D0233050

STRATHCLYDE UNIVERSITY LIBRARY
30125 00091506 5

ANDERSONIAN LIBRARY
WITHDRAWN
FROM
LIBRARY
STOCK
UNIVERSITY OF STRATHCLYDE

Macromolecular Syntheses

Macromolecular Syntheses

A Periodic Publication

of Methods for the Preparation

of Macromolecules

COLLECTIVE VOLUME I

1977

J. A. MOORE, Editor

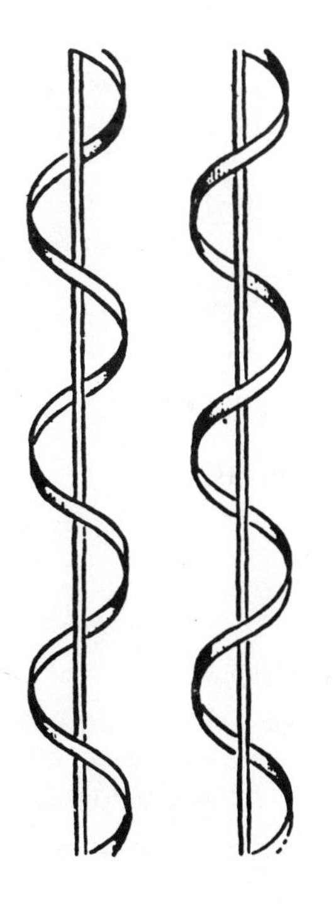

John Wiley & Sons
New York • Chichester • Brisbane • Toronto

Library of Congress Catalog Card Number: 63-18627

ISBN 0-471-61451-3

Printed in the United States of America

10 9 8 7 6 5 4 3 2 1

Submission of Preparations

All chemists are invited to submit procedures for the preparation of polymers to *Macromolecular Syntheses.* Preparations of both old and new polymers that are of general interest or that illustrate useful preparative techniques are desired. Chemists who plan to submit preparations are urged to write to the secretary concerning their intentions.

The author should carefully check his preparation before submitting it. Explicit directions, including elaboration in notes when necessary, for all the steps of the preparation, isolation, and characterization of the polymer should be given. The synthesis of the starting materials or their source along with criteria to be used in determining their purity should be given. Experience in checking procedures for preparing polymers has shown that impure or uncharacterized starting materials present the greatest difficulty in duplicating the author's results. The range of yields and polymer property values should be reported rather than the maximum values obtainable. The characterization of the polymer should include such information as molecular weight, softening or melting point, spectra, functional group analysis, and solubility data.

Authors should submit three copies of their preparation to the secretary of *Macromolecular Syntheses*

Professor James A. Moore
Department of Chemistry
Rensselaer Polytechnic Institute
Troy, New York 12181

Foreword

The essence of practical chemistry is synthesis—reproducible syntheses of strictly characterized substances.

In fact, chemistry developed and expanded as the methods for purification and characterization became more and more reliable, and premitted checking, with increasing sharpness, whether a molecule of desired structure and properties was in fact synthesized. Large parts of the chemical literature—articles, journals, reference books, handbooks, and encyclopediae—occupied themselves exclusively with problems of sythesis and characterization.

Polymer chemistry was plagued in its infancy—between 1920 and 1930—by an absence of knowledge concerning the chemical nature of macromolecules and by the insufficient reproducibility of standard characterization methods.

Professor C. G. Overberger, the Editor-in-Chief and originator of this series, emphasized in his 1963 preface to Volume I that "Since about 1950, however, new and improved methods have been developed and it could be demonstrated that it is quite possible to prepare and characterize a mixture of large molecules by other than conventional techniques. A number of chemists working in the area of organic and polymer chemistry in academic institutions and in industrial research laboratories were therefore convinced that the time was appropriate for the beginning of a series on macromolecular syntheses containing detailed directions for polymerization and characterization. At the same time, it was felt that these procedural methods should be checked in a rigorous way following the successful pattern of *Organic Syntheses,* so that they might be easily repeated by a student or a research worker."

At that time—in 1961—the first edition of *Preparative Methods of Polymer Chemistry* had just appeared and was an extremely valuable help for the initiation of *Macromolecular Syntheses.* Later—in 1968—the second, considerably expanded edition appeared. In 1966–Dietrich Braun, Harald Cherdron, and Werner Kern had published the *Praktikum der Makromolekularen Organischen Chemie*, which described the synthesis and characterization of numerous polymers together with chemical reactions of macromolecular materials. These books are a very welcome aid to anybody who wants to prepare a certain polymer but they, evidently, can only teach syntheses known before the texts were published. *Macromolecular Syntheses,* on the other hand, provides ongoing instructions as new synthetic efforts become available.

This book, edited and revised by Professor J. A. Moore, contains the first five volumes that appeared betwen 1963 and 1974, with Drs. C. G. Overberger, J. R. Elliot, N. Gaylord, W. J. Bailey, and E. L. Wittbecker as Editors. In each volume the editors particularly attempted to include the syntheses of new and, to a certain extent, somewhat extravagant materials, which are not yet produced on a commercial level but offer, because of their unusual properties, special interest of fundamental character.

We are all indebted to the editors and to the authors for their contributions to this highly useful and stimulating series.

H. F. MARK

New York, 1977

Preface to Vol. 1

The science of polymer chemistry—our knowledge and understanding of macromolecules—is relatively recent. Initial pioneering work in this area was carried out between 1920 and 1930. However, the preparation and characterization of large macromolecules did not become a part of organic chemistry until the 1930's. A major reason for this was ignorance of the chemical nature of macromolecules, as well as the inapplicability of standard methods for their characterization.

Soon the enormous industrial interest in the properties of these large molecules gave a strong impetus to research in this area and helped to elucidate their preparation and characterization, although even now many organic chemists have the strong archaic bias that non-crystalline products cannot be of interest because they cannot be characterized in conventional fashion.

Since about 1950, however, new and improved methods have been developed, and it could be demonstrated that it is quite possible to prepare and characterize a mixture of large molecules by other than conventional techniques. A number of chemists working in the organic polymer area in academic institutions and in industrial research laboratories were therefore convinced the time was appropriate for the beginning of a series on macromolecular syntheses, containing detailed directions for polymerization and characterization. At the same time, it was felt these procedural methods should be checked in a rigorous way following the successful pattern of *Organic Syntheses*, so they might be easily repeated by a student or a research worker.

One intent in this series is to provide trustworthy examples of polymer preparations for organic laboratory courses at the graduate and undergraduate levels, since there are few such examples in the usual organic laboratory textbook. In addition, it is well recognized that, unless an industrial laboratory has a staff with a background of working with macromolecules, it is sometimes difficult to obtain sufficiently detailed information for the preparation of polymers from the patent literature. Thus we hope this series will provide help to industrial concerns wishing to enter the field, by aiding the chemists in these organizations who have had no previous experience in the synthesis and characterization of macromolecules.

The projected series has been aided greatly by the recent publication of a book entitled *Preparative Methods of Polymer Chemistry* by Wayne Sorenson and Tod W. Campbell. In particular, we refer the readers to Chapter 2, "Preparation, Fabrication, and Characterization of Polymers." This excellent book, with a chapter clearly describing methods of characterization used in many of the preparations, greatly helps and strengthens our series.

While the first volume contains preparations submitted largely by members of the Editorial Board, we now encourage contributions from the scientific community at large. The preparation of monomers and polymers of all types will be included. Thus it is intended to include also macromolecules of biochemical origin, such as polypeptides, polysaccharides, and nucleic acids. In the preparation of the first volume it has been necessary to do most of the checking at the Polytechnic Institute of Brooklyn for the sake of expediency. Some of the preparations represent older, some newer, procedures. As the series proceeds, the older preparations will be mixed with the new, the main "driving force" being an attempt to represent all types of monomers and polymers. All useful methods of preparation will be represented, with the further goals that the procedures can be easily repeated, and that appropriate methods of characterization of these polymers are clearly described.

C. G. OVERBERGER

Brooklyn, New York
June, 1963

Preface to Vol. 2

The aim of *Macromolecular Syntheses* is to provide an expanding number of detailed syntheses of polymers that have been checked by other workers for reproducibility. Our experience in checking these syntheses has impressed us with the importance of using purified reagents and carefully following all the steps of the preparation. The synthesis of many polymers is extra-ordinarily affected by trace impurities in ordinary reagents. Wherever the trace impurity has been beneficial, we have tried to isolate it and call for its known inclusion in the preparation. In several cases these problems have required continued refinement of the preparation by submitter and checker until the directions are explicit and reproducible. The care required in making polymers successfully is much greater than the care required in making most low molecular weight organic compounds for which impurities and minor changes in preparative conditions seldom spell the difference between success and failure.

Volume 2 contains syntheses of both old and new polymers. By this approach in the first volumes of *Macromolecular Syntheses* we hope to provide syntheses of the older classes of organic polymers as well as syntheses of new polymers.

The ability to characterize polymers in molecular terms is improving with time. We plan to emphasize such characterization as part of *Macromolecular Syntheses* to improve the utility of the series for teaching, research, and other purposes. Thus we have included means of characterizing the degree of stereospecificity in appropriate polymers in this volume. Submitters of syntheses for future volumes are urged to keep this goal in mind.

J. R. ELLIOTT

Schenectady, New York
March, 1966

Preface to Vol. 3

Volume 3 of *Macromolecular Syntheses* continues the custom established in the earlier volumes and contains syntheses of both old and new polymers as well as descriptions of old and new techniques applicable to a broad class of polymers. In several instances a number of approaches to the same polymer by the use of different catalysts or monomers are described, and in one case different polymers are produced from the same monomer using different catalysts.

The extreme difficulty in reproducing polymer syntheses, pointed out in earlier volumes, was again confirmed with the procedures in this volume. This was as true with well-known polymers, which the submitters had made innumerable times, as with new polymers or procedures in which the submitter was the acknowledged expert. The procedures finally included in the volume were successfully checked by other workers, generally after several unsuccessful attempts and communication with the submitter to pinpoint small details not included by the submitter because they were common practice in the submitter's laboratory, or because the latter was not consciously aware of their importance.

In addition to failures resulting from the influence of trace impurities on the yield and molecular weight of the polymer, insufficient details for handling the product, for example, stabilization, resulted in failure to obtain confirmation of the characteristics described by the submitter, although the product actually reproduced the submitter's results.

With Volume 3 we have reached the stage in which we have become accustomed, but not reconciled, to the problems. Hopefully, these volumes will result in more detailed reporting not only in future submittals to *Macromolecular Syntheses* but also in the journal and patent literature.

NORMAN G. GAYLORD

Newark, New Jersey
October, 1968

Preface to Vol. 4

With the publication of Volume 4 of *Macromolecular Syntheses* the series has described the laboratory synthesis and characterization of a total of 124 polymers, including examples of the preparation of biopolymers, new exotic research polymers, and commercially important materials. An effort has been made to cover every major polymerization technique or method so these syntheses can serve as models for the preparation of a large variety of analogous materials. We hope that by making these very detailed and carefully checked procedures available in one series the whole will be more valuable than the sum of the parts.

As was pointed out in previous volumes, great difficulty is often encountered in duplicating even the most carefully described polymer synthesis. This arises from the fact that minute changes in the purity of the starting materials and the reagents or changes in experimental conditions often have a profound effect on the polymeric product. In an organic synthesis if the directions can be followed for the reaction, isolation, and purification of the product in the required yield, the preparation is a success. In a macromolecular synthesis, however, even if the reaction and isolation procedures give the reported yield, the product may differ in molecular weight, molecular weight distribution, stereochemistry, branching, and even composition. For these reasons it has been found that more carefully detailed directions are necessary for polymer syntheses and the resulting product must be characterized more extensively than in organic syntheses.

It has been the experience of the editors of the first four volumes of *Macromolecular Syntheses* that very few preparations have been spontaneously submitted for checking and publication, and that most contributions resulted from direct solicitation by the Editorial Board. We would like to encourage more polymer chemists to submit examples of their favorite polymerization technique or polymer synthesis to this series. Even if you do not submit a preparation, we welcome suggestions for preparations or techniques to be included in forthcoming volumes.

WILLIAM J. BAILEY

College Park, Maryland
February 1972

Preface to Vol. 5

Volume 5 of *Macromolecular Syntheses* represents another step toward the goal of providing an increasing number of syntheses that have been submitted and checked for reproducibility. The availability of detailed descriptions of the preparation and characterization of a wide variety of polymeric structures should be of considerable value of those already engaged in polymer work or those desiring to enter this important field. The syntheses in this volume can be classified into five different types. Polymers made by addition and condensation polymerization are the most numerous. However, there are some examples of cycloaddition and ring-opening polymerizations, and two-syntheses involve the chemical modification of natural polymers, that is, rubber and cellulose. The polymers described range from well-known structures to new and unusual ones, including some that are ionic, heterocyclic, optically active, network, and catalytic.

The inability of checkers to reproduce a synthesis on the first try continues to be a problem. This emphasizes the importance of submitting detailed, accurate directions and information that describe completely the purity of reagents and the steps required to run the polymerization. It is equally important that sufficient information on characterization (e.g., the method used for determining the melting point of the polymer) be included.

As was expected with a new series, most of the syntheses have been contributed by the members of the Editorial Board or have been submitted as a result of direct solicitation by them. Polymer chemists who have run syntheses suitable for checking and publication are urged to submit them. Suggestions for polymerizations to be included in future volumes are also desired by the Editorial Board.

EMERSON L. WITTBECKER

Wilmington, Delaware
June, 1974

Preface to Collective Volume 1

I was privileged to be "in on" the beginning of *Macromolecular Syntheses.* As a graduate student working with C. G. Overberger at "Brooklyn Poly" I heard of the nascent idea, saw the conception embodied, and made suggestions for which procedures should be considered for inclusion in some of the volumes. As a student of Polymer Science (then and now) I gained immeasurably by having access to the fund of reliable information assembled between the covers of Volumes 1-5. The reader can, based on this introduction, appreciate my satisfaction at having the opportunity of attempting to place the "gem" into a polished setting.

Solicitations of criticism and comment were made to all submitters and checkers of preparations to (as well as to the members of the Advisory and Editorial Boards of) Macromolecular Syntheses, Volumes 1-5. This endeavor resulted in almost four hundred inquiries, of which about 60% received replies. This represents an overwhelmingly positive response to the concept of *Macromolecular Syntheses.*

Thirteen years have passed since the first volume of *M.S.* appeared and, at this writing, Volume 6 has been completed and Volume 7 is well on its way. As Mark Twain said, "Recent reports of our demise are greatly exaggerated." In fact, Volumes 4, 5, 6 are arranged in a schedule of approximately one volume every two years—a slightly shorter and more regular interval than for Volumes 1, 2, and 3. We ask for the continued support of the scientific community in the form of willingness to submit and to check timely procedures so the recent publication schedule may be maintained.

Despite the realities of the employment marketplace (more than 50% of the chemists employed in industry work in some area of polymer science) the numbers of academic institutions offering comprehensive programs in polymer science is still surprisingly small. Because of this simple fact the central idea of

Macromolecular Syntheses remains valid and relevant. Not only may the syntheses described be used as the nucleus of a university laboratory course, but they may also be used as an independently pursued practical training program for chemists who find it necessary to learn how to synthesize specific classes of polymers.

Approximately 150 procedures are included here, covering *A*crylonitrile to *Z*iegler-Natta polymerization. Bringing these procedures to the reader has required the combined energies and resources of some 400 individuals. The enormous amount of labor freely donated by these people for the advancement of Polymer Science cannot be adequately compensated by having their names included on a title page, and we trust their selfless devotion to their chosen field of endeavor is fully appreciated by the recipients of this largess. Special efforts made by C. G. Overberger, N. Gaylord, and E. Wittbecker in the compilation of this collected work deserve commendation and are thankfully acknowledged. The assistance, cooperation and forbearance of the editor's wife in doing the "dog work" required to complete an undertaking of this magnitude constitute perhaps the largest single contribution to its success.

J. A. MOORE

Troy, New York
July, 1977

Contents

Macromolecular Syntheses

Crystalline Polystyrene

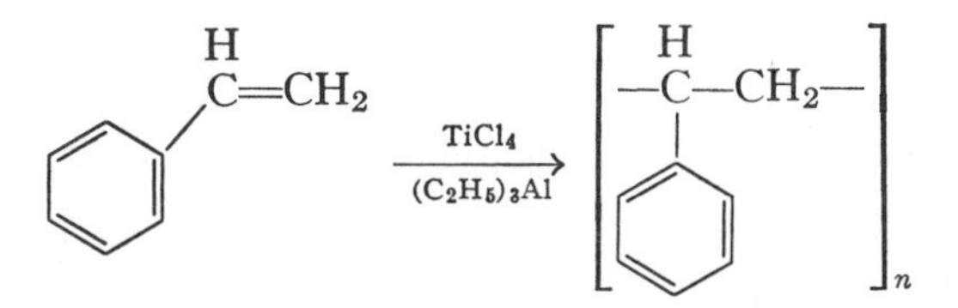

Submitted by R. J. Kern (1)
Checked by C. G. Overberger (2a, b) and E. Davidson (2c)

1. Procedure

Caution! Triethylaluminum is highly incendiary.

A 1-1. four-necked flask is fitted with a mechanical stirrer, a thermometer, and a nitrogen sweep system. The stirring shaft operates through a vacuum-tight bearing and has a blade of stainless steel or 1.125-in. Teflon® sheet cut to wipe the flask bottom closely. The assembly is baked for 1 hr. at 120° and then swept with purified nitrogen (Note 1) for 30 min. while cooling (Note 2). Titanium tetrachloride (0.9 ml., 1.55 g., 8 mmoles) is added with a hypodermic syringe. The flask is cooled externally to about 0°. Because of the extremely hazardous nature of the reaction between triethylaluminum and water, a Dry Ice-dimethoxyethane cooling bath is used. A solution of triethylaluminum (3.3 ml., 27 g., 24 mmoles) in 5 ml. of hexane is added dropwise with a small pressure-equilibrated addition funnel, over 20 min. with slow (40-60 r.p.m.) agitation. After addition is complete, the cooling bath is removed and stirring is continued for 30 min. at room temperature. Four hundred ml. of styrene (Note 3) is added from a pressure-equilibrated addition funnel. The temperature is then raised to 50° and agitation is increased to about 120 r.p.m. The black mixture soon becomes granular and steadily more viscous. Polymerization is allowed to proceed until a granular gel is obtained (2-4 hr.) without apparent free-flowing liquid (Note 4).

After removal of the heating unit, methanol is cautiously added dropwise with agitation. Considerable quantities of gas and heat are evolved because of decomposition of excess triethylaluminum. Local pools of methanol must not be allowed to form until this decomposition is completed. Addition of methanol is continued until 400 ml. has been mixed. Stirring is then continued for 10 min. longer. Solids are filtered from the black liquors, washed on a Büchner funnel with 400 ml. of acetone containing 2 ml. of conc. hydrochloric acid, and then reslurried in 400 ml. of acetone (Note 5). The solids are filtered and sucked nearly dry on a filter funnel and then stirred in refluxing, freshly distilled methyl ethyl ketone for 2 hr. The product is filtered, rinsed with acetone and then with distilled water, and dried (Note 6). Yield 30-50 g. of product, m.p. 233-236°.

2. Notes

1. Prepurified or lamp grade nitrogen containing not more than 10 p.p.m. of oxygen or water should be used.
2. Repeated evacuation with vacuum release to dry nitrogen is preferred.
3. Styrene containing no more than 10 p.p.m. of water should be used and may be obtained by distillation (50°/14 torr) after a fore-run equal to one-third of the charge has been discarded. Styrene should be distilled directly into a previously baked addition funnel.
4. The degree to which polymerization is allowed to proceed is arbitrary. As higher conversions are attained, the mass becomes stiffer and less tractable in subsequent operations. Reaction mixtures containing 8-12 wt. % of solids are about the maximum that can be conveniently handled in conventional laboratory glassware.
5. It is recommended that at this point the solids be vigorously stirred in acetone in an explosion-proof blender.
6. Drying temperatures up to 140° in air may be used.

3. Methods of Preparation

Crystallizable forms of polystyrene have also been prepared using heterogeneous (3, 4) and homogeneous (5-7) organo-alkali metal compounds.

4. References

1. Monsanto Co., St. Louis, Missouri 63166.
2a. Polytechnic Institute of New York, Brooklyn, New York 11201.
2b. Current Address: The University of Michigan, Ann Arbor, Michigan 48104.
2c. Current Address: Singer Co., 1225 McBride Ave., Little Falls, New Jersey 07424.
3. J. L. Williams, T. M. Laakso, and W. J. Dulmage, *J. Org. Chem.*, **23**, 628 (1958).
4. S. Smith, *J. Polym. Sci.*, 38, 259 (1959).
5. R. J. Kern, *Nature*, **187**, 410 (1960).
6. R. J. Kern, *Amer. Chem. Soc. Polym. Prep.*, 4, 324 (1963).
7. D. Braun, W. Betz, and W. Kern, *Makromol. Chem.*, **42**, 89 (1960).

Amorphous Polystyrene

A. Bulk Polymerization with Peroxide Catalysis

$$C_6H_5CH{=}CH_2 \xrightarrow{(C_6H_5CO_2)_2} \left[-CH_2-\underset{\displaystyle C_6H_5}{\underset{|}{CH}}- \right]_n$$

Submitted by C. G. Overberger (1a, b)
Checked by H. Friedman (1a, c)

1. Procedure

A glass polymerization tube (Note 1) is charged with 25.00 ml. (22.68 g.) of freshly distilled styrene (Note 2) and 0.50 g. of benzoyl peroxide. *Caution! Benzoyl peroxide is a strong oxidant and can ignite or explode if not properly treated.*

The tube is connected to a three-way stopcock and the other two inlets of the stopcock are connected to a vacuum pump and a balloon containing nitrogen, respectively. The contents of the polymerization tube are then degassed and sealed under vacuum (Note 3).

The monomer is polymerized by heating the sealed tube at 55-60° for 66 hr. (2.75 days).

At the end of the reaction period the plug of polymer is obtained by breaking the glass tube with a hammer (Note 4) and separating the broken glass from the plug.

The polymer is dissolved in 300 ml. of benzene at room temperature with stirring, and the solution (Note 5) is then added dropwise to 3 1. of methanol (Note 6) with vigorous stirring to precipitate the polymer. The polystyrene is collected by filtering through a sintered glass funnel (coarse porosity). The polymer is further purified by redissolving it in benzene and reprecipitating it by addition to methanol.

After drying overnight in a vacuum oven at 60°, the polymer weighed 22.5 g., η_{inh} = 0.37-0.38 dl./g. at 29.4° in a 0.5% benzene solution.

2. Notes

1. The polymerization tube has a bulb 12 cm. in length and 2.5 cm. in diameter. A constricted neck, 15 cm. long, is connected to the bulb.

2. It is necessary to distill the styrene under reduced pressure to free the monomer of inhibitor and other impurities. The first 5% of styrene distilled is discarded, and the middle fraction is saved. An alternate procedure is to wash the styrene with a 5% aqueous sodium hydroxide solution and to then dry the monomer over anhydrous calcium sulfate.

3. The contents of the tube are frozen by immersing the tube in a Dry Ice-acetone bath. After freezing, the tube is evacuated, nitrogen is introduced, the monomer is thawed, refrozen, and reevacuated. This procedure is repeated several times. Before the tube is sealed the monomer is refrozen and the tube is reevacuated.

4. The tube should be wrapped in a towel to prevent flying glass from being a hazard.

5. The solution may be freed of dust particles by filtering it through a sintered glass funnel (coarse porosity).

6. The non-solvent is used in a volume ten times that of the solvent to ensure complete precipitation of the polymer.

B. Solution Polymerization with Cationic Catalysis

$$C_6H_5CH{=}CH_2 \xrightarrow{SnCl_4} \left[-CH_2-\underset{\underset{C_6H_5}{|}}{CH}- \right]_n$$

Submitted by C. G. Overberger (1a, b)
Checked by H. Friedman (1a, c)

1. Procedure

The Catalyst. Into a dried 2-oz. screw-cap bottle (Note 1), which has been thoroughly dried in an oven, is pipetted 20 ml. of C.P. reagent grade carbon tetrachloride from a newly opened bottle (Note 2). The bottle and its contents are then accurately weighed.

Stannic chloride catalyst, 0.6 ml. (1.32 g.) (Note 3), is withdrawn with a 5-ml. hypodermic syringe and immediately injected into the 2-oz. bottle without unscrewing the cap. The bottle is then accurately reweighed to determine the exact amount of catalyst transferred. Sufficient carbon tetrachloride is added with a syringe to the bottle to bring the solution to a concentration of 2.14 wt. % in catalyst. The catalyst solution is immersed in an ice bath until needed (Note 4).

The Monomer. Freshly distilled styrene, 6.5 ml. (Note 2, Part A), is pipetted into a thoroughly dried 4-oz. screw-cap bottle (Note 1). From a pipet, 20 ml. of carbon tetrachloride and 25.5 ml. of nitrobenzene (Note 5) are added to the monomer. The bottle cap is closed tightly, and the bottle is shaken briefly. The bottle is then placed in the same ice bath with the catalyst solution.

After about 10 min. has been allowed for the contents of both bottles to cool to 0°, 4 ml. (6.5 g.) of the catalyst solution is withdrawn with a hypodermic syringe. The contents of the syringe are injected into the 4-oz. bottle of monomer solution (Note 6). After standing for a period of 5 hr. in an ice bath, the solution is added to 500 ml. of methanol (Note 6, Part A) with vigorous stirring to precipitate the polymer. The polystyrene is collected by filtering through a sintered glass funnel. The polymer is further purified by redissolving it in 50 ml. of methyl ethyl ketone and reprecipitating it from 500 ml. of methanol. This purification process is repeated twice.

After drying overnight in a vacuum desiccator (Notes 7, 8) the polymer weighed 5 g., η = 0.19-0.20 dl/g. at 29.2° in a 0.5% benzene solution.

2. Notes

1. The cap of the bottle should be perforated and fitted with two rubber gaskets. The lower gasket, exposed to the contents of the bottle, consists of vulcanized rubber (chemically inert to the solvents), and the upper gasket is a puncture-sealing rubber material such as butyl rubber.

2. The carbon tetrachloride should be stored over anhydrous calcium chloride if it is not used immediately after opening the bottle.

3. The main supply of stannic chloride catalyst should be stored in a screw-top bottle, fitted at the cap as described in Note 1 with two rubber gaskets. If the main supply of catalyst becomes contaminated, it may be purified by heating under reflux over phosphorous pentoxide for 1-2 hr. and then distilling under reduced pressure.

4. A more elaborate procedure than that described above for preparing the catalyst in such polymerization reactions is reported in the literature (2).

5. If the nitrobenzene is not obtained from a newly opened bottle of C.P. reagent grade material, it must be purified according to standard procedures (3).

6. The polymerization mixture consists of a 10.0 mole % solution of styrene in solvent (an equimolar mixture of carbon tetrachloride and nitrobenzene) containing 0.1 mole % of stannic chloride catalyst. The molar ratio of monomer to catalyst is 100:1.

7. Paraffin chips and phosphorous pentoxide are placed in the desiccator to absorb any solvent adhering to the polymer. The pressure in the desiccator was 0.5 torr.

8. Care must be taken when admitting air to the desiccator because the powdery polymer often acquires a static charge. It is helpful to cover the polymer container with a piece of porous filter paper, held in place with a rubber band.

3. Methods of Preparation

Polystyrene has also been prepared by suspension or pearl polymerization (4), emulsion polymerization (5), solution polymerization (6), and thermal polymerization (7).

4. References

1a. Polytechnic Institute of Brooklyn, Brooklyn, New York 11201.
1b. Current Address: The University of Michigan, Ann Arbor, Michigan 48104.
1c. Current Address: School of Medicine, University of Pennsylvania, Philadelphia, Pennsylvania 19174.
2. J. George, H. Mark, and H. Wechsler, J. Amer. Chem. Soc. 72, 3896 (1950).
3. A. I. Vogel, *Practical Organic Chemistry,* third edition, Longmans, Green and Co., New York, 1959, p. 175.
4. W. P. Hohenstein (Polytechnic Institute of Brooklyn,) U.S. Pat. 2,524,627 (1950) (C.A., 45, 903 (1951).
5. Combined Intelligence Objectives Sub-Committee Report 79-85 (C.A. Schildknecht, *Vinyl and Related Polymers,* John Wiley and Sons, Inc., 1952, p. 16).
6. M. Hunt (E. I. duPont de Nemours & Co.,) U.S. pat. 2,471,959 (1949) (C.A., 43, 6002 (1949).
7. Dow Chemical Company, Product Bulletin, "The Polymerization of Styrene,"

Bisphenol-A Polycarbonate (p,p′-Isopropylidene bisphenol polycarbonate)

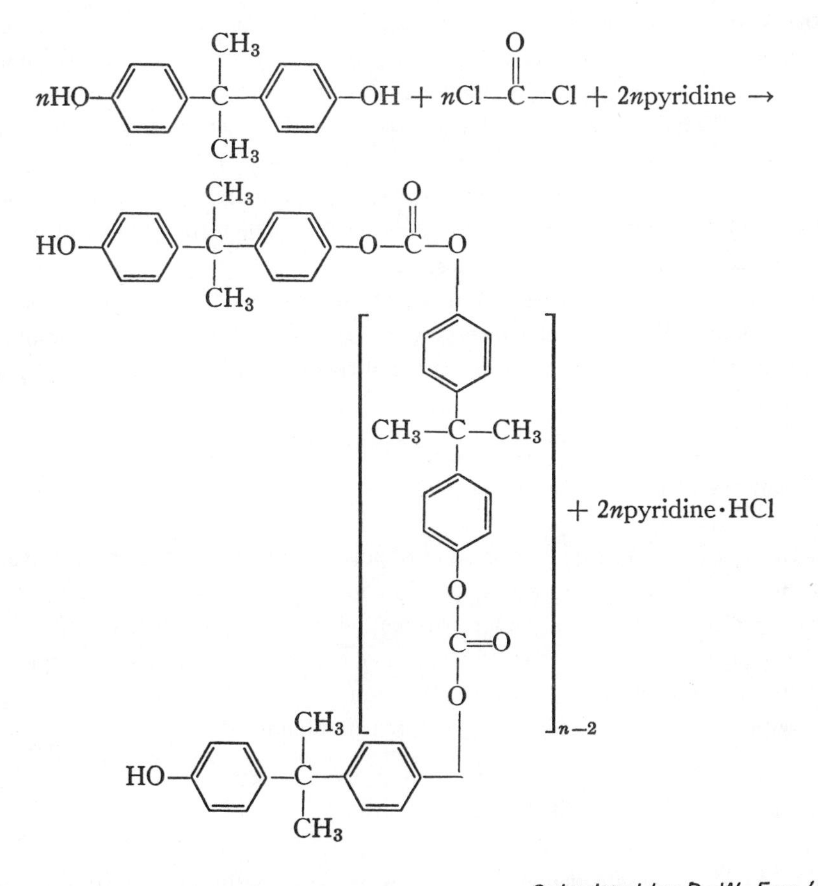

Submitted by D. W. Fox (1)

Checked by C. G. Overberger (2a, b) and B. Avchen (2a, c)

1. Procedure

Caution! The reaction should be conducted in a hood.

A 500-ml. four-necked flask (or resin pot) equipped with a stirrer, a thermometer, a wide-bore (Note 1) gas inlet tube, and a gas outlet (Note 2) is charged with 22.8 g. (0.10 mole) of bisphenol-A (Note 3) and 228 ml. of pyridine (Note 4). Phosgene, (*Caution! Phosgene is a toxic, irritating gas and should be used only in an efficient fume hood.*) at a rate of 0.25 g./min. (Note 5), is passed into the rapidly stirred reaction mixture, which is maintained at 25-30° with a water bath (Note 6). Pyridine hydrochloride will begin to separate from the reaction mixture after about 25 min. This is an indication that the reaction is about 60% completed. Approximately 15 min. later a marked increase in viscosity will be noted over a period of 2-3 min.; the polymerization is then essentially completed (Note 7).

The polymer may be precipitated directly in the reaction flask. A dropping funnel containing 250 ml. of methanol is substituted for the gas inlet tube, and methanol is added to the stirred reaction mixture over a period of 2-3 min. (Note 8). The polymer suspension is stirred for 5 min. after precipitation and recovered by filtration on a Büchner funnel. General purpose filter paper may be used. The polymer may be conveniently freed of residual pyridine and pyridine hydrochloride by two or three successive cycles of agitation in an explosion-proof blender with 250-ml. portions of methanol (Note 9). The filtered product may be dried overnight in a vacuum oven at 125°. The yield is 22 g.

2. Characterization

The intrinsic viscosity of bisphenol-A polycarbonate determined in dioxane at 30° is 0.46 dl./g., melting range 225-250° (determined on a melting point block) (Note 10). It is soluble in methylene chloride, chloroform, tetrachloroethane, chlorobenzene, dioxane, and pyridine. Colorless transparent films may be conveniently cast on glass from methylene chloride solution. The polymer may be pressed into films between aluminum foil at 240-260°.

3. Notes

1. A 7-mm. inside-diameter tube was used with the end positioned just above the stirring blade. A gas diffuser or smaller tube is undesirable because

pyridine hydrochloride and/or the pyridine-phosgene complex tends to clog the inlet.

2. A length of rubber tubing may be connected to the gas outlet to carry escaping phosgene to a scrubber consisting of an empty trap connected to a trap containing well-stirred ammonium hydroxide. The entrance to this trap should be above the liquid level to prevent formation of urea from clogging the opening.

3. The freezing point of bisphenol-A may vary from 148°-157°. The monomer used in the reaction described had a freezing point of 156.2°. Recrystallization from chlorobenzene may be used for purification.

4. Reagent grade pyridine was used. A technical grade may also be used. Dimethyl aniline may be substituted for pyridine, and part of the tertiary amine may be replaced with chlorinated aliphatic hydrocarbons or chlorobenzene.

5. The rate of phosgene addition is not critical and, in fact, need not be measured. A dry trap between the phosgene cylinder and the reaction vessel is recommended to prevent the reaction mixture from being drawn into the cylinder if the phosgene flow is stopped.

6. The temperature is not critical, but elevated temperatures may promote color formation.

7. The exact end point is difficult to characterize. Somewhat more than the theoretical 10 g. of phosgene is usually required. A marked increase in viscosity is a good indication. Overshooting of the end point may be visually determined if the yellow color of the transient phosgene-acid chloride complex does not dissipate rapidly. The reaction may essentially be back-titrated by adding a small quantity of the original reaction mixture or, alternatively, precipitated at, or past, the end point.

8. Some degradation of the polymer may occur if the precipitation with alcohol is too slow. A stringy, sticky ball of polymer may form if precipitation is too rapid. Other alcohols, acetone, and hydrocarbons may be used as precipitating solvents. Alternatively, the polymer may be precipitated in an explosion-proof blender by pouring the reaction mixture into stirred methanol.

9. An alternate purification procedure may be followed: it consists of one or more cycles of dissolution in methylene chloride, precipitation, and washing with methanol. The adequacy of washing in either procedure may be determined by checking the wash liquor for chloride ion with alcoholic silver nitrate.

10. The melting point is approximately 15° lower when measured in a capillary tube. The melting point is affected by the extent of crystallinity and

by the molecular weight of the polymer. The use of acetone to precipitate the polymer (Note 8) may promote crystallization and raise the melting point. Slow drying rates will also promote crystallinity. Polymer with an intrinsic viscosity of 1.0-3.0 dl./g. will not flow until the temperature is above 250°.

4. Methods of Preparation

Bisphenol-A polycarbonate may be prepared from bisphenol-A and diphenyl-carbonate by a thermally induced ester interchange.

5. References

1. General Electric Co., Pitsfield, Massachusetts 01201.
2a. Polytechnic Institute of Brooklyn, Brooklyn, New York 11201.
2b. Current Address: The University of Michigan, Ann Arbor, Michigan 48104.
2c. Current Address: 7402 Big Cypress Drive, Miami Lakes, Florida, 33014.

Poly(hexamethylenesebacamide) by Interfacial Polycondensation

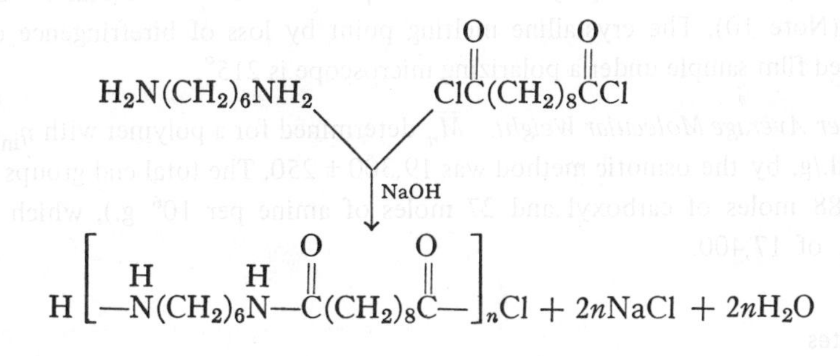

Submitted by P. W. Morgan and S. L. Kwolek (1)
Checked by C. G. Overberger (2a, b) and L. Herin (2a)

1. Procedure (Note 12)

In a 1-qt. explosion-proof blender (Note 1) is placed a solution of 2.32 g. (0.02 mole) of hexamethylenediamine and 1.60 g. (0.04 mole) of sodium hydroxide (Note 2) in 330 ml. of water. The blender is turned to high speed with a rheostat, and there is added (Note 3) over a period of about 15 sec. a solution of 4.78 g. (0.02 mole) (Note 4) of sebacoyl chloride in 250 ml. of tetrachloroethylene (Note 5).

After the mixture has been stirred for 2 min. (Note 6), the polymer is collected on a fritted glass filter (Note 7) and washed with water until free of alkali and salt (Note 8). The granular polymer is dried in air at a temperature below 100° or in a vacuum oven. The yield is about 4.8 g. (85%).

2. Characterization

Solubility. The polymer is insensitive to most common low-boiling organic solvents but is soluble in formic acid, concentrated sulfuric acid, trifluoroacetic acid, phenol, *m*-cresol, and hot benzyl alcohol. Viscous solutions are formed at 5% to 20% solids.

Dilute Solution Viscosity. The inherent viscosity (η_{inh}) is 1.0-1.8 dl./g. (Note 9) determined in *m*-cresol at 30° and 0.5 g. of polymer per 100 ml. of solution. The intrinsic viscosity and weight average molecular weight can be approximated from the equations $\eta_{sp/c} = [\eta] + 0.42\ [\eta]^2 c$ and $\overline{M}_n = 10{,}750\ [\eta]^{1.05}$, which were derived for polymer formed by melt polymerization.

Melt Temperature. The polymer melt temperature on a hot metal surface is 220° (Note 10). The crystalline melting point by loss of birefringence of an oriented film sample under a polarizing microscope is 215°.

Number Average Molecular Weight. $\overline{M}_n$ determined for a polymer with η_{inh} = 1.87 dl./g. by the osmotic method was 19,300 ± 250. The total end groups were 115 (88 moles of carboxyl and 27 moles of amine per 10^6 g.), which gives an $\overline{M}_n$ of 17,400.

3. Notes

1. Vigorous stirring and high shearing action are needed to disperse the reactants rapidly and break down the tough film of polymer as it forms. Inefficient stirring will reduce the molecular weight and the yield of polymer. Detergents may be added to assist stirring, but they may interfere with the polymerization and the final polymer will not be stable to melting. Flammable solvents should not be used in an electrically-driven blender unless the motor is isolated from the fumes, or the housing is modified as suggested in ref. 12, p.409.

2. Other strong inorganic bases may be used, although some adjustment in the ratio of the phase volumes may be necessary to obtain the highest polymers. Sodium carbonate is acceptable if 1 mole is used for each mole of hydrogen chloride. Excesses of inorganic base up to 10% can be used without detriment to the properties of the product.

3. A smooth addition without loss of product may be made by adding the solution through a powder funnel inserted in a hole made in the plastic cover of the blender. It is also helpful to place a foil of aluminum on top of the jar but under the cover.

4. Liquid acid chlorides are most easily measured out in graduated pipets or in calibrated syringes. The volume of sebacoyl chloride to be used is 4.28

ml. at 25°. Diamines and alkalies may be prepared as concentrated aqueous solutions (0.2 g./ml.) and dispensed from automatic burets equipped with stopcocks of Teflon® (Note 11) fluorocarbon resin. Intermediates of adequate purity may be obtained from reagent supply houses (3, 4).

5. For best results the solvent should be distilled to free it of traces of water and other minor impurities. Various other water-immiscible solvents such as carbon tetrachloride, dichloromethane, xylene, and benzene may be used, but adjustments in the reactant concentration ratios must be made to obtain the highest molecular weight polymers (4, 5).

6. Stirring may be continued for a longer time, but for an unhindered aliphatic polyamide there is little change after 5 min. The narrowest molecular weight distribution is obtained by stopping the polymerization at about 2 min. Cessation of polymerization may be assured by adding 100 ml. of 3% aqueous hydrochloric acid to the blender at 2 min. and continuing the stirring for 1 min. more.

7. Paper filters contaminate the product with cellulose fiber that degrades and discolors if the polyamide is later melted.

8. The washing may be hastened by the use of such organic solvents as acetone or alcohol or mixtures of these with water. They help to remove the water-immiscible solvent. Stirring in the blender will also speed the washing. A final rinse with ethyl ether yields a fluffier form of dried product.

Distilled or demineralized water should be used in the washing if end groups are to be determined and if the polymer is to be melted.

9. A lower viscosity number indicates a lower molecular weight. This will result from various causes, such as impure reactants or solvents, incorrect measurement of reactants, failure to stir rapidly, or degradation by careless after-treatment.

10. Moisture absorbed from the air acts as a plasticizer and may lower the melting point of the polymer.

11. Teflon is a trademark of the du Pont Company.

12. The transfer of solvents, the polymerization and the recovery of polymer should be carried out in a well-ventilated hood. The waste solvents should be collected and disposed of in an incinerator or by other suitable means.

4. Methods of Preparation

High molecular weight, aliphatic polyamides were first made by melt polycondensation (11). This is the commercial process; it is a useful laboratory procedure. Longer times and special equipment are required.

The preparation of aliphatic polyamides by an unstirred interfacial polycondensation has been described as a lecture demonstration experiment (3, 5). The unstirred system has also been used to study polymerization variables and mechanism (4). Variations in procedure that may be used or that may be necessary to apply interfacial polycondensation to other polyamides have been reported (6-10, 12, 13).

5. References

1. Pioneering Research Division, Textile Fibers Dept., E. I. du Pont de Nemours & Co., Wilmington, Delaware 19898.
2a. Polytechnic Institute of New York, Brooklyn, New York 11201.
2b. Current Address: The University of Michigan, Ann Arbor, Michigan 48104.
3. P. W. Morgan and S. L. Kwolek, *J. Chem. Educ.,* **36**, 182 (1959).
4. P. W. Morgan and S. L. Kwolek, *J. Polym. Sci.,* **40**, 299 (1959).
5. P. W. Morgan, *Soc. Plast. Eng. J.,* **15**, 485 (1959).
6. R. Beaman, P. W. Morgan, C. R. Koller, E. L. Wittbecker, and E. E. Magat, *J. Polym. Sci.,* **40**, 329 (1959).
7. M. Katz, *J. Polym. Sci.,* **40**, 337 (1959).
8. V. E. Shashoua and W. M. Eareckson, III, *J. Polym. Sci.,* **40**, 343 (1959).
9. C. W. Stephens, *J. Polym. Sci.,* **40**, 359 (1959).
10. J. R. Schaefgen, F. H. Koontz, and R. F. Tietz, *J. Polym. Sci.,* **40**, 377 (1959).
11. H. Mark and G. S. Whitby, *Collected Papers of Wallace H. Carothers on High Polymeric Substances,* Interscience Publishers, Inc., New York, 1940.
12. P. W. Morgan, *Condensation Polymers by Interfacial and Solution Methods,* Interscience Publishers, New York, 1965.
13. P. W. Morgan and S. L. Kwolek, *Macromolecules,* 8, 104 (1975).

Poly(ethylene terephthalate)

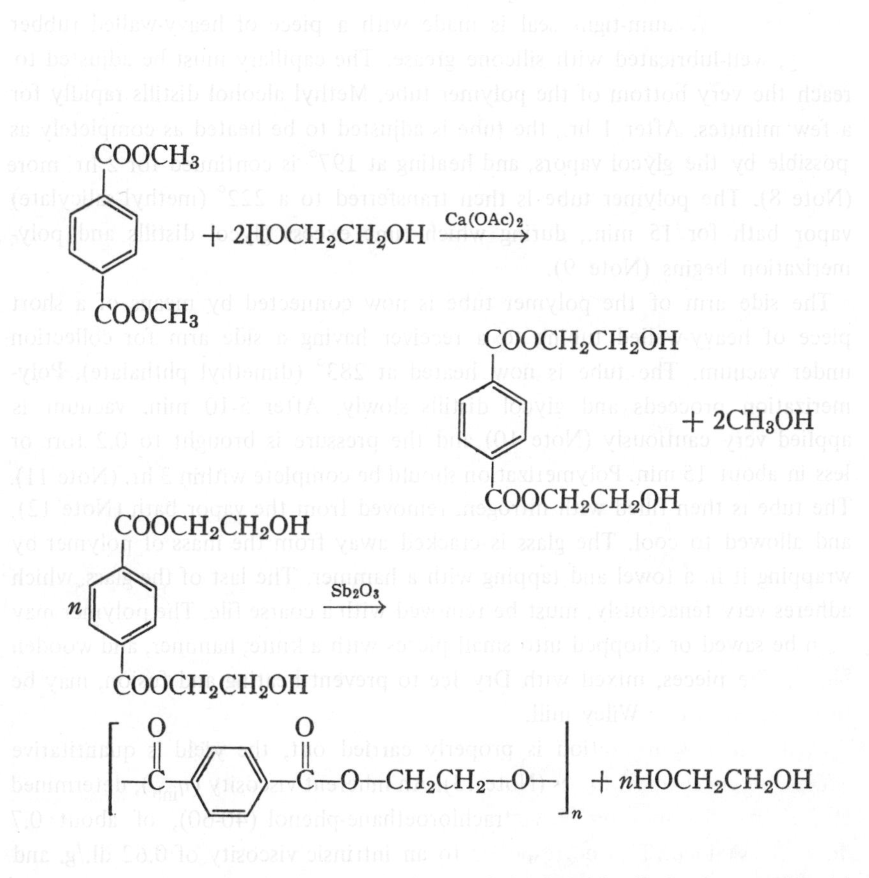

Submitted by F. B. Cramer (1a, b)
Checked by C. G. Overberger (2a, b) and A. M. Schiller (2a, c)

1. Procedure

The condensation may be conveniently carried out in a glass "polymer tube" about 25 mm. x 250 mm., sealed to a 10-mm. x 70-mm. neck carrying a side arm for distillation. Into the tube is charged 13.6 g. (0.07 mole) of dimethyl terephthalate (DMT) (Note 1), 10 g. (8.8 ml., 0.16 mole) of ethylene glycol

(Note 2), 0.022 g. (0.15% based on DMT) of calcium acetate dihydrate (Note 3), and 0.005 g. (0.035% based on DMT) of antimony trioxide (Note 4). The charge is melted by submerging the tube about halfway (Note 5) in the vapors (Note 6) of boiling ethylene glycol (197°), and a fine capillary (Note 7), connected to nitrogen under pressure, is introduced through the neck of the tube. A vacuum-tight seal is made with a piece of heavy-walled rubber tubing, well-lubricated with silicone grease. The capillary must be adjusted to reach the very bottom of the polymer tube. Methyl alcohol distills rapidly for a few minutes. After 1 hr., the tube is adjusted to be heated as completely as possible by the glycol vapors, and heating at 197° is continued for 2 hr. more (Note 8). The polymer tube is then transferred to a 222° (methyl salicylate) vapor bath for 15 min., during which time excess glycol distills and polymerization begins (Note 9).

The side arm of the polymer tube is now connected by means of a short piece of heavy-walled tubing to a receiver having a side arm for collection under vacuum. The tube is now heated at 283° (dimethyl phthalate). Polymerization proceeds and glycol distills slowly. After 5-10 min. vacuum is applied very cautiously (Note 10) and the pressure is brought to 0.2 torr or less in about 15 min. Polymerization should be complete within 3 hr. (Note 11). The tube is then filled with nitrogen, removed from the vapor bath (Note 12), and allowed to cool. The glass is cracked away from the mass of polymer by wrapping it in a towel and tapping with a hammer. The last of the glass, which adheres very tenaciously, must be removed with a coarse file. The polymer may then be sawed or chopped into small pieces with a knife, hammer, and wooden block. The pieces, mixed with Dry Ice to prevent heating and fusion, may be further ground in a Wiley mill.

When the polymerization is properly carried out, the yield is quantitative except for mechanical losses (Note 13). An inherent viscosity (η_{inh}), determined at 30° in 0.5% solution in *s*-tetrachloroethane-phenol (40-60), of about 0.7 dl./g. is obtained. This corresponds to an intrinsic viscosity of 0.62 dl./g. and a number average molecular weight of about 16,000. Polymers having an η_{inh} above about 0.5 dl./g. may be melt-spun or pressed into flexible films. The crystalline melting point (hot-stage, polarizing microscope) of polymer prepared as described is about 258°.

2. Notes

1. All ingredients must be of the highest purity. DMT may be purified by crystallization from methanol or dioxane.

2. Satisfactory glycol may be prepared by dissolving 1% of metallic sodium in the commercial product under nitrogen, heating under reflux for 1 hr., and distilling through a short Vigreux column.

3. At the temperature used for the first ester exchange reaction, basic catalysts are most effective. Calcium acetate is most satisfactory for laboratory work. Reagent grade chemicals should be used as catalysts to ensure best color in the final polymer.

4. Although the polymerization is also an ester exchange reaction, it cannot be driven to completion with calcium acetate. The simultaneous addition of the basic and acidic catalysts does not result in mutual interference, however (3). Zinc acetate alone may be used for both ester exchange and polymerization catalysts, but darker colored polymer usually results.

5. There is a strong tendency for DMT to codistill with methyl alcohol. This can lead to clogging of the side arm, and to altered composition when copolymers are made.

6. Pyrex® test tubes (50 x 400-mm.) make convenient vapor baths. In use, the upper wall acts as an air condenser to prevent loss of vapors. Electrical heating is safest and easiest to control. The reaction should be carried out so any of the refluxing vapors that do leave the tube are entrained by an exhaust system leading to a hood.

7. Commerical heavy-walled capillary tubing, small enough to fit inside the neck of the polymer tube, is drawn to a fine thread after softening in a gas-oxygen flame. The shoulder of the heavy capillary tubing is kept above the side arm of the polymer tube so the flow of methanol and glycol vapors is not impeded. The extremely fine capillary obtained by drawing makes additional control of the rate of nitrogen flow unnecessary.

8. It is essential to remove the last traces of methyl alcohol with the nitrogen sweep before proceeding, if high viscosity is to be obtained. With ethylene glycol terephthalate, this step can be allowed to proceed overnight. With more unstable monomers, the heating cycle can be shortened by using a larger excess of glycol, which often entrains the last traces of methyl alcohol in the next step.

9. If the charge solidifies at any of the intermediate stages, it will remelt without difficulty when transferred to the next higher temperature.

10. If vacuum is applied too rapidly, much material will splatter to the top of the tube and solidify, and bis(hydroxyethyl)terephthalate will distill in appreciable quantities.

11. The melt may be too thick to bubble at 283° when an η_{inh} of 0.8 dl./g. is approached. The separation of calcium terephthalate will cause the melt to become hazy, and decomposition products of antimony oxide may cause a

greyish green color. When cooling, the polymer should crystallize and become opaque. Later, the glass polymer tube may shatter with considerable violence because of differential shrinkage effects. Flying glass may be confined by wrapping the tube in a towel.

12. The glass capillary may be withdrawn from the melt before it cools. The "manual spinning" of a filament may thus be observed. The phenomenon of "necking down" may be demonstrated by hand-drawing the filament in contact with a warm surface (hot water pipe or desk lamp shade).

13. In general, this type of polyesterification is applicable to any system in which the monomers and polymers are thermally stable at temperatures above the polymer melting point, and the glycol is sufficiently volatile to permit the excess to be completely removed under vacuum.

3. Methods of Preparation

Polyethylene terephthalate has been prepared by direct esterification of ethylene glycol with terephthalic acid (4), by the condensation of bis(hydroxyethyl)terephthalate prepared from terephthalic acid and ethylene oxide (5), from terephthalic acid and ethylene carbonate (6), from monomethyl terephthalic acid ester and ethylene carbonate (7), from terephthalic acid and ethylene glycol (8), by self-condensation of methyl hydroxyethyl terephthalate (9), or mono(hydroxyethyl)terephthalic acid (10), from terephthalic acid and bis(hydroxyethyl)terephthalate (11), from disodium terephthalate and ethylene chlorohydrin (12), and from terephthaloyl chloride and ethylene glycol (13).

4. References

1a. Pioneering Research Division, Textile Fibers Dept., E. I. du Pont de Nemours & Co., Wilmington, Delaware 19898.
1b. The assistance of Dr. W. H. Watson and his associates at the Du Pont Dacron® Research Laboratory in revising this procedure is thankfully acknowledged.
2a. Polytechnic Institute of New York, Brooklyn, New York 11201.
2b. Current Address: The University of Michigan, Ann Arbor, Michigan 48104.
2c. Current Address: American Cyanamid Corp., Stamford, Connecticut.
3. W. R. Billica and J. T. Carriel (E. I. du Pont de Nemours & Co.), U. S. Pat. 2,739,957e (1956) [*C. A.*, 50, 11719 (1956)].
4. J. R. Whinfield and J. T. Dickson (E. I. duPont de Nemours & Co.), U. S. Pat. 2,465,319 (1949) [*C. A.*, 43, 4896g (1949)].
5. A. C. Farthing (Imperial Chemical Industries), Brit. Pat. 623,669 (1949) [*C. A.*, 44, 2028h (1950)].

6. J. G. N. Drewitt and J. Lincoln (British Celanese), Brit. Pat. 707,913 (1954) [*C. A.*, **48**, 12465a (1954)].
7. J. G. N. Drewitt and J. Lincoln (British Celanese), U. S. Pat. 2,802,807 (1957) [*C. A.*, **52**, 1643g (1958)].
8. Farbwerke Hoechst, Brit. Pat. 781,169 (1957) [*C. A.*, **52**, 1682a (1958)].
9. J. W. Batty, W. A. Cowdrey, C. Gardner, K. B. Wilson and N. Fletcher (Imperial Chemical Industries), Brit. Pat. 738,509 (1955) [*C. A.*, **50**, 13091h (1956)].
10. Gevaert Photo-Producten, Belg. Pat. 542,060 (1956) [*C. A.*, **52**, 5033f (1958)].
11. N. Munro (Imperial Chemical Industries), Brit. Pat. 775,030 (1957) [*C. A.*, **51**, 17237c (1957)].
12. J. Lincoln (British Celanese), Brit. Pat. 753,214 (1956).
13. I. Okamura, J. Shimeha, and K. Hashimoto, *Bull. Res. Inst. Teikoku Jinzo Kenshi Kaishi Ltd.*, **2**, 48-55 (1950) [*C. A.*, **45**, 388c (1951)].

Poly(methylmethacrylate) Suspension Polymer

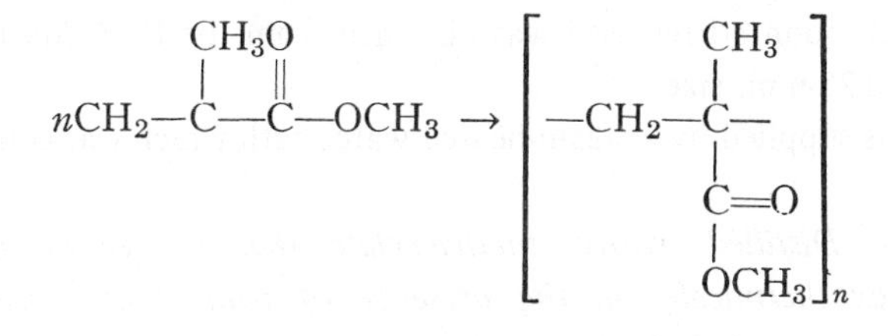

Submitted by D. P. Hart (1a, b)
Checked by M. Goodman (2a, b) and L. Levine (2a)

1. Procedure

Into a 3-1. three-necked flask are charged 1.5 1. of distilled water, 15 g. of Cyanamer A-370 (Note 1), 8.5 g. of disodium phosphate (Na_2HPO_4), and 0.5 g. of monosodium phosphate (NaH_2PO_4). The flask is fitted with a thermometer, a condenser, and a glass stirrer of the half-moon type (Note 2); the mixture is warmed to 30-35° (Note 3) and stirred until a clear solution is obtained.

In a 1-1. beaker are mixed 500 g. of distilled methyl methacrylate (Note 4) and 5 g. of benzoyl peroxide (Note 5). When the peroxide has been completely dissolved, the solution is added to the flask. The half-moon paddle is adjusted to about ½ in. below the top surface, and agitation is begun at about 400 r.p.m.

The reactor is flushed lightly with nitrogen gas for 1-2 min. to remove atmospheric oxygen. The agitator speed is adjusted to 250 r.p.m., and the reaction mixture is heated to 76-78°. This temperature is maintained for 2.5 to 3 hr. (Note 6).

The mixture is cooled to room temperature, and the polymer is recovered by filtration in a Büchner funnel (Note 7). The polymer is washed several times with water and dried at 65° for 5 to 10 hr.

The yield is 450-475 g. (90-95%) of clear polymer beads. Using chloroform, an intrinsic viscosity of 1.87-1.90 dl./g. is obtained (Note 8). This intrinsic viscosity may be lowered by the addition of tertiary dodecyl mercaptan to the monomer mixture (Note 9).

2. Notes

1. Cyanamer A-370 is a water-soluble modified polyacrylamide resin available from American Cyanamid Company as a free-flowing powder.

2. The half-moon stirrer used was obtained from the H. S. Martin Company and was the 125-mm. size.

3. Heat was supplied by a steam-heated water bath which was controlled manually.

4. *Caution! Distilled methyl methacrylate that is free of inhibitor can polymerize exothermically in the presence of light. Store distilled methyl methacrylate in a cool, dark place.*

5. *Caution! Dry benzoyl peroxide is known to spontaneously detonate on handling. To avoid this hazard a wet grade of benzoyl peroxide, e.g., Lucidol-78 (78% benzoyl peroxide), available from Lucidol Div. of Pennwalt Corp., may be used.*

6. A definite exotherm occurs after about 45 min. at reaction temperature. This exotherm should be controlled by the addition of cold water to the bath so that the *reaction temperature does not rise above 80°.*

7. Good filtration properties are obtained by using a Coors porcelain Büchner funnel with fixed plate (size No. 4 or 5). The best filter paper for this filtration was found to be the M grade from Sparkler Manufacturing of Mundelein, Ill. Volan 181 glass cloth or fine nylon cloth also serves well as the filtering diaphragm. Optimum filtering speed is attained if the cooled reaction mixture is diluted with an equal amount of cold water before filtering.

8. Using the relationship (3) $[\eta] = 2.8 \times 10^{-5} M^{0.8}$, the molecular weight of such polymers ranged from 1.07×10^6 to 1.09×10^6. Actual intrinsic viscosities of three different polymers were 1.89, 1.88, and 1.90 dl./g.

9. The addition of 3.75 g. of tertiary dodecyl mercaptan (0.75%) to the monomer-catalyst solution gave a polymer having an intrinsic viscosity of 0.47-0.49 dl./g. This is equivalent to a molecular-weight range of 1.90×10^5 to 2.01×10^5.

3. References

1a. Pittsburgh Plate Glass Co., Springdale, Pennsylvania 15144.
1b. The assitance of M. E. Hartman (PPG Industries, Allison Park, Pennsylvania 15101) in revising this procedure is thankfully achnowledged.
2a. Polytechnic Institute of Brooklyn, Brooklyn, New York 11201.
2b. Current Address: University of California at San Diego, LaJolla, California 92037.
3. Preprints, Division Paint, Plastics and Printing Ink Chemistry, American Chemical Society Meeting, Boston, April 1959, p. 136.

Isotactic Poly(isopropyl acrylate)

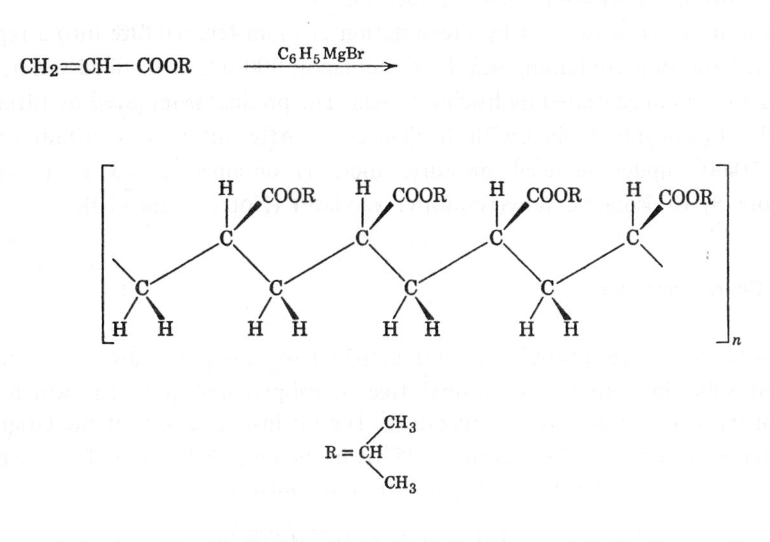

Submitted by W. E. Goode, R. P. Fellmann, and F. H. Owens (1a, b)
Checked by A. Katchman (2a, b) and E. E. Bostick (2a, c)

1. Procedure

A 1-l. four-necked round-bottomed flask is equipped with a mechanical stirrer, a thermometer, and a graduated addition funnel. One neck of the flask and the addition funnel are fitted with rubber serum caps (Note 1). In the flask is placed 525 ml. of toluene (Note 2); 43 ml. (38.3 g., 0.336 mole) of isopropyl acrylate (Note 3) is charged to the addition funnel. Oxygen is removed from the solvent and the monomer at room temperature by slowly bubbling a stream of nitrogen (Note 4) through the materials for at least 1 hr. Degassing is conveniently effected with hypodermic needles inserted through the serum caps. The nitrogen effluent should be vented through a mineral oil or mercury bubbler. Short hypodermic needles inserted through the serum caps serve as convenient vents.

To the toluene is added with stirring 0.024 eq. of phenylmagnesium bromide in ether (Note 5) by means of a hypodermic syringe inserted through the serum cap. The mixture is cooled to 0-5°. From the addition funnel is added 5 ml. of isopropyl acrylate, and the mixture is allowed to stir at 0-5° for 3 hr. (Note 6), whereupon it is cooled to -70 to -80° with a Dry Ice-acetone bath. The remaining monomer is added dropwise while the temperature is maintained at -70 to -80°. The mixture is stirred for 1 hr. and is then held at -70 to -80° overnight without stirring (Note 7).

The polymer is isolated by precipitation at room temperature into a rapidly stirred solution containing 4.5 l. of methanol, 800 ml. of distilled water, and 175 ml. of concentrated hydrochloric acid. The product is removed by filtration and is thoroughly washed with distilled water. After drying to constant weight at 50-60° under reduced pressure, there is obtained 29-33 g. (76-86%) (Note 8) of isotactic poly(isopropyl acrylate) (Notes 9 and 10).

2. Characterization

Isotactic poly(isopropyl acrylate) prepared by this procedure is a granular, hard solid in contrast to normal free-radical-prepared polymer, which is a rubbery material at room temperature. The intrinsic viscosity of the Grignard-initiated polymer in chloroform at 25° is in the range 5-10 dl./g. The viscosity average molecular weight is calculated from the relationship

$$[\eta] = 1.4 \times 10^{-4} \bar{M}_v^{0.72}$$

derived from measurements on fractions of conventional polymer. A preliminary investigation of tactic polymer fractions indicates this relationship to be at least approximately valid (3). It should be noted that the ratio of viscosity average to number average molecular weight for Grignard-initiated acrylic polymers is large. Mechanistic reasons for the broad molecular weight distribution in these polymers have been discussed (3).

Isotactic poly(isopropyl acrylate) shows a typical crystalline powder diffraction pattern with strong scattering from spacings at 8.4, 5.15, 4.85, and 4.20 Å. (4). Indeed it is difficult to obtain this polymer as an amorphous solid. Dilatometric measurements indicate a glass transition at -11° and a crystal melting point at 162°. Data now available indicate a fiber repeat distance of 6.32 Å. and a density of 1.08 g./cc. (5). In-depth characterization of the structure of isotactic isopropyl acrylate polymer was not carried out by the original submitters. Details of NMR spectra and of the mechanism of isotactic polymerization have been reported (12) as have studies of the basic physical properties of isotactic poly (isopropyl acrylate) (13). Films of normal free-radically-initiated poly(isopropyl acrylate) exhibit infrared maxima at 1265 and 920 $cm.^{-1}$, while crystalline isotactic polymer shows sharp maxima at 1305, 1245, 1210, and 912 $cm.^{-1}$. The maxima at 1305, 1245, and 1210 $cm.^{-1}$ are not present in spectra of solutions of isotactic poly(isopropyl acrylate).

3. Notes

1. The submitters used rubber stoppers, catalog No. 8826, supplied by the A. H. Thomas Co., Philadelphia, Pa. Size 15 stoppers are convenient for use with a 24/40 standard taper joint.

2. It is essential that this polymerization be conducted under anaerobic, anhydrous conditions. The toluene may be conveniently dried by azeotropically removing the water. The toluene should be stored over sodium metal or calcium hydride.

The attainment of very high molecular weights requires scrupulous purity of the reaction mixture and the absence of even small amounts of water, alcohol, oxygen, and peroxides. All transfers should be conducted under a nitrogen atmosphere. Merely pouring isopropyl acrylate through air is sufficient to introduce significant amounts of oxygen.

3. Isopropyl acrylate is most conveniently prepared by direct esterification of glacial acrylic acid (Rohm & Haas Co., Philadelphia, Pa.), with concentrated sulfuric acid or *p*-toluenesulfonic acid as the catalyst. The monomer may also be prepared by transesterification (6). Freshly distilled, uninhibited monomer should be used for the polymerization. Traces of alcohol are conveniently removed by distillation from boric acid (7). The monomer should be distilled from a good inhibitor, collected in an ice-cooled receiver, and stored under nitrogen over calcium hydride in a refrigerator. The monomer has the following properties: b.p. 110-111° (760 torr), 53-54° (95 torr); n_D^{20} 1.4059; d_4^{20} 0.893.

4. The submitters used prepurified nitrogen (Air Reduction Sales Co.) without further purification.

5. The Grignard reagent should be 3-6*N* phenylmagnesium bromide in ether. The submitters found that the best results were obtained when the Grignard reagent was prepared under helium (8). Commercial phenylmagnesium bromide

(Arapahoe Chemicals, Inc., Boulder, Colo.) or the reagent prepared under nitrogen may also be used.

6. The isotacticity and the viscosity average molecular weight are greatest when this prereaction technique is employed. Polymer of lower tacticity and molecular weight results when this step is omitted.

7. The viscosity of the reaction mixture increases markedly during the course of the reaction; therefore it is advisable to discontinue stirring.

8. The yield of isolated polymer is based on the total weight of monomer charged. In addition to the isolated solid polymer, the aqueous methanol contains polymer of very low molecular weight (3).

9. Isotactic poly(*sec*-butyl acrylate), poly(*tert*-butyl acrylate), and poly(cyclohexyl acrylate) can be prepared by the procedure described. These polymers are also crystalline as prepared (4).

10. Isotactic poly(methyl methacrylate) can be prepared by the method described; however, the prereaction technique appears to be much less effective for methyl methacrylate and is omitted. The following procedure is used to prepare isotactic poly(methyl methacrylate).

Freshly distilled, uninhibited, oxygen-free methyl methacrylate (32 ml., 30 g., 0.3 mole) is added to 0.024 eq. of phenylmagnesium bromide in 525 ml. of anhydrous, oxygen-free toluene at 0-5°. The mixture is stirred at 0-5° overnight. The polymer is isolated by precipitation at room temperature into 10 volumes of rapidly stirred petroleum ether. The residual solvent is removed from the polymer under reduced pressure, and inorganic salts are removed by suspending the product in a stirred mixture of 1.7 l. of water, 300 ml. of methanol, and 90 ml. of concentrated hydrochloric acid for 2 hr. The polymer is then thoroughly washed with distilled water and dried to constant weight at 50-60° under reduced pressure. There is obtained 23 to 25 g. (77-83%) of isotactic poly(methyl methacrylate).

Isotactic poly(methyl methacrylate) is usually amorphous as obtained by this procedure; suitable treatment is required to develop crystallinity (9). The properties of isotactic poly(methyl methacrylate) have been discussed in detail (9-11).

11. Solutions of lithium aluminum hydride in diethyl ether have been used to prepare isotactic poly(isopropyl acrylate) (14). Stereoregular polymers are obtained from acrylate esters with unbranched ester side chains by using calcium or strontium complexed with diethylzinc (15, 16).

4. References

1a. Rohm and Haas Co., Bristol, Pennsylvania, 19007.
1b. The assistance of Dr. R. K. Graham in revising this procedure is thankfully acknowledged.
2a. General Electric Co., Research and Development Laboratory, Schenectady, New York 12345.
2b. Current Address: General Electric Co., Selkirk, New York 12158.
2c. Current Address: General Electric Co., Mount Vernon, Illinois 62864.
3. W. E. Goode, F. H. Owens, and W. L. Myers, *J. Polym. Sci.,* **47**, 75 (1960).
4. B. S. Garrett, W. E. Goode, S. Gratch, J. F. Kincaid, C. L. Levesque, A. Spell, J. D. Stroupe, and W. H. Watanabe, *J. Amer. Chem. Soc.,* **81**, 1007 (1959).
5. Unpublished results.
6. C. E. Rehberg, Org. Syn. Coll. Vo. **3**, 146 (1955).
7. R. C. Lemon and W. E. Goldsmith (Union Carbide Corp.), Brit. Pat. 812,498 (April 29, 1959) [*C. A.,* **53**, 16968d (1959)].
8. F. H. Owens, R. P. Fellmann, and F. E. Zimmerman, *J. Org. Chem.,* **25**, 1808 (1960).
9. W. E. Goode, F. H. Owens, R. P. Fellmann, W. H. Snyder, and J. E. Moore, *J. Polym. Sci.,* **46**, 317 (1960).
10. T. G. Fox, B. S. Garrett, W. E. Goode, S. Gratch, J. F. Kincaid, A. Spell, and J. D. Stroupe, *J. Amer. Chem. Soc.,* **80**, 1768 (1958).
11. J. D. Stroupe and R. E. Hughes, *J. Amer. Chem. Soc.,* **80**, 2341 (1958).
12. W. Fowells, C. Schuerch, F. A. Bovey, and F. P. Hood, *J. Amer. Chem. Soc.,* **89**, 1396 (1967).
13a. J. E. Mark, R. A. Wessling, and R. E. Hughes, *J. Phys. Chem.,* **70**, 1895 (1966).
13b. R. A. Wessling, J. E. Mark, E. Hamori, and R. E. Hughes, *J. Phys. Chem.,* **70**, 1903 (1966).
13c. R. A. Wessling, J. E. Mark, and R. E. Hughes, *J. Phys. Chem.,* **70**, 1909 (1966).
14. T. Tsuruta, T. Makimoto, and Y. Nakayama, *Makromol. Chem.,* **90**, 12 (1966).
15. J. Furukawa, T. Tsuruta, and T. Makimoto, *Makromol. Chem.,* **42**, 165 (1960).
16. T. Makimoto, T. Tsuruta, and J. Furukawa, *Makromol. Chem.,* **50**, 116 (1961).

Syndiotactic Poly(isopropyl acrylate)

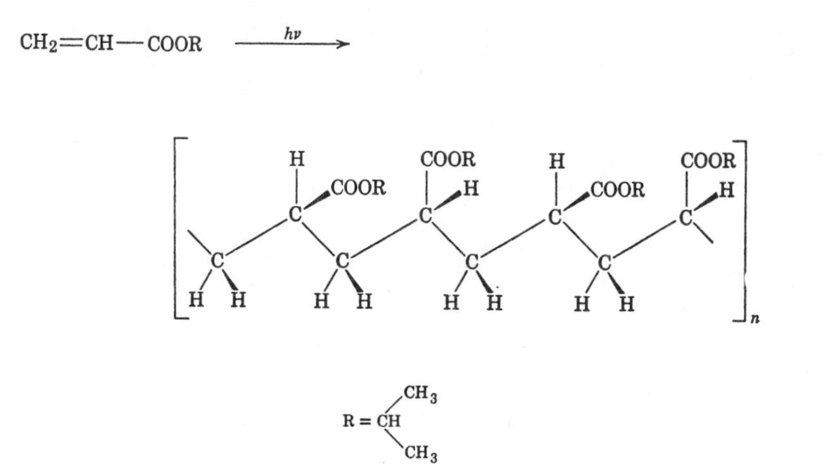

Submitted by C. F. Ryan and J. J. Gormley (1)
Checked by A. Katchman (2a, b) and E. E. Bostick (2a, c)

1. Procedure A

Apparatus (Fig. 1). An unsilvered Dewar flask (A) is placed on the platform (*B*) (Note 1). In the Dewar flask are placed a Pyrex® cylinder (*D*) (dia. = 8 cm.) open at the top, a mechanical stirrer (*E*), a pentane thermometer graduated from −200° to +30° (*F*), and the reaction vessel (*H*) (Note 2). 1-Propanol, which is used as the bath liquid, is placed in the Dewar flask to a level sufficient to cover the reaction vessel. Liquid nitrogen is placed in the Pyrex® cylinder (*D*) and serves as a coolant for the bath liquid. The temperature of the bath is adjusted by controlling the depth to which the cylinder (*D*) is inserted in the bath liquid. Constant temperature is maintained by repeatedly lowering and

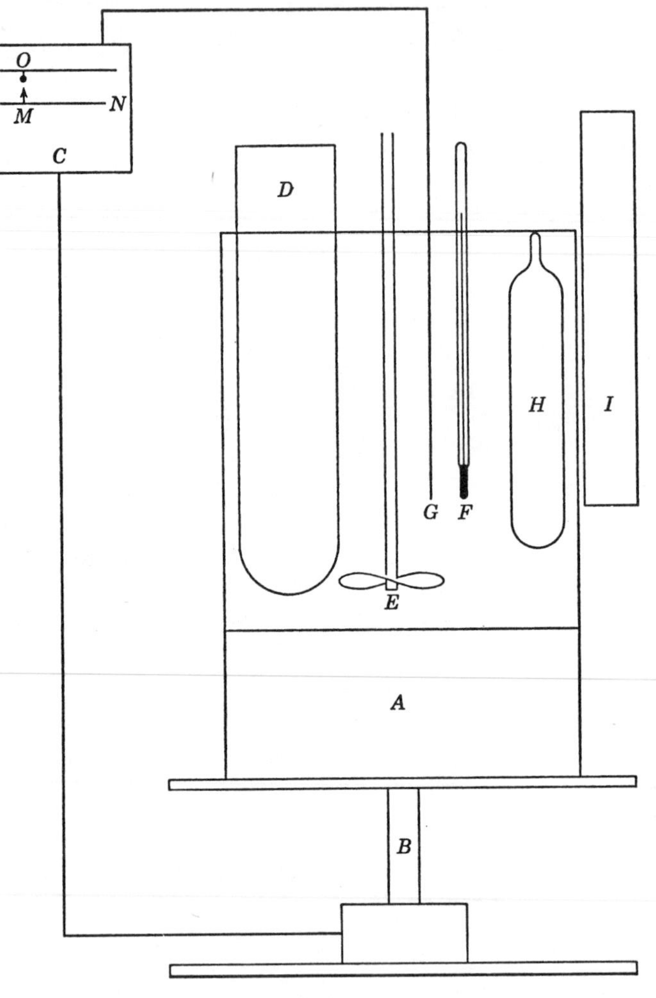

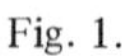
Fig. 1.

raising the platform when (*D*) is clamped in a fixed position. The platform can be lowered and raised either manually or automatically. An iron-constantan thermocouple (*G*) can be used in conjunction with a relay (*C*) to provide automatic movement of the platform (Note 3). An ultraviolet light source (*I*) (Note 4), mounted to within 2-3 in. of the reaction vessel, is used to initiate the reaction.

Polymerization. In the reaction vessel are placed 15 g. of reagent grade toluene, 15 g. (0.13 mole) of isopropyl acrylate (Note 5), and 0.124 g.

(6×10^{-4} mole) of benzoin (Note 6). The reactants are frozen in liquid nitrogen, placed under high vacuum (10^{-5} to 10^{-6} mm.) removed from the vacuum, and melted to liberate occluded gases. This degassing process is repeated several times, after which the reaction vessel is sealed undeı vacuum while the reactants are frozen (Note 7). The vessel is suspended in the Dewar flask, and the temperature of the bath is lowered to and maintained at $-105° \pm 7°$. The ultraviolet lights are turned on for 2 hr. The reaction vessel is removed, immersed in liquid nitrogen, and then opened. After thawing, the reaction mixture is poured slowly into ten times its volume of petroleum ether at $-20°$ to $-50°$. The precipitated polymer is isolated, dissolved in 20-30 ml. of reagent grade benzene, and this solution is filtered. The final syndiotactic polymer (1-2 g., which corresponds to an average rate of polymerization of approximately 5% per hour) is recovered by freeze-drying this solution (Notes 8 and 9).

2. Procedure B

An alternate procedure can be used to prepare larger quantities of material without the use of high vacuum techniques (Fig. 2). Here the reaction vessel (*P*) is a 500-ml., three-necked, round-bottomed Pyrex flask equipped with a mechanical stirrer (*Q*), a pentane thermometer (*F*), an iron-constantan thermocouple (*G*), and a Claisen adapter (*R*). A ground glass joint containing a gas inlet tube (*S*), is inserted in (*R*) below the surface of the reactants. An adapting tube (*T*), used to carry off the nitrogen effluent to a mercury or mineral oil bubbler (*U*), is inserted in the curved arm of (*R*). Temperature control and means of irradiation are accomplished by use of the same Dewar flask (*A*), platform (*B*), relay (*C*), and ultraviolet light source (*I*) described in Procedure A and illustrated in Fig. 1. (Note 1.)

In the reaction flask are placed 150 g. of reagent grade toluene, 150 g. (1.32 moles) of isopropyl acrylate (Note 5), and 1.25 g. (0.006 mole) of benzoin (Note 6). The reactants are cooled to $-95°$, and the solution is degassed by bubbling a stream of nitrogen through it for 2 hr. (Note 7). The temperature is lowered to and maintained at $-105° \pm 7°$ before the reaction is initiated by irradiation from the ultraviolet light source. Nitrogen bubbling and agitation are maintained throughout the entire reaction time.

After 6 hr. of irradiation, the ultraviolet lamps are turned off, and the reaction mixture is allowed to warm to room temperature. The polymer is isolated as indicated in Procedure A, dissolved in 500 ml. of benzene, and then this solution is filtered. The final syndiotactic polymer (23 g., which corresponds to an

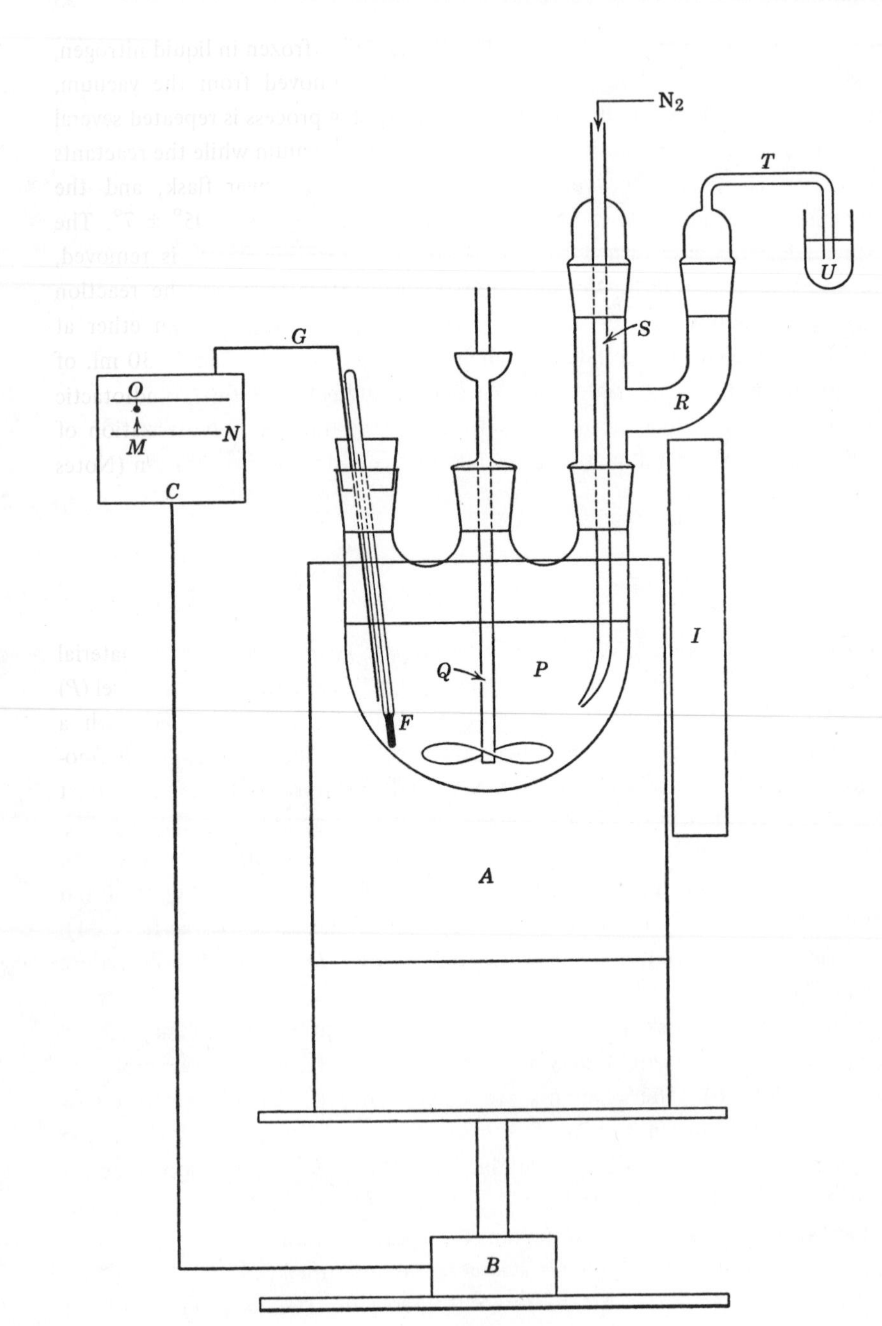

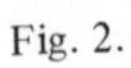

Fig. 2.

average rate of polymerization of approximately 2.6% per hr.) is isolated by freeze-drying this solution.

3. Characterization

The viscosity average molecular weight of the rubbery, partially crystalline, syndiotactic poly(isopropyl acrylate) prepared by the methods described was calculated from the intrinsic viscosity $[\eta]$ in benzene at 30° from the relationship

$$[\eta] = 1.18 \times 10^{-4} \overline{M}_V^{0.71}$$

derived from measurements on fractions of conventional polymer (3) prepared at 44.1°. For the material prepared by Procedure A, $[\eta] = 0.62$ dl./g.; for material prepared by Procedure B, $[\eta] = 1.4$ dl./g.

Syndiotactic poly(isopropyl acrylate) gives a typical crystalline powder diffraction pattern with strong scattering from spacings at 7.4 and 4.25 Å.; a fiber repeat distance of 5.18 Å.; a calculated density of 1.18 g./cc. (4); and a dilatometric glass transition temperature between 5° and 11° (Note 10). A crystalline melting point, $T_m = 115°$, has been found from dilatometric measurements for a sample of syndiotactic poly(isopropyl acrylate).

Films of conventional poly(isopropyl acrylate) exhibit infrared maxima at 1265 and 920 $cm.^{-1}$, whereas syndiotactic polymer exhibits sharp maxima at 1250 and 924 $cm.^{-1}$. The maximum at 1250 $cm.^{-1}$ is not present in spectra of solutions of syndiotactic polymer.

4. Notes

1. The submitters used a Jack-O-Matic, model HD-4, as the platform (available from Instruments for Industry and Research, 108 Franklin Avenue, Cheltenham, Pa.). The checkers reported satisfactory temperature control by manual adjustment and stirring liquid nitrogen into 1-propanol.

2. A Pyrex® ampoule (dia. = 2.5 cm., length = 17 cm.), fitted with a 10/30 standard taper joint and constricted below the joint, was used as the reaction vessel.

3. The submitters used the Gardsman, Model J, Indicating Pyrometric Controller (available from West Instrument Corporation, 525 North Noble Street, Chicago 22, Ill.) as a relay (shown in section *C* of Fig. 1). A manually operated setting arm (*M*) is set at the desired position on the temperature

scale (*N*). When a decrease in temperature occurs in the bath, the current generated by the thermocouple (the leads of which must be reversed to accommodate low temperature operation) activates the pointer (*O*). The relay is energized, and the Jack-O-Matic rises at the point where (*M*) and (*O*) coincide. When the pointer (*O*) exceeds the position of (*M*), the relay is de-energized, and the Jack-O-Matic falls to its normal position. The repetition of this action provides temperature control to within ±7° at very low temperature (−100°) and to within ±3° at higher temperature (−50°).

4. The submitters used two pairs of 15-watt Sylvania Blacklight bulbs (wavelength = 320-420 nm 100% transmission at 360 nm. The lights were placed in position so that their beams struck the reactants at an angle of 45° to each other. A Hanovia Inspectolite may also be used, but, because this is a 100-watt source, the amount of photosensitizer should be reduced proportionately.

5. Isopropyl acrylate was distilled from diphenylpicrylhydrazyl just before use; b.p. = 111° (760 torr), 53-54° (95 torr), n_D^{20} = 1.4059, d_4^{20} = 0.893.

6. Reagent grade benzoin, which serves as the photosensitizer, was used as received from Eastman Organic Chemicals Company.

7. It is essential that the polymerization be conducted in the complete absence of oxygen.

8. If the reaction mixture contains peroxidic impurities, a higher yield of lower molecular weight polymer is obtained. The checkers reported a yield of 4.5 g. of polymer with $[\eta]$ = 0.27 dl./g.

9. Syndiotactic poly(methyl methacrylate) can be prepared by Procedure A, but a Dry Ice-acetone mixture is used in the cylinder (*D*). Forty grams of freshly distilled uninhibited methyl methacrylate and 0.0002 g. of benzoin are placed in the reaction vessel, and the solution is degassed. The reactants are irradiated at a constant temperature of −48° ±3° for 5 hr. with a 100-watt Hanovia Inspectolite, or equivalent, mounted 6 in. from the reaction vessel. The average rate is approximately 0.3% per hour, and the viscosity average molecular weight is approximately 3×10^6. The properties of syndiotactic poly(methyl methacrylate) have been discussed in detail (5, 6).

10. Conventional polymers prepared at 44.1° have a glass transition temperature of less than −2°.

5. References

1. Rohm & Haas Co., Bristol, Pennsylvania 19007.
2a. General Electric Co., Research Laboratories, Schenectady, New York 12345.
2b. Current Address: General Electric Co., Selkirk, New York 12158.
2c. Current Address: General Electric Co., Mount Vernon, Illinois 62864.
3. E. S. Cohn, T. A. Orofino, and L. L. Scogna, Unpublished results.
4. H. S. Yanai, Unpublished results.
5. T. G. Fox, B. S. Garrett, W. E. Goode, S. Gratch, J. F. Kincaid, A. Spell and J. D. Stroupe, *J. Amer. Chem. Soc.*, **80**, 1768 (1958).
6. T. G. Fox, W. E. Goode, S. Gratch, C. M. Huggett, J. F. Kincaid, A. Spell and J. D. Stroupe, *J. Polym. Sci.*, **31**, 173 (1958).

Poly(acrylic anhydrides)

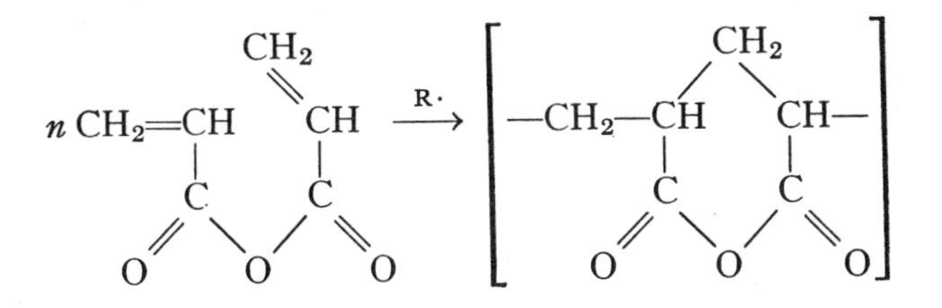

Submitted by G. B. Butler, A. Crawshaw, and W. L. Miller (1)
Checked by C. G. Overberger (2a, b) and H. Ringsdorf (2c)

Acrylic anhydride and substituted acrylic anhydrides may be polymerized *via* a cyclic chain propagation mechanism (3, 4) to yield linear, cyclic, polymeric anhydrides (5-7).

1. Procedure

A solution of 3.96 g. (4.0 ml.) of freshly distilled acrylic anhydride (Note 1) and 40 mg. of benzoyl peroxide (*Caution! Benzoyl peroxide is a strong oxidant and can ignite or explode if not properly treated.)* in 4.0 ml. of dry benzene (Note 2) is prepared in a screw-capped vial. The vial is flushed with dry nitrogen and capped. This solution is then placed in an oven at 65° for 2 hr. (Notes 3, 4). At the end of the 2-hr. period of heating, the whole mass has solidified.

The contents of the vial are dissolved in 100 ml. of dry dimethylformamide (Note 5) and precipitated by addition to 500 ml. of vigorously stirred, dry, ethyl ether.

The polymer is collected by filtration and dried under reduced pressure to yield 3.08-3.36 g. (78-85%). The polymer is a finely divided white solid, very susceptible to electrostatic charge, with a softening point of approximately 220°. The polymer does not add bromine in solution, and the infrared spectrum shows negligible absorption for both OH and C=C groups (Note 6).

The method of Kagawa and Fuoss (Note 7) is used to determine the molecular weight of the polymer, and a value of approximately 500 is obtained for the degree of polymerization of the poly(acrylic acid) obtained by hydrolysis of the poly(acrylic anhydride). This corresponds to a molecular weight of 95,000 for the poly(acrylic anhydride). A sample obtained from a reaction carried out with 2.5% of monomer in benzene gives a molecular weight of approximately 32,000.

The same general procedure (Note 8) may be used to prepare poly(methacrylic anhydride) in 85-95% yields with a molecular weight of approximately 62,000 (Note 9). This material shows the same solubility characteristics as poly-(acrylic anhydride) with the exception that it is more difficult to hydrolyze with water (Note 10).

2. Notes

1. The monomer obtained from the Borden Company contains an inhibitor, usually copper salts or hydroquinone. The distillation is carried out under reduced pressure, using a short fractionating column packed with copper beads or shot. The receiver is cooled with an ice-salt mixture, and the distilled monomer is kept cold and protected from light until used. The monomer, so treated, has a density of 1.0940 at 25° and is a clear, water-white liquid, b.p. 85-86°/17 torr., 30-31°/1 torr.
2. To protect the anhydride group from hydrolysis, all solvents must be dried prior to use. The benzene and ethyl ether are distilled and dried over sodium wire. The dimethylformamide is distilled and dried over Drierite®
3. Experiments carried out in an oil bath give a higher polymerization rate because of more efficient heat exchange.
4. After several minutes the polymerization mixture becomes turbid and solidifies before the end of the reaction period. The reaction is extremely exothermic.
5. It requires from 5-8 hr. to dissolve 5 g. of polymer in 100 ml. of dimethylformamide.
6. The infrared spectra of acrylic anhydride, poly(acrylic anhydride), and glutaric anhydride all contain two C=O bands at 5.5 and 5.8 μ as expected, and both poly(acrylic anhydride) and glutaric anhydride have a small shoulder at 6.0-6.15 μ which suggests the presence of a small amount of the enol form in both of these materials. The relative intensity of the two carbonyl bands in glutaric anhydride and poly(acrylic anhydride) is reversed as compared to that of acrylic anhydride.

7. The molecular weight of the polymer may be calculated from the intrinsic viscosity of a 2*N* sodium hydroxide solution at 30° by using the following relationship (8):

$$[\eta] = 4.27 \times 10^{-3} P^{0.69}$$

where P is the degree of polymerization.

8. The heating period is 12 hr. and the monomer/solvent ratio is 1/10, in the case of methacrylic anhydride.

9. This value was approximated from the curve given by Katchalsky and Eisenberg (9). The molecular weight of poly(methacrylic anhydride) was also determined by conversion of a sample of the anhydride (from a similar preparation, with a monomer/solvent ratio of ¼) to poly(methyl methacrylate) with diazomethane (9) and then using the relationship of Baxendale, Bywater, and Evans (10):

$$M_n = 2.81 \times 10^3 [\eta]^{1.32}$$

where M_n is the number average molecular weight, and the intrinsic viscosity is determined in benzene at 25°. This gave a value of 124,000 for the poly(methyl methacrylate), which corresponds to a value of 95,000 for the poly(methacrylic anhydride).

10. Both poly(acrylic anhydride) and poly(methacrylic anhydride) were found to be insoluble in all the common organic solvents such as ether, benzene, acetone, methanol, ethanol, chloroform, and carbon tetrachloride and ethyl acetate. They dissolve slowly in dimethylformamide and dimethyl sulfoxide. They dissolve with reaction in water, sodium hydroxide solution, and sodium bicarbonate solution. Poly(acrylic anhydride) dissolves in water (1 g. of polymer in 25 ml. of distilled water) at room temperature on standing for a period of about 24 hr. Poly(methacrylic anhydride), however, requires a longer period (50-100 hr.) and a more dilute solution (1 g. in 40 ml.).

3. References

1. The University of Florida, Gainesville, Florida 32601.
2a. Polytechnic Institute of Brooklyn, Brooklyn, New York 11201.
2b. Current Address: The University of Michigan, Ann Arbor, Michigan 48104.
2c. Current Address: Institute for Organic Chemistry, The University of Mainz, D-65 Mainz, West Germany.
3. G. B. Butler and R. J. Angelo, *J. Amer. Chem. Soc.*, 79, 3128 (1957).
4. G. B. Butler, A. Cranshaw, and W. L. Miller, *J. Amer. Chem. Soc.*, 80, 3615 (1958).

5. A. Cranshaw and G. B. Butler, *J. Amer. Chem. Soc.*, **80**, 5464 (1958).
6. J. F. Jones, *J. Polym. Sci.*, **33**, 15 (1958).
7. J. F. Jones (B. F. Goodrich Co.), Ital. Pat. 563,941 (June 7, 1957).
8. I. Kagawa and R. M. Fuoss, *J. Polym. Sci.*, **18**, 535 (1955).
9. A. Katchalsky and H. Eisenberg, *J. Polym. Sci.*, **6**, 145 (1951).
10. J. H. Baxendale, S. Bywater, and M. G. Evans, *J. Polym. Sci.*, **1**, 237 (1946).

Ethylene Maleic Anhydride Copolymer

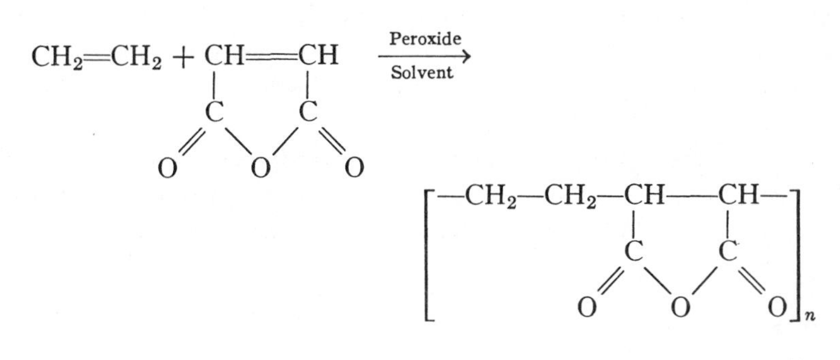

Submitted by J. H. Johnson (1)
Checked by J. R. Schaefgen (2)

1. Procedure

Maleic anhydride (*Caution! Maleic anhydride is a very toxic substance and gloves should be worn when using this material. Avoid breathing, or skin contact with, vapors or dust from this substance.*) (267 g., 2.72 moles) is dissolved in 2089 ml. of benzene (Note 1) at room temperature. The solution is filtered to remove any insoluble matter (Note 2) and charged to a standard 3-1. autoclave, with rocking or stirred agitation provided (Note 3). Benzoyl peroxide (*Caution! Benzoyl peroxide is a strong oxidant and can ignite or explode if not properly treated.*) (3.48 g. of 95% active, 0.50 mole % based on maleic anhydride) is added to the charge and the autoclave is closed. This charge fills the autoclave approximately one-half full. The system is then pressure flushed (Note 4) with high purity ethylene (Note 5) by alternately raising the pressure to 100 p.s.i.g. and then releasing it to a suitable exhaust system three times. The system is then pressurized to approximately 250 p.s.i.g., agitation is started, and the charge is heated to the polymerization temperature of 70° (Note 6).

The pressure is adjusted to 300 p.s.i.g. and this pressure is maintained throughout the reaction by continuously adding ethylene to replace that consumed in the polymerization. The reaction rate can be followed roughly by the rate of ethylene consumption (Note 7) with 1 mole of ethylene being consumed for each mole of maleic anhydride (Note 8). Polymerization time is 15-18 hr. (Note 9).

The pressure is reduced and the autoclave is opened. The polymer is present as a swollen gel. It is recovered by filtration, washed 4-6 times with ethylene dichloride (Note 10) and dried in a vacuum (28-29-in.) oven at 110° for 18-24 hr. The product is a fine white powder (yield 90-99% of theory based on maleic anhydride). Analysis indicates a copolymer composition of 1/1 mole ratio of reacting monomers (Note 8). The degree of polymerization is estimated by determination of the specific viscosity of a 1% solution of the copolymer in dimethylformamide at 25°. The product has a value of η_{sp} 0.5 dl./g. (Note 11). It should be stored in a dry atmosphere to prevent hydrolysis of the anhydride groups. The reaction described is applicable, with minor modifications, to the copolymerization of other olefins with maleic anhydride (Note 12).

2. Notes

1. Dry, thiophene-free benzene should be used. Ethylene dichloride is another preferred solvent which offers the advantage of a denser product. However, its successful use depends on the availability of good-quality maleic anhydride (essentially free of maleic acid). Ethylene dichloride should be anhydrous and free of hydrogen chloride.

2. Filtration of benzene-maleic anhydride solutions at room temperature removes a major portion of hydrolysis products which form during normal storage. Maleic acid markedly reduces polymerization rates, conversions, and polymer molecular weight. This filtration procedure does not completely compensate for the superior results obtained with high purity anhydride. Commercial anhydride that has been stored away from moisture is satisfactory, the critical test being production of clear solutions in benzene. For large scale production, maleic anhydride, shipped and stored in the molten state, provides good protection against maleic acid formation.

3. Standard pressure autoclaves with rated operating pressures of 650 p.s.i.g. or higher are suitable. Either rocking- or propeller-type agitation is satisfactory.

4. Oxygen present in the vapor space and dissolved in the charge interferes with polymerization. It is conveniently removed by this pressure flushing

technique. The gas is introduced through an inlet tube arranged to keep the opening above the liquid level to prevent plugging during reaction. A simple sweep of the vapor space with inert gas is insufficient to ensure good polymerization. A suitable alternate technique involves boiling the charge under reduced pressure.

5. Ethylene of high purity with low acetylene and oxygen contents is required. Saturated hydrocarbon impurities (methane, ethane, etc.) do not interfere with polymerization other than to reduce monomer concentration. A satisfactory ethylene stock has the following analysis: ethylene, 99.85; ethane, 0.1; methane, 0.01; acetylene, 40 p.p.m.; carbon dioxide, 50 p.p.m.; oxygen, less than 2 p.p.m.

6. The pressure is not adjusted to its final value until the reactor is at operating temperature to allow for pressure increase of the charge upon heating. Reaction conditions can usually be achieved in 20-30 min. after heating is begun.

7. A convenient method of following the extent of reaction, if the possibility of intermittent sampling is provided, consists of filtering an aliquot sample and titrating the maleic anhydride in the filtrate with standardized base using thymol blue indicator. Precise aliquots are unnecessary if uniform sample size is adopted and titration values are compared to an initial value determined for the charge immediately before initiation of polymerization.

8. Ethylene and maleic anhydride enter the polymer in a 1/1 ratio with alternating monomer units (3). The 1/1 composition is indicated by the following typical carbon-hydrogen analysis. Calcd. for $C_6H_6O_3$: C, 57.19; H, 4.80. Found: C, 57.18; H, 4.88.

9. The polymerization time can be varied from 4 to 30 hr. by the selection of varying combinations of temperature, pressure, and catalyst. Higher temperature and pressure favor faster rates. When lower temperatures are required to produce higher-molecular-weight product, more active catalysts such as 2,4-dichlorobenzoyl peroxide must be used to ensure practical rates.

10. Polymer is washed to remove unreacted maleic anhydride and catalyst residues. Ethylene dichloride is preferred to benzene.

11. The specific viscosity of the polymer can be varied over a range of 0.05-3.0 by the use of suitable regulators (*n*-butyraldehyde, ethylbenzene, etc.) and by appropriate choice of pressure and temperature. High pressures and low polymerization temperatures give polymer with higher specific viscosities, the highest value being obtained at about 600 p.s.i.g. ethylene pressure at 45°

12. This procedure, with minor modifications, has been utilized in preparing a large number of maleic anhydride copolymers including propylene, isobutylene, 1-dodecene, styrene, vinyl acetate, methyl vinyl ether, and vinyl chloride.

3. Methods of Preparation

Preparation of ethylene-maleic anhydride copolymers are described in the patent literature (4, 5). Ethylene/maleic anhydride copolymer is commercially available from Monsanto Industrial Chemicals Co., St. Louis, Missouri 63166.

4. References

1. Monsanto Chemical Company, St. Louis, Missouri 63166.
2. Experimental Station, E. I. du Pont de Nemours & Co., Wilmington, Delaware 19898.
3. P. J. Flory, *Principles of Polymer Chemistry,* Cornell University Press, 1953, pp. 187 ff.
4. W. E. Hanford (E. I. du Pont de Nemours & Co.), U.S. Pats. 2,378,629 (June 19, 1945) [*C.A.,* **39**, 4265 (1945)] and 2,396,785 (March 19, 1946) [*C.A.,* **40**, 3459 (1946)].
5. J. H. Johnson (Monsanto Chemical Co.), U.S. Pats. 2,857,365 (Oct. 21, 1958) [*C.A.,* **53**, 1853h (1959)], 2,913,437 (Nov. 17, 1959) [*C.A.,* **54**, 7126i (1959)], and 2,938,016 (May 24, 1960) [*C.A.,* **54**, 19024h (1960)].

Poly(propylene maleate phthalate)

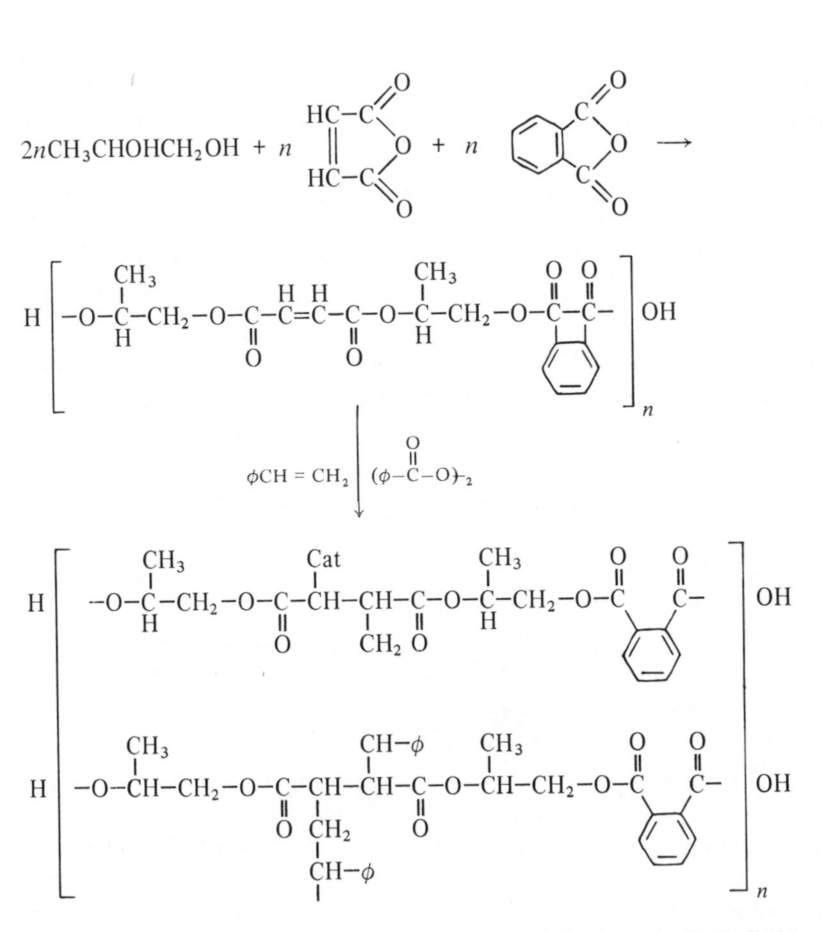

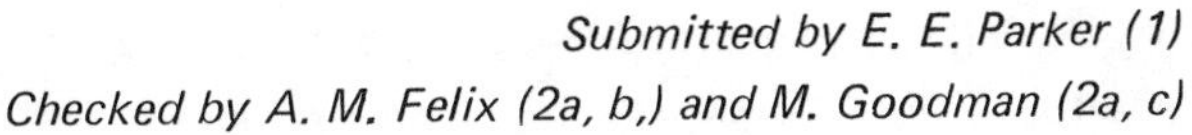

Submitted by E. E. Parker (1)

Checked by A. M. Felix (2a, b,) and M. Goodman (2a, c)

Ed. Note: In modern practice, no xylene is used and the reaction temperature is about 210°. This shortens the reaction time. Nitrogen is introduced at a rate of 0.02 in.^{3}/min. during the "cooking stage" and 0.04 in.^{3}/min. during the blowing stage. A mechanical stirrer is always used as suggested in Note 2. A fractionating column is used to prevent excessive losses of propylene glycol. An all-glass bulb (Liebig) condenser with boiling water or steam in the jacket works well in this capacity. A thermometer at the top of the column is used to control the rate of heat input to the reaction when the nitrogen sparge is increased to 0.04 in.^{3}/min. (usually when the top column temperature drops to 65-70°).

1. Procedure

A 5-1. three-necked flask is equipped with a thermometer, an inlet tube for inert gas (center neck) (Notes 1 and 2), a 150-ml. Dean-Stark trap with a bottom drain, and a reflux condenser. Into the flask are charged 1068 g. (11 moles) of maleic anhydride (Note 3), 1628 g. (11 moles) of phthalic anhydride (Note 4), 1839 g. (24.2 moles) of 1,2-propylene glycol (Note 5), 400 g. of xylene, and 0.91 g. of hydroquinone. Additional xylene is used to fill the trap. The flask is heated with a 5-1. hemispherical Glas-Col heating mantle attached to a 1-KVA variable transformer.

The charge is heated at 110 volts with occasional stirring to 100°. Inert gas (Note 6) is bubbled slowly through the batch as soon as it becomes sufficiently molten. At 100° heating is stopped until the temperature reaches a maximum and starts to fall (Note 7). Heating is then restarted. A setting of 90 volts on the transformer is usually sufficient to maintain a steady reflux. Water is removed through the bottom drain of the Dean-Stark trap as required. Water should not be allowed to run from the trap into the batch. A typical heating schedule is given in Table 1 (Note 8).

Xylene may be added or withdrawn from the batch to control the reflux rate and the temperature. The xylene can be added *slowly* through the top of the reflux condenser. The heating schedule can be varied considerably without materially affecting the results. When the acid number (Note 9) of the batch reaches 50-52, the contents of the trap are drained and the flow of inert gas is gradually increased to as high a rate as possible without entrainment of the resin to remove the xylene and other unreacted materials. The fumes from this procedure should be disposed of in a good fume hood. Blowing is continued until the viscosity of a 60% solution of polyester in ethylene glycol monoethyl ether (Cellosolve) reaches G-H on the Gardner-Holt scale (Note 10). At this point, the inert gas is reduced to a slow stream, and the batch is allowed to

TABLE 1. HEATING SCHEDULE

Time, hr.	Temperature	H_2O, total ml.	Remarks, Acid Number
0:00			Heat on
0:30	100		Heat off
1:00	160		Heat on
8:00	195	300	
10:00	195	380	74
11:00	195	410	63
12:00	195	420	57
13:00	195	430	52
13:10	Blow		Viscosity D+
13:40			Viscosity G+; heat off
13:50	Cool to 150° and thin		

cool to 150°. It is then weighed quickly into a previously tared 8-l. stainless steel beaker and cooled to 125°. (Note 11). The yield will be about 4 kg. The polyester should be stirred with a large stainless steel spatula or a mechanical agitator to prevent the formation of a hard layer on the sides of the beaker. Enough styrene (commercial rubber or polymer grade inhibited with 50 p.p.m. *p*-tertiary butyl catechol) should be added to make the resin solution 35% in styrene. It is preferable to get a small portion of the styrene to mix with the polyester before the remainder is added. After the mixture becomes homogeneous, it is cooled as rapidly as possible to room temperature with a cold water bath. Any lumps can be removed by filtration through cheese cloth.

Small castings (50-100 g.) can be prepared by dissolving 1% benzoyl peroxide (*Caution! Benzoyl peroxide is a strong oxidant and can ignite or explode if not properly treated.*) in the resin and heating the catalyzed resin for 1 hr. at 75° and 1 hr. at 125°. Larger castings tend to crack because of the heat of polymerization. This problem can be overcome, to some extent, by the use of a catalyst-accelerator system that will gel the resin at room temperature. A suitable recipe is:

100 parts resin
0.3 part cobalt octoate (12% Co) (Note 12)
1.0 part Lupersol DDM (Note 13)

The cobalt octoate is dissolved in the resin, and then the Lupersol DDM is added and dispersed. The casting is allowed to gel at room temperature and the exotherm to subside. It is then heated for 1 hr. at 75° and for 1 hr. at 125° as before. Typical properties of the cast material are:

Flexural strength, p.s.i. at 25°	14,000-15,000
Modulus of elasticity in flexural, p.s.i.	550,000
Heat distortion point	85-90°
Tensile strength, p.s.i. at 25°	7000-8000
Hardness: Barcol Impressor	45-50

2. Notes

1. Corks are more durable than rubber stoppers for this reaction. Ground glass joints can be used if care is taken to avoid contamination of the batch with stopcock grease.

2. A mechanical agitator can be used if desired. Then the inlet tube for inert gas and the thermometer are combined in one of the small necks of the flask.

3. Commercial grade, m.p. 52° (minimum).

4. Commercial grade, m.p. 130-132°.

5. Commercial grade, refractive index at 25° of 1.4300-1.4320 is satisfactory.

6. Oxygen-free nitrogen or carbon dioxide can be used. Care must be taken that the resin does not become superheated. It is vital that the stream of inert gas be maintained at all times. If it does stop for as long as a few minutes, the heat should be turned off and the batch allowed to cool before the flow of inert gas is started again. If a superheated batch is disturbed, a steam-boiler type of eruption will result.

7. The temperature in a freshly charged batch will rise about 60°C. because of the heat liberated by the exothermic reaction. It is reduced to some extent if the charge is stored overnight before heating is started.

8. If smaller batches are made, considerably shorter heating schedules may be used. Smaller batches are considerably more difficult to control, especially during the blowing cycle. Care should be taken with smaller batches that the acid number and viscosity of the batch do not go beyond the desired values.

9. Polyester (1-2 g.) is dissolved in 20 ml. of neutral acetone. The solution is titrated with approximately 0.2*N* potassium hydroxide dissolved in 90% ethanol with a phenol red indicator. Acid number = mg. of KOH/g. of sample. The KOH can be standardized against potassium acid phthalate.

10. If Gardner-Holt viscosity standards are not available, the batch can be

blown until phthalic anhydride starts to crystallize on the apparatus. Blowing should then continue for an additional 10 min.

11. Part of the batch can be thinned in 4-l. glass beakers or other suitable vessels. Iron containers should be avoided.

12. Obtained from Mooney Chemical Company.

13. Obtained from Lucidol Division, Wallace and Tiernan, Inc.

14. An alternate synthesis uses propylene oxide in place of propylene glycol (3, 4) requiring a shorter reaction time.

3. References

1. PPG Industries, Springdale, Pennsylvania 15144.
2a. Polytechnic Institute of Brooklyn, Brooklyn, New York 11201.
2b. Current Address: Hoffman-La Roche Inc., Nutley, New Jersey 07110.
2c. Current Address: The University of California, San Diego, La Jolla, California 92037.
3. J. A. Seiner and E. E. Parker, U.S. Pat 3,374,208, Mar. 19, 1968.
4. R. E. Carpenter and C. R. Peterson, U.S. Pat. 3,723,390, Mar. 27, 1973.

Polydisulfide of 1,9-Nonanedithiol by Catalytic Air Oxidation (1)

$$HS(CH_2)_9SH + (O) \xrightarrow[(OH^-)]{SeO_2} [-S(CH_2)_9S-]_n + nH_2O$$

Submitted by P. V. Bonsignorè (2a, b) and C. S. Marvel (2a)
Checked by C. G. Overberger (3a, b) and J. J. Ferraro (3a, c)

1. Procedure

In a 4-oz. screw-cap bottle is placed a solution of 4 g. of potassium hydroxide (85% C.P.) in 50 ml. of distilled water. Lauric acid, 1.5 g. (C.P.), is dissolved in the alkaline solution, and 25 mg. of selenious acid (*Caution! Many selenium compounds are extremely toxic and due care should be exercised in their use.*) is added (Note 1). Five ml. of 1,9-nonanedithiol (*Stench! All operations using this material should be conducted in a hood. The use of disposable gloves is advisable throughout the procedure when free mercaptan is present.*) (Note 2) is added and a stable solution or emulsion is formed by vigorous shaking (Note 3). Filtered, compressed air is bubbled through the solution—through a 6-mm. glass tubing that reaches almost to the bottom of the vessel—at a rate of about two bubbles per second for four to ten days (Note 4).

At the end of this time the polymer is coagulated by pouring the latex into about 500 ml. of methanol (Note 5). The precipitated polymer is separated by decantation, dissolved in approximately 50 ml. of chloroform, and reprecipitated into methanol (Note 6). It is collected and dried in a vacuum oven at 50°. The usual yield is between 60% and 90% of polydisulfide, inherent viscosity 0.2-0.6 (0.5% dl./g. (0.5% in chloroform) (Notes 7 and 8).

2. Notes

1. Of numerous catalysts tried for the oxidation of dimercaptans, selenious acid proved by far the most effective (4).

2. The structure of the dimercaptan can be varied over a wide range. Data have been published for the preparation of a variety of polydisulfides by oxidation of dimercaptans, e.g., $HS(CH_2)_nSH$ where n = 6, 7, 9, 10 (5), 2-mercaptoethylcyclohexyl mercaptan (5), and dimercaptans containing disiloxane groups (6):

$$HS(CH_2)_n\overset{\displaystyle CH_3}{\underset{\displaystyle CH_3}{\overset{|}{\underset{|}{Si}}}}-O-\overset{\displaystyle CH_3}{\underset{\displaystyle CH_3}{\overset{|}{\underset{|}{Si}}}}(CH_2)_nSH \qquad \text{where } n = 2, 3\ (6)$$

Two convenient methods are available for the formation of dimercaptans suitable for oxidation to polymeric disulfides. These are: (*a*) the replacement of the halogen atoms of α,ω-dihalides by the acition of sodium hydrosulfide or thiourea (isothiuronium salt method) (7), or (*b*) by the addition of thiolacetic acid to α,ω-diolefins, and the saponification of the resulting bis thioacetates (6, 8, 9). Numerous methods are available for the synthesis of α,ω-diolefins (10) that are suitable as intermediates in synthesis (*b*).

3. The addition of an antifoaming agent is usually necessary at this point to eliminate troublesome foam both in the preparation of the emulsion and in the subsequent air-bubbling treatment. Dow-Corning "Antifoam A" has been used satisfactorily.

4. The addition of distilled water from time to time is necessary to replace water lost by evaporation.

5. Attempts to coagulate the latex using alum coagulant (potassium aluminum sulfate in dilute hydrochloric acid) resulted in polymers which could not be entirely freed of inorganic impurities even after repeated precipitations from chloroform or benzene into methanol.

6. If the polymer obtained by coagulation of the latex into methanol assumes a pink or red coloration (from the selenious acid), the color can be removed by adding a small amount of sodium hydroxide to the methanol used in the second reprecipitation (about 1 g. per 500 ml.).

7. In some cases, higher temperatures for the oxidation resulted in higher inherent viscosities for the resultant polydisulfides (6). A convenient apparatus for carrying out the oxidative polymerization reaction of dimercaptans at

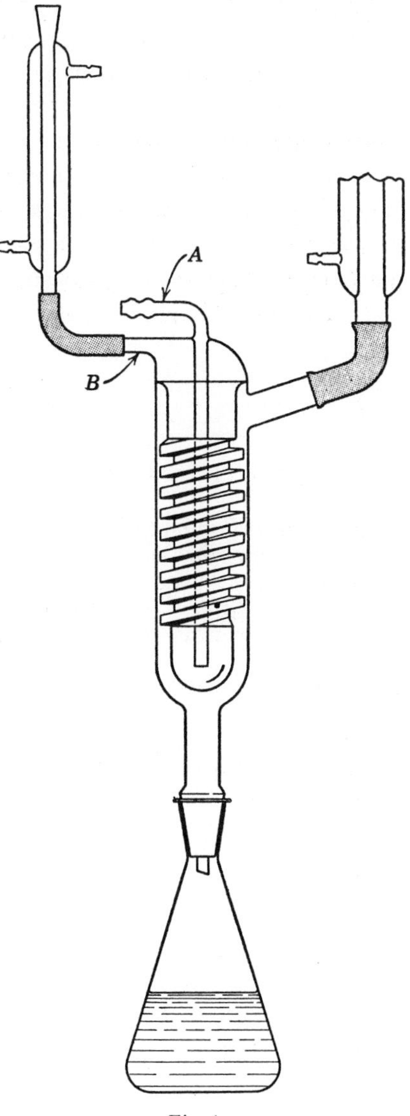

Fig. 1.

constant, controlled temperatures is a water condenser of the Friedrichs type, as illustrated in Fig. 1 (6).

A double batch of the emulsion is placed in the inner portion of the condenser. Filtered compressed air is led in through the inlet tube *A*. The outlet tube *B* leads through a short length of Tygon tubing to a small upright condenser that prevents excessive evaporation of the reaction solution. The reaction is kept at a constant specified temperature by refluxing a liquid of suitable boiling point through the outer portion of the condenser.

8. The yield and inherent viscosity depend on the purity of the monomer and the length and temperature of oxidation. The checkers obtained yields of 88-91% after an oxidation period of 7 days at room temperature.

9. Dimercaptans can also be oxidized to poly (disulfides) using dimethyl sulfoxide as solvent and oxidant (11). A novel system for the production of poly(disulfides) is by reaction of dithiols with oxycarbonyl disulfides (12).

3. References

1. J. C. Patrick (Thiokol Corp.), U.S. Pat. 2,142,144 (Jan. 3, 1939) [*C.A.*, **33**, 3027 (1939)]
2a. The University of Arizona, Tucson, Arizona 85721.
2b. Current Address: Aluminum Company of America, Alcoa Center, Pennsylvania 15069.
3a. Polytechnic Institute of Brooklyn, New York 11201.
3b. Current Address: The University of Michigan, Ann Arbor, Michigan 48104.
3c. Current Address: Long Island University, Zeckendorf Campus, Brooklyn, New York 11201.
4. G. H. Meguerian [Standard Oil Co. (Indiana)], U.S. Pat. 2,740,748 (April 3, 1956) [*C.A.*, **50**, 9737a (1956)].
5. C. S. Marvel and L. E. Olson, *J. Amer. Chem. Soc.*, **79**, 3089 (1957).
6. C. S. Marvel, P. V. Bonsignore, and S. Banerjee, *J. Org. Chem.*, **25**, 237 (1960).
7. W. P. Hall and E. E. Reid, *J. Amer. Chem. Soc.*, **65**, 1466 (1943).
8. C. S. Marvel and L. E. Olson, *J. Polym. Sci.*, **26**, 23 (1957).
9. C. S. Marvel and H. N. Cripps, *J. Polym. Sci.*, **9**, 53 (1952).
10. C. S. Marvel and W. E. Garrison, *J. Amer. Chem. Soc.*, **81**, 4737 (1959).
11. J. V. Carabinos and C. N. Yiannios, U.S. Pat. 3,513,088 (May 19, 1970) [*C.A.*, **73**, 34781c (1970)].
12. N. Kobayashi, A. Osawa, and T. Fujisawa, *J. Polym. Sci., Part B, Polymer Letters Ed.*, **11**, 225 (1973).

Crystalline Poly(vinyl chloride) Prepared in an Aldehyde Medium

$$CH_2{=}CHCl \xrightarrow[\text{AIBN}]{\text{Aldehyde}} \left[-CH_2-\underset{Cl}{\overset{H}{C}}-CH_2-\underset{H}{\overset{Cl}{C}}- \right]_n$$

Submitted by I. Rosen (1a, b)
Checked by E. A. Ofstead and R. M. Pierson (2)

1. Procedure

This method describes the preparation of a highly crystalline poly(vinyl chloride) by a simple solution polymerization. Into a thick-walled polymerization tube (Note 1) of 200-ml. capacity, which has been dried and flushed with lamp grade nitrogen, is distilled 62.5 g. (1.0 mole) of purified vinyl chloride (*Caution! Vinyl Chloride should be considered as toxic by inhalation and as a potential carcinogen. Distillation and subsequent venting should be conducted in a manner which avoids worker exposure.*) (Note 2). While the nitrogen atmosphere is maintained, 72 g. (1.0 mole) of freshly distilled *n*-butyraldehyde and 0.16 g. (0.001 mole) of azobisisobutyronitrile (AIBN) are added (Note 3). The tube is sealed and transferred to a heating bath maintained at 50°. (Note 4). The tube is agitated, or tumbled end over end in a bottle polymerizer, for 9 hr. (Note 5). The tube and its contents are cooled, the unpolymerized monomer carefully vented, and the contents poured, with stirring, into about 400 ml. of methanol. The precipitated polymer is filtered (Note 6), then dissolved in about 80 ml. of hot cyclohexanone (Note 7), and reprecipitated in methanol. After the polymer has been filtered and washed with methanol, the yield is about 9-15%. It is a white solid, difficulty soluble in tetrahydrofuran, and has a degree of polymerization of about 26. The polymer may be dissolved in hot cyclohexanone (120°). On cooling, the solution usually remains clear, and the viscosity can be determined.

The maximum melting point of 265° and the heat of fusion of 1180±90 cal/mole have been determined by differential scanning calorimetry. The powder x-ray diffraction pattern shows a highly crystalline structure (3). Refined x-ray crystal structure analyses have been done. The infrared absorption spectrum is obtained from a film cast on a KBr disk from a chlorobenzene or *o*-dichlorobenzene solution. The high 635 cm.$^{-1}$/692 cm.$^{-1}$ infrared absorbance ratio relative to that of conventional poly(vinyl chloride) indicates increased syndiotacticity in the polymer (Note 8). NMR spectroscopy indicates racemic sequences (corresponding to syndiotatic placements) equivalent to those observed in PVC prepared at −65° (9).

2. Notes

1. The polymerizations may also be conveniently carried out in 8-oz "pop" bottles. A common 4-oz., screw-cap bottle (able to withstand 100 p.s.i.) can be used with half the suggested quantities.

2. Vinyl chloride of adequate purity is obtained by distillation through an 18-in. packed column. The first 10% of the distillate, as well as the last 10% should be discarded. A good, polymerization grade vinyl chloride can be used directly without fractionation.

3. Increase of the mole ratio of aldehyde to vinyl chloride will decrease the yield of polymer without providing any significant increase in polymer crystallinity. The polymer crystallinity is so high as to be capable of little improvement.

4. Temperatures lower than 50° can be used, with consequent reduction in reaction rate and little improvement in polymer crystallinity.

5. Increase of polymerization time will not significantly increase the polymer yield.

6. Some polymer is dissolved in the filtrate. This material may be recovered by evaporation of the solvent. It has a lower molecular weight than the precipitated polymer and is fairly crystalline.

7. The hot cyclohexanone solution may require filtration prior to the methanol precipitation step.

8. To characterize the structure, the infrared spectrum of the sample in the tacticity-sensitive 700-600 cm.$^{-1}$ region was compared with those obtained from: (*a*) a "partly syndiotactic" sample prepared by free radical polymerization at −78°; (*b*) an "atactic" sample prepared by free radical polymerization at +50°; and (*c*) a "highly syndiotactic" sample prepared by the canal complex

polymerization technique. The submitter's measurements of the 635 cm.$^{-1}$/692 cm.$^{-1}$ absorbance ratio placed the aldehyde-prepared sample between the values exhibited by (*a*) and (*c*), usually close to (*c*). The checkers, however, found the spectrum to lie closer to (*a*). In any case, there is agreement that the procedure may be used for the preparation of at least a partly syndiotactic poly(vinyl chloride).

9. This method has been applied to the polymerization of vinyl trifluoroacetate and has provided a slight increase in polymer syndiotacticity (11).

3. Methods of Preparation

Three other methods of preparing crystalline poly(vinyl chloride) have been reported in the literature: (*a*) the low temperature, free radical polymerization of vinyl chloride, in which advantage is taken of the energetically favored syndiotactic propagation over the isotactic propagation (4); (*b*) the polymerization of vinyl chloride in a urea complex, in which the monomer is supposedly oriented prior to polymerization and then polymerized by radiation (5) (this polymerization takes place at $-78°$ and raises the question of the relative amounts of the crystallinity caused by the low polymerization temperature and the complexing); and (*c*) the polymerization of vinyl chloride initiated by the Ziegler type of catalyst (6). [There is some evidence that this polymerization is free-radical-initiated (7).]

4. References

1a. Diamond Alkali Company, Painesville, Ohio 44077.
1b. Current Address: Standard Oil Co., Cleveland, Ohio 44128.
2. Goodyear Tire and Rubber Co., Akron, Ohio 44309.
3. I. Rosen, P. H. Burleigh, and J. F. Gillespie, *J. Polym. Sci.*, **54**, 31 (1961).
4. J. W. L. Fordham, P. H. Burleigh, and C. L. Sturm, *J. Polym. Sci.*, **41**, 73 (1959).
5. D. M. White, *J. Amer. Chem. Soc.*, 82, 5678 (1960).
6. Hercules Powder Co., Australian Pat. 26,889 (April 8, 1957).
7. W. P. Baker, *J. Polym. Sci.*, **42**, 578 (1960).
8. C. E. Wilkes, V. L. Folt, and S. Krimm, *Macromolecules*, **6**, 235 (1974).
9. A. M. Hassan, *J. Polym. Sci.*, A-2, **12**, 655 (1974).
10. E. V. Gouinlock, *J. Polym. Sci.*, A-2, **13**, 1533 (1975).
11. S. Matsuzawa, K. Yamura, and H. Noguchi, *Makromol. Chem.*, **168**, 27 (1973).

Stereoregular Poly-(vinyl trifluoroacetate) and Poly-(vinyl alcohol)

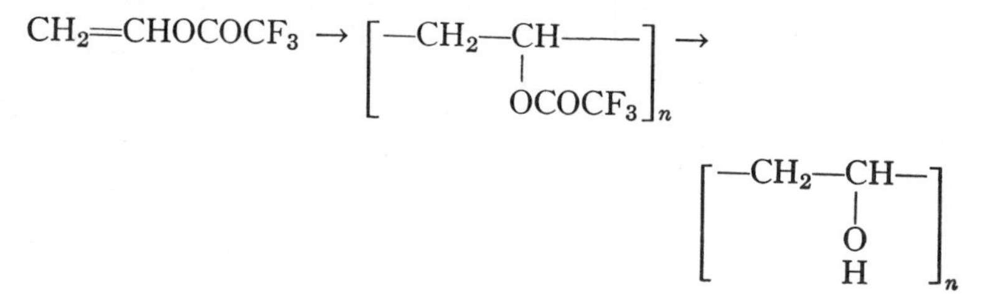

Submitted by G. H. McCain (1)
Checked by C. F. Hauser,(2a, b), J. Savory, and G. B. Butler (2a)

A. Polymerization of Vinyl Trifluoroacetate

1. Procedure

In each of four 200 x 25-mm. heavy-walled Pyrex® test tubes that have been narrowed at a point 1-1.5 in. from the top are placed 0.54 g. (0.6% by weight of monomer) of a 50% paste of 2,4-dichlorobenzoyl peroxide (*Caution! Acyl peroxides are strong oxidants and must be handled with care.*) in dibutyl phthalate (Note 1), and 45 g. of vinyl trifluoroacetate (Note 2). A small wad of glass wool is placed in the top of each tube and they are attached to a vacuum line, constructed as shown in Fig. 1, through which high purity nitrogen is flowing at a moderately rapid rate. The tubes are then immersed in Dry Ice-trichloroethylene or carbon tetrachloride/chloroform baths, (*Caution! Trichloroethylene, carbon tetrachloride and chloroform are toxic and should be handled with gloves in an efficient fume hood.*) under a slight positive nitrogen pressure.

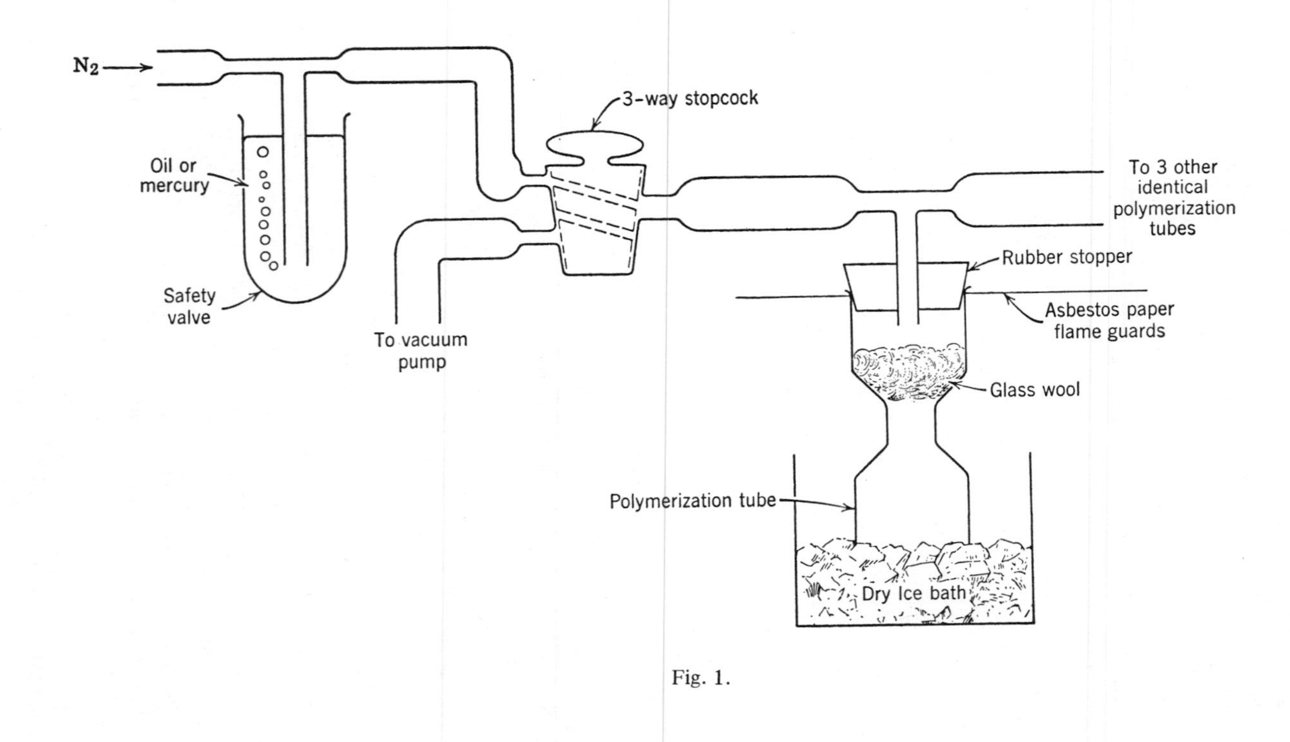
N_2
3-way stopcock
Oil or mercury
Safety valve
To vacuum pump
To 3 other identical polymerization tubes
Rubber stopper
Asbestos paper flame guards
Glass wool
Polymerization tube
Dry Ice bath

Fig. 1.

After cooling for about 15 min. the tubes are evacuated to a pressure of about 0.5 torr by reversing the three-way stopcock shown in Fig. 1. The system is then flushed with nitrogen by *cautiously* reversing the stopcock again until a slight positive pressure is attained. This procedure is repeated twice to ensure an inert atmosphere in the tubes. After the pressure is again reduced to 0.5 torr, the tubes are sealed at the constriction using a hand torch. When the glass seals have cooled, the tubes are permitted to warm to room temperature and are shaken vigorously to dissolve as much of the initiator as possible. The tubes are then inserted in metal shields (Note 3) and placed in a constant temperature bath at 40°. An occasional *cautious* (Note 4) agitation of the tubes during the first half hour will complete the dissolution of the initiator. After 1 day, the tubes are filled with purple to brown solid polymer. They are chilled in powdered Dry Ice to release the polymer from the walls, and the poly(vinyl trifluoroacetate) is removed by breaking the tubes. (*Caution! Wear heavy gloves to protect the hands and break the tubes behind a safety shield in case internal pressure has developed.*) Conversions to polymer of the order of 90-100% are achieved with an $\eta_{red.}$ in cyclohexanone of 3-4 dl./g.

2. Notes

1. Available from the Lucidol Division of Wallace and Tiernan, Inc., under the name Luperco CDB.

2. The vinyl trifluoroacetate can be purchased, or prepared by the reaction of acetylene with trifluoroacetic acid in the presence of mercuric oxide (6). *Caution! Vinyl trifluoroacetate should be handled in an efficient fume hood and should not come in contact with the skin.*

3. These can be conveniently made from standard 8-in. iron pipe nipples fitted with caps and perforated to permit the water to flow around the tube.

4. The agitation of the tubes (which are not removed from the metal shields) should be carried out behind a safety shield, heavy gloves being used to protect the hands. A gentle inversion of the tubes several times usually provides adequate stirring.

3. Methods of Preparation

Vinyl trifluoroacetate has been polymerized in bulk with benzoyl peroxide as the initiator (3) and also with ultraviolet radiation (4). In addition this monomer has been polymerized in acetone solution (3).

B. Conversion of Poly(vinyl trifluoroacetate) to Poly(vinyl alcohol)

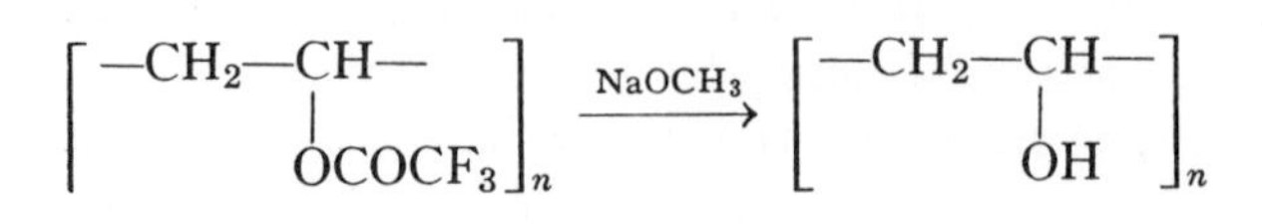

1. Procedure

Poly(vinyl trifluoroacetate) (ca. 180 g.) is dissolved in 1800 ml. of redistilled tetrahydrofuran (*Caution! The dissolving and alcoholysis steps must be carried out in an efficient hood to avoid building up a hazardous concentration of tetrahydrofuran vapor.*) (Note 1) with good agitation, at the reflux temperature of the solvent. The alcoholysis reaction (Note 2) is carried out in a 5-1., four-necked, round-bottomed flask equipped with a stirrer, a reflux condenser, a 150-ml. addition funnel containing 120 ml. of a 5% solution of sodium methylate in methanol, and a 1-1. addition funnel, which is either jacketed and heated by circulating hot water or wrapped with electrical heating tape (Note 2). In the flask are placed 660 ml. of redistilled tetrahydrofuran, 20 ml. of dry methanol, and 10 ml. of a 10% solution of sodium methylate in methanol. The mixture is heated to reflux and about one-half of the hot polymer solution is added through the large addition funnel over a period of 1.5 hr. This is repeated with the remainder of the polymer solution, using small additional quantities of fresh tetrahydrofuran to rinse residual poly(vinyl trifluoroacetate) from the dissolving flask and funnel into the alcoholysis vessel. Small amounts of the 5% sodium methylate solution (Note 3) are added as required to keep the reaction faintly basic, as determined with pH paper. After all the polymer solution has been introduced, the white to light yellow slurry is kept at reflux for an additional 2 hr. At the end of this period, the mixture is neutralized with glacial acetic acid and permitted to cool. The finely divided poly(vinyl alcohol) is separated by filtration, washed with several portions of methanol, and dried in a vacuum oven at 50°/20 torr. The yield of poly(vinyl alcohol) is about 50 g., 90% of theoretical, based on 180 g. of poly(vinyl trifluoroacetate).

2. Notes

1. Commercial tetrahydrofuran must be purified by distillation, because the inhibitor present will cause the poly(vinyl alcohol) to be discolored.

2. The polymer solution must be kept at about 50° to prevent gelation.

3. A lighter colored polyvinyl alcohol will result if an equivalent amount of a 1% sodium methylate solution is used, but a constant addition of this catalyst solution throughout the reaction period is required.

Ed. Note: Since this procedure was published controversy has arisen over the tacticity of Poly(vinyl trifluoroacetate) and the derived alcohol. The original submitter feels that some of the more recent work confirms the earlier thesis that these polymers are more syndiotactic than those made *via* vinyl acetate.

3. Methods of Preparation

A thin film of poly(vinyl trifluoroacetate) has been converted to a film of poly(vinyl alcohol) (PVA) by ammonolysis in dry, gaseous ammonia (3, 5). It has also been reported that sodium carbonate can be used as the catalyst for the alcoholysis (4).

The preparation of highly syndiotactic PVA *via* cationically polymerized vinyl trimethylsilyl ether has been reported (7). PVA has also been prepared from syndiotactic polyvinyl formate prepared by low temperature, free radical polymerization (8), and from free radical polymerization of vinyl esters of C_4 to C_6 ω-hydrofluoro-carboxylic acids (9); isotactic PVA has been obtained from cationically polymerized vinyl benzyl ether (10).

4. References

1. Diamond Shamrock Corp., Painesville, Ohio 44077.

2a. The University of Florida, Gainesville, Florida 32601.

2b. Current Address: Union Carbide Corp., South Charleston, West Virginia 25303.

3. H. C. Haas, E. S. Emerson, and N. W. Schuler, *J. Polym. Sci.,* **22**, 291 (1956).
4. C. R. Bohn, J. R. Schaefgen, and W. O. Statton, *J. Polym. Sci.,* **55**, 531 (1961).
5. J. Fordham, G. H. McCain, and A. Alexander, *J. Polym. Sci.,* **39**, 335.
6. B. W. Howk and R. A. Jacobson, U.S. Pat. 2,436,144 (Feb. 17, 1948) [*C.A.*, **42**, 3215i (1948)].
7. S. Murahashi, S. Nozakura, and M. Sumi, *J. Polym. Sci., Part B, Polymer letters Ed.,* **3**, 245 (1963).
8. K. Fujii, T. Mochizuki, S. Imoto, J. Ukida, and M. Matsumoto, *J. Polym. Sci., Part A,* **2**, 2327 (1964).
9. L. D. Budovskaya, E. N. Rostovskii, A. V. Sidorovich, A. I. Kol'tsov and E. V. Kuvshinskii, *Vysokomol. Soyed.,* B **11**, 850 (1969) [*C.A.*, **72**, 56143c (1970)].
10. S. Murahashi, H. Yuki, T. Sano, U. Yonemura, H. Tadokoro, and Y. Chatani, *J. Polym. Sci.,* **62**, S77 (1962).

Poly(N-hexyl 1-nylon)

$$\mathrm{R{-}N{=}C{=}O} \xrightarrow{\mathrm{CN^-}} \left[\begin{array}{c} \mathrm{O} \\ \| \\ -\underset{\displaystyle \mathrm{R}}{\underset{|}{\mathrm{N}}}-\mathrm{C}- \end{array} \right]_n$$

Submitted by V. E. Shashoua (1)
Checked by R. W. Lenz (2a, b) and G. M. Scharlach (2a)

1. Procedure

A 100-ml. three-necked flask is equipped with a stirrer and two Y side arm adapters. One side arm has a calcium chloride tube and a low temperature thermometer, suitably placed for immersion in the reaction medium. The other side arm is fitted with a nitrogen inlet tube, and the vertical arm is sealed with a rubber bulb. The flask is then flamed (Note 1) while being swept with nitrogen. This removes the traces of moisture adhering to the walls of the reaction vessel. Then 30 ml. of dry N,N-dimethylformamide (DMF, Note 2) is added, and the flask and its contents are cooled to −58°, approximately the melting point of DMF. Eight ml. of *n*-hexyl isocyanate (*Caution! Most isocyanates are potent lachrymators.*) (Note 3) is added to the reaction flask, and the mixture is stirred rapidly until the temperature reaches −58°. At this point, 4 ml. of a solution of sodium cyanide in DMF (*Caution! Sodium cyanide is extremely toxic. Solutions in solvents such as DMF are rapidly absorbed through the skin.*) (Note 4) is added dropwise during 2-3 min. with a hypodermic syringe, to the vigorously stirred reaction mixture. The addition is carried out by piercing the rubber bulb at the side arm with the hypodermic needle. The polymer precipitates immediately as a white solid. After stirring for 15 min. at −58°, 50 ml. of methanol is added to quench the reaction. The polymer is then collected on a filter and washed with a total of about 300 ml. of methanol. The product is then dried at 40° under vacuum, to give 5.9-6.5 g. of polymer (75-85% yield).

2. Characterization

Solubility. The polymer is soluble in most common aromatic and chlorinated hydrocarbons, such as benzene, chloroform, and methylene chloride. Films can be prepared by drying benzene solutions containing 2-5% polymer. The polymer degrades in concentrated sulfuric acid and trifluoroacetic acid (3, 4).

Viscosity Measurements. The inherent viscosity of the polymer should be 2.0-2.9 dl./g. determined in benzene at 30°, for a concentration of 0.1 g. in 100 ml. of solvent (Note 5).

Melt Temperature. The polymer melt temperature measured on a hot metal surface is 195°. The polymer softens at a temperature of 120°. This softening point is the temperature at which the polymer becomes plastic without sticking to the metal surface.

3. Notes

1. All flasks, pipets, and hypodermic syringes should be thoroughly dried and "flamed" before use. The "flaming" of the moisture traces adhering to the walls of the glass equipment can be readily carried out with a Bunsen burner. In the case of pipets, it is convenient to stopper the two ends with eye-dropper rubber bulbs to prevent readsorption of moisture on their internal walls before letting them cool to room temperature.

2. The DMF must be dried before use by distillation from phosphoric anhydride at atmospheric pressure. Polymerization grade solvent can be prepared by starting with 2.1 of DMF and 40 g. of phosphoric anhydride. A fore-cut of about 600 ml. is discarded and a middle fraction of about 800 ml. of pure solvent is collected for use. To avoid a vigorous reaction of the P_2O_5 with the DMF, which takes place in the pot residue, it is important not to allow the distillation to proceed to dryness.

3. The *n*-hexyl isocyanate can be synthesized by a method similar to that described by Boehmer (5). Freshly distilled heptanoyl chloride (100 ml.) (Eastman Organic Chemicals) was added to 150 ml. of toluene containing 46 g. of activated sodium azide (6). The mixture is heated under reflux until no further nitrogen is evolved. This generally takes about 4 hr. The toluene

solution is then decanted from the solid residue and distilled through a glass helices-packed column. The residue is fractionated twice, yielding 52 g. of a clear white liquid, b.p. 162-163°.

$$RCOCl + NaN_3 \rightarrow RCON_3 \rightarrow RNCO + N_2$$

It is found that this method of synthesis gives the best polymerization grade isocyanate. When the isocyanates were prepared from the amine hydrochloride and phosgene, extreme difficulty was experienced in obtaining polymerization grade product, probably because of the presence of traces of phosgene in the isocyanate.

4. The sodium cyanide catalyst solution is prepared by dissolving dry reagent grade compound in dry DMF to give a saturated solution containing 0.68%. The sodium cyanide may be dried over potassium hydroxide in a drying pistol for 2 days at 100° under vacuum.

5. The viscosity of the polymer and the molecular weight depend on the amount of catalyst used in the polymerization at a given temperature. Small amounts of catalyst give higher molecular weight polymers (4, 7).

4. References

1. Pioneering Research Division, Textile Fibers Department, E. I. du Pont de Nemours & Co., Wilmington, Delaware 19898.

2a. The Dow Chemical Company, Framingham, Massachusetts 01701.

2b. Current Address: The University of Massachusetts, Amherst, Massachusetts 01002.

3. V. E. Shashoua, *J. Amer. Chem. Soc.,* **81**, 3156 (1959).
4. V. E. Shashoua, W. Sweeny, and R. F. Tietz, *J. Amer. Chem. Soc.,* **82**, 866 (1960).
5. J. W. Boehmer, *Rec. trav. chim.,* **55**, 379 (1936).
6. P. A. S. Smith, *Org. Reactions,* **3**, 382 (1946).
7. V. E. Shashoua, U.S. Pat. 2,965,614 (Dec. 20, 1960) [*C.A.,* **55**, 10966c (1961)].

Poly(2,5-dimethyl-2,4-hexadiene)

$$(CH_3)_2C{=}CH{-}CH{=}C(CH_3)_2 \xrightarrow{BF_3} [-(CH_3)_2C{-}CH{=}CH{-}C(CH_3)_2-]$$

Submitted by F. B. Moody (1)
Checked by R. W. Lenz (2a, b) and G. M. Scharlach (2a)

1. Procedure

In a 1-l. four-necked flask equipped with an efficient stirrer, a thermometer, a dropping funnel, a gas inlet tube, and a gas outlet tube provided with a drying tube is placed 350 ml. of petroleum ether. The petroleum ether is cooled to -70° by stirring in a solid carbon dioxide-solvent bath. A gentle stream of boron trifluoride gas is introduced through the gas inlet tube for a few moments until fumes are seen escaping from the gas outlet tube. (*Caution! The operation should be carried out in a well-ventilated hood.*) The gas inlet tube is removed and a solution of 20 g. of 2,5-dimethyl-2,4-hexadiene (Note 1) in 35 ml. of petroleum ether is added with vigorous stirring over about 4 min. A pronounced yellow color rapidly develops, the polymerization mixture becomes a slush that is difficult to stir, and the temperature rises to about -60°. Stirring is continued for another minute or two, cooling is discontinued, and 25 ml. of denatured alcohol followed by 250 ml. of acetone is added with stirring, resulting in the disappearance of the yellow color. The white, finely-divided polymer is collected by suction filtration, washed with acetone, and dried (Note 2). The yield of poly(2,5-dimethyl-2,4-hexadiene) is 12-16 g. (60-80%). The inherent viscosity (0.5 g./100 ml. of decahydronaphthalene at 130°) varies from 0.8 to 1.5 dl./g. Melting, accompanied by some decomposition, occurs at 260-265°.

2. Notes

1. 2,5-Dimethyl-2,4-hexadiene, obtained from Benzol Products Company, was purified by a simplified zone melting process followed by distillation. The liquid diene was incompletely frozen in a bottle, either by immersing in ice water or by cooling in a refrigerator. The unfrozen core, amounting to about 15% of the total, was decanted, the solid was melted, and the process was repeated twice more. Distillation yields pure product, m.p. 14°, b.p. 133°, n_D^{25} 1.475, and gas chromatographic purity = 99.4%.

2. Storage of the polymer is sometimes accompanied by development of a violet or brown coloration. This can be prevented by stirring the polymer with very dilute ammonium hydroxide solution, followed by filtration, washing with water, and drying.

3. References

1. Textile Fibers Department, E. I. du Pont de Nemours & Co., Wilmington, Delaware 19898.

2a. The Dow Chemical Company, Framingham, Massachusetts 01701.

2b. Current Address: The University of Massachusetts, Amherst, Massachusetts 01002.

Poly(1,4-butylene hexamethylene carbamate)

$$OCN(CH_2)_6NCO + OH(CH_2)_4OH \xrightarrow{\text{Solvent}} \left[-O(CH_2)_4O\overset{O}{\overset{\|}{C}}\overset{H}{N}(CH_2)_6\overset{H}{N}\overset{O}{\overset{\|}{C}}- \right]_n$$

Submitted by W. R. McElroy (1a, b, c)
Checked by R. W. Lenz (2a, b) and G. M. Scharlach (2a)

1. Procedure

Hexamethylene diisocyanate (*Caution! Hexamethylene diisocyanate is a severe primary eye and skin irritant (3).*) (Note 1) (51.5 ± 0.1 g., 0.3065 mole) is weighed directly into the dry (Note 2) reactor, and the air in the vessel is again displaced by dry nitrogen. The reactor is a 1-1., three-necked, round-bottomed flask with ground glass joints, equipped with a thermometer, an agitator sealed to prevent access of air, a Friedrichs condenser with a drying tube on the outlet, and a heating mantle. Monochlorobenzene (0.2 1.) (Note 3) is added. Gentle agitation is started, and the mixture is heated to 100°. 1,4-Butanediol (27.0 ± 0.1 g., 0.3000 mole) is added (Note 4) by weighing in all but 1-2 g. by difference from a small glass-stoppered Erlenmeyer flask, then adding the remainder volumetrically (Note 5) from a hypodermic syringe calibrated in 0.1-cc. divisions. The mixture is heated to reflux; 132-134°, within about 20 min. after the 1,4-butanediol is added, and maintained under reflux. After about 30-40 min. from the time the reflux temperature is reached the solution becomes cloudy, indicating separation of polymer. Heating is continued for 75 min. at the reflux temperature after the cloudiness first appears. During this time the polymer precipitates. The slurry is filtered

(Note 6) while it is at a temperature above 100°. The polymer is collected in a 3-in. Büchner funnel fitted with Fisher Scientific Co. semicrimped, rapid, qualitative filter paper No. 9-795 or its equivalent. The polymer is compressed and sucked as dry as possible on the filter (Note 7). It is then washed twice with 0.2-1. portions of water (Note 8) to remove most of the remaining monochlorobenzene. The filter cake is returned to the reactor, 0.2 1. of water is added (Note 8), and the position of the condenser is changed for distillation.

The slurry· is agitated to prevent bumping and foaming, and the reactor is heated with an oil bath maintained at 115° to remove the remaining monochlorobenzene by steam distillation, until the distillate is not cloudy (Note 9). The polymer is filtered as before and dried under a pressure of 10 torr at 75° for 18 hr. The resulting fine white powder has a flow point in a capillary melting-point tube of 174-178° (Note 6). The intrinsic viscosity in *m*-cresol at 25° is in the range 0.57-1.26 (Note 10). The yield is 69-74 g., 88-95%.

2. Notes

1. Hexamethylene diisocyanate is purified by distillation, b.p. 127°/10 torr sp. gr. = 1.046 at 20/4°, crystallizing point −67°. (Hexamethylene diisocyanate is a reactive chemical and should be stored in tightly closed containers away from moisture and heat. Contact with water, alcohol, or strong bases should be avoided. In case of a spill, it should be covered with sawdust and treated with dilute ammonia-water, or alcohol and water in an open container, before disposal. It will cause irritation of the respiratory tract and headache if exposure is prolonged. Wash thoroughly with soap and water if it comes in contact with the skin, and flush with water for 15 min. if splashed into the eyes. Wear rubber gloves and safety glasses when handling this substance.) A slight excess, corresponding to an NCO/OH = 1.02, is used to make up for small, but inevitable, losses by reaction with moisture; using the stoichiometric amount produces a polymer with lower melting point.

2. The reactor components are dried at 110° for at least 1 hr., then quickly assembled while hot, and flushed with dry nitrogen to displace the air in the system.

3. The solvent is dried by distillation just before use.

4. The 1,4-butanediol must be anhydrous. It is purified by distillation, b.p. 107-108°/4 torr. The crystallizing point is a minimum of 19°.

5. The volume from the remaining weight is calculated by using the density 1.02 g./cc. at 20°.

6. Normally, the polymer precipitates in a finely divided form. However, if it should agglomerate because of variations in procedure, the polymer may be removed, after the solvent has been decanted, by dissolving it in *m*-cresol and precipitating it in methanol. In this case a melting point up to about 189° has been obtained between microscope slides on a Fisher-Johns melting point apparatus.

7. Lower melting polymer containing some cyclic urethane (4, 5) is dissolved in the solvent and is recoverable by evaporation. At this point the product may be dried under vacuum at 75° instead of proceeding to the steam distillation to remove the remaining solvent. However, a much longer time is required to remove the last traces of solvent than when steam distillation is used.

8. Distilled or deionized water is used.

9. The melting temperature is not depressed by this treatment.

10. The polymer is insoluble in acetone, dimethylacetamide, and toluene, but it is soluble in *m*-cresol.

3. Methods of Preparation

Solvent Method. The method described herein is an adaptation of that mentioned by Bayer (4) and described in greater detail by Müller (5). It is disclosed in a German patent (6). Lyman (7) describes a method for preparing ethylene glycol-diisocyanate polymers in solution. These references give the melting points of a great variety of polymers of this type.

Melt Method. A melt technique is mentioned by Bayer (4) and described by Müller (5); this is also disclosed in a German patent (6). This procedure was found by the submitter to be more difficult to reproduce in the laboratory than the solvent method. Furthermore, because of the high viscosity of the polymer melt it was difficult or impossible at times to remove the product from the reactor before it solidified. Somewhat higher melting points were observed in products made by the melt procedure.

4. References

1a. Mobay Chemical Co., New Martinsville, West Virginia.
1b. Current Address: Action Products Inc., Olean, New York 14760.
1c. The assistance of H. B. Staley and D. H. Chadwick (Mobay Chemical Co.) in the revision of this procedure is thankfully acknowledged.
2a. Dow Chemical Co., Framingham, Massachusetts.

2b. Current Address: The University of Massachusetts, Amherst, Massachusetts 01002.
3. "Toxicity and Safe Handling of Isocyanates", Mobay Chemical Corp.
4. A. Bayer, *Angew. Chem.*, **59**A, 257 (1947).
5. E. Müller, in Houben-Weyl, *Die Methoden der Organischen Chemie,* Vol. 14, part 2, pp. 71-75, Thieme Verlag, Stuttgart, 1963.
6. I. G. Farben., Germ. Pat. 728,981 (1937) [*C.A.*, **38**, 3816 (1944)].
7. D. J. Lyman, *J. Polym. Sci.*, **45**, 49 (1960); Rev. Macromol. Chem., **1**, 191 (1966).

Poly[*ethylene methylene bis(4-phenyl carbamate)*]

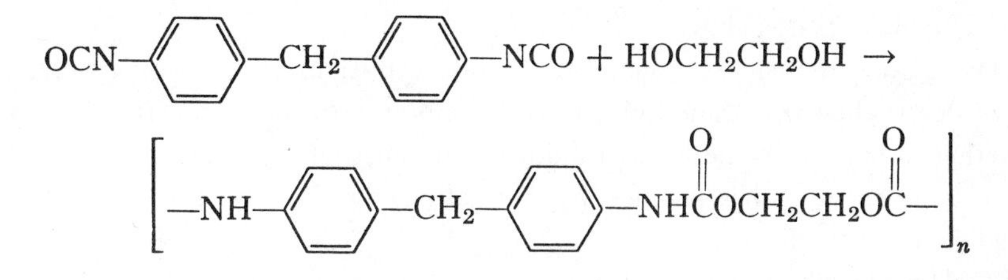

Submitted by V. S. Foldi, T. W. Campbell, and D. J. Lyman (1)
Checked by R. W. Lenz (2a, b) and G. M. Scharlach (2a)

1. Procedure

To a suspension of 25.02 g. (0.10 mole) of methylene *bis*(4-phenyl isocyanate) (*Caution! Most isocyanates are severe irritants and should be handled with appropriate precaution.*) (Note 1) in 40 ml. of 4-methylpentanone-2 (Note 2) in a 500-ml., three-necked, round-bottomed flask equipped with a stirrer and a condenser and protected from moisture with a drying tube is added, with rapid stirring, a solution of 6.20 g. (5.56 ml., 0.10 mole) of ethylene glycol (Note 3) in 40 ml. of dimethyl sulfoxide (Note 2). The reaction mixture is slowly stirred and heated at 115° for 1½ hr.

The polyurethane is isolated (Note 4) by pouring the clear viscous solution into water. The tough, white polymer is washed with water (three 200-ml. washes) and acetone (one 100-ml. wash) in an explosion proof blender and is then dried in a vacuum oven at 100°. The yield of polymer is 97-100% with an inherent viscosity of 1.0 dl./g. in dimethylformamide at room temperature (conc. 0.5%). The polymer melt temperature is about 250°, and the glass transition temperature is 90°.

2. Notes

1. A high-purity methylene bis(4-phenyl isocyanate) is available commercially from Mobay Chemical Co. It may be purified by distillation through an air-jacketed Vigreux column, b.p. 148-150°/0.12 torr, or by recrystallization in the following manner: the compound is dissolved in an equal volume/weight of hexane at the boil, the hot solution is treated with decolorizing charcoal and is filtered hot into an equal volume of ice-cold hexane. By this procedure separation of the diisocyanate as an oil is prevented, and the compound is obtained as a pure, white crystalline solid, m.p. 42°. The pure material should be stored under nitrogen in a refrigerator.

2. The solvents are purified by distillation: dimethyl sulfoxide, b.p. 66°/5 torr 4-methylpentanone-2, b.p. 115°.

Other solvents can be used in this polymerization. In decreasing order of usefulness they are dimethyl sulfoxide/carbon tetrachloride (50/50), tetramethylene sulfone, N-methylpyrrolidone, and dimethyl sulfoxide.

3. Ethylene glycol is purified by distillation; b.p. 79°/4.4 torr n_D^{25} 1.4300, and % H_2O = 0.05 or less.

4. The resulting viscous solution can be dry-spun or cast into films directly without isolation.

3. References

1a. Pioneering Research Division, Textile Fibers Department, E. I. du Pont de Nemours & Co., Wilmington, Delaware 19898.
2a. The Dow Chemical Company, Framingham, Massachusetts 01701.
2b. Current Address: The University of Massachusetts, Amherst, Massachusetts 01002.

Poly(2,6-dimethyl-1,4-phenylene ether)

A. By Oxidative Coupling

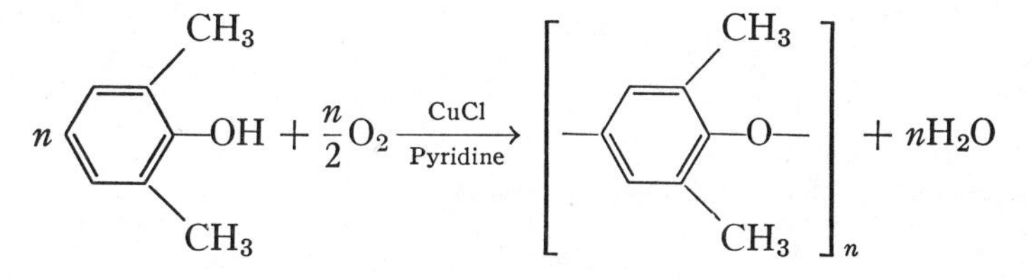

Submitted by A. S. Hay, H. S. Blanchard, G. F. Endres, and J. W. Eustance (1)
Checked by W. A. Butte, Jr., and C. C. Price (2)

1. Procedure

To a 500-ml. wide-mouthed Erlenmeyer flask in a water bath at 30° equipped with a Vibromixer stirrer, an oxygen inlet tube, and a thermometer are added 200 ml. of nitrobenzene, 70 ml. of pyridine, and 1 g. of copper(I) chloride. Oxygen (300 ml./min.) is bubbled through the vigorously stirred solution and then 15 g. (0.12 mole) of 2,6-dimethylphenol (Note 1) is added. Over a period of 16 min. the temperature rises to 33°, at which point the reaction mixture begins to get viscous. The reaction is continued for 12 min., then it is diluted with 100 ml. of chloroform and added to 1.1 l. of methanol containing 3 ml. of concentrated hydrochloric acid. The precipitated polymer is filtered and washed with 250 ml. of methanol, then with 250 ml. of methanol containing 10 ml. of conc. hydrochloric acid, and finally with 250 ml. of methanol. The polymer is dissolved in 500 ml. of chloroform, filtered and reprecipitated in 1.2 l. of methanol containing 3 ml. of conc. hydrochloric acid. After washing with methanol and drying at 110° (5 torr) for 3 hr. there is obtained 13.5 g.

(0.11 mole, 91%) of almost colorless polymer, $[\eta] = 0.96$ dl./g. ($CHCl_3$, at 25°) (Note 2).

2. Notes

1. 2,6-Dimethylphenol was purchased from Aldrich Chemical Co. and twice recrystallized from heptane. Copper(I) chloride was purified by dissolving it in concentrated hydrochloric acid, filtering, and precipitating with water. The copper(I) chloride was washed with alcohol and then with ether and dried under reduced pressure. Pyridine and nitrobenzene were purified grades.
2. This is equivalent to an osmotic molecular weight of 28,000.
3. An alternate method for producing the polymer from 2,6-xylenol which uses MnO_2 as oxidant has been reported. (3)

3. References

1. General Electric Company, Schenectady, New York 12345.
2. University of Pennsylvania, Philadelphia, Pennsylvania 19104.
3. E. McNelis, *J. Org. Chem.*, **31**, 1255 (1966).

B. By Oxidative Displacement of Bromine

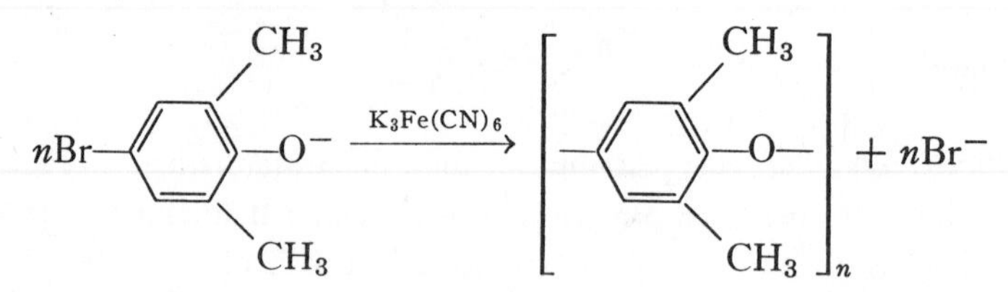

Submitted by W. A. Butte, Jr., N. S. Chu, and C. C. Price (1)
Checked by H. S. Blanchard and A. L. Klopfer (2)

1. Procedure

A 1-l. three-necked flask is fitted with an efficient stirrer, a dropping funnel, and a gas inlet tube connected to a stream of purified nitrogen (Note 1). A solution of 5 g. of potassium hydroxide in 200 ml. of water, 8 g. (0.04 mole) of 4-bromo-2,6-xylenol (Note 2), and 200 ml. of benzene is introduced. The stirrer is started and 1.3 g. of potassium ferricyanide in 20 ml. of water is added dropwise over a period of 30 min. After an additional 15 min. of stirring, the mixture is transferred to a separatory funnel and the lower aqueous phase is removed (Note 3). The yellow benzene solution is transferred to a 300-ml. distilling flask and concentrated to 50 ml. at a water aspirator. The concentrate is poured slowly, with stirring, into 250 ml. of methanol acidified with 2.5 ml. of concentrated hydrochloric acid. The precipitate is collected by suction filtration and washed by resuspending it in 150 ml. of methanol. It is then collected, redissolved in 50 ml. of benzene, and precipitated again as described above.

The reprecipitated polymer is collected and redissolved in 50 ml. of benzene contained in a 250-ml. round-bottomed flask. The flask is swirled in Dry Ice-acetone until the contents are frozen and it is quickly connected to a vacuum line protected by a large trap. The pump is started and a pressure of 1 torr is maintained for 5 hr. Yield: 4.6 g. (96%) of spongy, white solid, $[\eta]^{30°}_{C_6H_6}$ = 0.5-0.6 dl./g. (Note 4). The product darkens and softens above 270° (Note 5).

2. Notes

1. The nitrogen is freed of traces of oxygen by bubbling it through Fieser's solution (3).
2. 4-Bromo-2,6-xylenol is prepared by bromination of 2,6-xylenol (4).
3. Separation of the phases, if difficult, may be facilitated by the addition of 10 ml. of concentrated hydrochloric acid.
4. The intrinsic viscosity corresponds to a molecular weight of 20,000-25,000 as determined by an equation relating osmotic molecular weight to intrinsic viscosity ($[\eta] = 3.8 \times 10^{-4} M^{0.73}$) supplied by Dr. A. S. Hay, General Electric Co., Schenectady, N.Y. A product of higher molecular weight may be obtained by adding the ferricyanide solution more slowly. Thus when the reaction time was doubled, the product decomposed without softening above 270° and had a viscosity of 0.9 dl./g.
5. A sample of the polymer gave the following analysis: C, 79.19; H, 7.02; Br, 1.06. The amount of residual phenol group was negligible as determined by titration with tetrabutylammonium hydroxide. Exchange with tritiated water, however, indicated one exchangeable hydrogen per 19,000 molecular weight units.
6. An alternate method for producing the polymer from 4-bromo-2,6-xylenol in a homogeneous system has been reported (5).

3. References

1. University of Pennsylvania, Philadelphia, Pennsylvania 19104.
2. General Electric Company, Schenectady, New York 12345.
3. L. F. Fieser, *J. Am. Chem. Soc.*, **46**, 2639 (1924).
4. T. C. Bruice, N. Kharasch, and R. J. Winzler, *J. Org. Chem.*, **18**, 83 (1953).
5. C. C. Price, M. A. Semsarzadeh, and T. B. L. Nguyen, *Appl. Polym. Symp.* **26**, 319 (1975).

Poly-3,3-bis-(chloromethyl)oxetane

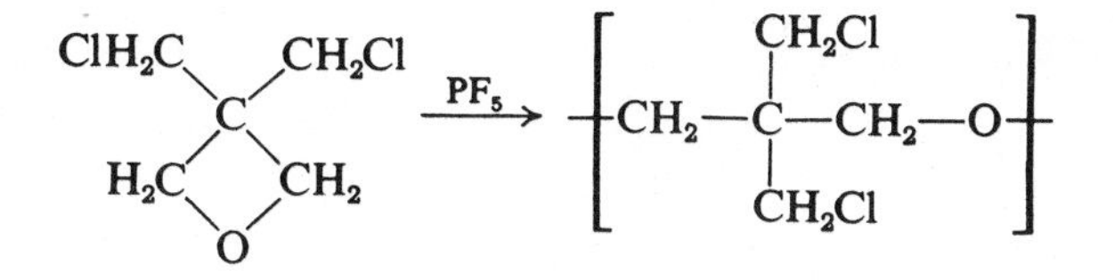

Submitted by T. W. Campbell (1)
Checked by K. D. Kopple (2a, 2b) and J. R. Ladd (2a)

1. Procedure

Anhydrous methyl chloride (100 ml.) (Note 1) is condensed into a 500-ml. three-necked flask equipped with a stirrer, a Dry Ice condenser, and a gas inlet tube. Pure 3,3-bis(chloromethyl)oxetane (25 g., Note 2) is then added all at once. The Dry Ice bath is removed, and the mixture is allowed to reflux (-25°). Into the refluxing mixture is introduced a trace of phosphorus pentafluoride gas (*Caution! This gas is very toxic and should be handled in an efficient hood.*) (Note 3); after a short induction period, polymerization takes place. Ordinarily this polymerization is uneventful, and the solid polymer precipitates during the course of 15-30 min. Occasionally, however, a polymerization is violent and the contents of the reaction flask are ejected through the top of the condenser.

The methyl chloride is allowed to evaporate, and the solid polymer is isolated and washed several times with methanol. Poly-3,3-bis(chloromethyl)oxetane is obtained as a spongy white solid with η_{inh} (Note 4) in the range 0.7-1.0 dl./g. The yield is 87-95%. It may be fabricated into films and fibers by the usual melt techniques at 175-200°. The polymer has a melt temperature (3) of 165° and a crystalline melting point of 177°.

2. Notes

1. The methyl chloride is purified by passing the gaseous product through a glass tube packed with silica gel. The silica gel becomes quite warm because of the heat of absorption.

2. 3,3-Bis(chloromethyl)oxetane is prepared by the reaction of methanolic potassium hydroxide on pentaerythritol trichloride (4). It is a stable compound. Polymerization grade material may be stored for reasonable periods of time at room temperature or below, if air and moisture are excluded.

3. Phosphorus pentafluoride gas may be obtained by the thermal decomposition of a diazonium fluoborate, such as Phosfluorogen A, obtainable commercially from Ozark-Mahoning, Tulsa, Oklahoma, or it may be purchased in cylinders from the Matheson Company, East Rutherford, New Jersey. Other Lewis acids such as boron trifluoride may be used; however, phosphorus pentafluoride is preferred because the speed of polymerization is much greater.

4. The polymer is soluble in hot cyclohexanone, hot DMF, and a number of similar solvents. However, the most satisfactory solvent is hexamethylphosphoramide. (*Caution! This substance has been shown to be carcinogenic in test animals.*) in which it is soluble to the extent of about 20%. All inherent viscosities have been measured in this solvent at 0.5% concentration (3).

3. Merits of the Preparation

Other oxetanes may be polymerized in a similar manner (4, 5).

4. References

1. Benger Research Laboratory, Textile Fibers Department, E. I. du Pont de Nemours & Co., Inc., Waynesboro, Virginia 22980.
2a. Research Laboratory, General Electric Company, Schenectady, New York 12345.
2b. Current Address: Illinois Institute of Technology Dept. of Chemistry, Chicago, Illinois 60616.
3. W. R. Sorenson and T. W. Campbell, *Preparative Methods of Polymer Chemistry,* Interscience Division, John Wiley and Sons, New York, 1961; see Chapter 2.
4. T. W. Campbell, *J. Org. Chem.,* **22**, 1029 (1957).
5. T. W. Campbell, *J. Org. Chem.,* **26**, 4654 (1961).

Sodium Carboxylmethylcellulose (CMC) (Note 1)

$$R_{cell}\text{-}OH + ClCH_2COONa + NaOH \longrightarrow R_{cell}\text{-}OCH_2COONa + NaCl + H_2O$$

$$Cl\text{-}CH_2COONa + NaOH \longrightarrow HO\text{-}CH_2COONa + NaCl \quad \text{(side reaction)}$$

Submitted by E. D. Klug (1)
Checked by A. B. Savage (2)

1. Procedure (3)

A slurry of 50 g. (0.3 mole) of chemical cellulose (cotton linters or wood pulp in shredded, chopped, or ground form; Note 2) is stirred mechanically in 1 l. of butanol in a 1500-ml. glass resin kettle (Note 3) fitted with a glass paddle stirrer with precision ground shaft and bearing, a thermometer, and a reflux condenser. The agitation is adjusted so that all the cellulose is in motion, but solid material is not splashed above the liquid level (Note 4). To the stirred slurry, cooled to 20-22°, is added 208 g. (1.55 moles) of 30% aqueous sodium hydroxide over a period of 30 min. (Note 9). The slurry is stirred for an additional 30 min., after which 73 g. (0.77 mole) of monochloroacetic acid (Notes 5 and 9) is added. The stirred slurry is heated in a water bath to 55° and kept at that temperature for 5 hr. During this period, the stirring speed must be decreased occasionally.

At the end of the reaction, any solid deposited on the walls of the reactor above the liquid level is discarded. The reaction liquid is removed by filtration with suction, using a rubber dam to prevent absorption of moisture. The filter cake is slurried in 80% methanol, and the pH is adjusted to 7.0 with acetic acid (Note 6). The product is given five 1-l. half-hour steep washes in 80% methanol and two such steep washes in anhydrous methanol. It is dried under reduced pressure at 70°. The resulting purified product has a degree of substitution (D.S.) of 0.88 (Note 7). Aqueous solutions of the product are substantially free of insoluble cellulose fibers.

The D.S. of the product may be varied by changing the ratio of monochloroacetic acid to cellulose, the molar ratio of NaOH to monochloroacetic acid being held at slightly greater than 2.0. The water in the system should be such that the NaOH concentration is 20-30%. CMC of D.S. appreciably greater than 1.2 may be prepared by repeating the carboxymethylation on the purified sample.

The viscosity of the product is measured on aqueous solutions in a Brookfield viscometer at 25° (4). The magnitude of this viscosity depends on the molecular weight of the starting cellulose. This molecular weight is expressed in terms of the viscosity of the cellulose (Note 8). CMC viscosities of 2000 cp. and higher in 1% aqueous solution may be obtained by using very high viscosity cotton linters (2000 sec. at 2.5 concentration), with rigorous exclusion of oxygen during the reaction. Wood pulp of 35 sec. viscosity gives CMC having viscosities of 500-1000 cp. in 2% solution. Solka-Floc BW-40, a wood pulp in powder form (Brown Company), gives a CMC with a viscosity of about 100 cp. in 2% solution. Lower viscosities may be obtained by adding hydrogen peroxide and controlled amounts of manganese or cobalt during the alkali cellulose stage (5).

2. Notes

1. Sodium carboxymethylcellulose is frequently called carboxymethylcellulose or CMC. Carboxymethylcellulose, strictly speaking, is the free acid that is insoluble in water and is of minor technological interest.
2. Chemical cotton may be purchased in the United States from the Buckeye Cellulose Corporation, Hercules Powder Company, or The Southern Chemical Cotton Company. Chemical wood pulp (dissolving pulp) may be purchased from ITT Rayonier, Inc., Brown Company, or several other manufacturers of dissolving pulps.
3. The resin kettle is obtainable from Ace Glass, Inc., or Labglass, Inc. A three-necked flask and a half-moon stirrer may also be used.
4. Satisfactory agitation can generally be obtained by placing the paddle close to the bottom of the vessel and adjusting the stirring speed.
5. Reagent grade monochloroacetic acid, 98% assay or better, was used. No correction is made for the impurities.
6. The pH of neutrality of CMC is about 8.25 (6). Neutralization to phenol-

phthalein is satisfactory. If the product is dried at a pH below 7, the solubility in water is impaired.

7. The degree of substitution is the average number of hydroxyl groups that have been carboxymethylated per anhydroglucose unit. It is determined by titrating a weighed sample of the pure free acid. The free acid is prepared by washing the sample with 80% methanol acidified with nitric acid (4, 7). Other methods of measuring the D.S. are also available (4).

8. The molecular weight of the cellulose is expressed in terms of its viscosity in cuprammonium hydroxide or cupriethylene diamine (8-10).

9. Safety Hazards: When preparing and handling 30% aqueous sodium hydroxide and monochloroacetic acid, safety glasses and soft-sided chemical workers' goggles should be worn. Any spill on skin or clothing should be washed with large amounts of water, and contaminated clothing should be removed.

3. Methods of Preparation

The classical methods involve the use of specialized equipment such as a sigma-bladed mixer, preferably with serrated blades (11). Generally, no inert organic liquid is added. If the product is water-soluble, it may be obtained as a dough that is transformed into granules by precipitation with methanol.

4. Merits of the Preparation

Compared with the classical method, this procedure has these advantages: all required equipment is readily available in most chemical laboratories; and the procedure is less laborious because the precipitation step is avoided.

5. References

1. Hercules Research Center, Hercules Powder Company, Wilmington, Delaware 19898.
2. The Dow Chemical Company, Midland, Michigan 48640.
3. E. D. Klug and J. S. Tinsley (Hercules Powder Comapny), U.S. Pat. 2,517,577 (Aug. 8, 1950).
4. Tentative Methods for Testing Carboxymethylcellulose D1439-61T, *ASTM Std.*, 1961, Pt 8.
5. E. D. Klug and H. M. Spurlin (Hercules Powder Company), U.S. Pat. 2,512,338 (June 20, 1950).
6. G. Sitaramaiah and D. A. I. Goring, *J. Polym. Sci.*, **58**, 1107 (1962).
7. R. W. Eyler, E. D. Klug, and F. Diephuis, *Ind. Eng. Chem. Anal. Ed.*, **19**, 24 (1947).

8. J. A. N-C 206A Amend. I (for cotton).
9. J. A. N-C 216A (for wood pulp).
10. *Tappi* T230 su-63.
11. A. B. Savage, A. E. Young, and A. T. Maasberg, p. 939 in E. Ott, H. M. Spurlin, and M. W. Grafflin, eds., *Cellulose and Cellulose Derivatives,* Interscience Division, John Wiley and Sons, New York, 1954.
12. E. D. Klug, p520-539, Vol. 3, N. G. Gaylord Ed. *Encyclopedia of Polymer Science and Technology,* Interscience-Wiley, New York, 1965.

Poly-ϵ-caprolactam

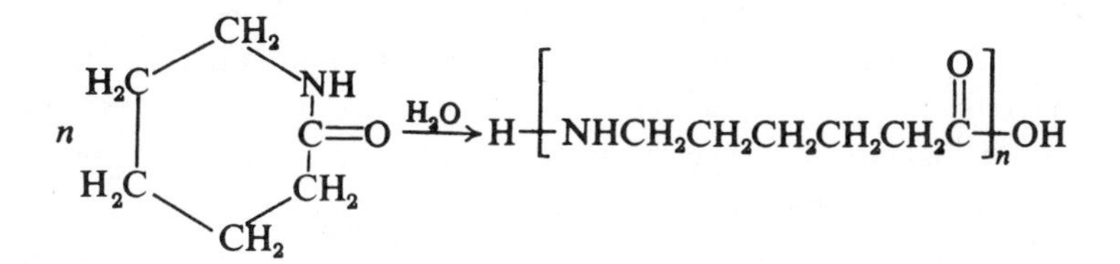

Submitted by M. I. Kohan (1)

Checked by J. Brandrup (2a, b) and M. Goodman (2a, b)

1. Procedure

ϵ-Caprolactam 50 g. (Note 1) and 2.0 g. of a 50 wt. % aqueous solution of catalyst (Note 2) are placed in a 38 x 300-mm. Pyrex® test tube fitted with 8-mm. Pyrex® inlet and exit tubes. The tube is swept with nitrogen for 5 min. before adjusting the rate to 350 cc./min. (Note 3). The tube is immersed to a depth of 9 in. in a vapor bath at 280-285° (Note 4). After 4 hr. (Note 5) the test tube is removed from the bath and allowed to cool to room temperature, under nitrogen. The test tube is then broken away from the polymer. The polymer plug can be cut to the desired particle size (Note 6). After extraction, the polymer has a crystalline melting point of 227° and a density at 23° of 1.13 g./ml. (Note 7).

2. Notes

1. Caprolactam is commercially available in a pure form. It is a white crystalline substance with a melting point of 68-69° when completely dry. One of the easiest ways to ascertain whether the lactam is of adequate purity is to carry out the polymerization and note whether a high molecular weight polymer of good color is obtained. This is fundamentally the simplest analysis

not only because removal of trace water is unnecessary, but, more importantly, because attainment of high molecular weight in a trial polymerization is an extremely sensitive test of purity.

2. Amine salts are effective catalysts leading to a polymer with a Flory distribution after extraction. Because this polymerization is carried out at an elevated temperature in an atmosphere constantly swept with fresh nitrogen, it is important to use high-boiling components. It is also important to use difunctional components to avoid introduction of non-reactive ends that limit the molecular weight. The product is normally a copolymer comprising approximately 98 wt. % polycaprolactam and 2 wt. % catalyst residues, unless the catalyst chosen is ϵ-aminocaproic acid, which can be obtained by hydrolysis of the lactam (3). The 50 wt. % solution is convenient for preparation of the catalyst. The amine and acid (stoichiometric equivalents or a 1% excess of amine if somewhat volatile) are dissolved or dispersed in water, and one solution is then slowly added to the other while stirring. The resulting solution can then be treated with Darco G-60 to remove impurities, and the filtrate can be used directly as the catalyst.

Suitable amines are hexamethylene diamine, piperazine, 3,3′(methylimino)-bispropylamine, 3,3′-iminobispropylamine, and *m*-xylylene diamine. Representative acids are adipic and sebacic acids. Also suitable are amino acids, which have little tendency to participate in ring closure reactions such as ϵ-aminocaproic acid or 11-aminoundecanoic acid. Salts from these amines and acids should give equivalent products.

3. Within reasonable limits the nitrogen flow rate is not critical.

4. A suitable vapor bath is made from an 80-mm. Pyrex® glass tube 13 in. long, rounded at one end, and fitted with a side arm of 15-mm. tubing located 4 in. below the open end and bent at a 90° angle so it extends 4 in. above the open end. This tube can be heated with a Glass-Col 400-ml. beaker heating mantle. The rest of the tube should be wrapped with Fiberglass or other insulating material. A sheet of asbestos or other suitable material can be cut to sit on top of the bath with a hole just big enough to accommodate the 38-mm. test tube. Diphenylene oxide is placed in the bath so its liquid level is just below the top of the heating mantle.

5. Sublimate of monomer will accumulate in the exit tube but should not be enough to cause plugging. This should be checked. Any accumulation is readily removed by wiping with a damp swab or pipe cleaner.

A number average molecular weight of about 11,000 is reached in 2 hr.; 16,000 in 4 hr. Longer times result in only slightly higher molecular weight and cause significant variation in molecular weight from the top to the bottom of the plug.

The number average molecular weight of extracted polymer can be determined directly by end group analysis using the method of Waltz and Taylor (4), or it can be estimated from solution viscosity (5). A convenient empirical relationship observed for extracted resin is

$$\overline{M}_n = 15{,}600 \times \eta_{\text{inh}}^{1.49}$$

where η_{inh} is the inherent viscosity, at 25°, of 0.5 g. of resin in 100 ml. of *m*-cresol (98%, freshly distilled).

6. If a uniform particle size is not required, the plug can be fragmented for convenient handling by freezing in Dry Ice and compressing the cold plug in a hydraulic press or simply wrapping it in a towel and striking it with a hammer. Polycaprolactam in equilibrium at 280° contains over 10 wt. % of a water-extractable fraction made up principally of monomer and cyclic oligomers (6). Under the conditions of synthesis described here, some of the relatively volatile monomer is removed, but the higher cyclic oligomers are unaffected. Essentially complete extraction of the polymer can be accomplished after 16 hr. in ten times its weight of boiling water if the particles have one dimension not over one-sixteenth of an inch. Drying to a sufficiently low moisture content for melt pressing is achieved without discoloration or change in molecular weight by heating overnight at 100° under nitrogen or under reduced pressure.

7. The crystalline melting point can be determined microscopically on a Kofler hot stage (7). Melting points are often approximated by the simpler ASTM method (8), but the criterion of melting in this method is less certain and the values obtained are less reliable. The density is readily measured in a gradient tube (9) made from solutions of carbon tetrachloride (C) and toluene (T): lower layer = 46.3/53.7 C/T (v./v.); upper layer = 25.9/74.1 C/T (v./v.).

3. Methods of Preparation

The procedure described for the preparation of polycaprolactam is extremely simple in that it avoids special equipment and sealed systems and yields a polymer of normal Flory distribution. Hydrolytic polymerization under pressure and catalytic polymerization in evacuated sealed tubes using alkali or alkaline earth metal salts of ε-caprolactam as catalyst have been described (10). The latter method yields a polymer whose molecular weight decreases on continued heating (11). High viscosity polycaprolactam has been obtained with very short reaction times by using an alkali metal hydride as catalyst, but the subsequent decrease in viscosity with time, denoting a peculiar initial distribution of molecular weights, was again observed (12). The imide-promoted anionic polymerization of caprolactam takes place at relatively low temperatures but yields products of uncertain molecular weight distribution (13). The non-hydrolytic polymerization of caprolactam via acid catalysis has been carried out in sealed tubes, but the molecular weights achieved were below 5000 (14). A two-step process involving alkali catalysis in a nitrogen atmosphere followed by catalyzed redistribution in a sealed tube has been reported (15). Elimination of the amine and acid end groups to reduce the rate of depolymerization to monomer at melt temperatures can be accomplished by reaction with an isocyanate in dimethyl formamide (16). Additional information on the characterization and utility, as well as on the synthesis of a variety of nylons, is now available (17).

4. References

1. Plastics Products and Resins Dept., Experimental Station, E. I. du Pont de Nemours & Company, Wilmington, Delaware 19898.
2a. Polytechnic Institute of New York, Brooklyn, New York 11201.
2b. Current Address: Farbwerke Hoechst, Frankfurt am Main, West Germany.
2c. Current Address: Department of Chemistry, The University of California/San Diego, La Jolla, California 92037.
3. J. C. Eck, *Org. Syn.*, Coll. Vol. **2**, 28 (1943).
4. J. E. Waltz and G. B. Taylor, *Anal. Chem.*, **19**, 448 (1947); see also ref. 5, and J. R. Schaefgen and P. J. Flory, J. Amer. Chem. Soc., **72**, 689 (1950).
5. F. Wiloth, *Makromol. Chem.*, **27**, 37 (1958); O. Fukumoto, *J. Polym. Sci.*, **22**, 263 (1956).
6. P. H. Hermans, D. Heikens, and P. F. Van Velden, *J. Polym. Sci.*, **16**, 451 (1955); P. F. Van Velden, G. M. Van Der Want, D. Heikens, Ch. A. Kruissink, P. H. Hermans, and A. J. Staverman, *Rec. trav. chim.*, **74**, 1376 (1955); H. H. Schenker, C. C. Casto, and P. W. Mullen, *Anal. Chem.*, **29**, 825 (1957).
7. L. Kofler, *Mikrochem.*, **15**, 242 (1934); R. D. Evans, H. R. Mighton, and P. J. Flory, *J. Chem. Phys.*, **15**, 685 (1947).
8. ASTM Std. D-789-59T.
9. R. F. Boyer, R. S. Spencer, and R. M. Wiley, *J. Polym. Sci.*, **1**, 249 (1946).
10. W. E. Hanford and R. M. Joyce, *J. Polym. Sci.*, **3**, 167 (1948); see also ref. 5.
11. J. Kralicek and J. Sebenda, *J. Polym. Sci.*, **30**, 493 (1958).
12. V. W. Griehl and S. Schaaf, *Makromol. Chem.*, **32**, 170 (1959).
13. O. Wichterle, *Makromol. Chem.*, **35**, 174 (1960).
14. G. M. Van der Want and Ch. A. Kruissink, *J. Polym. Sci.*, **35**, 119 (1959).
15. J. Saunders, *J. Polym. Sci.*, **30**, 479 (1958).
16. G. M. Van der Want, *J. Polym. Sci.*, **37**, 547 (1959).
17. M. I. Kohan, *Nylon Plastics*, Wiley Interscience, New York, 1973.

Polymerization of Acrylamide to Poly-β-alanine

$$n\mathrm{CH_2{=}CHCONH_2} \xrightarrow{t\text{-}\mathrm{C_4H_9ONa}} \mathrm{-\!\!\!+\!NHCH_2CH_2CO\!+\!\!\!-}_n$$

Submitted by D. S. Breslow (1),
G. E. Hulse (1a), and A. S. Matlack (1)
Checked by A. Abe (2a, b) and M. Goodman (2a, 2c)

1. Procedure

In a 250-ml., three-necked, round-bottomed flask equipped with a mechanical stirrer, a reflux condenser carrying a drying tube, and a stopper are placed 0.02 g. of *N*-phenyl-β-naphthylamine (Note 1) and 100 ml. of dry pyridine (Note 2). Stirring is started, the mixture is heated to 95-100° on a steam bath, and 10.0 g. of dry acrylamide (Note 3) are added. As soon as the acrylamide dissolves, a solution of 0.1 g. of sodium in 10 ml. of t-butyl alcohol (Note 4) is added. Polymer begins to form on the walls and stirrer in 3-10 min. (Note 5). After 16 hr. of heating, the polymer is removed by filtration, extracted with 200 ml. of water on the steam bath for an hour (Note 6), and dried overnight in a vacuum oven at 80° and 1 torr. The yield of water-insoluble polymer melting at about 335-340° (dec.) is 2.8-4.8 g. (28-48% of the theoretical amount). A 1% solution in 90% formic acid has a specific viscosity of 0.47-0.66 dl/g. at 25° (Note 7).

After neutralization with acetic acid, the aqueous extract is evaporated to dryness to recover 3.2-3.6 g. (32-36% of the theoretical amount) of water-soluble polymer melting at about 300-305°(dec.) (Note 8). A 1% solution in 90% formic acid has a specific viscosity of 0.19 dl/g. at 25° (Note).

2. Notes

1. This is added as an inhibitor to vinyl polymerization. In small runs it may be unnecessary. This procedure has been used with up to 500 g. of acrylamide.
2. Reagent grade pyridine allowed to stand over barium oxide for several days is suitable.
3. Commercial material may need to be sublimed at 0.1 torr before use. The checkers sublimed at 50° and 0.5-1.0 torr.
4. Because sodium dissolves slowly in boiling *t*-butyl alcohol, this solution must be prepared in advance.
5. If too much polymer forms on the stirrer, stirring can be discontinued without harm.
6. Because much of the polymer may adhere to the walls of the flask, it is convenient to carry out the extraction in the same flask.
7. The solution should be prepared just before the viscosity is determined. The strength of the acid used affects the viscosity considerably, stronger acids giving higher values.
8. During the evaporation some additional polymer may precipitate. After its removal by filtration or centrifugation, the evaporation is continued. If purification of the water-soluble polymer is desired, the evaporation should be stopped at a small volume. This solution is poured into a large volume of methanol to precipitate 2.6 g. of polymer.

3. Methods of Preparation

The method is that discovered by Breslow, Hulse, and Matlack (3) and studied further by Ogata (4). Low molecular weight poly-β-alanine has been prepared by the elimination of HX from a β-alanine derivative, $NH_2CH_2CH_2COX$ [where X is NH_2 (5), OC_2H_5 (6), Cl (7), or OH (8)] ; by the elimination of thiophenol and carbon dioxide from N-carbothiophenyl-β-alanine (9); by base-catalyzed polymerization of 4,5-dihydro-1,3-oxazine-2,6-dione (10) or the corresponding 1,3-thiazine (11); and by polymerization of β-isothiocyanatopropionic acid. Molecular weights similar to those obtained by this procedure have been reported for the polymerization of perhydro-1,5-diazocine-2,6-dione (12). Completely linear, high molecular weight poly-β-alanine has been prepared by polymerization of azetidine-2-one (13), poly-β-alanine from acrylamide has been reported to contain some branching (14).

4. References

1. Research Center, Hercules Powder Company, Wilmington, Delaware 19899.
2a. Polytechnic Institute of Brooklyn, Brooklyn, New York 11201.
2b. Current Address: Showa Denko Co., Tokyo, Japan.
2c. Current Address: Dept. of Chemistry, The University of California/San Diego, La Jolla, California.
3. D. S. Breslow, G. E. Hulse, and A. S. Matlack, *J. Amer. Chem. Soc.*, **79**, 3760 (1957).
4. N. Ogata, *Bull. Chem. Soc. Japan,* **33**, 906 (1960).
5. A. P. N. Franchimont and H. Friedman, *Rec. trav. chim.,* **25**, 80 (1906).
6. E. Abderhalden and F. Reich, *Z. Physiol. Chem.,* **178**, 169 (1928).
7. M. Frankel, Y. Liwschitz, and A. Zikha, *Experientia,* **9**, 179 (1953); *J. Amer. Chem. Soc.,* **76**, 2814 (1954).
8. J. S. Chirtel and A. M. Mark, U.S. Pat. 2,691,643 (1954).
9. J. Noguchi and T. Hayakawa, *J. Amer. Chem. Soc.,* **76**, 2846 (1954).
10a. L. Birkhofer and R. Modic, *Ann.* **628**, 162 (1969).
10b. H. R. Kricheldorf, *Makromol. Chem.* **173**, 13 (1973).
11. H. R. Kricheldorf, *Makromol. Chem.,* **175**, 3345 (1974).
12. Y. Iwakura, K. Uno, M. Akiyama, and Haga *J. Polym. Sci.,* (A-1) **7**, 657 (1969).
13. H. Bestian, *Angew Chem. Int. Ed.,* **7**, 278 (1968).
14. J. D. Glickson and J. Applequist, *Macromolecules,* **2**, 628 (1969).

Polyhydroxymethylene Films via Poly(vinylene carbonate)

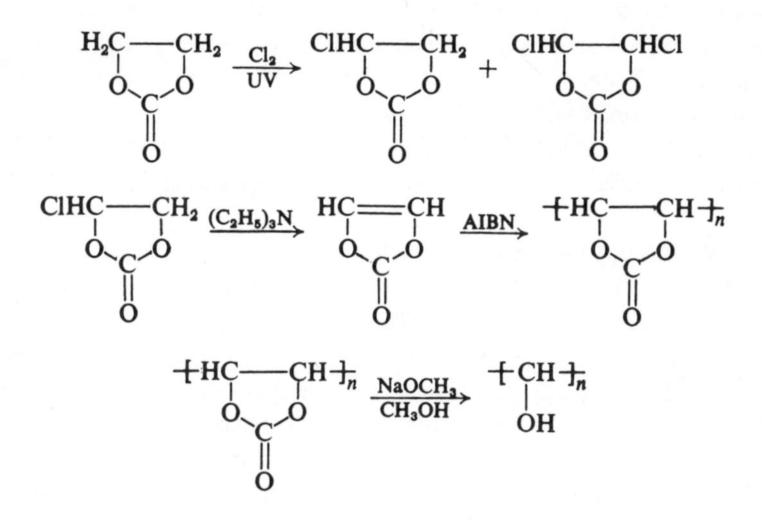

Submitted by J. R. Schaefgen (1a) and N. D. Field (1a, b)
Checked by J. B. Clements (2)

A. Poly(vinylene carbonate)

1. Procedure

Chloroethylene Carbonate. Ethylene carbonate (Note 1) (500 g., 5.7 moles), and carbon tetrachloride (1 1.) are added to a 2-1., three-necked, standard taper flask equipped with an ultraviolet lamp in a quartz jacket immersed in the flask, a gas inlet tube, and an efficient condenser (Note 2). The two-phase system is heated with a heating mantle to reflux. Heating is then stopped, the ultraviolet lamp is turned on, and chlorine from a cylinder is introduced at a rate sufficient to maintain vigorous refluxing (Note 2). The ethylene carbonate-rich phase gradually disappears and a homogeneous solution forms. Chlorination is continued beyond this point until the total weight of chlorine added is about 600 g. (8.5 moles). This requires 3-5 hr. (Note 3). The product is isolated by fractional distillation through an efficient column (Note 4). After removal of the solvent and a low-boiling solid impurity (Note 5), 1,2-dichloroethylene carbonate, 114 g. (13%), b.p. 91°/30 torr, n_D^{25} 1.4606, and chloroethylene carbonate, 420-475 g. (60-68%), b.p. 102°/8 torr, n_D^{25} 1.4525, are isolated.

Vinylene Carbonate. The chloroethylene carbonate from the preceding step (450 g., 3.65 moles), 450 ml. of anhydrous ether (Note 6), and 4 g. of di-*t*-butyl-*p*-cresol (Note 7) are added to a 2-1. three-necked, standard taper flask equipped with a precision-ground stirrer, dropping funnel, and reflux condenser to which a drying tube filled with Drierite is attached. Triethylamine (560 ml., 4.1 moles) is added slowly over about 4 hr. to the stirred refluxing solution (Note 8). Gentle refluxing is maintained for 2 days. A copious precipitate of amine salt forms, and the color of the solution becomes dark brown. The precipitate is collected and washed four times, with 400 ml. of a mixture of 50/50 vol. % of benzene and ether. The second and third washings are carried out by slurrying the solid precipitate with the solvent mixture in a beaker. The filtrate and washings are combined, and most of the ether and some of the benzene are removed by simple distillation. Distillation of the remainder of the solution at reduced pressure through an efficient column (Note 4) yields 200-230 g. (63-73%) of vinylene carbonate, b.p. 74°/30 torr. This material rapidly colors (brown) on standing. It is further purified by refluxing for 1 hr. over 1.5% by weight sodium borohydride and then distilling (Note 9). A second treatment with sodium borohydride is recommended to obtain a color-stable purer product, n_D^{25} 1.4185; m.p. 20.5°; d, 1.35 (27°).

Polymerization. A thick-walled test tube is narrowed near the top and, after cooling, 0.01 g. of azobisisobutyronitrile (AIBN) (Note 10) is introduced. By use of a hypodermic syringe, 5 ml. of the sodium borohydride-treated vinylene carbonate is added. The tube is cooled in ice water to freeze the monomer and is evacuated through a stopcock to 1 torr. The stopcock is closed and the monomer is then degassed by melting and refreezing. The evacuation and degassing procedure is then repeated, after which the system is sealed under reduced pressure (Note 11). The sealed tube is placed in a bath thermostated at 60-65°. Polymerization takes place slowly to give a clear solid resin in 18-72 hr. The tube is broken, and the tough plug of polymer is dissolved in 50 ml. of N,N-dimethylformamide (DMF) at room temperature (Note 12) and reprecipitated as a white fibrous solid by adding this solution slowly with stirring to 200 ml. of methanol. The polymer is collected by filtration and washed repeatedly by slurrying with methanol until the filtrate is nearly clear. The yield of polymer is 3.7-5.6 g. (57-87%); the inherent viscosity at 30° of a 0.5% solution of polymer in N,N-dimethylformamide is 2.0-3.5 dl./g. (Note 13).

B. Polyhydroxymethylene

1. Procedure

Poly(vinylene carbonate) is dissolved in DMF to form a 10% solution. This is cast as a 10-mil film on a glass plate by use of a doctor knife. After drying overnight at room temperature, the clear film is removed from the plate and hydrolyzed. To accomplish this, film is suspended in a 1% sodium methylate (Note 14) solution in methanol in a covered beaker. Hydrolysis to clear but crinkled films of polyhydroxymethylene is complete after 24 hr. at 50-60° or after 3-5 days at room temperature. The progress of hydrolysis may be followed conveniently by noting the disappearance of the carbonyl bond at 5.5 μ in the infrared spectrum.

The films of polyhydroxymethylene (Note 15) are stiff and brittle when thoroughly dry but become limper and tougher in moist air. They are insoluble even in boiling water and retain a moderate amount of strength when wet. The wet film can be cut into narrow strips (about 5 mm. wide), and these can be oriented by drawing them quickly (Note 16) over a rod heated to 200°. Such films are quite strong [>6 g./denier (g.p.d.)] and very stiff (~300 g.p.d. initial modulus). Although measurements of inherent viscosity or molecular weight are precluded by the insolubility of the polymer, the retention of good mechanical properties is indicative of a high molecular weight.

2. Notes

1. Material obtained from the Jefferson Chemical Company is sufficiently pure to use as received. It was found convenient to add this chemical to the reaction flask as a liquid (m.p. 36°).

2. Two Liebig or Allihn condensers, one on top of the other, were found

convenient. The top of the upper condenser should be connected to a scrubber to trap the evolved hydrogen chloride, excess chlorine, and any carbon tetrachloride which fails to condense. The rate at which chlorine can be added depends on the efficiency of the condenser system. The checker found that an external sun lamp (G.E. RS type) was also satisfactory, but the time of reaction was longer, i.e., 9 hr.

3. Near the end of the reaction, the solution turns green and this color change may be used as an indication of completion of the reaction if it is not convenient to weigh the chlorine cylinder.

4. A 3-ft. spinning band column in which the spinning element is a tungsten wire supporting a platinum gauze was used by the submitter. The checker found that a 30-cm. Vigreux column gave sufficiently pure chloroethylene carbonate for dehydrohalogenation.

5. This impurity was not identified. If the take-off head clogs, an infrared lamp may be used to maintain fluidity. It may also be necessary to clean the column before collecting the pure 1,2-dichloroethylene carbonate.

6. The commercial purified product is suitable.

7. This is added to prevent polymerization of the monomer by adventitious impurities during either reaction or the subsequent distillation.

8. A water bath should be used to avoid overheating the reaction. If convenient, a thermometer should be used in the flask, and a temperature of about 45° should be maintained.

9. A simple Vigreux column is sufficient. Nearly all the product is recovered.

10. Du Pont Vazo vinyl polymerization catalyst is a convenient product to use.

11. An alternate procedure (3) is to repeatedly (5 or 6 times) alternately evacuate to about 30 torr pressure and flush with purified nitrogen, and then seal the tube under reduced pressure.

12. Hot DMF degrades the polymer (3). Solution requires about 8-16 hr. (i.e., overnight) with occasional shaking by hand.

13. The rate of polymerization is quite low, necessitating the long polymerization time. The inherent viscosity of the polymer is a function of the initiator to monomer ratio, but it is greater than 2 even at five times the ratio used in the example. The preparation also has been carried out at up to five times the scale indicated here with corresponding yields of polymer in the designated range of inherent viscosity.

14. Commercially available sodium methylate may be used. The reaction probably involves hydrolysis by small amounts of water present in the methanol

rather than alcoholysis, because a white precipitate, presumably sodium carbonate, forms. The presence of a few per cent of water does not cause difficulty.

15. The checker was able to make 12% solutions of polyhydroxymethylene in dimethyl sulfoxide from which 5-mil films were cast. The submitter was not able to obtain gel-free solutions of high molecular weight polymer in this solvent.

16. The films must be cut when wet because the dry films are brittle and may shatter. Some practice is required to orient the films uniformly without breaking them. The films may be drawn 600-800% when the technique is perfected. Such films are smooth and are oriented, as shown by their wide angle x-ray diffraction patterns. Alternatively, the films can be stretched to a lower degree of orientation in warm water (60°). The great difference in stiffness of the dry and wet films is intriguing.

3. Methods of Preparation

This preparation is based on the article by Field and Schaefgen (3). Other authors have used similar methods to prepare vinylene carbonate (4) but have used more involved and less effective purification methods (5). The hydrolysis of poly(vinylene carbonate) (6) has usually been carried out in aqueous solution. In aqueous hydrolysis, polyhydroxymethylene precipitates as a completely intractable white powder.

4. References

1a. Pioneering Research Division, Textile Fibers Department, E. I. du Pont de Nemours & Co., Inc., Wilmington, Delaware 19898.
1b. Current Address: Internaltional Playtex Co., 215 College Rd. Paramus, New Jersey 07652.
2. Chemstrand Research Center, Inc., Durham, North Carolina 27709.
3. N. D. Field and J. R. Schaefgen, *J. Polym. Sci.,* **58**, 533 (1962).
4. M. S. Newman and R. W. Addor, *J. Amer. Chem. Soc.,* **75**, 1263 (1953); **77**, 3789 (1955).
5. See, for example, R. M. Thomas, U.S. Pat. 2,873,230 (1959).
6. See ref. 4, and H. C. Haas and N. W. Schuler, *J. Polym. Sci.,* **31**, 237 (1958).

Propylene Sulfur Dioxide Copolymer

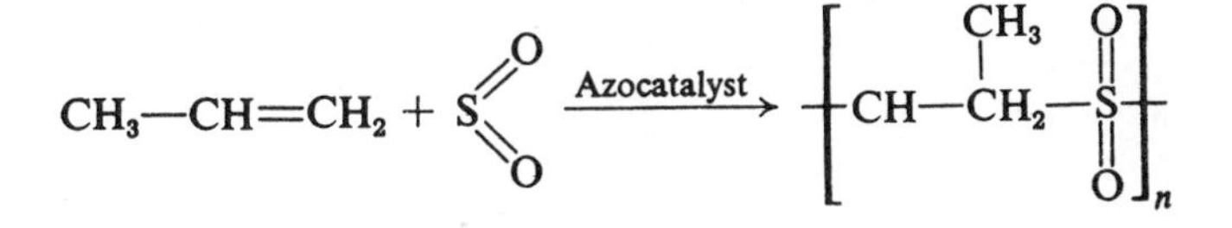

Submitted by A. H. Frazer and W. P. Fitzgerald, Jr. (1)
Checked by N. L. Zutty (2a, b)

1. Procedure

Into a 1-l. rocker bomb is charged 0.50 g. of α,α′-azo*bis*(α,α-dimethylvaleronitrile) catalyst (Note 1). The bomb is pressure-tested with nitrogen at 400 p.s.i. to detect leaks and then evacuated to less than 1 torr for 3 or 4 hr. The bomb is chilled in Dry Ice-acetone, and 42 g. (1 mole) of propylene (Note 2) and 256 g. (4 moles) of sulfur dioxide (Notes 2 and 3) are distilled into it. The bomb is sealed and the polymerization is carried out at 40-45° for 8 hr. Vent the excess reactants from the bomb through a system designed to prevent contact with laboratory personnel. The product (ca. 110 g.) is removed from the bomb and washed twice with ethanol. The dried polymer is ground in a Wiley mill through a 20 mesh screen and rewashed with alcohol (Note 4). After drying at 80° for about 15 hr., 96 g. (88%) of a hard white product (Note 5) with η_{inh} (Note 6) of 3.3 analyzing for 30.5% sulfur (theory, 31.5%) is obtained. The polymer melt temperature is 300° (dec.).

2. Notes

1. The catalyst should be kept under refrigeration prior to use. In checking

this procedure, α,α'-azo*bis*isobutyronitrile was used, and essentially the same results were obtained.

2. The propylene and sulfur dioxide are purified products offered by the Matheson Company (both analyze over 99% pure).

3. Unless this excess of sulfur dioxide is employed, diminished yields of the desired copolymer are obtained.

4. The solubility of the resultant product in sulfuric acid is markedly affected by its state of subdivision.

5. In a similar run, on the same scale, 89 g. of polymer, η_{inh} 3.15 dl/g. (sulfuric acid) was obtained.

6. The η_{inh} is determined in sulfuric acid *at room temperature.* An η_{inh} dl./g. of 3.3 corresponds to a $\overline{M}_w$ of 250,000.

3. Methods of Preparation

Preparation of olefin-sulfur dioxide copolymers is described in the patent literature (3-10) in the open literature (11-17) and in several review articles (18-20). Areas of application for this class of polymers are varied and include blends with PVC (8, 9), biomedical applications (21), and electron-beam resists for microelectronic circuitry (22).

4. References

1. Pioneering Research Division, Textile Fibers Department, E. I. du Pont de Nemours & Co., Inc., Wilmington, Delaware 19898.
2a. Union Carbide Corporation, Chemicals Division, South Charleston, West Virginia 25303.
2b. Current Address: Union Carbide Corp., Bound Brook, New Jersey 08805. The assistance of R. J. Cotter in revising this procedure is thankfully acknowledged.
3. J. T. Rivers and A. H. Frazer (E. I. du Pont de Nemours & Co., Inc.), U.S. Pat. 2,684,950 (July 26, 1954) [*C.A.*, 48, 13131 (1954)].
4. F. E. Frey and R. D. Snow, U.S. Pat. 2,112,986 (April 5, 1938) [*C.A.*, **32**, 3858 (1938)]; Phillips Petroleum Co. (Brit. Pat. 480,777 (Feb. 24, 1938) [*C.A.*, **32**, 6360 (1938)]; French Pat. 808,580 (Feb. 10, 1937) [*C.A.*, **31**, 6373 (1937)].
5. J. Harmon (E. I. du Pont de Nemours & Co. Inc.), U.S. Pat. 2,190,836 (Feb. 20, 1940) [*C.A.*, **34**, 4183 (1940)].
6. C. S. Marvel and D. S. Frederick (C. S. Marvel), U.S. Pat. 2,136,389 (Nov. 15, 1939) [*C.A.*, **33**, 1343 (1939)].
7. F. E. Frey, R. D. Snow, and L. H. Fitch, Jr. (Phillips Petroleum Co.) U.S. Pat. 2,136,389 (April 30, 1940) [*C.A.*, **34**, 5972 (1940)].
8. C. F. Hammer and J. J. Hichman (E. I. du Pont de Nemours & Co.) U.S. Pat. **3,657,394** (April 18, 1972).

9. C. F. Hammer and T. F. Sashihara (E. I. du Pont de Nemours & Co.,) U.S. Pat. 3,657,202 (April 18, 1972).
10. B. L. Atkins, W. M. Welsh, and W. R. Moore (Dow Chemical Co.,) U.S. Pat. 3,792,026 (February 12, 1974).
11. M. Hunt and C. S. Marvel, *J. Amer. Chem. Soc.*, **57**, 1691 (1935).
12. R. D. Snow and F. E. Frey, *J. Amer. Chem. Soc.*, **65**, 2417 (1943).
13. C. S. Marvel and E. D. Weil, *J. Amer. Chem. Soc.*, **76**, 61 (1954).
14. R. D. Snow and F. E. Frey, *Ind. Eng. Chem.*, **30**, 176 (1938).
15. W. W. Crouch and J. E. Wicklatz, *Ind. Eng. Chem.*, **47**, 160 (1955).
16. N. L. Zutty and C. W. Wilson, *Tetrahedron Letters*, **30**, 2181 (1963).
17. J. E. Crawford and D. N. Gray, *J. Appl. Polym. Sci.*, **15**, 1881 (1971).
18. O. Grummit and A. Ardis, *J. Chem. Educ.*, **23**, 73 (1946).
19. E. J. Goethals, *J. Macromol Sci., Rev Macromol. Chem.*, **C2** (1), 105 (1968).
20. E. M. Fettes and F. O. Davis, in "High Polymers", Vol. XIII (N. G. Gaylord, Ed.).
21. Anon., *Chem. Eng. News*, December 15, 1975, p. 24.
22. M. J. S. Bowden and E. A. Chandross (Bell Telephone Lab.) U.S. Pat. 884,696 (May 20, 1975).

Poly-p-phenylene

$$nC_6H_6 + (2n-2)CuCl_2 \xrightarrow{AlCl_3} C_6H_5\text{-}[C_6H_4]_{n-2}\text{-}C_6H_5 + (2n-2)CuCl + (2n-2)HCl$$

Submitted by P. Kovacic and J. Oziomek (1)
Checked by D. Hoeg and E. P. Goldberg (2)

1. Procedure

A 1-l. three-necked flask is fitted with a thermometer, a reflux condenser, a gas inlet tube, and a paddle stirrer. After the apparatus is swept with dry nitrogen (Note 1), the gas flow is adjusted to a moderate rate.

While stirring is maintained, the flask is charged at 25° with 469 g. (6 moles) of benzene (Note 2), 200 g. (1.5 moles) of anhydrous aluminum chloride (Note 3), and 101 g. (0.75 mole) of anhydrous cupric chloride (Note 4) in the indicated order. The temperature is increased to 31-32° and held in this range. After an induction period of about 25 min. the reaction proceeds with good rapidity. During 2 hr. 0.64-0.65 mole (85-87%) of hydrogen chloride is evolved (Note 5).

The dark viscous mixture is cooled to 15° and filtered through an 8.5-cm. fritted glass filter (Note 6). After being washed with benzene, the residue is carefully added, with stirring, to 1.4 l. of ice-cold 18% hydrochloric acid, and then heated at the boiling point of the mixture (Note 7). The filtered solid is pulverized in the presence of water with a blender and triturated three times with a total of 4.2 l. of boiling 18% hydrochloric acid (Note 8). Washing is effected by several triturations with 1.4 l. of boiling water and the drying by storage overnight at 110°. The yield of light brown product is 25-26 g. (89-91%) (Notes 9 and 10).

2. Notes

1. Concentrated sulfuric acid is used as the drying agent.

2. Benzene, thiophene-free, is distilled from sodium and then stored over sodium.

3. The powder form works well.

4. Commercial anhydrous cupric chloride is dried at 110° for 24 hr.

5. The reaction can be followed by titration of the generated gas with standard sodium hydroxide solution.

6. The benzene can also be removed by steam distillation after careful addition of the reaction mixture to dilute hydrochloric acid with good cooling.

7. The necessary precautions must be taken in relation to the small amounts of benzene present in the residue.

8. This degree of purification should be satisfactory for most purposes.

9. The same yield is obtained when the reaction is carried out under air. The evolved acid gas is passed into a trap containing caustic solution.

10. Analysis of the polymer shows 88.7-92.6% carbon, 4.9-5.3% hydrogen, 1.6-4.8% chlorine, and 0-1.7% residue.

3. Methods of Preparation

In addition to the method described (3, 4), poly-*p*-phenylene has been prepared by dehydrogenation of poly-1,3-cyclohexadiene (5).

4. References

1. Department of Chemistry, Case Institute of Technology, Cleveland, Ohio 44101.
2. Roy C. Ingersoll Research Center, Borg-Warner Corporation, Des Plaines, Illinois 60016.
3. P. Kovacic and A. Kyriakis, *Tetrahedron Letters,* 467 (1962).
4. P. Kovacic and J. Oziomek, unpublished work.
5. C. S. Marvel and G. E. Hartzell, *J. Amer. Chem. Soc.,* 81, 448 (1959).

Polymerization of Acrylamide in the Crystalline State

$$\underset{\displaystyle CONH_2}{CH_2{=}CH} \longrightarrow \left[CH_2{-}\underset{\displaystyle CONH_2}{CH}\right]_n$$

Submitted by H. Morawetz (1a) and T. A. Fadner (1a, b)
Checked by D. Ballantine (2)

1. Procedure

Reagent grade acrylamide is recrystallized from chloroform (m.p. 85-85.7°). A test tube of about 1 cm. dia. containing a 4-g. sample of powdered monomer crystals is sealed under reduced pressure (Note 1). It is then exposed to a dose rate of 4 x 10^5 rad of gamma-rays from a Co^{60} source at Dry Ice temperature (-78°) (Note 2). The irradiated test tube is placed in a thermostat at 35° for 24 hr. It is then opened, and the unreacted monomer is extracted from the sample by two portions of 30 ml. of dry butanone containing 0.1% of hydroquinone. The polymer is filtered, washed with butanone, and dried. The indicated conditions lead to a monomer conversion to polymer of 8.5-11% and a weight average molecular weight ($\overline{M}_w$) of about 600,000 (Note 3).

2. Notes

1. The reaction is insensitive to oxygen, but the rate is affected somewhat by water vapor. It is therefore desirable to carry out the process under reduced pressure.

2. The polymerization of crystalline acrylamide by high energy ionizing radiation may be carried out either by exposing the crystals to the radiation

at the polymerization temperature, or by polymerizing at a suitable temperature crystals that have been preirradiated at a much lower temperature, at which chain propagation is not significant. The latter procedure is preferred because it leads to a process that is better controlled and avoids exposure of the macromolecules to the radiation, which leads to the production of branched structures. The polymerization is exothermic and, because heat transfer is inefficient in a polycrystalline sample, the recommended conditions are such as to lead to a very slow reaction, so thermal control may be maintained.

3. Polymerization times of 5 hr. and 12 days yield about 5% and 25% of polymer with weight average molecular weights around 400,000 and 900,000, respectively.

3. Methods of Preparation

Irradiation of monomer crystals formed by shock-cooling molten monomer leads to polymerization rates about twice as high as those given above, but the molecular weights are similar for the same polymerization times. Polymerization in solid solutions containing acrylamide and propionamide proceeds almost at the same rate as in pure acrylamide, but the molecular weight of the product decreases with increasing propionamide content (3).

4. References

1a. Department of Chemistry, Polytechnic Institute of New York, Brooklyn, New York 11201.
1b. Current Address: Graphic Arts Technical Foundation, 4615 Forbes Ave., Pittsburgh, Pennsylvania 15213.
2. Brookhaven National Laboratory, Upton, Long Island, New York.
3. T. A. Fadner and H. Morawetz, *J. Polym. Sci.*, **45**, 475 (1960).

Stereopolymerization of Isobutyl Vinyl Ether

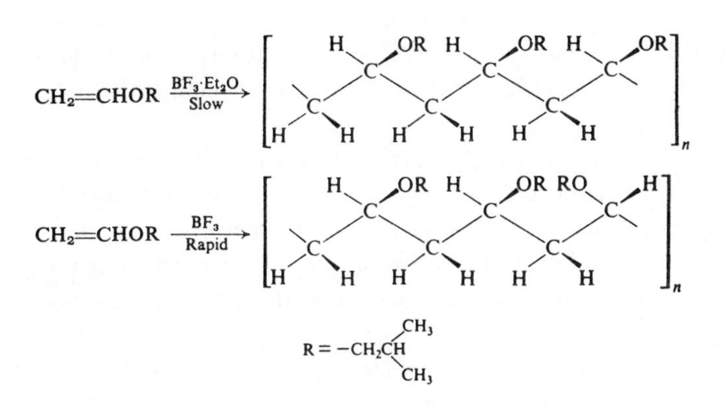

Submitted by C. E. Schildknecht, C. H. Lee, and W. E. Maust (1)
Checked by J. R. Schaefgen (2)

A. Relatively Crystalline Isobutyl Vinyl Ether Polymers

1. Procedure

Clean, granular Dry Ice (35 g.) is placed in a 250-ml. Erlenmeyer flask. The Dry Ice is in the form of granules of 1/8 to 1/4 inch rather than a powder. To the Erlenmeyer flask is added 10 g. of freshly distilled isobutyl vinyl ether (Notes 1 and 2). Liquid propane is condensed from a cylinder of Matheson C.P. propane into 1-l. Erlenmeyer flask surrounded by Dry Ice-acetone. To the cold 250-ml. Erlenmeyer flask containing Dry Ice and isobutyl vinyl ether is added approximately 40 g. of liquid propane. This reaction flask is cooled by immersing the lower half in a mixture of Dry Ice and petroleum ether in a Dewar flask. The flask may be plugged with cotton or glass wool.

In a 50-ml. beaker, 0.20 ml. of distilled boron fluoride-diethyl ether complex from a micropipet is mixed with 5 g. of granular Dry Ice. This Dry Ice containing adsorbed boron fluoride etherate catalyst is transferred in portions during 10 min. to the flask containing monomer, propane, and Dry Ice (Notes 3 and 4). In the reaction flask is placed a low temperature thermometer, and the plug again is put in place. From time to time the flask may be removed from the Dewar flask for observation of the growth of polymer masses. The temperature within the reactants should not rise above $-70°$. If the liquid becomes viscous or if the temperature rises with rapid evolution of carbon dioxide, the experiment must be repeated with purer materials and precautions taken to avoid local high temperatures and rapid reaction. If the polymerization is of the desired proliferous type, during 1-2 hr. relatively hard, irregularly shaped polymer masses will have grown in the liquid medium, which retains its low viscosity.

After 1-2 hr. of reaction the catalyst is deactivated or quenched by adding 10 ml. of a mixture cooled below $-70°$ and consisting of 4 parts by volume of methanol and 1 part by volume of 28% ammonium hydroxide containing 0.5% antioxidant such as Solux (*p*-hydroxy-*N*-phenyl morpholine) or thymol. After thorough stirring, the flask is removed from the cold bath, the temperature is allowed to rise slowly, and the propane is allowed to evaporate. Precautions are taken against fire by working in an adequate fume hood. Sufficient methanol is added to cover the white polymeric mass, and the mixture is allowed to stand overnight to facilitate removal of catalyst residues and to permit penetration of the polymer by the antioxidant. The polymer is washed twice with 100-ml. portions of methanol and is dried to constant weight in a forced draft oven at 50°. Depending on the monomer purity, the temperature, and the mode of contact with the catalyst, isobutyl vinyl ether polymers of reduced viscosity, η_{sp}/C 1.0-8.0, dl./g. can be prepared (solutions of 0.100 g. per 100 ml. of benzene at 25°) (Notes 5 and 6). Yields of 80% to substantially quantitative are obtained.

Molded films of these relatively crystalline isotactic isobutyl vinyl ether polymers are non-tacky, show cold-drawing, and have crystal melting ranges by birefringence of 90-120°. Molded films show crystallinity by x-ray diffraction in both the drawn and the undrawn states. Quenching films by rapid cooling from above 100° to 0° results in lowered crystallinity as indicated by Shore A hardness. However, after 24 hr. hardness and crystallinity are restored (Note 7).

B. Relatively Amorphous Isobutyl Vinyl Ether Polymers

1. Procedure

Caution! These flash polymerizations in liquid propane diluent on addition of boron fluoride occur so rapidly with a burst of flammable gas that it is recommended that they be carried out only on a small scale in an efficient hood or, preferably, outdoors. Closed reactors are hazardous on a laboratory scale, but the open reactor may be covered with a screen to prevent ejection of the polymer by the violent reaction. A commercial process has been operated in Germany (Note 8).

In a 250-ml. Pyrex® beaker cooled with Dry Ice-petroleum ether are placed purified isobutyl vinyl ether 10 g. (Note 1) and 20 g. of liquid propane. The beaker is removed from the cold bath, and the temperature of the monomer-propane mixture is allowed to rise to -60°. Into the middle of the liquid, gaseous boron fluoride is passed from a 6-mm. I.D. glass tube so rapidly that separate bubbles do not show distinctly. The boron fluoride is not cooled before introduction. Almost at once the polymerization occurs with a puff of gas that may throw some of the polymer into the air. The quenching, washing, and drying steps are the same as in procedure A, except that, because the soft product is not granular, it is cut into pieces about ¼ in. before drying.

If the reaction requires long bubbling with boron fluoride, or if the mixture only slowly becomes viscous, the procedure must be repeated with purer materials. Depending on the monomer purity, temperature, and exact mode of introducing the catalyst, rubbery amorphous polymers of reduced viscosity, η_{sp}/C 0.5-5.0, dl/g. can be prepared (0.10 g. per 100 ml. of benzene at 25°) (Note 9). Polymerizations with boron fluoride gas that occur less violently, e.g., by a growth of polymer during 1 min., show more evidence of crystallinity but less than by procedure A. The amorphous polymers show halo-type x-ray patterns. Shore A hardness values from 0 to 5, pressure-sensitive adhesion, rubber-like extension with slow retraction, and films cannot be cold drawn.

2. Notes

1. Commercial isobutyl vinyl ether monomer can be used as prepared from

acetylene and isobutanol by the Reppe process (General Aniline and Film Corporation). Purer monomer may be obtained by washing commercial material four times with an equal volume of water until substantially free of aldehyde (Tollens test). The monomer is dried by shaking it with pellets of potassium hydroxide and is allowed to stand overnight over fresh potassium hydroxide pellets. The monomer may be stored over solid potassium hydroxide or with a small amount of an amine as stabilizer against hydrolysis (3). For greatest monomer purity and polymers of highest molecular weight, the isobutyl vinyl ether may be refluxed for 1 hr. with metallic sodium wire before distillation. The monomer is distilled using a 2-ft. Vigreux column; a fore-run of 10% and tails of 10% are discarded.

2. Closed agitated resin flasks may be used for polymerization, but evaporating Dry Ice inside the Erlenmeyer flask provides agitation. Some contact with moisture and air can be tolerated. Polymerizations of isobutyl vinyl ether without a solid phase of Dry Ice normally give less isotactic polymer. However, boron fluoride in solid benzene and certain other catalyst-solid combinations give isotactic polymer without excess Dry Ice.

3. Instead of adding the etherate catalyst absorbed in a small amount of Dry Ice, the liquid etherate may be added directly if precautions are taken to prevent premature local reaction with rise in temperature. The cold droplets can be added in portions from a buret cooled to its tip by Dry Ice. If a homogeneous solution polymerization in propane begins, the viscosity and temperature will rise throughout and the reaction may run away in a few minutes.

4. The checker suggests that the addition of the catalyst adsorbed on Dry Ice can be made as fast as polymerization can be conducted without rise of temperature to higher than $-70°$. The checker obtained nearly quantitative yields in 20-30 min.

5. In these so-called proliferous ionic polymerizations, values of η_{sp}/c rise during 10-15 min. [see C. E. Schildknecht, A. O. Zoss, and F. Grosser, *Ind. Eng. Chem.*, **41**, 2891 (1948)]. The seat of polymerization seems homogeneous, although the system is heterogeneous [see C. E. Schildknecht and P. H. Dunn, *J. Polymer Sci.*, **20**, 597 (1956)].

6. The checker obtained reduced viscosities of 3.47 and 3.04 dl./g. for two proliferous polymers.

7. The checker found little difference in crystallinity as measured by x-ray between films quenched in ice water after removal from the mold and those not quenched.

8. This flash type of polymerization of isobutyl vinyl ether has been used in a continuous process by Badische Anilin und Soda Fabrik to make Oppanol C

[see M. Otto, H. Gueterbock, A. Hellemanns, U.S. Pat. 2,311,567 (Jasco); A. O. Zoss and D. L. Fuller, U.S. Department of Commerce, Office of Technical Service PB 67, 694].

9. The checker found reduced viscosity of 1.28 dl./g. for an amorphous polymer.

3. Methods of Preparation

Cationic polymerization conditions varying from those above have been observed to give polyvinyl isobutyl ethers of intermediate crystallinity. Thus slower polymerization by diluted boron fluoride gas gave moderate crystallinity (4), and proliferous polymerization using boron fluoride dissolved in methylene chloride as an immiscible catalyst gave crystalline polymers similar to those obtained using boron fluoride etherate (5). That the more crystallizable isobutyl vinyl ether polymers have the more regular DDD structure was confirmed by Natta and co-workers (6). Lower monomer concentrations were found to favor formation of more isotactic polyvinyl isobutyl ethers (7), and polymerizations in homogeneous solution gave isotactic fractions of polyvinyl isobutyl ether (8). Aluminum sulfate-sulfuric acid catalysts, which have been used commercially to prepare amorphous polyvinyl ethyl ethers (9), were found to give partially isotactic polyvinyl isobutyl ethers (10). The submitters found that isopropyl vinyl ether, which requires lower catalyst-to-monomer ratios than does isobutyl vinyl ether, gives substantially amorphous polymers by both procedures A and B.

Only the type A slow reaction gives high polymers from *n*-butyl vinyl ether. These are normally amorphous, but are substantially stereoregular because films and filaments crystallize on streching. Normally crystaline polymers have been prepared by proliferous polymerization of 2,2,2-trifluoroethyl vinyl ether from CH_2Cl_2 solution using BF_3 etherate as catalyst (11). Polyvinyl ethyl ethers of moderate degrees of normal crystallinity were prepared by procedure A, except for the use of an excess of fine granular Dry Ice and a monomer of exceptionally high purity (12).

Vinyl isobutyl ether has been polymerized to isotactic polymer at −78° using modified Ziegler-type catalysts obtained by reaction of titanium tetrachloride with an aluminum alkyl (13). Polyvinyl isobutyl ether of crystalline melting point 117° was obtained using $AlCl(C_2H_5)_2$ as a catalyst (14). Polyvinyl isobutyl ethers and some other polyvinyl ether fractions of higher degrees of tacticity than those discussed above have been obtained by use of complex catalysts without refrigeration (15). Reaction products of vanadium chlorides with aluminum alkyls (16), metal sulfates with aluminum alkyls or alkoxides (17), and aluminum alkoxides with sulfuric acid (18) have been used for slow stereoregular polymerizations of vinyl ethers at 0-30°. Additional references and applications of alkyl vinyl ether polymers have been discussed (19).

4. References

1. Gettysburg College, Gettysburg, Pennsylvania 17325.
2. E. I. du Pont de Nemours and Company, Pioneering Research Division, Textile Fibers Department, Wilmington, Delaware 19898.
3. C. E. Schildknecht, "Vinyl Ethers" in *Monomers,* H. Mark et al., eds., Interscience Division, John Wiley and Sons, New York, 1951.
4. C. E. Schildknecht, A. O. Zoss, and F. Grosser, *Ind. Eng. Chem.,* **41**, 2891 (1949); cf. **40**, 2014 (1948).
5. C. E. Schildknecht, *Ind. Eng. Chem.,* **50**, 107 (1958).
6. G. Natta, P. Corradini, and I. W. Bassi, *Makromol. Chem.* **18-19**, 445 (1955).
7. M. Takeda, Y. Imamura, S. Okamura, and T. Higashimura, *J. Chem. Phys.,* **33**, 631 (1960).
8. S. Okamura, T. Higashimura, and H. Yamamoto, *J. Polym. Sci.,* **33**, 510 (1958).
9. S. A. Moseley, U.S. Pat., 2,549,921.
10. S. Okamura, T. Higashimura, and T. Watanabe, *Makromol. Chem.,* **50**, 137 (1961).
11. C. E. Schildknecht, U.S. Pat. 2,820,025; c.f. J. A. Manson and H. Sorkin, U.S. Pat. 3,365,433.
12. C. E. Schildknecht, C. H. Lee, K. P. Long, and L. C. Rinehart, *Poly. Eng. Sci.,* 7 (4), 257 (1967).
13. J. Lal, *J. Polym. Sci.,* **31**, 179 (1958); G. Natta et al., *Angew. Chem.,* **71**, 205 (1959).
14. G. Dall'Asta and N. Oddo, *Chim. Ind. (Milan)* **42**, 1234 (1960).
15. E. J. Vandenberg, R. F. Heck, and D. S. Breslow, *J. Polym. Sci.,* **41**, 519 (1959); Ital. Pat. 571,741 (1958); E. J. Vandenberg, *J. Polym. Sci.,* Part C, No. 1, 207 (1963).
16. E. J. Vandenberg (Hercules), Brit. Pat. 820,469; Ger. Pat. 1,033,413.
17. R. F. Heck and E. J. Vandenberg (Hercules), U.S. Pat. 3,025,283.
18. R. Chiang (Hercules), U.S. Pat. 3,025,281.; D. L. Christman and E. J. Vandenberg (Hercules), U.S. Pat. 3,025,282.
19. C. E. Schildknecht, in *Kirk-Othmer Encyl. Chem. Tech.* 2nd ed. **21**, 412 (1970), John Wiley & Sons, New York.

High Density Polyethylene

A. $n\text{CH}_2{=}\text{CH}_2 \xrightarrow{\text{TiCl}_4 + \text{Al(C}_2\text{H}_5)_3} {+}\text{CH}_2{-}\text{CH}_2{+}_n$

B. $n\text{CH}_2{=}\text{CH}_2 \xrightarrow{\text{AlBr}_3 + \text{VX}_n + \text{Sn(C}_6\text{H}_5)_4} {+}\text{CH}_2{-}\text{CH}_2{+}_n$

Submitted by W. L. Carrick (1)
Checked by R. R. Jones and A. W. Anderson (2)

A. $TiCl_4 + Al(C_2H_5)_3$ Catalyst

1. Procedure

A 2-1., three-necked, round-bottomed flask is suspended in a water (or oil) bath held at ~60° and then fitted with a reflux condenser, mechanical stirrer, and gas inlet tube reaching to the bottom of the flask with one opening sealed with a serum stopper to permit injecting the catalyst slurry (Note 1). The flask is then charged with 1 1. of Phillips ASTM grade *n*-heptane and purged with dry nitrogen for 1 hr. to remove water and other volatile impurities (Note 2).

While the reaction flask is being purged, a 100-ml. bottle is charged with about 25 ml. of dry *n*-heptane and a magnetic stirring bar, purged with dry nitrogen, and sealed with a serum stopper. By way of hypodermic syringes, 4.6 mmoles of titanium tetrachloride and 2.0 mmoles of aluminum triethyl are added to the bottle as dilute solutions in heptane (Note 3). The total volume of heptane should now be about 40 ml. A colloidal brown precipitate deposits as the co-catalysts interact, and the slurry is stirred on a magnetic stirrer for 30 min. at room temperature.

The nitrogen purge through the reactor is discontinued, and ethylene is admitted at the rate of 3-4 1./min. The aged catalyst slurry described above is then injected into the reactor with a hypodermic syringe to start the reaction. Polymerization begins immediately with the evolution of heat, and particles of polyethylene may be observed. After addition of the catalyst, the ethylene flow is gradually reduced to 2 l./min. (Note 4). At the end of 1 hr. the reaction is stopped by the addition of 50 ml. of isopropanol to inactivate the catalyst. The resulting white polymer is separated by filtration, boiled in 1 l. of isopropanol, recovered by filtration, and dried overnight in a vacuum oven at 70°. The yield is 65 g. of polyethylene in the form of a fluffy white powder. This polymer has a melting point, as determined by disappearance of birefringence, of 128°. Melt Index (ASTM D-1238-52T) 1-2 (Note 5), and intrinsic viscosity of 1.5-2.0 dl./g. The polymer is soluble in boiling xylene or other hydrocarbon solvents at temperatures above 130°. Above its melting point it can be compression-molded into films suitable for physical testing or infrared spectra. In the subsequent discussion of the infrared spectrum this polymer will be designated Sample A.

2. Notes

1. The glassware should be dried in an oven at >100° assembled while warm, and immediately purged with nitrogen. Joints should be greased with a minimum of lubricant to prevent contamination, and the stirrer must be sealed to prevent air from entering the reactor.

2. A desirable additional step is to purge the diluent overnight in a separate flask with a slow stream of dry nitrogen and transfer it to the reactor through a glass or polyethylene tube without exposure to air.

If ASTM grade *n*-heptane is not available, heptane of unknown purity can be refined by stirring with concentrated sulfuric acid overnight, washing several times with water, drying over calcium hydride, and passing it through alumina. Cyclohexane of equivalent purity may be substituted for *n*-heptane.

Pure ethylene is also desirable. For this experiment Phillips ethylene (99 mole % minimum) was used.

Caution! Aluminum triethyl spontaneously ignites in air and must always be handled in an inert atmosphere. Dilute solutions of aluminum triethyl and titanium tetrachloride should be prepared in a dry box at a concentration of 0.5 mmole/ml. and then handled with hypodermic syringes.

4. When the catalyst is first added, polymerization may be so rapid that ethylene concentration in the reactor is depleted and air is sucked in. A bubbler may be attached to the exit from the condenser to show that an excess of ethylene is passing through the system.

Caution! The reaction should be carried out in a good hood to carry away the excess ethylene without danger of explosion.

5. The polymer Melt Index is inversely proportional to the Al/Ti ratio, and the Melt Index may be varied over a considerable range by fairly small changes in the Al/Ti ratio.

B. $AlBr_3 + VX_n + Sn(C_6H_5)_4$ Catalyst

1. Procedure

The polymerization apparatus described for part A is assembled and the reactor is charged with 1 l. of purified cyclohexane (or *n*-heptane) and 2 mmoles of tetraphenyl tin and purged with dry nitrogen as before (Note 1). A flow of Matheson C.P. ethylene (Important! see Note 2) is started at 3-4 l./min., and then 5 mg. of vanadium tetrachloride (or vanadium oxytrichloride) and 4 mmoles of aluminum bromide (Notes 3 and 4) are added as dilute solutions in cyclohexane with hypodermic syringes. Polymerization begins as soon as the last catalyst component is added, and white polyethylene particles may be observed almost immediately. The ethylene flow is then reduced to 2 l./min. over an interval of 3-5 min. The reactor contents should remain white or very light tan throughout the experiment. A brown color indicates contamination.

At the end of 1 hr. the reaction is stopped by the addition of 50 ml. of isopropanol. The polymer is separated by filtration, boiled in 1 l. of isopropanol, recovered by filtration (a second extraction may be required), and dried overnight in a vacuum oven at 70°. The yield is 40 g. of white particulate polyethylene having a melting point of 136°, Melt Index (ASTM D-1238-52T) of 0.0-0.1, and an intrinsic viscosity of 2-3 dl./g. in tetralin at 130°. This polymer is called sample B in the discussion of polymer structure.

2. Notes

1. This catalyst is more sensitive to contamination than the titanium catalyst, and the precautions mentioned in part A should all be carefully followed.

2. This particular catalyst *requires* a trace of oxygen in the ethylene for sustained polymerization. The reason for specifying Matheson C.P. ethylene is that this gas consistently contains about 400-600 p.p.m. of oxygen which is about the optimum range for the $AlBr_3$-VX_n-$Sn(C_6H_5)_4$ catalyst. Significantly higher

or lower oxygen concentrations lead to reduced yields of polymer and changes in the polymer end groups (3). Ethylene from some sources does not contain enough oxygen for good polymerization results with this catalyst. In these cases 400-600 p.p.m. of oxygen can be metered into the ethylene to achieve good polymerization activity; however, this technique is mechanically difficult and requires specialized equipment.

3. The vanadium halide and aluminum bromide solutions should be prepared in a dry box and handled by hypodermic syringes. Tetraphenyl tin is stable in air and may be added to the reactor as a solid.

4. Aluminum chloride is equivalent to aluminum bromide as a cocatalyst and may be substituted for it. A saturated solution of aluminum chloride in boiling cyclohexane contains about 6 g./1. of $AlCl_3$ and may be prepared by refluxing a ten-fold excess of aluminum chloride in cyclohexane under a nitrogen atmosphere. Solutions prepared this way usually contain some colloidal aluminum chloride which need not be separated from the solution. An aliquot of the hot solution is used directly for the polymerization.

3. Polymer Structure

The polyethylene from part A [$Al(C_2H_5)_3$ + $TiCl_4$ catalyst] is more branched than the polyethylene from part B [$AlBr_3$ + VX_n + $Sn(C_6H_5)_4$ catalyst]. Curve 1 on Fig. 1 is the infrared spectrum of sample A. Curve 2 is the spectrum of sample A compensated against sample B from 7 to 7.5 μ, which shows the substantially deeper methyl group absorption at 7.25 μ in sample A. Curve 3 gives the spectrum of sample B from 8 to 13 μ and shows that the double bonds found in sample A are missing in sample B.

These data clearly illustrate the subtle changes in polymer structure caused by different catalyst formulations and show the utility of infrared spectroscopy in detecting them.

4. Methods of Preparation

The Ziegler-type catalysts (4) normally consist of a reactive metal alkyl in combination with a transition metal compound of Groups IV to VI of the Periodic Table (Ti, V, Cr). Some other typical combinations that are active catalysts for olefin polymerization are $AlR_3 + TiCl_3$, $AlR_3 + VCl_3$, $AlR_3 + VOCl_3$, $R_2AlCl + VCl_4$, $R_2AlCl + VOCl_3$. More extensive lists of catalytically active systems are given elsewhere (5-8). The simple procedures described above are suitable for qualitative work with many of these systems.

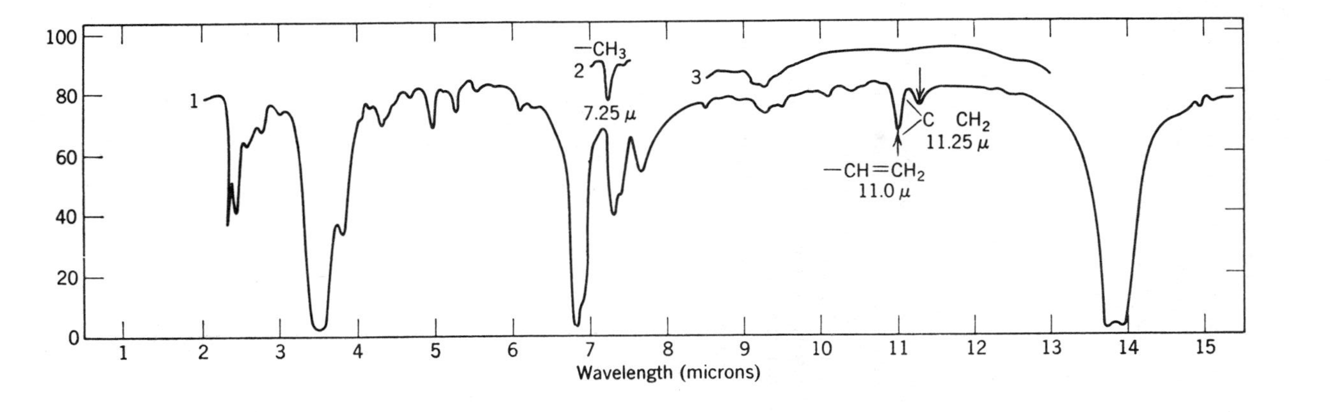

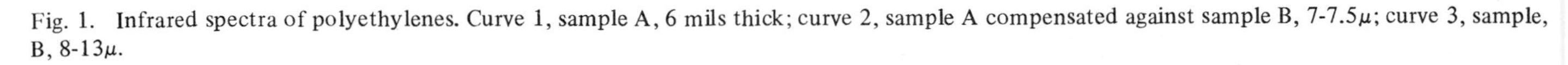

Fig. 1. Infrared spectra of polyethylenes. Curve 1, sample A, 6 mils thick; curve 2, sample A compensated against sample B, 7-7.5μ; curve 3, sample, B, 8-13μ.

Generally, the yield or molecular weight, or both, of the polymer are significantly influenced by polar impurities, such as water or oxygen, which may be present in the system in the parts per million range (3,9). A quantitative study, such as a kinetic analysis, is meaningless unless all polar impurities in the system are rigorously monitored and controlled.

5. References

1. Union Carbide Plastics Company, Bound Brook, New Jersey 08805.
2. Plastics Department, Experimental Station, E. I. du Pont de Nemours & Co., Wilmington, Delaware 19898.
3. G. W. Phillips and W. L. Carrick, *J. Polym. Sci.,* **59**, 401 (1962).
4. K. Ziegler, E. Holzkamp, H. Breil, and H. Martin, *Angew. Chem.,* **67**, 541 (1955).
5. N. G. Gaylord and H. F. Mark, *Linear and Stereoregular Addition Polymers,* Interscience Division, John Wiley and Sons, New York, 1959.
6. *Rept. Progr. Appl. Chem.,* **42**, 436 (1957).
7. J. K. Stille, *Chem. Rev.,* **58**, 541 (1958).
8. C. E. H. Bawn and A. Ledwith, *Quart. Rev.,* **16**, 361 (1962).
9. A. Orzechowski, *J. Polym. Sci.,* **34**, 65 (1959).

trans-1,4-Polybutadiene by Rhodium Salt Catalysis in Emulsion

$$nCH_2{=}CH{-}CH{=}CH_2 \xrightarrow{RhCl_3} \left[CH_2 \diagup CH \diagdown\!\!\diagdown CH \diagup CH_2 \right]_n$$

Submitted by R. E. Rinehart and H. P. Smith (1)
Checked by T. M. Wathen (2)

1. Procedure

In a 24-oz. carbonated beverage bottle are placed 200 ml. of distilled water, 5 g. of sodium dodecylbenzene sulfonate (Note 1), and 0.5 g. of "rhodium tichloride trihydrate" (Note 2). When this mixture has dissolved, the bottle is cooled in ice, then weighed on a balance in an efficient hood. Slightly more than 100 g. of freshly distilled butadiene is carefully added. The excess butadiene is allowed to evaporate until the increase in weight is exactly 100 g. The bottle is quickly sealed (Note 3) with a metal cap provided with a small hole for venting, and fitted with a self-sealing rubber gasket and a nylon or Teflon liner.

The bottle and contents are warmed to room temperature before being placed on a rocker in a regulated bath at 50°. The emulsion begins to form almost immediately, and there is no induction period for the polymerization.

The bottle is removed from the bath after 20 hr. (Note 4). Unreacted butadiene is vented (*Hood!*) through a syringe needle through the hole in the cap. The cap is removed, and the emulsion is poured, with stirring, into 500 ml. of methanol to which has been added about 0.5 g. of N-phenyl-β-naphthylamine.

The polymer is removed by filtration, washed with more methanol, and dried under vacuum at room temperature. Yield 25-30 g. (Note 5).

The polymer may be further purified by dissolving it in 300 ml. of chloroform. The solution is filtered or centrifuged and the polybutadiene reprecipitated by pouring it into a stirred solution of 0.5 g. of N-phenyl-β-naphthylamine in 500 ml. of ethanol. The purified polymer is filtered and dried.

The intrinsic viscosity of the product in chloroform at 30° is 0.4 dl./g. The *trans*-1,4 configuration represents more than 98% of the total unsaturation. The per cent crystallinity determined by x-ray diffraction is about 40%.

2. Notes

1. Suitable emulsifiers include alkali metal sulfonates and sulfates, such as sodium lauryl sulfate and sodium tetrahydronaphthalene sulfonate. Cationic or non-ionic emulsifiers, or other types of anionic dispersing agents, cannot be used.

2. Available from Engelhard Industries, Newark, New Jersey: 40% rhodium.

3. *Caution! The bottle should be protected with a metal screen or shield after sealing. The pressure from the butadiene can cause the bottle to explode as the contents warm to the temperature of the polymerization bath.*

4. A longer reaction time will lead to higher yields, but the emulsion becomes unstable. A considerable amount of precipitate will be formed after about 20 hr. This does not, however, slow the polymerization appreciably, nor does it appear to affect the polymer properties adversely.

5. Differences in the rate of polymerization may be observed, depending, in part, on the source of the rhodium chloride used. Rhodium salt from a source other than that indicated in Note 2 gave 15-22 g. of *trans*-1,4-polybutadiene.

3. Methods of Preparation

trans-Polybutadiene can be prepared by Ziegler-Natta catalysts (3). Certain transition metal salts may be used in an emulsion recipe, rhodium being an outstanding example (4, 5).

4. References

1. Research Center, United States Rubber Company, Wayne, New Jersey 07470, Contribution No. 234.
2. The Goodyear Tire & Rubber Company, Akron, Ohio 44309.

3. N. G. Gaylord and H. F. Mark, *Linear and Stereoregular Addition Polymers,* Interscience Division, John Wiley and Sons, New York, 1959, p. 368.
4. R. E. Rinehart, H. P. Smith, H. S. Witt, and H. Romeyn, Jr., *J. Amer. Chem. Soc.,* **83**, 4864 (1961), **84**, 4145 (1962).
5. A. J. Canale, W. A. Hewett, T. M. Shryne, and E. A. Youngman, *Chem. & Ind. (London),* 1054 (1962).

cis-1,4-Polybutadiene by Cobalt-Aluminum-Water Catalysis

$$x\mathrm{CH_2{=}CH{-}CH{=}CH_2} \xrightarrow[\text{+ cobalt octoate}]{\mathrm{Et_2AlCl + H_2O}} \mathrm{{+}CH_2{-}CH{=}CH{-}CH_2{+}_x}$$

Submitted by M. Gippin (1)
Checked by G. H. Smith and R. M. Pierson (2)

1. Procedure

Two clean 12-oz. and three 28-oz. beverage bottles are dried in a 110° oven overnight. While purging the bottles with nitrogen (Note 2) (Fig. 1), they are capped with suitable unlined beverage bottle crown caps having fitted rubber liners (Note 3). The metal caps have a center puncture to admit a hypodermic needle. The bottles are pressurized to 15 p.s.i.g. with nitrogen (Note 4) and set aside until needed.

Benzene (2.5 1.) (Note 5) is distilled at 10 torr pressure (Note 6) through a closed, nitrogen-purged apparatus having ground glass joints throughout (Fig. 2). The first 10% by volume of the distillate is discarded (Note 7). Distillation under nitrogen is continued, 2.0 1. of the benzene being collected in a 3-1. nitrogen-purged round-bottomed flask of the type shown in Fig. 2. The receiver is disconnected and attached to a 12 x 1.5 in. column filled with activated silica-gel (Note 8) through which nitrogen was previously passed for about 15 min. The first 500 ml. of the percolate is discarded, the next liter of benzene is collected in another flask of the type shown. The silica-gel column is

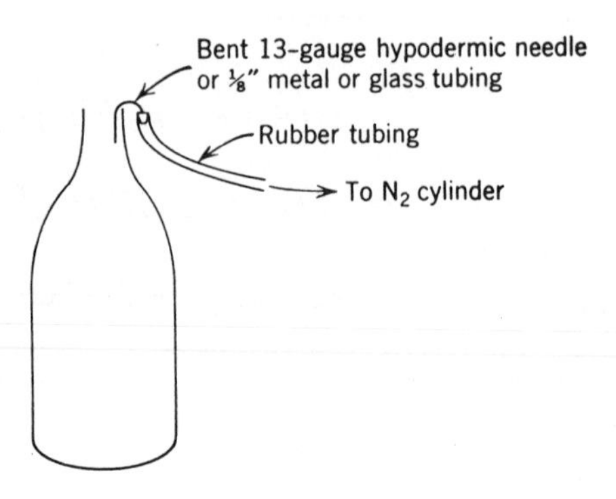

Fig. 1. Bottle purging method.

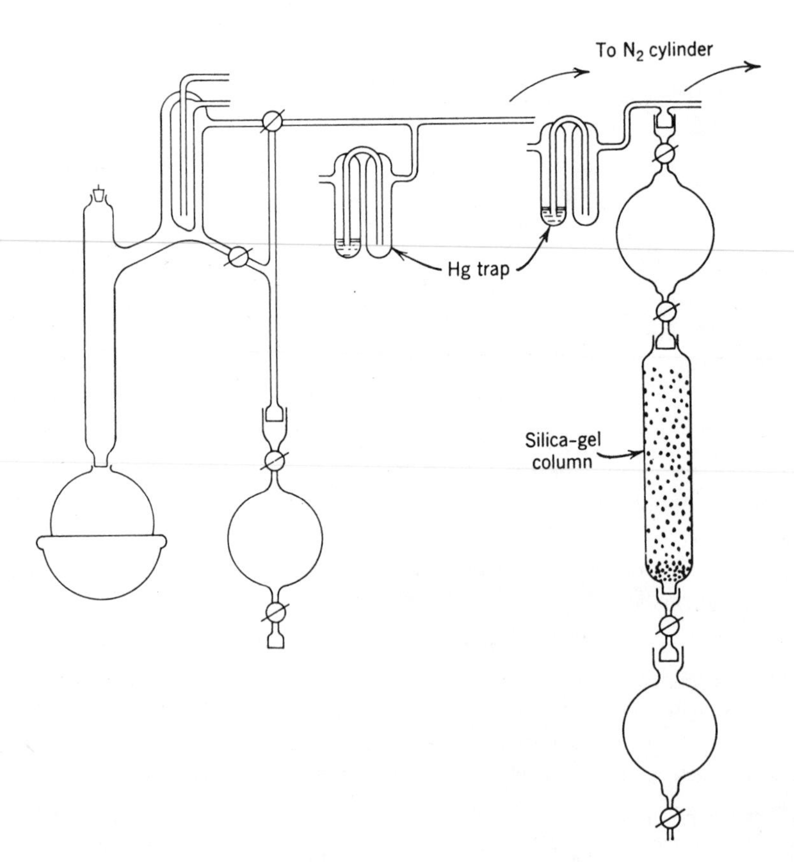

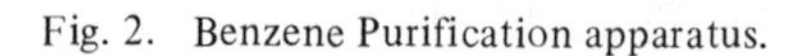

Fig. 2. Benzene Purification apparatus.

disconnected, and the flask containing the purified benzene is connected to the Hg trap. Under continuous nitrogen pressure, 250 ml. of the purified benzene is collected in one dry 28-oz. bottle (Note 9) to which about 15 g. of granular CaH_2 of 4-40 mesh is added. While purging the bottle with nitrogen, a bake-dried, glass-wool plug is inserted into the neck of the bottle, which is then capped with a rubber liner cap, pressurized to 15 p.s.i.g., and let stand overnight (bottle A). In a second dry 28-oz. bottle, 518 ml. of the benzene is collected, about 15 g. of CaH_2 is added, and the bottle is capped while purging (bottle B). In a clean, but not necessarily baked, 12-oz. beverage bottle about 0.1 1. of the benzene is collected, 2 ml. of distilled H_2O is added, the bottle is capped, pressurized, and set aside overnight (bottle C).

One-half liter of at least 99 mole % pure butadiene (Note 10) is distilled (using solid CO_2-acetone in the condensing head), and a 100-g. fore-run collected in a hydrocarbon solvent is discarded. The receiver is now removed, and bottle B, containing the benzene solvent and CaH_2, is uncapped and inserted into the receiver position. Distillation is continued until 88 ml. of butadiene (50 g.) is collected. The bottle is recapped, pressured, and set aside overnight.

The next day, an approximately 1*M* solution of diethylaluminum chloride (Note 11) is prepared as follows: A dried, pressurized 12-oz. bottle is punctured through the rubber gasket with an 18-gauge needle until the pressure is vented. Dry benzene (87.4 ml.) from bottle A is withdrawn by means of a carefully inserted hypodermic syringe and needle through the cap of the inverted bottle (Note 12) and injected through the rubber liner into the 12-oz. bottle. This bottle is now weighed to the nearest 0.1 g. Next, approximately 13 ml. of diethylaluminum chloride is transferred (Note 13) to the bottle, which is reweighed and repressured. The molarity is calculated from the weight of added diethylaluminum chloride and the total volume of solution.

A solution approximately 0.02 *M* in cobalt octoate is prepared as follows: If the commercial preparation of 12% cobaltous octoate (Note 14) is used, 1 ml. of the solution contained in a 2-ml. hypodermic syringe fitted with a needle and needle valve, is weighed on an analytical balance and set aside. A dry 12-oz. bottle is punctured through the rubber liner with an 18-gauge needle until the pressure is relieved. Dry benzene (0.1 1.) from bottle A is injected into the bottle with a 50-ml. syringe. The cobaltous octoate is then injected, the bottle capped and repressurized. The syringe is weighed again, the weight of cobaltous octoate taken, and the molarity for the given volume of solution is calculated.

The bottle containing the benzene solution of butadiene is vented and opened, and, while purging with nitrogen, a bake-dried glass-wool plug is packed into the

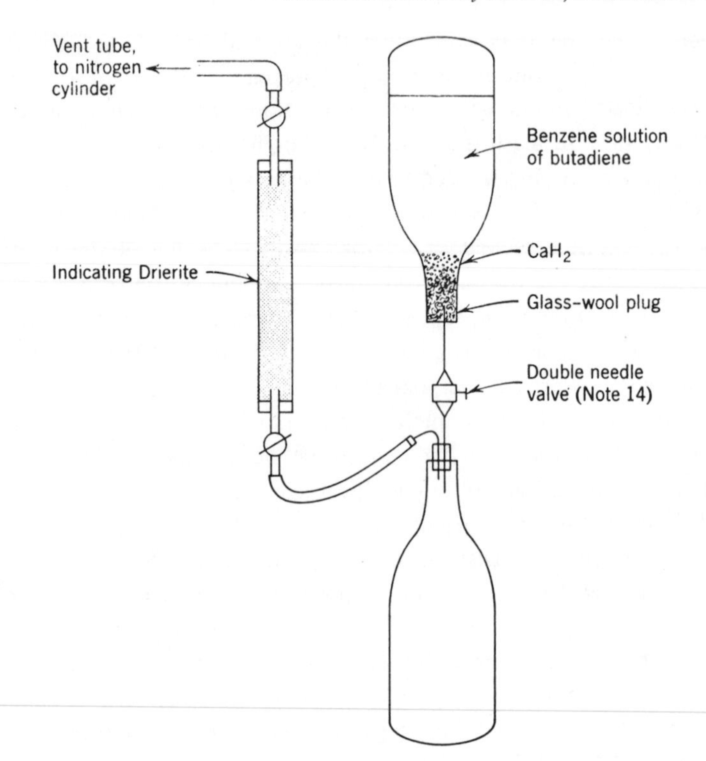

Fig. 3. Solution-transfer apparatus.

neck of the bottle. The bottle is capped and set into position in the apparatus shown in Fig. 3. A dry 28-oz. receiver bottle is vented and, while being purged with nitrogen, is forced into the rubber stopper attached to the lower needle (Note 15). With the needle valve open, a positive nitrogen pressure is admitted into the bottle of solution via the vent tube. The tube is disconnected from the nitrogen cylinder, and the open end is placed in the hood. The solution will flow into the receiver as long as a slight pressure of nitrogen is maintained above it.

After 551 ml. of the solution has been transferred (Note 16), the needle valve is closed, the nitrogen purge needle is quickly inserted into the receiver bottle, which is then capped with a cork-filled cap to which an aluminum disk 0.004 inch thick is snugly fitted. The bottle containing the solution is prechilled in a 5° bath for 1-2 hr. It is then removed, uncapped, and, while being purged with nitrogen, 10 mmoles of diethylaluminum chloride is added, using a syringe and needle to make the transfer. Similarly, 30 ml. of the benzene solution of H_2O from bottle C (Note 17) is added, followed by 0.02 mmole of the cobaltous octoate. The bottle is capped with an aluminum disk-containing, cork-filled liner cap, placed in a bottle guard, and rotated end over end in a constant temperature bath at 5°.

Although polymerization may be complete in 4-6 hr., it is best to allow the reaction to proceed overnight. The bottle is scored and carefully cracked open. The viscous polymerizate is transferred with a minimum of exposure to air to a 3-1. beaker or gallon jar filled with about 2 1. of isopropanol containing 1.0 g. of antioxidant (Note 18). After several minutes of agitation in the alcohol, the polymer is removed piecewise and placed in a screw-cap quart jar half-filled with isopropanol in which 1.0 g. of the antioxidant is dissolved. The next day the polymer is removed, blotted dry between absorbent paper towels or clean cloth, and then dried in a 50° vacuum oven overnight. Yield of polymer approximately 50 g.

2. Characterization

The dilute solution viscosity of *cis*-1,4-polybutadiene determined in toluene at 25° is 4.5-6.0 dl./g., which corresponds to a molecular weight of about 500,000. The polymer is free of any gel, as determined by filtering through a 100-mesh wire screen a solution of 0.4 g. of polymer in 100 ml. of toluene. The *cis*-1,4 content of the polybutadiene is 98% as measured by the film technique described in the literature (3). The influence of polymerization variables upon the polymer and procedure have been described in the published literature (4).

3. Notes

1. The preparation of polybutadiene having the characteristics described here requires manipulative procedures that rigorously exclude access of atmospheric moisture and oxygen to the reacting materials. The diethylaluminum chloride must be pure and free of any previous contact with moisture, or air, or aluminum metal. Small variations in the water content of the system will affect the *cis*-1,4 content and polymer viscosity. A large excess of water leads to highly gelled polymer. Insufficient or complete absence of water results in little or no polymer formation.

2. Lamp grade nitrogen of maximum 5 p.p.m. H_2O and oxygen impurity, or equivalent nitrogen, should be used.

3. Standard rubber cap liners, No. 3582-11-1579, are available from Firestone

Industrial Products Company, Noblesville, Indiana, to be used with the unlined crown caps.

4. The hot bottles are pressurized with an 18-gauge hypodermic needle attached to rubber pressure tubing connected to the nitrogen cylinder. Pressure adjustment to 15 p.s.i.g. at intervals during cooling of the bottles to room temperature will be necessary.

5. ACS reagent grade, thiophene-free benzene is used.

6. The Hg traps shown in Fig. 2 contain mercury to a distance of 10 mm. above the tip of the inner tube.

7. This contains the major amount of H_2O contaminant.

8. Mesh size 28-200.

9. Volumes to which bottles are to be filled can be estimated with sufficient accuracy by measuring the height of a required volume of the same liquid contained in another bottle of the same size, and making a wax pencil mark on the receiver bottle.

10. Butadiene obtained from Phillips Petroleum Company was used. It must be relatively dry even before distillation. It is recommended that a flask containing the butadiene be chilled in solid CO_2 to freeze most of the contaminant H_2O, and that the butadiene then be transferred to the distillation flask through cotton gauze.

11. The product of Ethyl Corporation or Texas Alkyls, Inc., is suitable.

12. Care must be taken not to penetrate the glass-wool plug beyond the point where particles of CaH_2 may enter the syringe. Using a 50-ml. syringe, it will be necessary to make two transfers.

13. Technique of transferring the alkylaluminum should conform to the instructions issued by the manufacturer. Twenty-five per cent by weight diethylaluminum chloride in toluene is commercially available and may be used without dilution; 5.4 ml. of this solution contains 10 mmoles of the alkylaluminum.

14. Cobaltous octoate from Witco Chemical Company, Inc., was used; per cent based on cobalt.

15. A 1/8 in. bar stock stainless steel needle valve to which a 1/8-in. brass close nipple is silver-soldered at each end. A hypodermic syringe lock tip is silver-soldered to each nipple. A suitable equivalent valve assembly may be used.

16. The transferred solution contains 50 g. of butadiene.

17. This contains 1.0 mmole of dissolved H_2O at room temperature. A valve-fitted needle is inserted through the rubber liner of bottle C, the bottle is carefully inverted so the excess H_2O flows down past the needle into the neck of the bottle. After allowing about 30 min. for the settling of finely dispersed

H_2O, the needle valve is opened to permit 5-10 ml. of the solution to be discarded. A 50-ml. syringe is attached to the needle valve, and 30 ml. of the solution is withdrawn.

18. Phenyl-β-naphthylamine is suitable. Other known rubber antioxidants may be used.

4. Methods of Preparation

cis-1,4-Polybutadiene can also be made with Et_2AlCl, H_2O, and other types of cobalt compounds, such as cobalt acetylacetonate, cobalt naphthenate, cobalt stearate, or cobalt chloride, the latter preferably rendered soluble through its dipyridyl complex (4). Another suitable catalyst consists of Et_2AlCl and cobalt octoate or other cobalt compound with certain organic compounds as substitutes for H_2O, such as cumene hydroperoxide, *t*-butanol, ethylene, chlorohydrin, allyl chloride, *t*-butyl chloride (5). Still another method uses mixtures of Et_3Al and $AlCl_3$, Et_2AlCl and $EtAlCl_2$ (5).

5. References

1. Central Research Laboratories, The Firestone Tire and Rubber Company, Akron, Ohio 44309.
2. The Goodyear Tire and Rubber Company, Akron, Ohio 44309.
3. Montecatini Soc. Gen. per l'Ind. Min. e Chim., Ital. Pat. 592,477 (June 12, 1957).
4. M. Gippin, *Ind. Eng. Chem., Prod. Res. Develop.*, **1**, 32 (1962).
5. M. Gippin, *Ind. Eng. Chem., Prod. Res. Develop.*, 4, 160 (1965).

cis-1,4-Polyisoprene by Aluminum-Titanium Coordination Catalysis

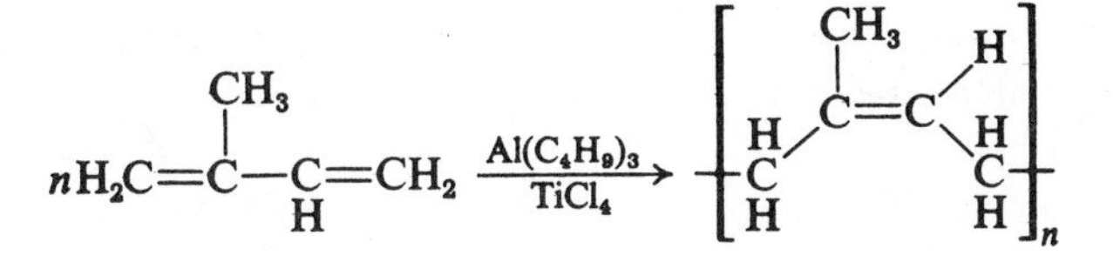

Submitted by W. M. Saltman and E. Schoenberg (1)
Checked by W. J. Bailey (2)

1. Procedure

To a carefully cleaned and dried 4-oz., narrow-mouthed, screw-cap bottle (Note 1) is added 55 g. of a 15% by weight isoprene (Note 2) in pentane (Note 3) solution. The solution should be taken directly from the exit of a jacketed column packed with silica gel (Note 4). Approximately 5 g. of the solution is vented by heating the uncapped bottle on a hot sand bath (Note 5). After 0.200 mmole of aluminum triisobutyl and 0.185 mmole of titanium tetrachloride in heptane (Notes 3 and 6) are added to the bottle with hypodermic syringes (Notes 7 and 8), the bottle is sealed with a cap having an inner Teflon lining and is allowed to rotate for 16 hr. in a 50° water bath. The bottle is removed, allowed to cool, and 10 ml. of a dilute solution of antioxidant (Note 9) is added to it. The contents of the bottle are then poured into a beaker containing 200 ml. of a 1% solution of antioxidant in isopropyl alcohol. The coagulated material is dried at 40° under vacuum to give a polymer having an inherent viscosity (3, 4) of 2.0-3.5 in 70-80% yield. The *cis* content of the polymer should be better than 94% when analyzed according to the method of Binder (5).

2. Notes

1. All apparatus should be clean and thoroughly dry.

2. Phillips polymerization grade isoprene is used after flash distillation under nitrogen just before use. The isoprene should be low in α-acetylenes, cyclopentadiene, 2-butyne, and any oxygen- or sulfur-containing compounds. Even within the specification limits for this high purity isoprene there may be variations in the isoprene quality that lead to induction periods and slow reaction rates. The isoprene may require preliminary treatment to remove these adventitious impurities. After distillation, isoprene should not be stored for long periods without addition of about 200 p.p.m. of *t*-butylcatechol as oxidation inhibitor.

3. Phillips pure grade (99+ %) pentane or heptane should be washed with concentrated sulfuric acid until no further discoloration takes place. After neutralizing and drying, the solvent is flash-distilled and stored over sodium until used.

4. A column ¾ in. in diameter and 13 in. long containing about 60 g. of silica gel is suitable; a cooling jacket is necessary to avoid polymerization in the column. It is convenient to use a graduated, cylindrical separatory funnel at the top of the column. Activated silica gel, 28-200 mesh size from the Davison Chemical Division of W. R. Grace and Co., is satisfactory. The column should be flushed with nitrogen and rinsed with 100 ml. of dry pentane before use.

5. A glass rod is placed in the bottle to prevent bumping.

6. Dilute solutions (0.2-0.5*M*) of aluminum triisobutyl and titanium tetrachloride are conveniently prepared in heptane and stored under nitrogen in 4- or 8-oz., narrow-mouthed bottles fitted with serum caps or with holed metal caps having an inner self-sealing gasket made of butyl rubber. The quantities needed may then be conveniently withdrawn and transferred by syringe. Triisobutyl aluminum from Ethyl Corporation or Stauffer Chemical Comapny is satisfactory without further purification. *Caution! Both the aluminum alkyl and titanium tetrachloride react vigorously with oxygen and water and should be handled in a dry box in a nitrogen or other inert gas atmosphere.*

7. A separate syringe should be used for each solution.

8. Slight variations in the Al/Ti mole ratios may be necessary to compensate for the aluminum triisobutyl used to destroy residual air and water in the reaction mixture. The total catalyst corresponds to about 0.8 g. per 100 g. of monomer.

9. This is to protect the polymer and destroy the catalyst without coagulation. Any hydrocarbon solvent may be used in which the antioxidant is soluble or in

which it may be suspended. Di-(*t*-amyl)-hydroquinone or other standard rubber antioxidant may be used. A solution containing 10 ml. of isopropyl alcohol and 2.0 g. of antioxidant per 100 ml. of benzene is satisfactory.

3. Methods of Preparation

cis-Polyisoprene has been prepared with this catalyst system (6) and with a lithium or lithium alkyl catalyst (7). The polymers prepared with lithium catalysts have somewhat lower *cis* contents and may be readily prepared to higher inherent viscosities and lower gels.

4. References

1. Research Laboratory, Goodyear Tire & Rubber Company, Akron, Ohio 44309, Contribution No. 261.
2. University of Maryland, College Park, Maryland 20740.
3. W. K. Taft and G. J. Tiger, in G. S. Whitby, ed., *Synthetic Rubber,* John Wiley and Sons, New York, 1954, p. 685.
4. W. K. Taft and G. J. Tiger, in G. S. Whitby, ed., *Synthetic Rubber,* John Wiley and Sons, New York, 1954, pp. 335-338; A. L. Back, *Ind. Eng. Chem.*, **39**, 1339 (1947).
5. J. L. Binder and H. C. Ransaw, *Anal. Chem.*, **29**, 503 (1957).
6. S. E. Horne, Jr., U.S. Pat. 3,114,743 (December 17, 1963); W. M. Saltman, E. W. Gibbs & J. Lal, *J. Amer. Chem. Soc.*, **80**, 5615 (1958); E. Schoenberg, D. L. Chalfant, and R. H. Mayor, *Rubber Chem. Technol.*, **37**, 103 (1964); D. C. Perry, F. S. Farson, and E. Schoenberg, *J. Polym. Sci.*, Polymer Chemistry ed. 13, 1071 (1975).
7. R. S. Stearns and L. E. Forman, *J. Polym. Sci.*, **41**, 381 (1959).

trans-1,4-Polyisoprene by Aluminum-Titanium-Vandium Catalysis

$$CH_2{=}\overset{\displaystyle CH_3}{\overset{|}{C}}{-}CH{=}CH_2 \xrightarrow{VCl_3-Al(i\text{-}C_4H_9)_3-Ti(OC_2H_{13})_4} \left[-CH_2-\overset{\displaystyle CH_3}{\overset{|}{C}}{=}CH-CH_2- \right]_n$$

Submitted by J. S. Lasky (1)
Checked by T. G. Mastin (2)

1. Procedure

Caution! Triisobutyl aluminum is pyrophoric and reacts violently with water. The vanadium compounds are toxic.

The Catalyst. In a 250-ml., three-necked flask fitted with a condenser, mechanical stirrer, and inert gas inlet (Notes 1 and 2) are placed 8.0 g. of kaolin (Note 3) previously dried at 120° for 16 hr., 70 ml. of dry benzene (Note 4), and 2.72 g. (1.5 ml.) of vanadium tetrachloride in 15 ml. of dry benzene (Note 5). The mixture is stirred gently and refluxed for 3 hr. in a fume hood. All operations are conducted under flowing inert gas. The mixture is allowed to cool to room temperature, filtered on a sintered glass funnel, washed with several volumes.of dry benzene, transferred to a suitable container, and dried under vacuum (1 torr for 24 hr.) (Note 6). The supported vanadium trichloride is analyzed for chlorine by the Volhard method. The product should have a VCl_3 content of 18-20%.

Polymerization. In a dry, heavy-walled glass container (Note 7) of at least 600-ml. capacity are placed, in the order given, 150 ml. of dry benzene, an amount of the VCl_3 on clay equivalent to 70 mg. (0.445 mmole) of VCl_3, 0.1 ml. (0.22 mmole) of tetra-2-ethylbutyl titanate in 10 ml. of dry benzene (Note 8), 1.15 ml. (4.45 mmoles) of triisobutyl aluminum in 10 ml. of dry benzene (Note 9), and 150 ml. (100 g.) of isoprene (Notes 10 and 11). The container is sealed (Note 7) vigorously shaken by hand a few times and then agitated (by rocking or shaking in a 50° bath for 6 hr. (Note 12). The container is allowed to cool to room temperature, the highly swollen polymer is removed (Note 13), cut into small chunks, and placed in 1 1. of methanol containing 2 g. of a phenolic antioxidant. The polymer is then shredded in a large explosion proof blender in the presence of methanol plus antioxidant. The polymer is soaked in 1 1. of fresh methanol plus antioxidant for a few hr. filtered, and dried in a vacuum oven at 40° overnight. The yield of *trans*-polyisoprene is 90-100 g. (90-100%).

The polymer is identical with natural balata in its x-ray diffraction pattern and dilatometric melting points (56° and 64°). The infrared spectrum of the synthetic polymer is almost identical with that of natural balata, the former sometimes having some *cis*-1,4 content. The polymer as obtained is slightly gelled. This gel can be broken down easily by milling on a two-roll rubber mill for a few minutes at 110-120° roll temperature. The polymer after milling is completely soluble in solvents such as benzene and is of far higher molecular weight than natural balata. Its intrinsic viscosity in benzene at 30° equals 3-5 dl./g. versus 0.8-1.0 dl./g. for the natural polymer. Mooney viscosity ML-4 at 100° equals 80-100 versus 10-20 for natural balata.

2. Notes

1. All glassware and auxiliary equipment used in the preparation of the catalyst and in the polymerization should be dried at 120° for at least 4 hr. before use and cooled in a dry inert atmosphere.

2. Dry argon or nitrogen is suitable.

3. The kaolin, Continental Clay, was obtained from R. T. Vanderbilt Company. The specific surface area is 10 $m.^2/g$.

4. Reagent grade benzene containing less than 10 p.p.m. of water was used. Passage through a column of silica gel surmounted by a layer of coconut charcoal and collection and storage under a dry inert atmosphere over sodium or Linde 4A Molecular Sieve suffices.

5. *Caution! VCl_4, if stored in glass, should be kept cool and in the dark. Storage of the glass container in a metal can containing limestone or the like is*

suggested, because samples of VCl_4 have been known to decompose in storage to vanadium trichloride and chlorine rapidly enough to shatter glass containers.

6. The filtration and transfer are conveniently carried out in a dry, inert atmosphere glove box. The vacuum is broken by admitting a dry inert gas. Storage is under a dry inert gas.

7. An ordinary soda bottle to be sealed with a crown cap is suitable. The bottle should be unscratched. The gasket in the crown cap may be nylon, polyethylene, or Teflon of suitable thickness.

8. Equimolar amounts of any tetraalkyl titanate may be used. The isopropyl and higher esters are used as commercially available. The ethyl and methyl ester should be distilled before use.

9. The triisobutyl aluminum is commercially available and used without purification. Twenty per cent solutions of triisobutyl aluminum in benzene are commercially available.

10. Polymerization grade isoprene (Phillips Petroleum Company, Special Products Division) distilled from sodium or purified by passage through a 3/1 mixture of silica/alumina is suitable. In the silica/alumina purification, care should be taken to avoid excessive heat on first wetting the absorbent. Inert gas blankets are used throughout purification and storage.

11. Additions to the polymerization vessel are made under an inert atmosphere with hypodermic syringes. The solid VCl_3 is conveniently weighed into a small, dry Erlenmeyer flask and transferred to the polymerization vessel with a dry powder funnel.

12. As short a time as is practicable should elapse between addition of the trialkyl aluminum and the isoprene. The container should be sealed very shortly after the isoprene addition and placed in the 50° bath immediately thereafter.

13. It is necessary to break the soda bottle to remove the polymer.

3. References

1. Research Center, United States Rubber Company, Wayne, New Jersey 07470, Contribution No. 249.
2. The Goodyear Tire & Rubber Company, Akron, Ohio 44309.
3. J. S. Lasky (United States Rubber Company, U.S. Pat. 3,054,754 (1962).
4. J. S. Lasky (United States Rubber Company), U.S. Pat. 3,114,744 (1963).

Emulsion Copolymerization of Butadiene and Styrene

$$CH_2{=}CH{-}CH{=}CH_2 + CH_2{=}\underset{\displaystyle C_6H_5}{\underset{|}{CH}} \longrightarrow \left[Bd{-}CH_2{-}\underset{\displaystyle C_6H_5}{\underset{|}{CH}} \right]$$

$$Bd = -CH_2{-}CH{=}CH{-}CH_2-$$

and/or

$$-CH_2{-}\underset{\displaystyle CH{=}CH_2}{\underset{|}{CH}}-$$

Submitted by J. A. Rozmajzl (1)
Checked by T. A. Jones (2)

1. Procedure

A. *At 50°*

The polymerization ingredients (Note 1) are added in the following manner (Note 2). The distilled water is heated to 50°. Polymerization grade hydrogenated tallow fatty acid soap is added with stirring to the hot water until dissolved (Note 3). The solution is then cooled to 35°, and the $K_2S_2O_8$ is added, with stirring, until dissolved. This solution is then added to the polymerization bottle (Notes 4 and 5). The *n*-dodecylmercaptan (*Stench!*) is dissolved in the styrene, and the solution is added to the bottle (Note 4). Finally the distilled butadiene is added to the bottle in slight excess on a triple beam balance and allowed to vent (*Hood!*) to the desired weight before capping. The bottle is placed in a constant temperature water bath at 50° in which it is tumbled at about 40 r.p.m. Latex samples may be taken for solids determinations at various times during the polymerization to determine the rate of polymerization (Note 6). The bottle is rotated until the desired conversion is reached. The polymerization is terminated at this time by adding a "shortstop" solution (Note 1) with a hypodermic syringe. The bottle is then allowed to rotate for an additional 15 min. It is removed from the bath, allowed to cool to room temperature, wiped dry with a towel, and weighed. A latex sample is then taken to determine solids content, and the final per cent conversion is calculated (Notes 6 and 7).

An open syringe needle is used to vent excess butadiene in a hood, and the cap is removed (Note 8). After most of the butadiene has vented, the emulsion is broken by pouring the latex with stirring (Note 9) into a 0.20% solution of N-phenyl-β-naphthylamine in 2-propanol (Note 10). The polymer is separated from the alcohol and water by decantation. It is then washed under agitation with a similar amount of 2-propanol containing N-phenyl-β-naphthylamine. The polymer is again separated from the alcohol by decantation and dried in an air oven at 80° (Notes 11 and 12).

B. At 5°

The polymerization ingredients (Note 13) are added in the following manner (Note 2): 195 parts of distilled water are heated to 50°, and Dresinate 214 (disproportionated rosin acid soap) and trisodium phosphate are added with stirring until dissolved (Note 3). The solution is allowed to cool to room temperature. The pH is then adjusted to 10.0 (Note 14). The soap solution is added to the polymerization bottle (Notes 4 and 5). The *t*-dodecyl mercaptan (*Stench!*) is dissolved in part of the available styrene (Note 15). The solution is then added to the bottle (Note 4). The distilled butadiene is added to the bottle on a triple beam balance and allowed to vent (*Hood!*) to the desired weight before capping. The Versene Fe-3, ferrous sulfate, and sodium formaldehyde sulfoxylate are added in that order to the remaining five parts of distilled water. This solution is added to the bottle with a hypodermic syringe. The bottle is allowed to rotate in a 5° constant temperature water bath for 5 min. at 40 r.p.m. *p*-Menthane hydroperoxide (*Caution! Strong oxidant*) is added to the remaining styrene (Note 16). This solution is added to the bottle with a hypodermic syringe. The bottle is then tumbled in the constant temperature water bath at 5° and 40 r.p.m. Latex samples may be taken for solids determinations at various times during the polymerization to determine the rate of polymerization (Note 17). The bottle is rotated until the desired conversion is reached. The polymerization is terminated at this time by adding a shortstop solution (Note 13) with a hypodermic syringe. The bottle is then allowed to rotate for an additional 15 min. The bottle is removed from the bath and allowed to come to room temperature. The bottle is wiped dry with a towel and weighed.

A latex sample is then taken to determine solids content, and the final per cent conversion is calculated (Notes 17 and 18). The butadiene is allowed to vent as before (Note 8) and the cap is removed. After most of the butadiene has vented, the emulsion is broken by pouring the latex with stirring (Note 9) into a 0.20% solution of N-phenyl-β-naphthylamine in 2-propanol (Note 10). The polymer is separated from the alcohol and water by decantation. It is then washed under agitation with a similar amount of 2-propanol containing N-phenyl-β-naphthylamine. The polymer is again separated from the alcohol by decantation and dried in an air oven at 80° (Notes 12 and 19).

2. Notes

1. Recipe for butadiene/styrene emulsion copolymer.

Polymerization temperature: 50°

Distilled water	180.0 parts
Sodium soap of hydrogenated tallow fatty acid (anhydrous basis)	4.3 parts
$K_2S_2O_8$	0.23 part
n-Dodecyl mercaptan	0.50 part
Styrene (polymerization grade)	29.0 parts
Butadiene (polymerization grade)	71.0 parts
Shortstop solution	
Sodium dimethyldithiocarbamate	0.32 part
Distilled water	10.0 parts

2. *Caution! For greater safety the bottle can be placed in a close-fitting perforated metal container after being charged. Safety glasses or a face shield should be worn in all operations. All work with butadiene and styrene is done in a hood. All sources of ignition should be eliminated when the butadiene is being charged or discharged such as during sampling or venting steps.*

3. For the purposes of accuracy and convenience the solutions are prepared in excess of the amounts actually required.

4. The requisite amount of solution can be added either by weight or by volume.

5. A 4-oz., Boston round bottle capable of withstanding a hydrostatic pressure of about 125 p.s.i.g., may be used. The 4-oz. bottle is sealed with a screw cap containing a two-ply gasket that is self-sealing [butyl (3)] and chemically resistant [Neoprene (3)]. The gasket should be positioned in the cap so

that the chemically resistant portion of the gasket will come in contact with the contents of the bottle when it is inverted. The caps contain two holes, about $\frac{1}{8}$ inch in diameter, through which solutions can be added by means of a hypodermic needle, 20-23 gauge, and syringe assembly or through which latex samples can be withdrawn with a hypodermic needle, 20-23 gauge. A 16-oz. bottle with a Crown pressure sealing cap containing a two-ply gasket may also be used. A 20-g. monomer charge is used in the 4-oz. bottle. A 100-g. charge is used in the 16-oz. bottle.

6. Latex samples are removed by shaking the bottle, inverting it, and inserting a hypodermic needle, 20-23 gauge, through one of the openings in the cap and through the self-sealing gasket. This is known as the ejection method (4) of determining solids. The bottle should be shaken thoroughly and the latex removed immediately, particularly at the lower conversion levels, to prevent a phase separation in the bottle that would result in a nonrepresentative sample. The latex (1.0-2.0 g., obtained by difference in the weight of the bottle before and after taking the latex samples) is transferred *directly* from the bottle to a weighed aluminum foil cup containing a few ml. of 2-propanol. The cup is placed on a hot plate at a medium heat, and the contents are evaporated to dryness. (*Hood!*) It is then weighed. The conversion is calculated from the per cent solids:

$$\% \text{ solids} = \frac{\text{weight of dry solids}}{\text{weight of latex}} \times 100$$

$$\% \text{ conversion} = \frac{(\% \text{ solids}) \times 285}{100} - 5 \text{ (parts of non-rubber)}$$

A factor of 295 is used after the shortstop solution is added.

7. Using this procedure, a conversion in the 65-75% range should be reached in 10 hr.

8. Venting is necessary because the bottle is normally under pressure.

9. *Caution! For safety an air-driven stirrer or explosion-proof blender should be used.*

10. N-Phenyl-β-naphthylamine is used as an antioxidant to protect the polymer against degradation. For coagulation 0.20% N-phenyl-β-naphthylamine in 2-propanol solution is used. The volume of this solution used is equal to 4 times the volume of the latex to be coagulated.

11. A 4-oz. bottle run was made to 67% conversion. The polymer had an inherent viscosity (5) of 1.94 in benzene at 30° at a concentration of 0.0703 g. of polymer per deciliter of solution.

12. A drying time of 2 hr. is normally adequate. The drying time will vary with the size of the polymer particles.

13. Recipe for butadiene/styrene emulsion copolymer.

Polymerization temperature: 5°

Butadiene (polymerization grade)	71 parts
Styrene (polymerization grade)	29 parts
Distilled water	200 parts
p-Menthane hydroperoxide (100% active) (Note 20)	0.08 part
Sodium formaldehyde sulfoxylate ($NaO_3SCH_3 \cdot 2H_2O$)	0.08 part
Ferrous sulfate ($FeSO_4 \cdot 7H_2O$)	0.01 part
Versene Fe-3 (100%)	0.02 part
Dresinate 214 (anhydrous basis) (Notes 20 and 21)	4.50 parts
Trisodium phosphate ($Na_3PO_4 \cdot 12H_2O$)	0.50 part
t-Dodecyl mercaptan	0.20 part
Shortstop solution	
Sodium dimethyldithiocarbamate	0.20 part
Distilled water	4.80 parts

14. If the pH is below 10.0, it should be adjusted with potassium hydroxide to 10.0. If the pH is between 10.0 and 10.5, it need not be adjusted.

15. The *t*-dodecyl mercaptan is added to 24.5 parts (5.40 ml. where a 4-oz. polymerization bottle is used) of the styrene.

16. The *p*-methane hydroperoxide is added to 4.5 parts (1.0 ml. where a 4-oz. polymerization bottle is used) of the styrene.

17. Latex samples are removed as described in Note 6. The conversion is calculated from the per cent solids:

$$\% \text{ solids} = \frac{\text{weight of dry solids}}{\text{weight of latex}} \times 100$$

$$\% \text{ conversion} = \frac{(\% \text{ solids}) \times 306}{100} - 5 \text{ (parts of non-rubber)}$$

A factor of 311 is used after the "shortstop" solution is added.

18. Using this procedure a conversion of about 60% should be reached in 9 hr.

19. A 4-oz. bottle run was made to 58% conversion. The polymer had an inherent viscosity (5) of 1.93 dl./g. in benzene at 30° at a concentration of 0.0681 g. of polymer per deciliter.

20. Obtained from Hercules Powder Company.
21. Potassium soap of disproportionated rosin acid.

3. References

1. Goodyear Tire and Rubber Company, Akron, Ohio 44309.
2. Marbon Chemical Division of Borg-Warner Corporation, Washington, West Virginia 26181.
3. S. A. Harrison and E. R. Meinke, *Anal. Chem.*, **20**, 47 (1948).
4. D. E. Williams and W. F. Johnson, *J. Polym. Sci.*, **2**, 346 (1947).
5. L. H. Cragg, *J. Colloid Sci.*, **1**, 266 (1946).

Emulsion Copolymerization of Butadiene and Acrylonitrile

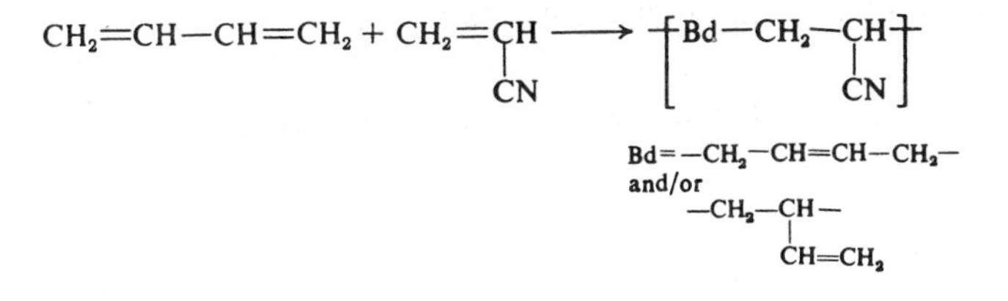

Submitted by R. J. Coleman (1)
Checked by T. A. Jones (2)

1. Procedure

The polymerization ingredients (Note 1) are charged according to the following procedure (Note 2). Polymerization grade tallow fatty acid soap is dissolved with stirring in distilled water warmed to 50° (Note 3). The solution is then cooled to 35°, and the $K_2S_2O_8$ is added, with stirring, until dissolved. This solution is added to the polymerization bottle (Notes 4 and 5). The *t*-dodecyl mercaptan (*Stench!*) is dissolved in the acrylonitrile and the solution added to the bottle (Note 4). Finally the distilled butadiene (in slight excess) is added to the bottle on a triple beam balance and allowed to vent (*Hood!*) to the desired weight before capping (Note 6). The bottle is placed in a constant temperature water bath at 38°, in which it is tumbled at about 40 r.p.m. Latex samples may be withdrawn for solid determinations at various times during the polymerization to determine the rate of polymerization (Note 7). When the required conversion has been reached, the reaction is terminated by adding the "shortstop" solution (Note 1) with a hypodermic syringe. The bottle is then rotated for an additional 15 min. The bottle is allowed to cool to room temperature. A latex sample is then taken to determine the solids content, and the final per cent conversion is calculated (Notes 7 and 8). The unreacted butadiene is vented through a syringe needle (Note 9). After most of the butadiene has evaporated, the emulsion is broken by pouring the latex, with stirring (Note 10), into a 0.2% solution of phenyl-β-naphthylamine in 2-propanol (Note 11). The polymer is separated from the alcohol and water by decantation. It is then washed, with agitation, using a similar amount of 2-propanol containing phenyl-β-naphthylamine. The polymer is again separated from the alcohol by decantation and dried in an air oven at 80° (Notes 12 and 13).

2. Notes

1. Recipe for butadiene/acrylonitrile emulsion polymerization.

Polymerization temperature: 38°

Distilled water	200	parts
Sodium soap of hydrogenated tallow fatty acid	3	parts
$K_2S_2O_8$	0.3	part
t-Dodecyl mercaptan	0.75	part
Acrylonitrile (practical grade)	25	parts
Butadiene (distilled polymerization grade)	75	parts
Shortstop solution		
Hydroquinone	0.1	part
Water	10	parts

2. *Caution! Safety glasses or face shield should be worn in all operations. Butadiene and styrene should be handled in a hood. All sources of ignition in the room should be eliminated when butadiene is being charged or vented. After charging, the bottle should be immediately placed in a close-fitting perforated metal container. This is not removed until after polymerization and after the bottle has been cooled.*

3. For the purposes of accuracy and convenience in charging, it has been found easier to make up two solutions, one containing the soap, $K_2S_2O_8$, and one containing the acrylonitrile and mercaptan.

4. The required amount of solution can be added by weight or volume.

5. A 4-oz., Boston round bottle capable of withstanding a hydrostatic pressure of about 125 p.s.i.g., may be used. The 4-oz. bottle is sealed with a screw cap containing a two-ply gasket which is self-sealing [butyl(3)] and chemically resistant [Neoprene (3)]. The gasket should be positioned in the cap so

that the chemically resistant portion of the gasket will come in contact with the contents of the bottle, when inverted. The caps should contain at least two holes, about $\frac{1}{8}$ in. in diameter, through which solutions can be added with a hypodermic needle (20-23 gauge) and syringe assembly, or through which latex samples can be withdrawn. A 20-g. monomer charge is used in 4-ox. bottles. An 8-oz. (50 g. of monomer) or 16-oz. (100 g. of monomer) bottle, with a screw or Crown pressure sealing cap containing a two-ply gasket, may also be used.

6. Venting of the small excess of butadiene helps to assure that the space above the contents of the bottle does not contain oxygen.

7. Latex samples are removed by shaking the bottle, inverting it, and then inserting a hypodermic needle (20-23 gauge) through one of the openings in the cap and through the self-sealing gasket. This is known as the ejection method (4) of determining solids. Because of phase separation the contents of the bottle should be shaken thoroughly and the latex removed immediately, particularly at low conversions. The latex (2 ± 0.5 g., obtained by difference in weight of the bottle before and after taking the latex samples) is transferred *directly* from the bottle into a weighed aluminum foil cup. The cup is placed on a hot plate at a medium heat, and the contents are evaporated to dryness (*Hood!*). Through weighing, the per cent solids is calculated as:

$$\% \text{ solids} = \frac{\text{weight of dry solids}}{\text{weight of latex}} \times 100$$

$$\% \text{ conversion} = \frac{\% \text{ solids} \times 304}{100} - 4 \text{ (parts non-rubber) or}$$

$$= \frac{\% \text{ solids} \times 314}{100} - 4 \text{ (after "shortstop" has been added)}$$

8. Using this procedure 90-95% conversion should be reached in 16 hr. or less.

9. The butadiene should be vented in an efficient hood.

10. An air-driven or explosion-proof stirrer should be used.

11. Phenyl-β-naphthylamine is used as an antioxidant to protect the polymer against degradation. For coagulation a 0.2% phenyl-β-naphthylamine in 2-propanol solution is used. The volume of this solution used is equal to 4 times the volume of the latex to be coagulated.

12. A 95% conversion polymer prepared by this formula had an inherent

viscosity (5) of 1.50 in methyl ethyl ketone when a 0.175 g./dl. solution is measured at 30°.

13. A drying time of 2 hr. is normally adequate. The size of the polymer crumb will usually determine the time needed.

3. References

1. Goodyear Tire and Rubber Company, Akron, Ohio 44309.
2. Marbon Chemical Division of Borg-Warner Corporation, Washington, West Virginia 26181.
3. S. A. Harrison and E. R. Meinke, *Anal. Chem.*, **20**, 47 (1948).
4. D. E. Williams and W. F. Johnson, *J. Polym. Sci.*, **2**, 346 (1947).
5. L. H. Cragg, *J. Colloid Sci.*, **1**, 266 (1946).

Butyl Rubber

$$CH_2{=}C(CH_3)_2 + CH_2{=}CH{-}C(CH_3){=}CH_2 \xrightarrow[-78^\circ]{AlCl_3 \cdot CH_3Cl}$$
$$\sim CH_2{-}C(CH_3)_2{-}CH_2{-}CH{=}C(CH_3){-}CH_2{-}CH_2{-}C(CH_3)_2 \sim$$

Submitted by J. P. Kennedy (1a, b)
Checked by R. F. Foerster (2)

1. Procedure

Preferably in a dry box (Note 1) filled with nitrogen, a 1-1. resin flask fitted with a thermocouple, stirrer, and two 250-ml. jacketed dropping funnels is immersed in a cooling bath maintained at -78°, and to it is added 600 ml. (at -78°) of methyl chloride (Notes 2 and 3). The two dropping funnels contain: (*a*) a mixture of 97 ml. (-78°) of isobutene and 3 ml. (-78°) of isoprene (Note 4), and (*b*) $AlCl_3$ in CH_3Cl solution prepared by refluxing 500 ml. of CH_3Cl and 2-3 g. of anhydrous sublimed $AlCl_3$ for 3 hr. (Note 5), and filtering the obtained solution through a jacketed, medium-coarse fritted glass funnel (Notes 6, 7, and 8). The contents of the dropping funnels are added dropwsie to the vigorously stirred CH_3Cl (Note 9). Introduction rates: monomers 1-2 ml./min., catalyst 0.25-0.5 ml./min. After a short induction period, the polymerization starts (haziness). Temperature control is extremely important; the catalyst addition rate is reduced if temperature rises above -78°. After the introduction of the monomer feed, an excess of about 0.25 ml. of catalyst is added and the slurry is stirred for 5 min. more. The polymerization is terminated by adding 100 ml. of pre-cooled methanol and stirring for 5 min. The unreacted gases are evaporated, the polymer is washed by kneading with methanol and dried under reduced pressure at 45-50° for 48 hr. Yield 75-85%. Intrinsic viscosity 0.640-0.840, corresponding to 130,800-200,200 molecular weight (Note 10); mole % isoprene content 1.3-1.5 (Note 11); gel content less than 5%; second order transition point $\sim -73^\circ$.

2. Notes

1. The checker found that satisfactory results can be obtained in the absence of a dry box with oven-dried glassware and connecting tubing and by flaming the assembled apparatus while purging with dry nitrogen. All operations should be conducted in an efficient hood.

2. The methyl chloride gas is passed through two columns (1 m. long, 3 cm. radius filled with BaO) and condensed at $-78°$ (Dry Ice and solvent). The jacketed dropping funnels are cooled with a Dry Ice-pentane mixture.

3. The molecular weight of butyl rubber is strongly influenced by the reaction temperature. A plot of the logarithm of the molecular weight versus inverse temperature gives a straingt line with a positive slope (3). Thus higher molecular weight butyl rubber is obtained at $-100°$, the temperature of commercial operations. Convenient laboratory equipment for carrying out polymerization at extremely low temperatures (down to $-190°$) has been described (4).

4. The isobutene is scrubbed with BaO (see Note 2 for methyl chloride); the isoprene is distilled.

5. Refluxing at room temperature using a Dry Ice-pentane cold finger reflux condenser.

6. The funnel is cooled with powdered Dry Ice in the jacket.

7. The concentration of the $AlCl_3$ catalyst solution should be 0.1-0.5 wt. %. For the analysis of $AlCl_3$ solutions see ref. 5.

8. Molecular weights are unaffected by catalyst concentrations in this catalyst concentration range.

9. *Caution! Operate safely! Because of the low flash points of isobutene and methyl chloride, particularly when working without a dry box, use sparkless motors, ground wires, exhaust hood, and Dry Ice and non-combustible solvent cooling baths.*

10. As a typical cationic polymerization system, butyl rubber molecular weights are extremely sensitive to reaction conditions (e.g., purity of chemicals, moisture, temperature fluctuations, and homogeneity of stirring among others) during synthesis. Using reasonably purified chemicals and care during polymerization, the specified molecular weight range should be attained. Molecular weights lower than about 130,000 are unacceptable when working at $-78°$ and indicate inadequate care in preparation. Molecular weights are affected little by conversions within a wide conversion range (10-90%). The molecular weights are calculated from intrinsic viscosities obtained from single point measurements of 0.1% polymer solutions in di-isobutene at 20° by Flory's equation (6). The checker found that cyclohexane can be used if di-isobutene causes difficulties.

11. The isoprene content of butyl rubber is determined as mole per cent unsaturation by the drastic iodine-mercuric acetate method (7). A solution of butyl rubber in CCl_4 is reacted for 30 min. with iodine in the presence of mercuric acetate and trichloroacetic acid, and the excess iodine is titrated with sodium thiosulfate. The iodine number (centigrams of iodine per gram of polymer) is multiplied by the factor 0.1472 to give mole per cent unsaturation and, consequently, mole per cent isoprene content.

3. Methods of Preparation

Preparation of butyl rubber by this method was first described in 1944 (8). A convenient laboratory set-up has been described in detail (4).

4. References

1a. Chemicals Research Division, Esso Research and Engineering Company, Linden, New Jersey 07036.

1b. Current Address: Institute of Polymer Service, University of Akron, Akron, Ohio 44309.

2. Thiokol Chemical Corporation, Trenton, New Jersey 08608.
3. P. J. Flory, *Principles of Polymer Chemistry,* Cornell University Press, Ithaca, New York, 1953, p. 218; J. P. Kennedy and R. G. Squires, *Polymer,* **6**, 579 (1965).
4. J. P. Kennedy and R. M. Thomas, Chapter 7 in *Polymerization and Polycondensation Processes,* Advances in Chemistry Series 34, A.C.S., Washington, 1962.
5. J. P. Kennedy and R. M. Thomas, *J. Polym. Sci.*, **46**, 481 (1960).
6. P. J. Flory, *J. Amer. Chem. Soc.*, **65**, 372 (1943).
7. S. G. Gallo, H. K. Wiese, and J. F. Nelson, *Ind. Eng. Chem.*, **40**, 1277 (1948).
8. R. M. Thomas and W. J. Sparks, U.S. Pat. 2,356,128 (1944); U.S. Pat. 2,356,130 (1944).

Chlorinated Butyl Rubber

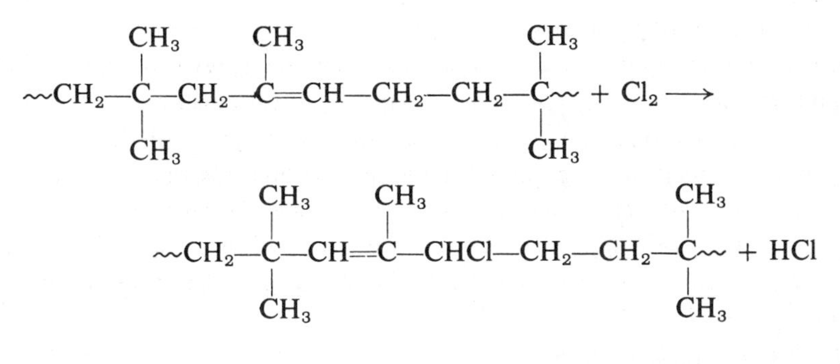

Submitted by F. P. Baldwin (1)
Checked by R. F. Foerster (2)

1. Procedure

A 1-1., three-necked round-bottomed flask is fitted with a stirring paddle, a gas dip tube (or dropping funnel), and a gas exit tube. Into the flask are placed 500 ml. of an inert solvent (Note 1) and 50 g. of characterized butyl rubber (Note 2) cut into small cubes. The mixture is stirred until the rubber has completely dissolved.

The reaction vessel is now shielded from light and the flask and contents preferably, though not necessarily, swept with nitrogen or other inert gas (Note 3). To the flask there is then carefully added 1 mole of chlorine (Note 4) for each mole of double bonds present in the reaction mixture (Note 5). The evolved HCl is vented to the hood or caught in a caustic or soda lime trap. After addition of all the chlorine, stirring is continued for a period of 1-2 min.

The contents of the flask are now slowly poured into 1-2 1. of vigorously agitated acetone or methanol containing about 1 g. of a phenolic oxidation inhibitor (e.g., 2,5 di-*t*-butyl-4-methylphenol) and the precipitated polymer is collected and dried in a vacuum oven for 12-16 hr. at 50° (Note 6).

The yield of polymer is 95-100%. Analysis of the weight per cent of chlorine indicates the equivalent of 1 atom of Cl for each double bond originally present in the polymer. The molecular weight is substantially identical with that of the original polymer. The value for the amount of unsaturation depends on the method of analysis (Note 7).

2. Notes

1. The reaction of interest is very fast and very probably of an ionic nature (3, 4). Suitable inert solvents are *n*-hydrocarbons (pentane, hexane, heptane), benzene, carbon tetrachloride, chloroform. If benzene is used, the temperature should not be allowed to fall below about 23° because high molecular weight polymer will tend to separate from the solution, resulting in a jelly-like mass.

2. Butyl rubber may be prepared by the method described by Kennedy (p. 155), or it may be purchased. The polymer may be characterized by viscosity average molecular weight by the method of Flory (5) and by mole per cent unsaturation by the method of Gallo, Wiese, and Nelson (6) or Rehner and Gray (7).

3. This is a precaution to preclude the onset of scission reactions. Although such reactions are much slower than the reaction of interest, they can, under the right conditions (excess chlorine, long reaction times, etc.), lead to molecular weight degradation (3, 4). For example, if a 100% excess of chlorine over that indicated by the stoichiometry is added and neither light nor oxygen rigorously excluded, the molecular weight of a typical commercial butyl polymer can be roughly halved.

4. Chlorine may be introduced as a gas or may be added as a solution in an inert solvent. Chlorine may be replaced by an equivalent amount (based on Cl_2) of sulfuryl chloride *but only if chloroform is used as the polymer solvent.*

5. In connection with the indicated reaction stoichiometry and the desirability of repressing scission reactions, it is wise not to add appreciable excess of halogen. The amount of chlorine to be added can be calculated as follows:

$$X = 0.186 \times \text{iodine number}$$

where X is the weight per cent of chlorine, based on the polymer, to be added to the reaction vessel.

While recent work has indicated that under carefully controlled conditions two atoms of halogen may be combined for each double bond originally present in the polymer (8), the product containing one atom of halogen per double bond is the one of interest and of commercial importance.

6. The addition of oxidation inhibitor is necessary to prevent oxidative degradation if the product is to be stored for long periods in the open atmosphere. Care should be taken not to heat the product excessively because it can crosslink on heating (3, 4). The commercial product (*Chlorobutyl*) contains calcium stearate as an additive to prevent premature crosslinking during processing and compounding.

7. The usual iodine-mercuric acetate method used for unsaturation measurements of butyl rubbers (6) is unsatisfactory as it stands because the conversion factor involving 3I/olefin link is unsuitable for the halogenated polymer (3, 4). If it is used, one can anticipate an indicated reduction in unsaturation of about 50% as compared to the unchlorinated polymer. Nor can the ozonolysis technique as described (7) be used. The external double bonds in the reaction product preclude clean cleavage of the main chain by ozone. Chain cleavage is a fundamental requirement for the ozonolysis technique.

3. Methods of Preparation

The preparation of chlorinated butyl rubber by this method was first reported in 1958 (9) and its properties and preparation described in succeeding articles and patents (3, 4, 10).

4. References

1. Exxon Chemical Co., P.O. Box 45, Linden, New Jersey 07036.
2. Thiokol Chemical Corporation, Trenton, New Jersey 08608.
3. F. P. Baldwin and I. Kuntz, in *Encyclopedia of Chemical Technology,* Second Supplement, The Interscience Encyclopedia, Inc., New York, 1960.
4. F. P. Baldwin, D. J. Buckley, I. Kuntz, and S. B. Robison, *Rubber Plastics Age,* **42**, 500 (1961).
5. P. J. Flory, *J. Amer. Chem. Soc.,* **65**, 372 (1943).
6. S. G. Gallo, H. K. Wiese, and J. F. Nelson, *Ind. Eng. Chem.,* **40**, 1277 (1948).
7. J. Rehner, Jr., and P. Gray, *Ind. Eng. Chem. Anal. Ed.,* **17**, 367 (1945).
8. I. C. McNiell, *Polymer,* 4, 15 (1963).
9. R. M. Thomas and F. P. Baldwin, French Pat. 1,154,488 (1958).
10. R. M. Thomas and F. P. Baldwin, U.S. Pat. 2,944,578 (1960).

Polycyclopentadiene by Solution Polymerization with Cationic Catalysis

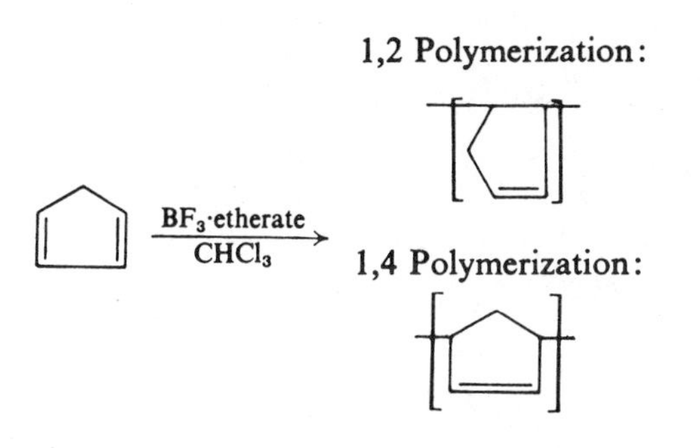

Submitted by M. Wismer and H. P. Doerge (1)
Checked by H. Roth and H. A. Gawel (2)

1. Procedure

Reaction should be conducted in a hood.

A 1-l. four-necked flask equipped with stirrer, thermometer, connecting tube with parallel side arm, gas inlet tube for prepurified nitrogen (Note 1), gas outlet tube connected to a gas bubble counter (Note 2), reflux condenser, and dropping funnel is charged with 300 g. of chloroform (Note 3) and 0.7 ml. of boron trifluoride etherate (Note 4). The flask is purged with prepurified nitrogen for 10 min. and 100 g. of monomeric cyclopentadiene (Note 5) is added dropwise over a period of 1 hr. and 15 min. at 20-30° Nitrogen flow rate is 1 bubble per second and is maintained throughout the reaction. After the addition is completed, the reaction is continued for five additional hours at 20-30°. During the addition and throughout the reaction it is necessary to maintain the temperature at 20-30° using external cooling (Note 6) because the polymerization is exothermic. The reaction mixture is then neutralized with a 20% aqueous solution of ammonium hydroxide (Note 7). Ten grams of anhydrous magnesium sulfate (Note 8) and 10 g. of Celite filter aid (Note 9) are added with stirring. The reaction flask is purged with prepurified nitrogen for several minutes, and the reaction mixture is filtered through a layer of Celite in a pressure filter press (Note 10). The solid polymer is obtained by means of freeze-drying.

Forty grams of the chloroform solution of the polymer is weighed into a 500-ml., one-necked, round-bottomed flask. To this solution is added 40 g. of benzene (Note 11). This solution is immediately purged with prepurified nitrogen, stoppered, and then swirled in a Dry Ice-acetone bath so that a thin film of the polymer solution will freeze on the inner walls of the flask. After the solution is thoroughly frozen, the flask is clamped onto a 90-105° angle vacuum adapter that is connected to a 250-ml., 1-necked, round-bottomed flask. This receiving flask is completely immersed in a Dry Ice-acetone bath. The system is then subjected to a vacuum of 0.4-0.7 torr for 3-4 hr., after which time the polymer-containing flask has reached room temperature. The product is a light yellow, fluffy material. The yield of the polymer after freeze drying is 60-80%, depending on the efficiency of the filtration step.

2. Characterization

Soluble polycyclopentadiene prepared by this method of cationic polymerization is a highly unsaturated type of resinous polymer containing approximately one double bond per cyclopentadiene unit. It is not known how much 1,2 or 1,4 polymerization occurs. The polymer is a light yellow solid soluble in aromatic and chlorinated hydrocarbons. Films may be cast from benzene or chloroform solution. The polymer itself oxidizes easily to an insoluble orange solid and should therefore be stored under prepurified nitrogen.

Some of the polymer (1.0-1.2 g.) was immediately weighed into a 100-ml. volumetric flask, diluted and dissolved in chloroform, and purged with prepurified nitrogen. The solution was conditioned at 25° for several hours in a constant temperature bath and the intrinsic viscosity determined. An intrinsic viscosity of 0.2 dl./g. was obtained.

The unsaturation in the polymer was determined by iodine number. Two solutions of 20-25% solids in chloroform were prepared and purged with nitrogen. Diphenylamine (0.1% by weight of polymer) was added to the one solution to stabilize the polymer. Iodine numbers of both solutions were then determined. The iodine numbers obtained are as follows: iodine number of (*a*) unstabilized solution = 263 at 100% solids; (*b*) of stabilized solution = 312 at 100% solids; (*c*) of stabilized solution after aging at 0° for 3.5 days = 290 at 100% solids (theory = 385).

3. Notes

1. Prepurified nitrogen was supplied by the Matheson Company, East Rutherford, New Jersey.

2. The gas bubble counter, consisting of a small suction flask in which a tube was immersed in glycerin, leads to the top of the condenser.

3. The chloroform used was ACS reagent grade supplied by Fisher Scientific Company, Pittsburgh, Pennsylvania.

4. The boron trifluoride-diethyl ether complex used is technical grade supplied by General Chemical Division, Allied Chemical Corporation, New York, New York.

5. Cyclopentadiene monomer was prepared by the thermal cracking of dicyclopentadiene. Dicyclopentadiene 96% pure was obtained from Enjay Chemical Company, New York. Twice-distilled monomer should be used in the polymerization. The monomer has the following properties: b.p. 41.5°/760 torr n_D^{20} = 1.4430.

6. External cooling of the flask during the exothermic polymerization is best afforded by a Dry Ice-acetone bath that is raised and lowered as desired.

7. The ammonium hydroxide used was ACS reagent grade, supplied by Fisher Scientific Company, Pittsburgh, Pennsylvania.

8. The anhydrous magnesium sulfate was reagent grade, supplied by Fisher Scientific Company, Pittsburgh, Pennsylvania.

9. Celite filter aid is Hyflo Super-Cel, a diatomaceous silica product supplied by Johns-Manville Celite Division, New York , New York 11216.

10. The pressure filter press, Model L-1000, was obtained from the Sparkler Manufacturing Company, Mundelein, Illinois.

11. The benzene used was analytical reagent grade supplied by Mallinckrodt Chemical Works, St. Louis, Missouri. Benzene was added to the chloroform solution to freeze the polymer solution sufficiently. It is difficult to freeze a chloroform solution sufficiently in a Dry Ice-acetone bath because of the low freezing point (−63°) of the chloroform.

4. Methods of Preparation

The polymerization of cyclopentadiene is described in a review by Wilson and Wells (3). Bruson and Staudinger (4) describe the use of metal halides as catalyst. The catalytic polymerization of cyclopentadiene using aluminum chloride complexes is also described in the patent literature (5).

5. References

1. Pittsburgh Plate Glass Company, Research and Development Center, Springdale, Pennsylvania 15144.
2. Inmont Corporation, Central Research Laboratories, Clifton, New Jersey 07015.
3. P. J. Wilson and J. H. Wells, *Chem. Rev.*, **34**, 1 (1944).
4. H. A. Bruson and H. Staudinger, *Ind. Eng. Chem.*, **18**, 381 (1926).
5. F. J. Soday (The United Gas Improvement Company), U.S. Pat. 2,314,904 (1943).

Polyacrylonitrile from a Slurry Polymerization

$$CH_2{=}CH{-}CN \xrightarrow[Fe^{+2},\ pH\ 3.2]{K_2S_2O_8{-}NaHSO_3} \left[CH_2\underset{\displaystyle CN}{\underset{|}{C}}H \right]_x$$

Submitted by W. K. Wilkinson (1)
Checked by N. G. Gaylord, (2a) S. Gottfried, (2a) and N. M. Bikales (2b)

1. Procedure

Deaerated distilled water (0.4 1.) and distilled acrylonitrile (50 ml., 40 g.) are placed in a 1-1. three-necked flask equipped with a stirrer, a stopper, and a reflux condenser (Notes 1, 2, and 3). The flask is placed in a thermostat in the hood at 50 ± 1°, and a constant pressure of nitrogen equal to 1 in. of water is applied to the top of the condenser. A solution of 20-25 ml. of water and 4.0 ml. of 0.1*N* sulfuric acid and 0.001 g. of ferrous ammonium sulfate hexahydrate is added quickly through the stopper (Note 4). When the contents of the flask are at 50°, 25 ml. of water containing 0.10 g. of potassium persulfate and then 50 ml. of water containing 0.50 g. of sodium metabisulfite are added quickly through one of the necks. Within 1 min. the appearance of opalescence indicates that the reaction has started. Stirring is continued at 50-200 r.p.m. for 1 hr., during which time a thick slurry develops. A 1% solution of sodium carbonate to reach pH 7-10 is added to stop the reaction (Note 5). The slurry is filtered by vacuum and washed with 200 ml. of water followed by 200 ml. of acetone or ethanol. The polymer is crumbled into a drying tray and dried below 100°. A yield of 65-75% is obtained. The intrinsic viscosity (in dimethylformamide) is 1.5-1.7 dl./g. (Note 6).

Copolymers may also be made by this procedure if there is not too great a deviaiton from acrylonitrile either in solubility characteristics or copolymerization reactivity ratios (Note 7).

2. Notes

1. The water is deaerated by bubbling nitrogen through it for 2-3 min. at a rate of at least 50 ml./min. Failure to deaerate may cause low conversions and erratic molecular weights. Alternatively, freshly boiled distilled water may be used.

2. Larger runs should be attempted only with great caution, because the heat of polymerization may exceed the cooling capacity of the thermostated bath, causing violent boiling of the acrylonitrile.

3. Distilled acrylonitrile is prepared by adding 1 ml. of 85% phosphoric acid per liter of monomer and distilling rapidly through a simple Claisen head. This removes trace amines sometimes used as inhibitors, and also removes phenolic inhibitors.

It is not necessary to distill most commercial grades of acrylonitrile to override the inhibitors with this redox recipe. However, the resulting product would have a slightly lower molecular weight, which may be compensated for by using less persulfate.

4. The sulfuric acid must be added at this step to obtain optimum rate and molecular weight control. The optimum range is pH 2.4-4.0 with best reproducibility at pH 2.8 ± 0.05. If the sulfuric acid is added after the bisulfite, the molecular weight control will be erratic because of rapid bisulfite addition to acrylonitrile at pH 6 and above. The iron salt is a catalyst for persulfate decomposition and must be present above 0.2 p.p.m. Fe^{2+}, based on the total volume. Often the water and raw materials provide the required iron, but, if a known amount is added, improved reproducibility results.

5. Adjustment to pH 7 or higher stops the polymerization by allowing bisulfite addition to acrylonitrile and by lowering ionizable iron content. Alternatively, sequestering agents may be used as well as the usual phenolic inhibitors.

6. Raising the molecular weight may be simply achieved by lowering the amount of persulfate or both persulfate and bisulfite used; concurrently, a lower rate of polymerization will be observed, especially if less persulfate is used. The maximum range of intrinsic viscosity attainable is 0.05-20 dl./g.

7. A mixture of comonomers may be used. Some of those that can be prepared by the procedure include vinyl acetate, acrylamide, methyl vinyl ketone, and methyl acrylate. Others may be used by slight adjustments (e.g., vinylidene chloride at 30° or 2-vinyl pyridine with enough acid to reach pH 3.2).

3. References

1. Textile Fibers Department, Benger Laboratory, E. I. du Pont de Nemours & Company Inc., Waynesboro, Virginia 22980.
2a. Gaylord Associates, Inc., Newark, New Jersey 07102.
2b. Rutgers University, New Brunswick, New Jersey 08901.

Calcium Oxide as an Initiator for Acrylonitrile Polymerization

$$CH_2{=}CH{-}CN \xrightarrow[DMF]{CaO} \left[CH_2{-}\underset{\displaystyle CN}{\underset{|}{C}H} \right]$$

Submitted by J. R. Schaefgen (1)
Checked by J. Cazes (2a, b)

1. Procedure

A 100-ml. three-necked flask is equipped with a precision-ground stirrer (glass paddle) and a nitrogen inlet. The third opening is stoppered and used for introducing the reactants. To provide a constant positive nitrogen pressure, the inlet line is attached to the flask via a T-tube, the third end of which is attached to a vent trap containing an inert dry liquid hydrocarbon such as cyclohexane. The flask is flamed and purged with nitrogen, and 1 g. of calcined, free-flowing, powdered calcium oxide is introduced (Note 1). The flask is then cooled. N,N-Dimethylformamide (DMF) (Note 2) (30 ml.) and acrylonitrile (Note 3) (10 ml.) are introduced by pipet. The reaction is stirred under nitrogen overnight (16 hr.). As polymerization proceeds, the suspended particles of calcium oxide become coated with white to slightly yellow polymer. The suspension stays quite fluid and is milky in appearance.

To isolate the polymer, the contents of the flask are added to 100 ml. of 1*N* HCl to give a copious, white, coherent precipitate. This mixture is then agitated in a blender for 5 min. to free the polymer of calcium oxide. The polymer is collected by filtration and is retreated with an additional 100 ml. of 1*N* HCl, agitated for 5 min. in the blender, and collected. The filter cake is washed with water until the filtrate is neutral, and then with acetone. After drying in an oven at 80° for 1 hr. under reduced pressure (<100 mm.), a yield of 2.5-6 g. (31-75%) of polyacrylonitrile is obtained (Note 4).

2. Characterization

The inherent viscosity of the polyacrylonitrile varies from 1 to 2 or higher in DMF solution at 30° at a concentration of 0.5%. Concentrated solutions of polyacrylonitrile to be used for forming films may be prepared from a solution of 1-2 g. of polymer in 8-9 ml. of DMF. Ice-cold DMF is added to the finely divided polymer in a screw cap vial, which is then stoppered and warmed on the steam bath for several hours to effect solution. Film is cast from the viscous solution by use of a 10- or 20-mil doctor knife. After evaporation of the solvent, the brittle films (1-4 mils thick) may be cut into strips on a warm (90-100°) glass plate without fracturing, and oriented by drawing 6-10 times at 160° on a heated rod or bar. The film still contains residual DMF, most of which may be removed by soaking overnight in methanol. The authenticity of the sample may be checked by the infrared spectrum (3). The material cannot be melted but undergoes an exothermic condensation on heating (4).

3. Notes

1. A reagent grade or technical grade calcium oxide (e.g., Mallinckrodt N.F. lumps) is suitable, but the latter is preferred. The material is ground to a fine powder with a mortar and pestle. Directly before use, the oxide sample is heated strongly with a Fisher burner in a platinum crucible until a *free-flowing* dry powder is obtained. This takes 10-60 min. The contents of the hot crucible are added directly to the reaction flask.

2. The commercial material (200 ml.) is refluxed at reduced pressure over 1-5% by weight of phosphoric anhydride for 30 min. and then distilled, b.p. 63°/30 torr. A simple Vigreux column is adequate. The center fraction is stored under dry nitrogen and is used soon after distillation because the product can absorb water.

3. The commercial product (100 ml.) is freed of inhibitor and traces of water by refluxing over, (for about 30 min.) and distilling from, 1-5% by weight of phosphoric anhydride through a short Vigreux column. The pure, constant-boiling center fraction is stored under dry nitrogen and is used promptly to avoid spontaneous polymerization.

4. Variations in yield and viscosity depend on the grade of calcium oxide used (see Note 1), the fineness of the particles, and the extent to which the oxide has been dried before using.

4. Methods of Preparation

Acrylonitrile polymerizes by free radical or anionic initiation in suspension or in solution. Detailed procedures for polymerization are given by Sorenson and Campbell (4). A number of other types of initiators, including calcium oxide, are described in Table I of the article by Bohn, Schaefgen, and Statton (5).

5. References

1. Pioneering Research Division, Textile Fibers Department, E. I. du Pont de Nemours & Co., Inc., Wilmington, Delaware 19898.
2a. Research and Development Division, Mobil Chemical Company, Metuchen, New Jersey 08840.
2b. Current Address: Waters Associates, Milford, Massachusetts 01757.
3. S. Krimm and C. Y. Liang, *J. Polym. Sci.,* **31**, 513 (1958).
4. W. R. Sorenson and T. W. Campbell, *Preparative Methods of Polymer Chemistry,* Interscience Division, John Wiley and Sons, New York, 1961, pp. 168-170, 200.
5. C. R. Bohn, J. R. Schaefgen, and W. O. Statton, *J. Polym. Sci.,* **55**, 531 (1961).

Graft Copolymers of Acrylamide on Poly(vinyl alcohol)

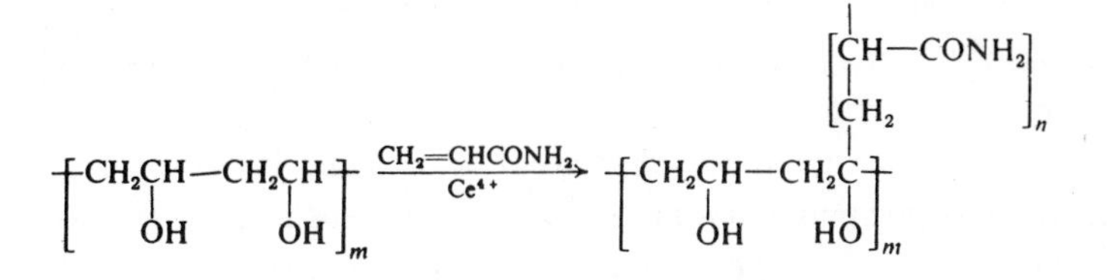

Submitted by G. Mino and S. Kaizerman (1)
Checked by W. L. Hahn (2)

1. Procedure

A 500-ml. three-necked flask is fitted with a stirrer, thermometer, and gas inlet tube. Water (100 ml.) is placed in the flask, and 15 g. of poly(vinyl alcohol) (Elvanol 51-05) is added and dissolved with rapid stirring (Note 1); then 15 g. of recrystallized acrylamide (*Caution! Acrylamide is toxic and should be handled with care.*) (Note 2) is added. The solution is flushed slowly with carbon dioxide (Note 3) for about 10 min. and the temperature is adjusted to 25°. Ceric ammonium nitrate solution (12 ml.), 0.1*M* in 1*M* nitric acid (Note 4), is added to initiate polymerization. Polymerization is allowed to proceed for 45 min. at 25°, during which a slow stream of carbon dioxide is bubbled through the solution. The yellow color of the ceric ion fades during the reaction, and the final solution is essentially colorless.

The clear polymer solution is poured slowly, with rapid stirring, into 1.2 1. of acetone to precipitate the polymer. The polymer is separated on a sintered glass filter, washed with acetone, and dried in a vacuum oven at 70°. The yield is 27-28 g., which represents 80-86% conversion of the monomer. The gross polymer contains 45-48% of combined acrylamide, as determined by nitrogen analysis. The grafting efficiency is close to 100% because no free polyacrylamide can be detected in the polymer (Note 5).

2. Notes

1. Elvanol 51-05 is a partially hydrolyzed poly(vinyl acetate) with about 12% of the acetate groups still present and is readily soluble in cold water. Poly(vinyl alcohols) of lower acetyl content (e.g., Elvanol 70-05) must be dispersed in cold water and heated to 85-90° to effect complete solution.

2. Commercial grade acrylamide cannot be used because it contains impurities which interfere with the ceric ion reaction. Acrylamide is conveniently recrystallized from either chloroform or ethyl acetate. The hot solutions should be filtered.

3. A simple carbon dioxide generator can be made by placing a few lumps of Dry Ice in a loosely stoppered side-arm flask. Nitrogen may also be used to purge the system of oxygen.

4. The ceric ammonium nitrate reagent is prepared by dissolving 5.5 g. of reagent grade ceric ammonium nitrate in sufficient 1*M* nitric acid to make 100 ml. of solution. At this acid concentration, the reagent may be kept indefinitely.

5. Grafting efficiency is determined by fractionation of the polymer. Five grams of the polymer is dissolved in 100 ml. of a 50:50 water-methanol mixture, and acetone is added slowly until a white opaque suspension is obtained. The suspension is stirred for 1 hr. at room temperature, then kept at 0° for 24 hr. and centrifuged. The supernatant liquid is decanted, the polymer is dissolved in a few milliliters of water and reprecipitated in acetone. It is then filtered, dried under reduced pressure, weighed, and analyzed for nitrogen. Several fractions are obtained by repeating the procedure with further acetone addition to the same polymer solution. The polymer is found to contain less than 10% of free poly(vinyl alcohol) and no free polyacrylamide.

3. Methods of Preparation

Other preparations of graft copolymers from acrylamide and other vinyl monomers by the ceric ion technique have been published (3-5). A partial elucidation of the initiation mechanism also appears in the literature (6-8).

4. References

1. American Cyanamid Company, Bound Brook, New Jersey 08805.
2. Textile Fibers Department, E. I. du Pont de Nemours & Company, Inc. Waynesboro, Virginia 22980.
3. G. Mino and S. Kaizerman, *J. Polym. Sci.*, **31**, 242 (1958).

4. G. Mino and S. Kaizerman (American Cyanamid Company), U.S. Pat. 2,922,768.
5. S. Kaizerman, G. Mino, and L. F. Meinhold, *Textile Res. J.,* **32**, 136 (1962).
6. G. Mino, S. Kaizerman, and E. Rasmussen, *J. Polym. Sci.,* **39**, 523 (1959).
7. G. Mino, S. Kaizerman, and E. Rasmussen, *J. Amer. Chem. Soc.,* **31**, 1494 (1959).
8. G. Mino, S. Kaizerman, and E. Rasmussen, *J. Polym. Sci.,* 38, 393 (1959).

Free Radical Telechelic Polystyrene with Amine End Groups

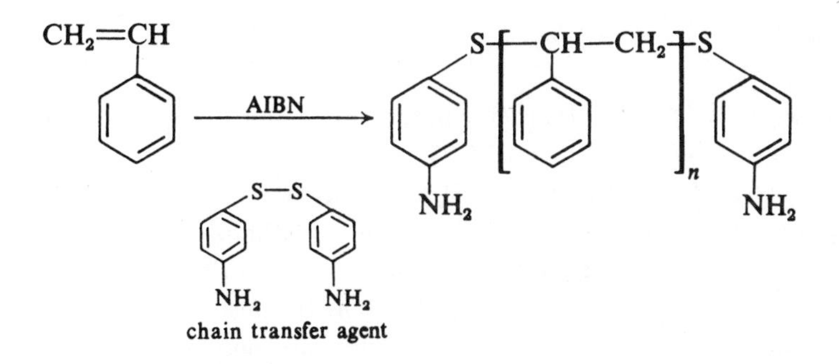

Submitted by A. J. Costanza (1)
Checked by L. Levine (2)

1. Procedure

Distilled styrene (100 g.) is weighed into a 4-oz. screw cap bottle with a cap provided with an aluminum foil or Teflon liner. Bis-(4-aminophenyl) disulfide (3.68 g.) (Note 1) and 0.05 g. of azobisisobutyronitrile (Note 2) are dissolved in the styrene, and the free space over the styrene is flushed with lamp grade nitrogen before and during capping. The bottle is transferred to a 50° constant temperature bath for a period of 16 hr. This will give a monomer to polymer conversion of about 7% (Note 3). The styrene cement is poured into 10 volumes of methanol with rapid stirring (Note 4). The precipitated polymer is allowed to settle until the supernatant liquid is clear. The polystyrene is decanted or filtered and redissolved in 50 ml. of benzene or methyl ethyl ketone and reprecipitated in 10 volumes of rapidly stirred methanol. The polystyrene is allowed to settle as above and is redissolved and reprecipitated twice more. The reprecipitation frees the polymer of the unreacted bis-(4-aminophenyl) disulfide. The polymer is dried under vacuum at 60° to constant weight. The intrinsic viscosity in benzene at 30° of the polystyrene so prepared is about 0.24. The number average molecular weight calculated from this intrinsic viscosity is 25,000 (Note 5). Titration of the telechelic polystyrene for combined amine groups employing the perchloric acid in nonaqueous solvent technique (3) indicates that the polystyrene contains 2.16 amine groups per polymer molecule after correcting for non-reactive end groups conferred by the initiator (Note 6). Because polystyrene prepared at this temperature is substantially linear and unbranched, the value of approximately two amine groups per mole demonstrates its telechelic character.

2. Notes

1. This disulfide has a transfer constant of 0.24 (4), and its synthesis is reported in the literature (5), m.p. 76-77°.
2. The amount of initiator can be varied according to the polymerization temperature and rate desired. The level of initiator used at any particular temperature determines the proportion of inert ends contributed by it.
3. Depending on the magnitude of the transfer constant, the conversion is kept within the limits of low depletion of monomer and modifier so that the M_n versus $[\eta]$ relationship for polystyrene is valid (6). One method of determining conversion is the total solids method using ethanol, containing a trace of polymerization inhibitor, to aid in the volatilization of residual styrene.
4. As an aid in precipitating low molecular weight polymer, a few grams of $Ca(NO_3)_2$ may be dissolved in the methanol.
5. Calculated from the relationship (7) $\overline{M}_n = 184{,}000[\eta]^{1.40}$.
6. The lower the molecular weight of the polystyrene prepared by mass polymerization through use of disulfide transfer agent, the lower is the ratio of inert ends produced by the initiator. However, to use the $\overline{M}_n$ versus $[\eta]$ relationship in the molecular weight region in which it is known to be valid (above about 20,000), preparations are preferably at $\overline{M}_n$ above 20,000. If reliable means are available for measuring $\overline{M}_n$ of polystyrene below 20,000, such telechelic polymers can be expected to possess the functionality of two. Because it is well established (6) that disulfides do cleave in such a manner as to impart one-half of the modifier to each end of polystyrene, very low molecular weight polymers (<5000) can be prepared by either mass or emulsion polymerization through use

of appropriate disulfides. The $\overline{M}_n$ of such polymers can be established through end group analysis.

3. Methods of Preparation

The commercially available bis-(2-amino-phenyl)disulfide

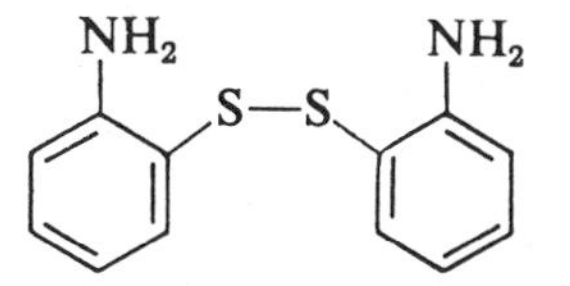

from American Cyanamid, if recrystallized, may be used to prepare telechelic polystyrene. This compound, with a relatively high transfer constant of 3 (6), can be used to produce very low molecular weight product at relatively low charge levels. However, it does retard polymerization rate. The amine groups in the polystyrene prepared with it cannot be titrated very readily by the perchloric acid method because of the absence of a sharp end point, but its end group can be analyzed by acetylation with excess acetic anhydride in pyridine.

Other telechelic polystyrenes containing terminal hydroxyl, carboxyl, or halogen can be prepared by the use of the appropriate disulfides (4, 5). A solvent such as pyridine must be used with disulfides containing hydroxyl or carboxyl groups to solubilize them in styrene. Functional group analyses can be made by the standard methods.

4. References

1. Goodyear Tire & Rubber Company, Akron, Ohio 44309.
2. Foster Grant Company, Leominster, Massachusetts 01453.
3. J. S. Fritz, Acid-Base Titrations in Non-aqueous Solvents, G. Frederick Smith Chemical Company.
4. A. J. Costanza, R. J. Coleman, R. M. Pierson, C. S. Marvel, and C. King, *J. Polym. Sci.*, 17, 319 (1955).
5. C. C. Price and G. W. Stacy, *Org. Syn.*, Coll. Vol. 3, 86 (1955).
6. R. M. Pierson, A. J. Costanza, and A. H. Weinstein, *J. Polym. Sci.*, 17, 221 (1955).
7. R. A. Gregg and F. R. Mayo, *J. Amer. Chem. Soc.*, 70, 2375 (1948).

Polyphosphonitrilic Chloride

$$(PNCl_2)_3 \xrightarrow[(NH_4)_2S_2O_8]{\text{Heat}} \underset{\displaystyle Cl}{\overset{\displaystyle Cl}{|}}{\!\!\!\!\!P}\!=\!N\!-\!\left[\underset{\displaystyle Cl}{\overset{\displaystyle Cl}{|}}{\!\!\!\!\!P}\!=\!N\right]_n\!-\!\underset{\displaystyle Cl}{\overset{\displaystyle Cl}{|}}{\!\!\!\!\!P}\!=\!N\!-$$

Submitted by R. L. Vale (1)
Checked by H. R. Allcock (2a, b) and K. J. Valan (2a)

1. Procedure

Phosphonitrilic chloride trimer (*Caution! This material is toxic and should be handled in a well-ventilated hood, while wearing protective clothing.*) (Note 1) is ground in a mortar and pestle with ammonium persulfate in a ratio of 100 g. of trimer to 2 g. of persulfate (mole ratio = 2.87:0.0875) to give an intimate mixture of the two compounds. Four grams of the mixture is sealed in a 10-mm. diameter glass tube 10-15 cm. long under vacuum (Note 2), and the tube is wrapped in a sheet of copper gauze and placed in an air oven at 210°.

After heating for 24 hr. (Note 4) the tube is allowed to cool, and the polymerized material is separated from the unchanged trimer by swelling the contents of the tube in 50 ml. of sodium-dried benzene for 24 hr. The insoluble $(PNCl_2)_n$ rubber is removed by filtration on a sintered glass crucible (porosity No. 3). The swollen rubber is dried in a vacuum oven at 60-80° until a constant weight is obtained. The yield is 3-3.4 g. representing a conversion of 75-85%. The rubber is a pale yellow rubbery solid (Note 3) that hydrolyzes slowly in moist air. It should be stored in a dry atmosphere, preferably in a small stoppered vessel.

2. Notes

1. The trimer, obtained from Albright and Wilson, Birmingham, England, as a commercial grade melting at 111.5-112.5°, was purified by the following procedures. (Ed. Note: Also available from Ethyl Corp., as "hexachlorocyclotriphosphazene".)

(*a*) Recrystallized from isopropyl alcohol, product m.p. 113-113.5°. Ligroin may also be used and may be preferred because it is quite inert with the trimer, whereas alcohols do react under certain conditions.

(*b*) Zone-refined using six passes at 1 in./hr. down a tube 12 mm. in diameter, 15 cm. long; product m.p. 113.5°. The rate of conversion of trimer to polymer is not greatly affected by the purity of the trimer. Conversions of 75% were obtained after 24 hr. of heating with both crude and recrystallized trimer. Zone-refined material gave 83% polymer.

2. The polymerization is not seriously retarded by air, and conversions of 70-75% are obtained if the polymerization is done in tubes sealed under 1 atm. of air. The vacuum used in the experimental procedure was 1 torr, obtained by using a rotary oil pump. Although there is no pressure generated during polymerization, it is always considered good practice to enclose a sealed-tube reaction vessel in a protective sheath. The phosphonitrilic chloride will sublime at 210° if a sealed tube is not used.

3. The color of the final rubber is very dependent on the water content of the trimer. The trimer may be conveniently dried by heating it at 140° in an oven with a small addition of anhydrous silica gel or by zone refining. If the material is not dried, the polymer will have a very dark brown to black appearance, presumably because of some dehydrochlorination. This is not shown, however, by elemental analysis.

4. The conversion time curve for the polymerization reaction depends on the method of determining the polymer yield. If the benzene extraction technique is used, the low molecular weight oils that are present at low conversions are not isolated because they dissolve in the benzene with the trimer. If, however, the unchanged trimer is removed by vacuum sublimation, the total yield of polymer is obtained. When using the latter technique, it is important to reevacuate the reaction tube, after heating at 210°, because the vapor pressure in the tube is too high to allow sublimation at a suitable temperature (140°). By using the sublimation method, it is shown that the polymerization is retarded during the first 4 hr. of heating. Between 4 and 20 hr. the conversion is linear with time, and a maximum is reached after approximately 24 hr. of heating.

3. Methods of Preparation

The polymerization of phosphonitrilic chloride trimer has been the subject of a considerable amount of research. The polymerizations can be conveniently divided into catalyzed and non-catalyzed reactions. The trimer polymerizes if heated alone in the range 250-350°, although, if the materials are carefully purified, the reaction is very slow. There are large discrepancies between the rates of polymerization obtained by a purely thermal reaction at 350°C., and it has been suggested that these discrepancies are caused by the surfaces of the glass vessels. Above 350° depolymerization occurs. The tetramer is considerably more expensive than the trimer and polymerizes at a much lower rate.

The advantage of using a catalyzed reaction is that polymerization is possible at lower temperatures, and the results are much more reproducible. Konecny and Douglas (3) reported the polymerization of both trimer and tetramer at 211° using a variety of catalysts such as benzoic acid and diethyl ether and obtained results very similar to those given in the procedure above. Ammonium persulfate gives very consistent results and a lighter-colored product than most other catalyst systems. The reaction is now believed to be ionic in character.

4. Characterization of Phosphonitrilic Chloride Rubbers

Rubbers obtained at very low conversions (<10%) are low molecular weight oils that are benzene-soluble and have intrinsic viscosities in the range 0.05-0.15. dl./g. At higher conversions insoluble rubbers are formed that have a swelling index (ml. of benzene imbibed per gram of rubber) of 3.7-3.9. If the rubbers are allowed to stand in moist air, this value falls, as a result of hydrolysis of the phosphorus-chlorine bonds, to 2.0-3.0. The insolubility and swelling of these polymers suggest that they are highly cross-linked structures.

The analysis of the rubber is given below and compared with that of the trimer. The zone-refined trimer gave almost the same analysis as the commercial material.

	% P	% N	% Cl
$PNCl_2$ (theory)	26.72	12.09	61.19
$(PNCl_2)_3$ trimer	26.92	12.06	61.00
$(PNCl_2)_n$ rubber	27.35	11.98	61.00

5. Use of Phosphonitrilic Chloride Rubbers

This inorganic rubber is potentially stable at high temperature, but it has not been possible to develop it commercially because of the ease with which it undergoes hydrolysis. Attempts have been made to prepare derivatives and also to graft non-polar side chains onto the rubber (4), but no satisfactory material appears to have been produced so far.

6. References

1. Adhesive Tapes Ltd., Borehamwood, Herts, England.
2a. American Cyanamid Company, Stamford Research Laboratories, Stamford Conneticut 06904.
2b. Current Address: Pennsylvania State University, University Park, Pennsylvania 16802.
3. J. O. Konecny and C. M. Douglas, *J. Polym. Sci.*, **36**, 195 (1959).
4. M. W. Spindler and R. L. Vale, *Makromol Chem.*, **43** (3), 237 (1961).

Poly[1,3-bis(p-carboxyphenoxy)-propane anhydride]

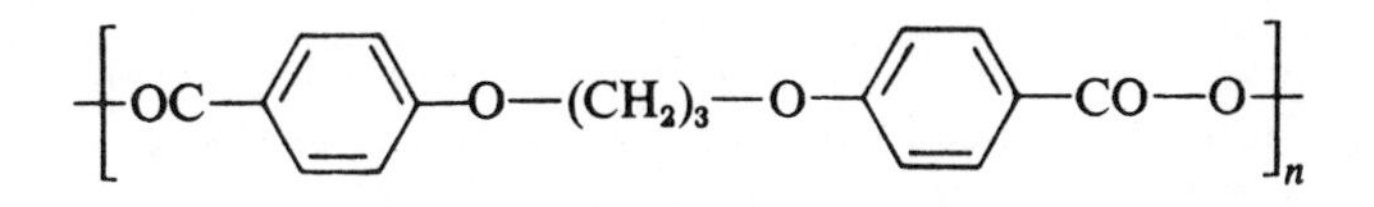

Submitted by A. Conix (1)
Checked by P. W. Morgan and S. L. Kwolek (2)

1. Procedure

1,3-Bis(p-carboxyphenoxy)propane.

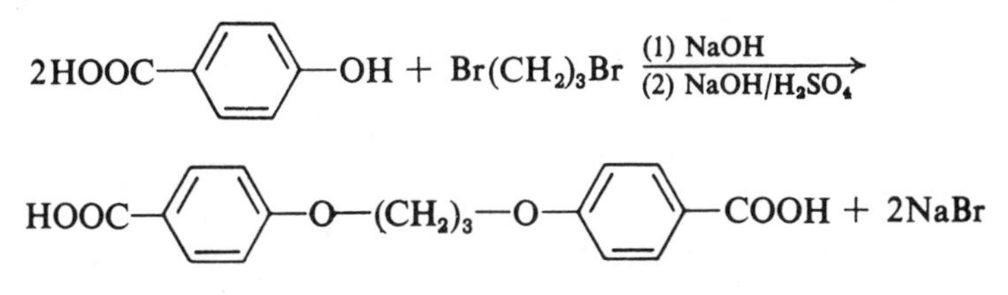

In a 1-l. three-necked flask equipped with a stirrer, a condenser, and a dropping funnel is placed a solution of 138 g. (1.0 mole) of *p*-hydroxybenzoic acid (Note 1) and 80 g. (2.0 moles) of sodium hydroxide in 400 ml. of water. Through the funnel, 102 g. (0.5 mole) of 1,3-dibromopropane (Notes 2 and 3) is added over a period of 1 hr. while the contents of the flask are stirred and kept at reflux temperature. After the addition of the 1,3-dibromopropane the reaction mixture is refluxed for 3.5 hr. Then 20 g. (0.5 mole) of solid sodium hydroxide is added to the mixture, which is refluxed for an additional 2 hr. Heating is discontinued, and the reaction mixture is left to stand overnight. The fine, powdery, white precipitate of the disodium salt is isolated by filtration and washed with 200 ml. of methanol. The still wet precipitate is dissolved in 1 l. of distilled water. The solution is warmed to 60-70° and acidified with 6*N* sulfuric acid. While the mixture is still warm, the dibasic acid is isolated by filtration and dried in a vacuum oven at 80° (Note 4). The yield is 79 g. (50%). The neutralization equivalent is 157 (calc. 158).

The Mixed Anhydride of 1,3-Bix(p-carboxyphenoxy)propane and Acetic Acid.

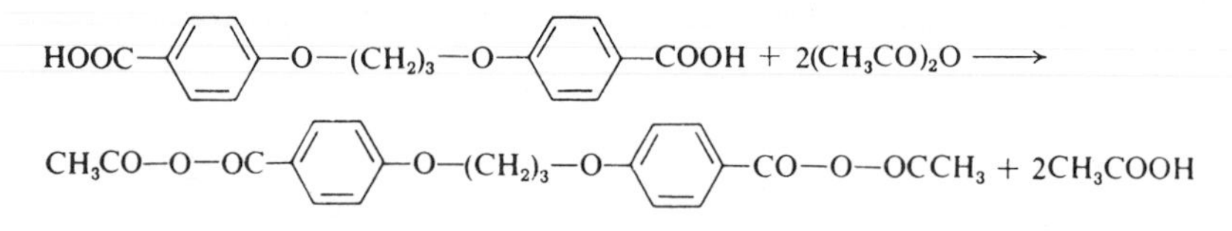

In a 1-1. three-necked flask fitted with a stirrer, a condenser, and a gas inlet tube are placed 60 g. (0.19 mole) of 1,3-bis(*p*-carboxyphenoxy)propane and 650 ml. of acetic anhydride (Note 5). A slow stream of dry nitrogen (Note 6) is bubbled through the mixture, which is refluxed. After 30 min. almost all the dibasic acid is dissolved. The hot mixture is filtered, and the slightly yellow-colored filtrate is concentrated to a volume of about 150 ml. by distilling acetic anhydride (contaminated by acetic acid) under vacuum at a temperature not higher than 65°. The remaining reaction mixture is stored in a refrigerator overnight. The white needle-like crystals formed are isolated by filtration, washed with dry ether, and dried in a vacuum oven at 70°. The yield is 66 g. (87%), and the melting point is 102-103°.

Poly[1,3-bis(p-carboxyphenoxy)propane anhydride].

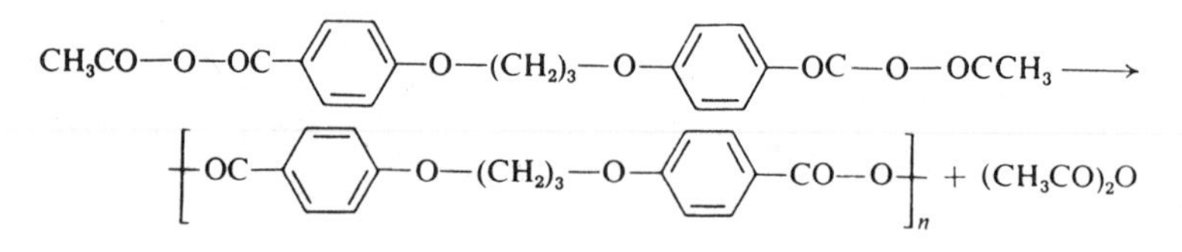

In a glass polymerization tube equipped with a side arm for distillation is placed 20 g. of the mixed anhydride of 1,3-bis(*p*-carboxyphenoxy)propane and acetic acid. A capillary reaching to the bottom of the tube is inserted. The polymerization tube is heated at 282° in a dimethyl phthalate vapor bath, brought to that temperature after the tube is in position. Dry nitrogen is passed through the mixture and acetic anhydride distills.

After 30 min. a vacuum of about 1 torr, or less, is applied. A slow stream of nitrogen is continually passed through the melt, which becomes more and more viscous. Periodically, the vacuum may be released and a strong current of nitrogen flushed through the viscous melt for additional mixing. After 30 min.

the polycondensation is terminated. The tube is then filled with nitrogen and removed from the vapor bath.

Caution! On cooling, adhesion of the polymer to the walls of the vessel and shrinkage during crystallization may cause the tube to shatter. As a safety measure the cooling tube should be handled with leather gloves, wrapped in a cloth towel, and kept behind a shield. If the tube does not shatter spontaneously, it is cracked and separated from the mass of polymer by tapping with a hammer while within a towel or other protective device, and the last traces of glass are removed by filing.

The polymer is obtained as a yellowish, opaque, hard block, which can be further crystallized by annealing at 130° in an oven for about 30 min. The crystalline melting point (hot stage polarizing microscope) is about 267°. From the melt, yellowish lustrous fibers can be drawn which show the typical phenomenon of cold-drawing (Notes 7 and 8). The intrinsic viscosity measured in a 60:40 phenol-tetrachloroethane solution is in the range 0.1-0.3 dl./g.

2. Notes

1. Commercial grade *p*-hydroxybenzoic acid (m.p. 216°) was used.
2. Commercial grade distilled 1,3-dibromopropane (b.p. 167°) was used.
3. The 1,3-dibromopropane may be replaced by 1,3-dichloropropane; the same procedure is used but longer reaction time (6 hr.) is required. The yield remains approximately the same.
4. Filtration of the diacid requires 2-5 hr. Care should be taken to filter the mixture while it is still warm, otherwise filtration requires even more time.
5. The acetic anhydride must be of highest purity. Reagent grade may be used directly, but a commercial product (b.p. 135-140°) should be fractionated, using a Todd distillation assembly with a 90-cm. column. The fraction at 140-142° should be used.
6. Pure nitrogen, containing a maximum of 5 p.p.m. oxygen, is dried with $CaCl_2$.
7. The procedure described is a general one for the preparation of aromatic polyanhydrides. Conix (4) has described polymers of the general formula

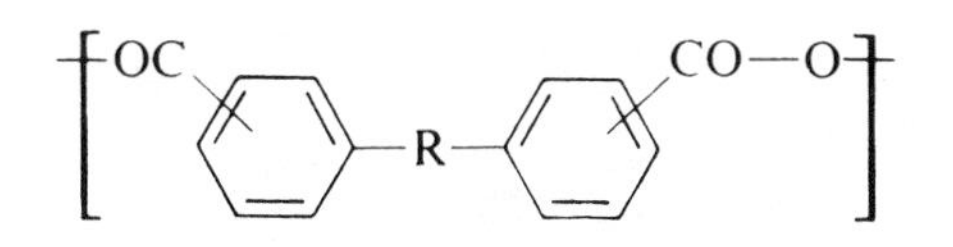

where the carbonyl functions are in the para- or meta-position and R represents an aliphatic residue, such as $-O-(CH_2)_n-O$ and $-(CH_2)_n-$ (n = 1-6).

8. A similar procedure was used by Yoda (5) to prepare a series of aromatic, heterocyclic, and heterochain polyanhydrides and mixed polyanhydrides.

3. Methods of Preparation

Poly(terephthalic acid anhydride) has been prepared by partial hydrolysis of terephthaloyl chloride with stoichiometric amounts of a hydrolysis agent (5); mixed polyanhydrides of regular structure were prepared by the reaction of acid chlorides on acids in pyridine-ether medium (6). An extensive review of aliphatic aromatic and heterocyclic polyanhydrides has been published (8).

4. References

1. Gevaert-Agfa N. V., Mortsel (Antwerp), Belgium.
2. Pioneering Research Division, Textile Fibers Department, E. I. du Pont de Nemours & Co., Inc., Wilmington, Delaware 19898.
3. W. K. Sorenson and T. W. Campbell, *Preparative Methods of Polymer Chemistry*, Interscience Division, John Wiley and Sons, New York, 1961, p. 141.
4. A. Conix, *Makromol. Chem.*, **24**, 76 (1957); *J. Polym. Sci.*, **29**, 343 (1958).
5. N. Yoda, *Makromol. Chem.*, **32**, 1 (1959); **55**, 174 (1962); **56**, 10, 36 (1962); *Chem. High. Polymers (Tokyo)*, **19**, 495, 553, 603, 613 (1962); *J. Polym. Sci.*, **A1**, 1323 (1963).
6. Gevaert Photo-Producten N.V., Belg. Pat. 545,313 (1956) [*C.A.*, **54**, 20316 (1960)].
7. N. Yoda and A. Miyake, *Bull. Chem. Soc. Japan*, **32**, 1120 (1959).
8. N. Yoda, *Encyl. Polym. Sci. Technol.*, Vol 10, Interscience Division, John Wiley and Sons, New York, 1969, pp. 630-653.

Polynorbornene
Polybicyclo[2.2.1]hept-2-ene

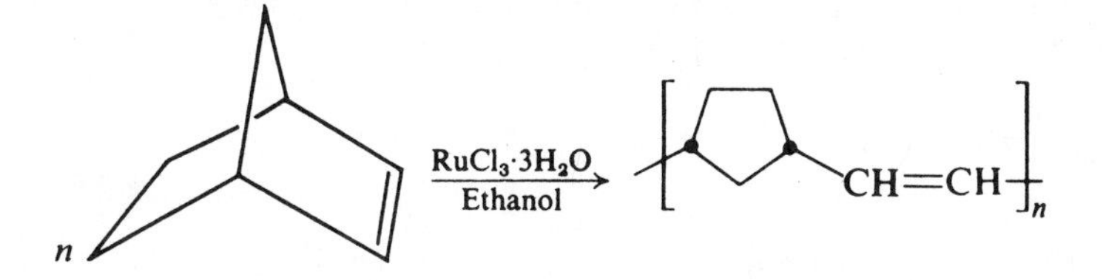

Submitted by F. W. Michelotti (1a, b) and W. P. Keaveney (1a)
Checked by J. J. Pellan and R. L. Kugel (2)

1. Procedure

A 500-ml. one-necked flask, fitted with a reflux condenser, is charged with 1 g. of hydrated ruthenium trichloride (Note 1), 50 g. (0.53 mole) of norbornene (Note 2), and 200 ml. of Synasol (Note 3). The resulting highly-colored solution is magnetically stirred and heated at reflux for 6 hr. (Note 4), cooled, and the supernatant liquid decanted. The dark mass remaining in the flask is washed several times with Synasol until the wash liquid is colorless. The product is pressed free of excess solvent as rapidly as possible to avoid undue exposure to air, and is then dissolved in several portions of hot benzene under a nitrogen blanket to prevent oxidation of the polymer (Note 5). The dark viscous solution is then poured with stirring into a large volume of methanol, from which the product immediately precipitates as a greenish-grey mass. The polymer is reprecipitated twice more, again taking precautions to exclude oxygen from the polymer, and is then vacuum-dried under nitrogen for 24 hr. The yield of purified polymer is about 30 g. (60%).

2. Characterization

Analysis. Calcd. for $(C_7H_{10})_n$: C, 89.29; H, 10.71. Found: C, 88.50; H, 11.23.

The infrared spectrum of the product (film from benzene dried at 100° and 0.05 torr) showed bands at 3.45 μ and 3.53 μ (S), 6.85 and 6.93 μ (M), and at 10.37 μ (S), and 13.5 μ (W) (Note 6). Bromination of this polymer in carbon tetrachloride resulted in an insoluble material, the spectrum of which showed no absorption in the 10.37 μ and 13.5 μ regions, respectively. The intrinsic viscostiy [η] of polynorbornene in benzene at 25° was 0.69 dl./g.; melting range (Fisher-Johns melting point block) 72-90°. A sample containing small amounts of Ional (Note 7) was milled and then pressed at 82° to a pliable plastic sheet with a tensile strength of 1275 p.s.i.

A sample of pure polymer was ozonized, hydrolyzed, and treated with *p*-bromophenacylbromide. The only isolable product was the bis-*p*-bromophenacyl ester of *cis*-1,3-cyclopentanedicarboxylic acid.

3. Notes

1. Ruthenium trichloride trihydrate, 38.60% ruthenium (theory, 38.79%) was obtained from Engelhard Industries, Inc. The corresponding osmium and iridium salts were also found effective for the polymerization of norbornene under identical conditions. The relative rates of polymerization were in the order Ir > Os > Ru.

2. Norbornene was obtained from the Aldrich Chemical Company, and was distilled prior to use, b.p. 95-96°/760 torr, single peak in VPC.

3. Synasol is the trade name for Union Carbide's denatured ethanol; 95% ethanol may also be used. The checkers employed absolute alcohol, and after heating at reflux for 1.5 hr. found that the growth of the polymer had apparently ceased. The polymer was in the form of a large white ball, having completely absorbed all the liquid in the mixture. The flask was broken to remove the polymer which was then cut into small pieces (1 x ½ in.) and kneaded with absolute alcohol until the alcohol was free of color.

4. As the polymerization progresses, a spongy precipitate forms. Heating is terminated when the growth of this precipitate apparently ceases or all the liquid is imbibed by the polymer (Note 3).

5. The process of dissolving the polymer requires about a day.

6. The salient features of this spectrum were the absorption bands at 10.37 μ and 13.5 μ, which indicated the presence of *trans* and *cis* double bonds, respectively, in the polymer.

7. This antioxidant is available from Shell Chemical Corporation.

4. Methods of Preparation

Polynorbornene has also been prepared by a Ziegler-type catalyst system derived from lithium aluminum tetraheptyl and titanium tetrachloride at room temperature (3).

5. References

1a. Interchemical Corporation, Central Research Laboratories, Clifton, New Jersey 07015
1b. Current Address: J. T. Baker Chemical Co., Phillipsburgh, Pennsylvania 16866.
2. American Cyanamid Company, Stamford, Connecticut 06904.
3. W. L. Truett, D. R. Johnson, I. M. Robinson, and B. A. Montague, *J. Amer. Chem. Soc.,* **82**, 2337 (1960).

Suspension Polymerization of Vinyl Chloride

$$nCH_2{=}CHCl \xrightarrow[\text{Peroxide}]{\text{Gelatin}} \left[-CH_2-\underset{H}{\overset{Cl}{\underset{|}{\overset{|}{C}}}}- \right]_n$$

Submitted by C. L. Sturm (1a) and I. Rosen (1a, b)
Checked by J. H. Dunn (2)

1. Procedure

In a 500-ml. beaker 0.3 g. of type A high bloom gelatin (Note 1) is slowly added to 300 ml. of distilled water with stirring. The mixture is heated to 70-80° to dissolve the gelatin. After cooling, 0.10 g. of ammonium bicarbonate is added.

Into a 1-qt. Duraglas beverage bottle (Note 2) are charged 0.25 g. of Alperox-C (*Caution! Peroxides are strong oxidants.*) (Note 3) and the prepared gelatin solution. The bottle is stoppered and cooled in a deep freeze unit to freeze the gelatin-catalyst mixture (Note 4). To the frozen mixture is slowly added slightly more than 100 g. of vinyl chloride (*Caution!*) (Note 5). The vinyl chloride in excess of 100 g. is allowed to evaporate, sweeping out the air in the bottle. The bottle is sealed with an aluminum foil lined cap using a beverage bottle capper. The bottle is placed behind a safety shield and allowed to thaw (Note 6).

After thawing, the bottle is agitated, or tumbled end over end in a bottle polymerizer at 50° for 14-18 hr. (Note 7). The bottle and its contents are cooled and the unpolymerized monomer is carefully vented with a hypodermic needle inserted through the cap. The polymer is filtered in a Büchner funnel, washed with hot distilled water during the filtration, and dried overnight at 40-50°. The conversion is 85-95% of granular white polymer.

The polymer produced has an intrinsic viscosity in the range 1.0-1.2 dl./g., measured in cyclohexanone at 25° (Note 8). The infrared spectrum and x-ray diffraction pattern of the polymer are similar to those reported for conventional PVC (3).

2. Notes

1. This gelatin is available from Atlantic Gelatin Company. In place of the gelatin-ammonium bicarbonate combination, 0.5 g. of Elvanol 50-42 may be used, with about equivalent results, except that a fine polymer powder is produced.

2. Duraglas bottles, available from the Owens-Illinois Glass Company, are able to withstand a pressure of 100-150 p.s.i.

3. Alperox-C, or lauroyl peroxide, is available from the Lucidol Division of Wallace and Tiernan, Inc.

4. Bottle breakage during freezing of the contents can be minimized by tipping the bottle at a 45° angle during the freezing operation.

5. About a 10% excess of vinyl chloride is recommended. to ensure good purging. The vinyl chloride may be distilled from shipping cylinders, condensed with a Dry Ice-acetone "cold finger" condenser, and collected in a flask cooled with Dry Ice. Rubber tubing should be avoided because it will lead to inhibition of the polymerization. Polyethylene or Teflon tubing is satisfactory.

Caution! Vinyl chloride boils at −14.6° and is highly flammable. All work should be conducted in a well-ventilated hood. Vinyl chloride is also a suspected carcinogen (5).

6. *Caution! Placing a bottle with frozen contents in a warm bath can cause a cracked bottle and an explosion. The bottle should be kept in a perforated metal cage or wrapped in wire mesh to contain glass fragments in the event that the bottle breaks.*

7. The purity of the vinyl chloride will affect the length of the polymerization period. The time for polymerization may be reduced by increasing the catalyst concentration, but with a concomitant small reduction in polymer molecular weight.

Too slow or too rapid agitation may lead to agglomeration. The submitters' 1-qt. bottle polymerizer rotates at 24 r.p.m. The checker's 1-pt. bottle polymerizer rotates at 32 r.p.m.

8. The molecular weight of the polymer may be estimated from the expression $[\eta] = 2.4 \times 10^{-4} M_n^{0.77}$, where $[\eta]$ is in deciliters per gram (4).

3. References

1a. Diamond Shamrock Corp., Painesville, Ohio 44077.
1b. Current Address: Standard Oil Co., Cleveland, Ohio 44128.
2. Ethyl Corporation, Baton Rouge, Louisiana 70821.
3. G. Natta and P. Corradini, *J. Polym. Sci.,* **20**, 251 (1956); S. Krimm, *Fortschr. Hochpolymer.-Forsch.,* **2**, 51 (1960); E. Treiber, W. Berndt, and H. Toplak, *Angew. Chem.,* **67**, 69 (1955).
4. F. Danusso, G. Moraglio, and S. Gazzera, *Chim. Ind. (Milan),* **36**, 883 (1954).
5. *Federal Register,* Vol 39, Oct. 4, 1974, p35890 OSHA "Regulation on Exposure to Vinyl Chloride".

Thermosetting Acrylic Polymer

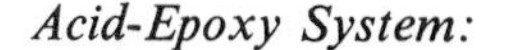

Acid-Epoxy System:

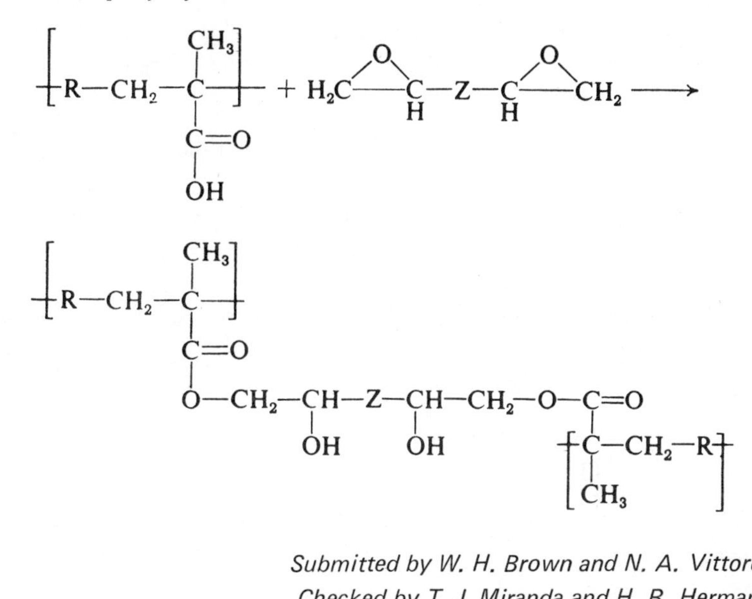

Submitted by W. H. Brown and N. A. Vittore (1)
Checked by T. J. Miranda and H. R. Herman (2)

1. Procedure

A 1-l. three-necked flask is equipped with a stirrer, nitrogen inlet, thermometer, and reflux condenser. Heating is accomplished with a Glass-Col heating mantle. The flask is charged with 191 g. of dry distilled xylene and heated to 136° under a slow stream of nitrogen. A mixture of 135 g. of methyl methacrylate (Note 1), 135 g. of ethyl acrylate, 30 g. of methacrylic acid, 6 g. of di-*t*-butyl peroxide (*Caution! Peroxides are strong oxidants.*) (Note 2), and 3 g. of *t*-dodecyl mercaptan (*Stench!*) (Note 3) is added over a period of 1.75 hr. The temperature is controlled between 136° and 143° by adjustment of the mantle temperature or by blowing an air stream over the surface of the flask. After the monomer mixture has been added, the temperature is maintained at 136-143° for an additional 5 hr., cooled to 65°, and 100 g. of anhydrous ethanol is then added.

2. Characterization

The polymer solution has a solids content of 51.3%, a Gardner-Holdt viscosity of U+, and a specific gravity of 0.988. The acid number of the polymer is 63.8. The acrylic polymer was cast on glass and dried in a circulating air oven at 150°. The film was removed by scraping and taken up in chloroform, filtered, and a number average molecular weight of 6780 was obtained (Note 4).

3. Conversion of Acrylic Polymer to Thermosetting Acrylic

Thermosetting films may be obtained by mixing 100 g. of the polymer solution with 10.9 g. of Epon 828 (Note 5); 0.6 g. of DMP-30 (Note 6) may be added to catalyze the reaction. Films are cast on tin plate and baked for 30 min. at 150°. The cured films were removed by immersing the tin plate in a mercury (*Caution! Mercury vapor is hazardous.*) bath.

The physical characteristics of the cured film were:

Tensile strength	5520 p.s.i.
Elongation	4.1% (Note 7)

4. Notes

1. Monomers used without removal of inhibitors:

Methyl methacrylate	10 p.p.m. MEHQ (Rohm & Haas Company)
Ethyl acrylate	200 p.p.m. MEHQ (Rohm & Haas Company)
Methacrylic acid	100 p.p.m. MEHQ (Rohm & Haas Company)

2. Di-*t*-butyl peroxide from Shell Chemical Corporation was used.
3. The *t*-dodecyl mercaptan was obtained from Pennsalt Chemicals Corporation.
4. This figure was determined on the Vapor Pressure Osmometer Model 301A of Mechrolab, Inc.

5. Epon 828 (epichlorohydrin-bisphenol A epoxy resin), a product of the Shell Chemical Corporation, was used.

6. DMP-30 [2,4,6-*tris*-(N,N′-dimethylaminoethyl)phenol] was obtained from the Rohm and Haas Company.

7. By partially substituting ethyl acrylate with 2-ethylhexyl acrylate the following data are obtained (3) (see also Note 8):

Composition	Wt. %	Wt. %
Methyl methacrylate	45	45
2-Ethylhexyl acrylate	20	10
Ethyl acrylate	25	35
Methacrylic acid	10	10
Tensile strength, p.s.i.	3635	5750
Elongation, %	24.1	5.0

8. Variations of the system in Note 7 can be made by substituting other functional monomers:

A 1-l. three-necked flask is equipped with stirrer, nitrogen inlet, thermometer, and reflux condenser. Heating is accomplished with a Glass-Col heating mantle. The flask is charged with 245 g. of distilled dry xylene and heated to 136° under a slow stream of nitrogen.

A mixture of 135 g. of methyl methacrylate (Note 1), 100 g. of 2-ethylhexyl acrylate, 25 g. of hydroxyethyl methacrylate, and 5.0 g. of di-*t*-butyl peroxide (Note 2) is added over a 1-hr. period. The temperature is controlled by adjustment of the mantle temperature or by air blowing and held between 136° and 140°.

After the monomer mixture has been added, the temperature is maintained at 136-140° for 2 hr. and cooled.

Characterization. The polymer has a solids content of 50%, a Gardner-Holdt viscosity of W^+, specific gravity of 0.978, and a number average molecular weight of 10,000 (Note 4).

Thermosetting films may be prepared by dissolving 12.5 g. of Cymel 301 (Note 9) in 100 g. of polymer solution, casting films on a tin plate, and baking for 30 min. at 150°

The physical characteristics of the film are:

Tensile strength	4766 p.s.i.
Elongation	4.45%

9. Cymel 301, a liquid form of hexakis-methoxymethyl melamine, was obtained from the American Cyanamid Company.

5. References

1. DeSoto Chemical Coatings, Inc., Chicago, Illinois 60018.
2. The O'Brien Corporation, South Bend, Indiana 46624.
3. W. H. Brown and T. J. Miranda, *Offic. Dig. J. Paint Technol. Eng.*, 36, No. 475, Pt. 2, 92 (1964).

Diels-Alder Polymer of 2,5-Diemethyl-2,4-diphenylcyclopentadieneone Dimer and N,N'-4,4'-3,3'-Dimethylbiphenyl-bis-maleimide

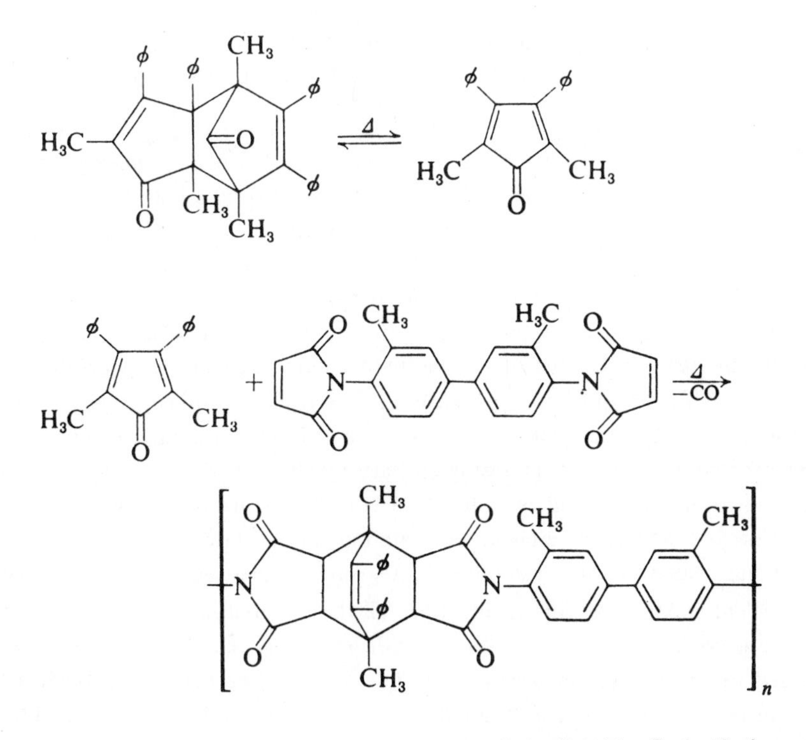

Submitted by E. A. Kraiman (1)
Checked by C. G. Overberger (2a, b) and I. Cho (2a, c)

1. Procedure

2,5-Dimethyl-3,4-diphenylcyclopentadieneone Dimer. Benzil (84 g., 0.4 mole) and diethyl ketone (68.8 g., 0.8 mole) are added to a stirred solution of 10 g. of sodium hydroxide in 250 ml. of absolute methanol at room temperature. Stirring at ambient conditions is continued for 1 hr. The solution is then poured into about 1 l. of cold water with stirring to precipitate the reaction product which is then separated by filtration, washed with water, and dried. The yield of the crude intermediate 2,5-dimethyl-3,4-diphenylcyclopenten-2-ol-4-one-1 is 110 g. (99% of theory) (Note 1).

The crude intermediate (35 g.) is added to a stirred mixture of 50 g. of acetic anhydride and 1.2 g. of concentrated sulfuric acid. A strong mechanical stirrer is necessary for efficient mixing. Ice and water cooling is applied immediately after the addition to control the temperature which ranges from 25° to 45°. When the reaction temperature has returned to the original ambient temperature (25°), about 200 ml. of water is added to decompose the excess acetic anhydride. After stirring for about 1 hr. at room temperature (cool if required), the resulting precipitate is separated by filtration, washed with cold methanol, and recrystallized from *n*-propanol. The yield ranges from 20 to 27.5 g. (60-83% of theory), m.p. 190-191°.

N,N'-4,4'-(3,3'-dimethylbiphenyl)-bis-maleimide. (*Caution! This material may be toxic and appropriate care should be exercised.*) *o*-Tolidine (100 g., 0.47 mole) (Note 2) is added to a solution of 98 g. (1.0 mole) of maleic anhydride (*Caution! Maleic anhydride is toxic. Do not inhale dust from, or allow skin contact with, this material.*) in 1 l. of methylene chloride (*Caution! Toxic vapor.*) with stirring at the reflux temperature. The rate of addition is regulated to maintain steady refluxing of the methylene chloride. Stirring and refluxing is continued for 30 min. after the addition. The resulting solid is separated by filtration, washed with methylene chloride, and dried. This intermediate *bis*-maleamic acid is obtained in almost quantitative yield.

All the *bis*-maleamic acid is added to a stirred mixture of 500 g. of acetic anhydride and 100 g. of potassium acetate (sodium acetate gives much poorer yields) and heated rapidly to about 80°. A strong mechanical stirrer is necessary for efficient mixing. At this temperature all the materials are in solution. The mixture is cooled to room temperature with ice water, and about 1 l. of water is added to decompose the excess acetic anhydride. After stirring for about 1 hr. at room temperature (cool if required), the resulting precipitate is separated by filtration, washed with water, dried, and then recrystallized twice from *s*-

tetrachloroethane. (*Caution! Toxic vapor.*) A yield of 116 g. (62% of theory) is obtained. This compound does not melt when heated gradually to 500°. A melting point of 286-289° can be obtained if a capillary is inserted in a preheated melting point apparatus (Note 3).

Polymer. A mixture of 32.5 g. (0.0625 mole) of 2,5-dimethyl-3,4-diphenylcyclopentadieneone dimer, 46.5 g. (0.125 mole) of N,N′-4,4′-(3,3′-dimethylbiphenyl)-*bis*-maleimide, and 250 ml. of α-chloronaphthalene is stirred and heated in a flask equipped with a large-bore (6-mm.) stopcock on the bottom. The temperature is raised to the reflux temperature (about 260°) as rapidly as possible and held at reflux for about 20 min. (Note 4). The stopcock is then opened to drain the reaction mixture into a 1-gal. explosion-proof blender containing about 3 l. of cold (0-5°) methanol, and running at maximum speed (Note 5). About 25 ml. of hot α-chloronaphthalene is used to wash any remaining polymer from the flask into the blender. The precipitated polymer is separated by filtration on a coarse fritted glass filter, washed several times with fresh methanol, and dried in an oven at 100° for 12 hr.

The polymer (an almost colorless, fluffy powder) amounts to 74.5 g. (99% of theory). The reduced viscosity is 1.61 in dimethylformamide at 25° (Note 6).

2. Notes

1. This crude product can be recrystallized from cyclohexane using a small amount of activated charcoal to obtain needle-like crystals melting at 113-115° in a yield of 83 g or 71% of theory.

2. The *o*-tolidine should be reasonably fresh material. If it is not, it can be purified by recrystallization from methanol using activated charcoal.

3. A modified process using N,N-dimethylformamide in the second step with acetic anhydride is described in a patent (3). Yields in the 80% range in the case of *m*-phenylene-*bis*-maleimide are reported.

4. At about 150° there is a noticeable evolution of carbon monoxide. After 5-10 min. at the reflux temperature there is a visible increase in viscosity. The reaction is stopped before the viscosity becomes too great to allow stirring and draining. The molecular weight of the polymer obtained will depend on the time at which the reaction is stopped as well as the purity of the monomers.

5. The reaction mixture is drained into a small opening in the stainless steel cover of the mixing jar. The cold methanol absorbs the heat of the reaction mixture.

6. The reduced viscosity is determined on a solution of about 0.4 g. of polymer in 100 ml. of dimethylformamide at 25°, using a Cannon-Fenske viscometer.

3. Methods and Merits of Preparation

No other methods of preparation for this particular polymer have been described. This method (4) has been used for similar polymers that have also been prepared by polymerizing *bis*-maleimides with α-pyrones (5) and thiophene dioxides (6).

4. References

1. Sun Chemical Corp., Carlstadt, New Jersey 07072.
2a. Polytechnic Institute of Brooklyn, Brooklyn, New York 11201.
2b. Current Address: Dept. of Chemistry, The University of Michigan, Ann Arbor, Michigan 48104.
2c. Current Address: The Korea Advanced Institute for Science, Seoul, Korea.
3. H. N. Cole and W. F. Gruber, U.S. Pat. 3,127,414 (March 31, 1964).
4. E. A. Kraiman, U.S. Pat. 2,890,206 (June 9, 1959).
5. E. A. Kraiman, U.S. Pat. 2,890,207 (June 9, 1959); Sui-Wu Chow, U.S. Pat. 3,074,915 (January 22, 1963).
6. S. W. Chow and J. M. Whelan, Jr., U.S. Pat. 2,971,944 (February 14, 1961).

Poly-p-xylylidene

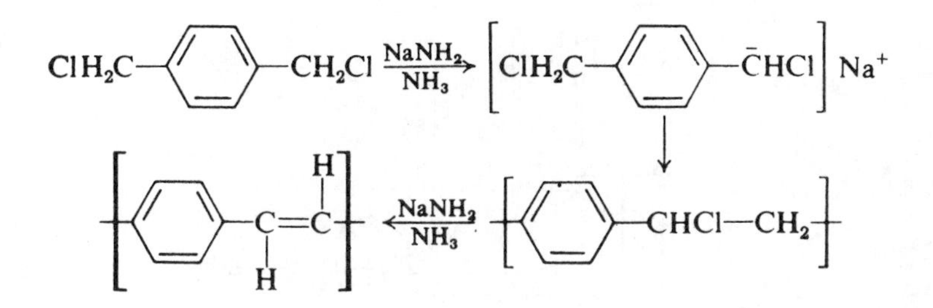

Submitted by D. F. Hoeg, D. I. Lusk, and E. P. Goldberg (1)
Checked by J. M. Hoyt and D. R. Fitch (2)

1. Procedure

The equipment used to carry out this preparation is shown in Fig. 1. The glass equipment is flamed gently under a purge of dry argon and allowed to cool under an argon atmosphere. A Dry Ice-acetone bath is then placed around the 500-ml. three-necked flask, and Dry Ice and acetone added to the reflux condenser. Liquid ammonia (300 ml.) is then added *via* the flexible connection shown (Note 1). Small pieces of sodium metal are dropped into the liquid ammonia until the blue color of dissolved sodium persists (Note 2). The temperature is raised by lowering the cooling bath enough to just maintain the ammonia near reflux conditions (−35°). Thereafter 4.8 g. (0.21 g. atom) of sodium is added (Note 3), and, after dissolving, a few crystals of ferric nitrate (Note 4) are dropped into the reaction to catalyze the formation of sodium amide. Within 1 hr. the steel-grey suspension of sodium amide has formed.

To this suspension α,α'-dichloro-*p*-xylene (Note 5) (17.5 g., 0.1 mole), dissolved in 80 ml. of dry tetrahydrofuran (*Caution! Toxic vapor.*) (Note 6) is added dropwise through the dropping funnel (Note 7). The reaction mixture becomes dark purple and quite thick. After 4 hr. the viscous, purple reaction

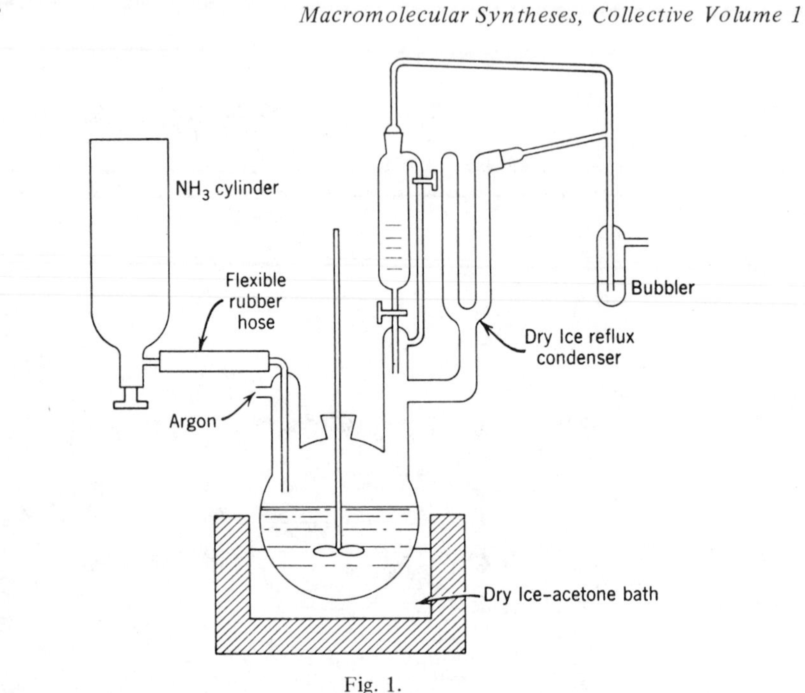

Fig. 1.

slurry is quenched with ammonium chloride and approximately 50 ml. of methanol. After evaporating the ammonia (Note 8), the yellow residue is triturated repeatedly with methanol, aqueous acid, aqueous base, and water until the filtrate is neutral (Note 9). It is then triturated repeatedly with benzene and then methanol, filtered, and dried. The product, approximately 9.2 g. (90% yield) (Note 10), is bright yellow and fluoresces brilliantly green-yellow under ultraviolet illumination. Yields are generally between 90 and 98% of theory. The shift in the fluorescence from the blue observed with stilbene to longer wavelengths observed with this product is consistent with previous sutdies of small molecules containing *p*-linked xylylidene groups and indicative of an extension in the degree of conjugation (3).

Analysis. Elemental analysis gave the following results (wt. %): C, 88.3; H, 6.10; N, 1.9; Cl, 2.7. In view of the chemistry of the synthesis, the structure that gives the best fit to these data is

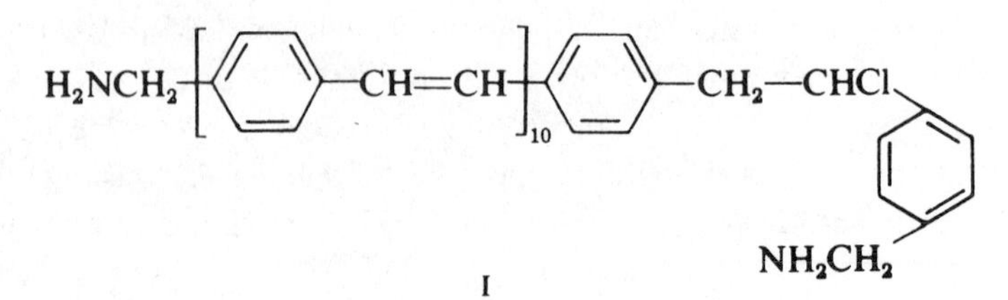

I

(Calcd. for I: C, 89.0; H, 6.14; N, 2.2; Cl, 2.7). The chain-limiting reaction is postulated to be the slow but competitive aminolysis of the chloromethyl end groups on the growing macromolecule. The chlorine content is thought to result from incomplete dehydrochlorination.

The poly-*p*-xylylidene resin is infusible and insoluble in all the solvents at their boiling points (including those found best for linear poly-*p*-xylylene, such as boiling Aroclor 1248 and benzyl benzoate). Preliminary electrical property measurements have also detected a low level of photoconduction. The infrared spectrum exhibited as its main distinguishing features absorption at 825 cm^{-1} (*p*-disubstituted benzene rings) and 960 cm^{-1} (*trans*-ethylenic unsaturation).

Bromination was accomplished by shaking a small amount of the xylylidene oligomer (I) with bromine/CCl_4 (~24 hr.). The product was cream-colored. *Analysis* (wt. %): C, 47.2; H, 3.4; N, 1.3; Br 45. Calc. for oligomer (I): C, 40; H, 2.7; N, 1.0; Br, 55. Moreover, the infrared spectrum indicated almost complete disappearance of the 960 cm^{-1} band. Iodine monobromide addition occurred slowly and gave results similar to the bromination and indicated the presence of approximately 8 to 10 xylylidene units per chain.

2. Notes

1. The checkers distilled the ammonia into the reaction flask because direct liquid transfer carried some foreign solid material into the reaction flask.
2. A stable blue color is obtained with approximately 0.1 g. of sodium.
3. The sodium was added over a period of 5-15 min.
4. $Fe(NO_3)_3 \cdot 9H_2O$ (0.10 g.) was used by the checkers. Ferric chloride, ca. 0.1 g., works as well.
5. The submitters found unpurified Eastman white label α,α'-dichloro-*p*-xylene satisfactory. The checkers used material from Columbia-Southern Chemical Corporation, recrystallized from acetone-hexane, m.p. 98-101°.
6. The tetrahydrofuran was obtained from Fisher Scientific Company and further purified by refluxing over potassium hydroxide for 8 hr. under argon, followed by distillation from calcium hydride and storage over sodium wire.
7. The addition required approximately 30 min., which is included in the total 4-hr. reaction time. Good stirring at this stage is desirable.
8. The evaporation of the ammonia is rather slow. After waiting 1 hr., the checkers poured the contents of the reaction flask cautiously into 1 l. of *ice-cold* (0°) methanol in a 2-l. Erlenmeyer flask in a hood. The product was subsequently filtered.

9. Approximately 5% aqueous sodium hydroxide and 5% sulfuric acid solution is satisfactory.

10. The checkers obtained 10.3 g. (101% yield) of product containing 4.3% ash. Soxhlet extraction of this material with water for 5 hr. gave a corrected yield of 97% of polymer—emphasizing the need for good washing.

3. Methods of Preparation

Poly-*p*-xylylidene may also be prepared by the method of McDonald and Campbell via the Wittig condensation of terephthalaldehyde and *p*-xylylene-bis-(triphenylphosphonium)-chloride (4). The polymer is similar to that derived via the sodamide route with minor structural differences (5). Another new route to poly-*p*-xylylidene is from p-phenylene dimethylene bis-(diethyl sulfonium chloride) with uses described for films, fibers coatings and foams (6). Alkali metal amalgam treatment of $\alpha,\alpha.\alpha',\alpha'$,-tetrabromo-p-xylylene has also afforded the title polymer (7).

4. Merits of the Preparation

The facile formation of poly-*p*-xylylidene from α,α'-dichloro-*p*-xylene described here suggests that this method is broadly applicable to the synthesis of new xylylidene homo- and copolymer structures from a wide variety of available aromatic-bis-halomethyl compounds. Polyxylylidene structures (including several copolymers) have been prepared from α,α'-dibromo-*m*-xylene, 2,4-di-(chloromethyl)-toluene, 2,4-di-(chloromethyl)-ethylbenzene, di-(*p*-chloromethylphenyl)-ether, 2,4-di-(chloromethyl)-anisole, 4,4′-di-(chloromethyl)-biphenyl, and 9,10-di-(chloromethyl)-anthracene. Semi-conductivity and photo-conductivity studies have been reported on a series of thermoplastic poly-p-xylyidenes that have been prepared by the Witting route and by basic dehydrohalogenation of 1,4-*bis* α-chlorobenzylbenzenes. (8)

5. References

1. Borg-Warner Corp., Roy C. Ingersoll Research Center, Des Plaines, Illinois 60016.
2. U.S. Industrial Chemicals Co., Research Department, Cincinnati, Ohio 45202.
3. J. F. Schmitt, J. Boitard, M. Suquet, and P. Comoy, *Compt. Rend.*, **242**, 649 (1956).
4. R. N. McDonald and T. W. Campbell, *J. Amer. Chem. Soc.*, 82, 4669 (1960).
5. D. F. Hoeg, D. I. Lusk, and E. P. Goldberg, *J. Polym. Sci.*, **B2**, 697 (1964).
6. H. Saikachi, *Chem. Pharm. Bull.*, **19**, 959 (1971).
7. U.S. Pat. 3,706, 677 (1972).
8. Russ. Pat. 421,708 (1974).

Phenol-Formaldehyde (Resol) Resin, One-Stage, Aniline-Modified

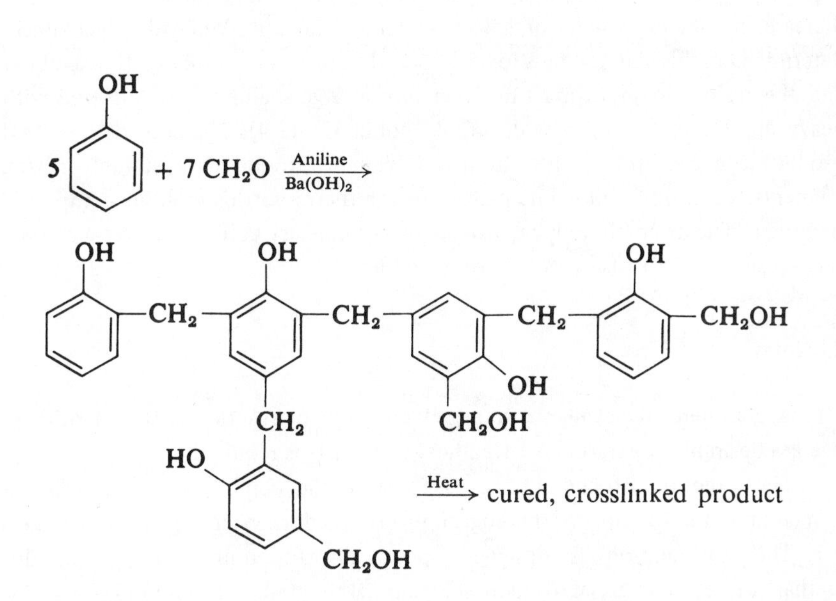

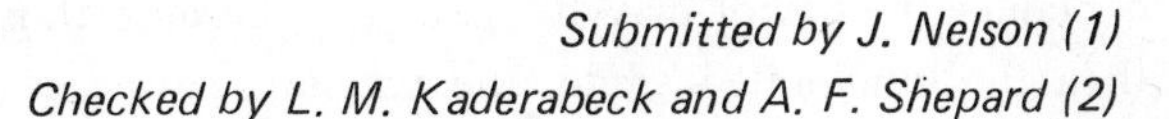
Submitted by J. Nelson (1)
Checked by L. M. Kaderabeck and A. F. Shepard (2)

1. Procedure

A 3-l., three-necked, round-bottomed flask (Note 1) equipped with a Teflon or stainless steel paddle-type stirrer about 3-4 in. long rotating at 300-500 r.p.m., thermometer, efficient bulb-type reflux condenser, and heating mantle is charged with 940 g. (10 moles) of phenol (99% purity), 70 g. (0.75 mole) of aniline (Note 2), 1130 g. of 37.2% formaldehyde solution (14 moles) (Note 3), and 110 g. of a 28.5% hot water solution of barium hydroxide octahydrate. The pressure is reduced to 300-350 torr and the reactants are heated slowly to a reflux temperature of 80° and maintained there for 50 min. The reflux condenser is then replaced with a condenser set for distillation, and the resin is dehydrated at 10-20 torr to a final temperature of 80-90°. As the dehydration proceeds, the molecular weight and viscosity of the condensate increase progressively and the resin becomes increasingly sensitive to further heating. A practical test to determine the extent of condensation is the so-called stroke cure test. A small amount of the resin (ca. 0.5 g.) is removed, spread on a metal hot plate maintained at 150° surface temperature, and stroked with a spatula until the sol-gel transition occurs. The sol-gel transition is indicated by a rapid increase in the rubberiness or gel character of the film. When the "gel time," determined in this manner, falls to 65-85 sec. the apparatus is quickly disassembled and the resin is poured in a thin layer into a large shallow vessel covered with heavy aluminum foil to provide rapid cooling (Note 4). The cooled, solidified product can be broken into lumps or can be ground with suitable fillers, pigments, lubricants, and pigments for use as a thermosetting molding compound (Note 5). The resin should be broken or ground just before use, because one-stage resins tend to fuse together on standing.

2. Notes

1. A commercially available standard resin pot or reaction kettle of suitable size is a desirable apparatus. A fitted heating mantle is required.

2. The function of the aniline is to slow the condensation, to improve grindability, and to improve rheological properties during subsequent molding.

3. This formulation is designed specifically for uninhibited, essentially methanol-free, commercial formalin solution. Other grades of inhibited, formalin-containing methanol may be used, but, because those grades are less reactive than the uninhibited grade, the reaction exotherm and final resin properties will not be the same. Normally, some increase in the weight of formalin charged would be required.

4. The resin should be transferred from the resin flask in a good hood. About 850-900 g. of aqueous distillate will have been collected at time of transfer.

5. One-stage phenol-formaldehyde resin molded products have better chemical resistance, better solvent resistance, and less odor than two-stage molded phenolic resin products. One-stage resins are often soft, and mixing with additives and fillers to make a molding compound generally requires mixing on a roll mill.

3. Methods of Preparation

A variety of alkaline catalysts may be used for preparation of one-stage resins. They include alkali metal hydroxides (NaOH, KOH, LiOH), alkaline earth hydroxides [($Ca(OH)_2$, $Ba(OH)_2$, etc.], weaker alkalies such as magnesium and zinc oxides and hydroxides, ammonium hydroxide, quaternary ammonium hydroxides, and various organic amines (alkyl and alkanol amines). Each one produces a somewhat different rate of reaction and somewhat different properties in the product.

Phenolic resins may also be prepared in essentially non-aqueous media by the use of polyformaldehyde (paraform or trioxane). Such procedures are limited in terms of the variety of catalysts that may be used, but they are more efficient in that the dehydration stage is either eliminated or greatly shortened.

Water-soluble phenolic resins are prepared by using formaldehyde in moderate excess, a strong aqueous alkali as catalyst, and a short reflux time. The dehydration step is either shortened or dispensed with entirely.

Phenolic resins suitable for casting are prepared by using a large excess (2-3:1) of formaldehyde and, an alkali metal hydroxide as catalyst. After reaction the mixture is neutralized to about pH 6, usually with a weak organic acid, and a polyol such as glycerol or glycol is added to improve the clarity of the product. Curing normally involves heating in an open mold at about 75° for extended periods of time.

Many alkylated or arylated phenols, or mixture of phenols (e.g., cresols, xylenols, phenyl phenols, alkoxyphenols) can be used to prepare phenol-formaldehyde resins. Optimum conditions for making the resins vary from one phenol (or mixture) to the next, as do their properties. If the substituents are large alkyl groups, or aryl groups, resins soluble in organic solvents can be produced.

4. References

1. General Electric Company, Plastics Department, Pitsfield, Massachusetts 01201.
2. Hooker Research Center, Niagara Falls, New York 14302

Phenol-Formaldehyde (Novolak) Resin, Two-Stage, and Molding Compound

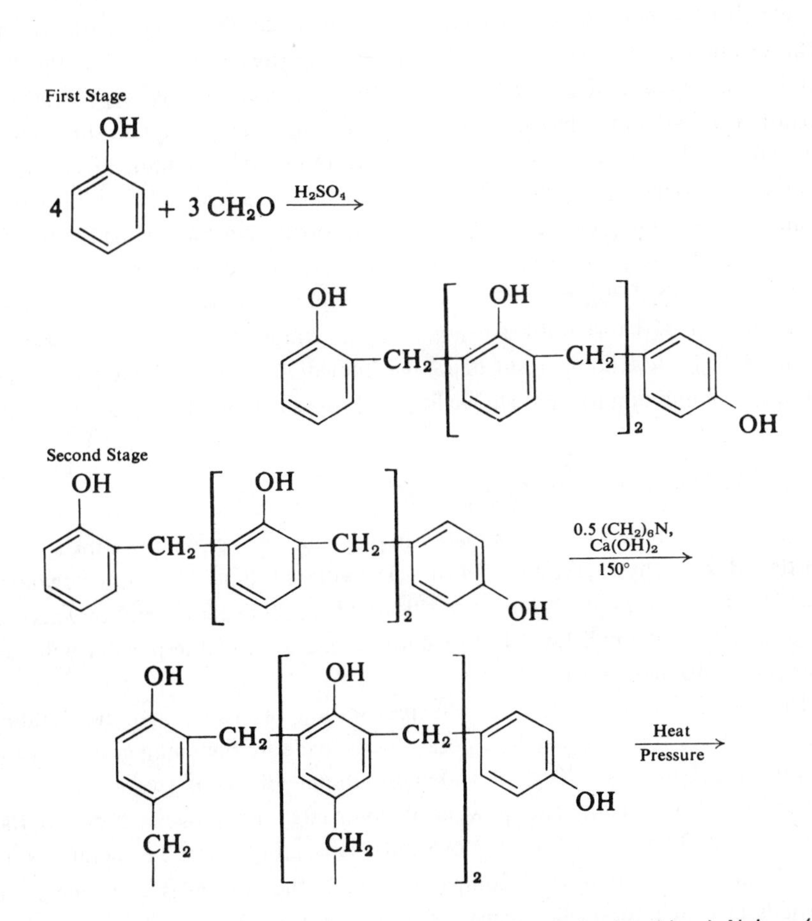

Submitted by J. Nelson (1)

Checked by L. M. Kaderabeck and A. F. Shepard (2)

1. Procedure

A. The Resin

A 3-1., four-necked, round-bottomed flask (Note 1) equipped with a Teflon or stainless steel paddle-type stirrer about 3-4 in. long set to rotate at 300-500 r.p.m., and a water-cooled, bulb-type reflux condenser, is charged with 940 g. (10 moles) of phenol (99+ % purity) and heated to 90°. Concentrated sulfuric acid, 5.6 g., is added cautiously through the dropping funnel (Note 2). Heating is continued to raise the temperature of the contents of the flask to 100-102°, at which time the mantle is removed. Formaldehyde, 630 g. of 37.2% solution (Note 3) (7.8 moles of formaldehyde), is added during 75 min. with stirring. The reaction is exothermic, and the temperature rises to 115-120° at the start of formaldehyde addition. After the exothermic reaction subsides, the heating mantle is raised and external heat is added as required to maintain the reaction at reflux. The reactants are maintained at reflux for 15 min. after all the formaldehyde has been added. The reflux condenser is replaced by a distillation condenser, and the pressure is reduced to 350 torr to remove H_2O and unreacted formaldehyde (<5% by weight). When the temperature of the reactants drops to 80°, 28 parts of 20% hydrated lime slurry is added and vacuum dehydration is continued at 350 torr until a reactant liquid temperature of 108° is reached (Note 4). The hot liquid resin is poured immediately into a shallow container lined with aluminum foil and allowed to cool to a solid, brittle resin.

B. The Molding Compound

The Novolak resin (100 g.) prepared in procedure A is ground in a mortar and pestle. Hexamethylenetetramine (16 g.), hydrated lime (1.5 g.), powdered stearic acid (0.25 g.), and aluminum silicate (2.0 g.) are added, and the mixture is ground in a ball mill for 3 hr., or until at least 95% of the powder will pass through a 140-mesh screen.

This powder can be used for the preparation of many phenolic molding compounds. A typical general-purpose type is made by ball-milling 50 g. of the powder for 3 hr. with 50 g. of 100-mesh ground soft wood (wood flour) and 1.5 g. of stearic acid. The powder is compacted by pressing between flat platens at 2000 p.s.i. in a Carver or similar laboratory press. Temperatures up to 70° may be used to assist compaction. After the powder is preformed, the sheet is broken into smaller pieces of about 1/8 in. size. This stock may then be compression-molded at 165° and 2000 p.s.i. for 5 min.

2. Notes

1. A commercially available standard resin pot or reaction kettle of suitable size is a desirable apparatus to use, if available. If used, an appropriate heating mantle is also required.

2. Sulfuric acid is added at 90° to assure formation of phenolsulfonic acid, a soluble, active catalyst.

3. Formulation is designed specifically for uninhibited, essentially methanol-free, commercial formalin solution. Other grades of inhibited formalin containing methanol may be used but, because those grades are less reactive than the uninhibited grade, the reaction exotherm and final resin properties will not be the same. Normally, some increase in the weight of formalin charged would be required.

4. The amount of distillate that should be collected is 560 g.

3. References

1. General Electric Company, Plastics Department, Pittsfield, Massachusetts 01201.
2. Hooker Research Center, Niagara Falls, New York 14302.

Poly(nonamethylene urea)

$$NH_2—(CH_2)_9—NH_2 + COS \longrightarrow$$

$$[NH_2—(CH_2)_9—NH—CO—SH] \rightleftharpoons$$

$$[\overset{+}{N}H_3—(CH_2)_9—NHCOS^-] \xrightarrow{120°C.}$$

$$NH_2—(CH_2)_9\!\left[NH—CO—NH—(CH_2)_9\right]_m\!NH_2 + H_2S$$

$$\downarrow 180° C.$$

$$NH_2—(CH_2)_9\!\left[NH—CO—NH—(CH_2)_9\right]_n\!NH_2 + NH_2—(CH_2)_9—NH_2$$

$$(m < n)$$

Submitted by H. G. J. Overmars (1)
Checked by W. Memeger, Jr., and A. H. Frazer (2)

1. Procedure

A. Carbonoxysulfide (COS)

(*Caution! Carbonoxysulfide is a toxic gas. All operations should be conducted in an efficient hood*).

A 2-l., four-necked flask (Fig. 1) is provided with a stirrer (*A*) (conveniently a Vibromixer stirrer), a specially designed (because of slight overpressure in vessel) dropping funnel (*B*), a gas inlet tube (*C*), and a gas outlet tube (*D*). The flask is immersed in a water bath maintained at about 30-40° (*E*). The gas outlet is connected to a gas purification train, which consists successively of a tube (*F*) filled with mercuric oxide dispersed on pumice (Note 1), a horizontal tube (*G*) filled with phosphorus pentoxide on pumice, a tube (*H*) packed with soda asbestos (Note 2), and finally a bubbler containing kerosine (*J*) (Note 3).

The flask is charged with a mixture of 400 ml. of concentrated sulfuric acid

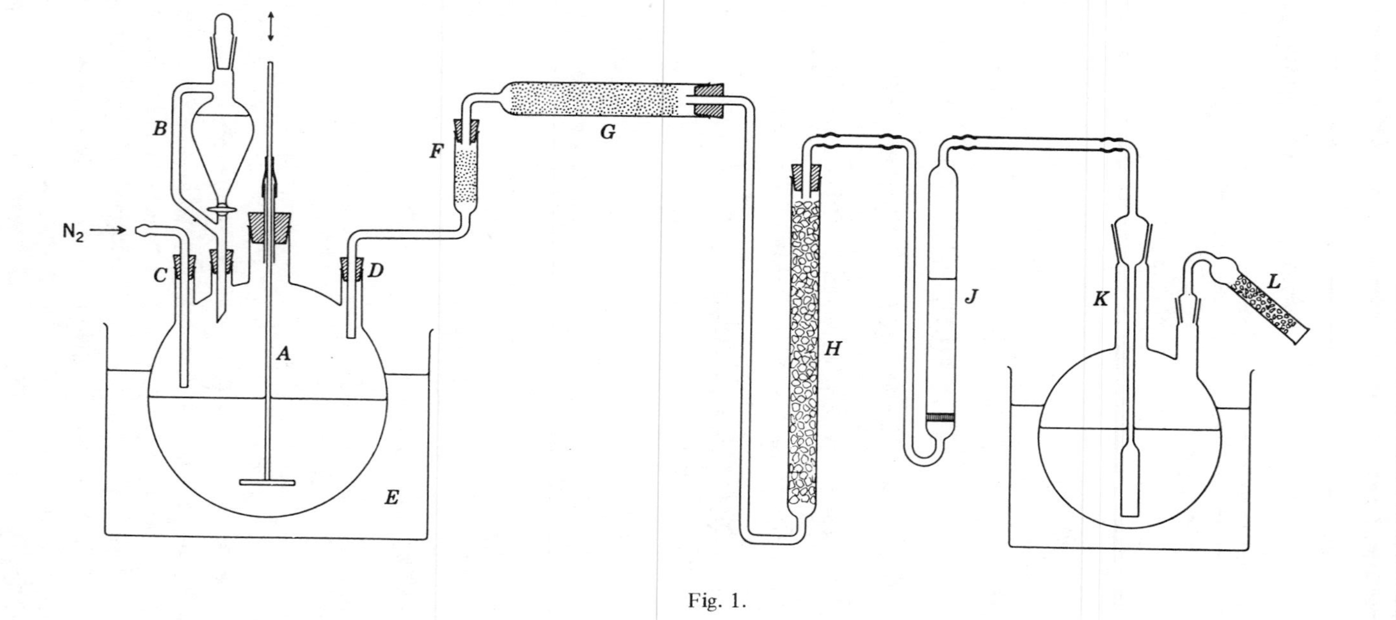

Fig. 1.

and 700 ml. of water previously prepared and cooled to room temperature. The air in the flask and purification train is replaced with nitrogen. The purification train at its other end is connected to a receiver (*K*) filled with a solution of nonamethylene diamine in methanol as described in procedure B.

A saturated aqueous solution of KCNS (20 ml.) is added all at once through the dropping funnel and, subsequently, the mixture is stirred for a short time to start the evolution of carbonoxysulfide (Note 4). A gentle stream of nitrogen is led through the generator during its operation. Then 60 ml. of the KCNS solution is added dropwise at such a rate that a gentle stream of gas is evolved. After this addition the stirrer is started again, at first slowly, to complete the reaction. In all, 20-25 g. of carbonoxysulfide is generated.

B. COS-Salt of Nonamethylene diamine (ω-Ammonium-Nonamethylenethiolo-carbamate) (4) (Note 5)

Caution! H_2S is a toxic gas. All precautions necessary to prevent exposure of laboratory workers to its effects should be strictly followed.

In an efficient hood a 1-l. two-necked flask (Fig. 1, *K*) is equipped with an inlet tube reaching to the bottom of the flask (Note 6) and protected from CO_2 by a drying tube (*L*) filled with soda lime. The flask is filled with nitrogen and charged with 50 g. of nonamethylenediamine (Note 7) and 500 ml. of purified methanol. As soon as the diamine has dissolved, the flask is cooled externally to about 5° and connected to the purification train (see procedure A). A gentle stream of carbonoxysulfide is passed into the solution while constantly tilting the flask (Note 8), until the soda lime turns to bluish green.

The flask is then left standing for 30 min. at room temperature. Finally the precipitate is filtered by suction on a fritted glass filter and rinsed with methanol. The white granular salt is dried at room temperature with a vacuum rotating evaporator. The yield is about 95%

C. Polycondensation of the COS-Salt of Nonamethylenediamine to a Semipoly-mer (Note 9)

A 750-ml. reaction flask with a standard-taper joint is connected with a sublimation tube constructed as in Fig. 2. The tube is fitted with a clamp that carries a rod that can be inserted in a vibrator (Vibromixer). The flask is filled with 50 g. of the COS-salt of nonamethylenediamine and, the joints are coated with silicone grease.

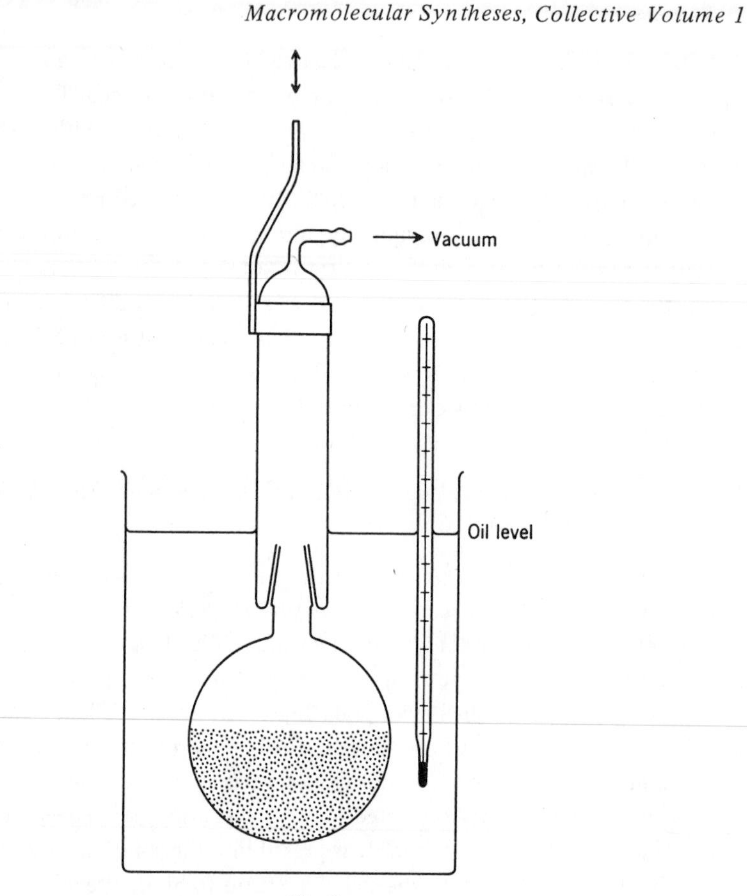

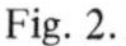
Fig. 2.

The air in the flask is carefully replaced by oxygen-free, dry nitrogen by repeated evacuation and vacuum release. The flask is finally evacuated, and a vacuum of about 12 torr is maintained.

The apparatus is immersed in an oil bath so that the oil level reaches well above the joint of the reaction flask, and the oil bath is heated to 80° (Note 10). Under moderate vibration (Note 11) the salt is then heated for 2 hr. at 80°, subsequently for 2 hr. at 100°, and finally for 3 hr. at about 120°. The vacuum is then lowered to just below 1 torr and heating is continued at 120° for 8 hr. During the polycondensation, apart from the evolved hydrogen sulfide, a small amount of sublimate (about 5%) is deposited on the walls of the tube just above the oil bath level (Note 12).

After cooling, pressure is released using nitrogen and the semipolymer is collected as a pure white, freely-flowing powder. It is completely devoid of detectable sulfur-containing products. The yield is about 95% (Note 13).

D. Polycondensation of the Semipolymer

The semipolymer (50 g.) is transferred to a 2-1. flask, constructed as in Fig. 3, that is connected to a rotating vacuum evaporator, and the flask is immersed

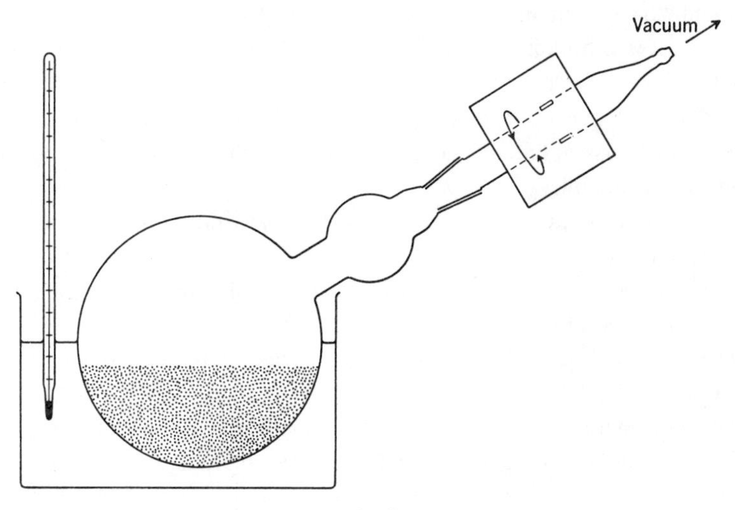

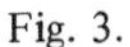

Fig. 3.

in an oil bath. The air in the flask is carefully replaced with oxygen-free, dry nitrogen by repeated evacuating and refilling. Finally a vacuum of about 0.5 torr is applied and the oil bath is heated to 120°. The mass is heated for 1 hr. while continuously rotating the flask. Subsequently the temperature is raised to 150° and maintained at that temperature for 2 hr. and finally at 180° for 10 hr. During the polycondensation a white sublimate is formed in the bulb which consists mainly of pure nonamethylenediamine. At the end, while still under vacuum, the flask is cooled to room temperature and air is then admitted. The yield of the freely flowing, white polymer is about 95%. The intrinsic viscosity in 95.0% sulfuric acid measured at 25° is 0.45 (Note 14). Its crystalline melting point is 210-220° (hot stage polarizing microscope).

2. Notes

1. Hydrogen sulfide, which usually is present as an impurity, is trapped by the mercuric oxide. The tube has a length of a few centimeters.

2. Soda asbestos is prepared as follows: Sodium hydroxide (20 g.) is melted in a nickel crucible after addition of a few drops of water. Asbestos (*Caution! Exposure to the dust–consisting of fine fibrous particles–from asbestos should be avoided. The use of a filter mask is a necessity to prevent inhalation of the dust.*) (3 g.) is carefully added to the melt and stirred with a glass rod until the melt is a thick mass. The melt is poured on an iron plate and after solidification immediately put in a well-stoppered flask. Just before filling the tube the treated asbestos is broken into small lumps. Soda asbestos is a very hygroscopic product.

3. The boiling range of this petroleum fraction is 100-120°/9 torr. The bubbler, which has a sintered glass inlet tube, is filled with about 20 ml. of kerosine or mineral oil.

4. If gas evolution does not start readily, the temperature of the water bath is temporarily increased.

5. Because carbonoxysulfide is a hazardous gas, all handling should be done in a well-ventilated hood.

6. The inlet tube must have a bore 2 cm. wide at its end because the salt tends to clog the inlet.

7. The nonamethylenediamine is freshly distilled over sodium (b.p. about 116°/9 torr) and transferred to the flask with rigorous exclusion of carbon dioxide.

8. A conventional stirrer can be used also.

9. This polycondensation is carried out in the solid phase and is an example of powder polycondensation.

10. *Caution! The water pump must be connected to the apparatus via a tube filled with ferruginous earth or other suitable scavenger to absorb the hydrogen sulfide.*

11. The apparatus should be connected to the vibrator so that the granular powder is slowly but evenly turning in the flask when the vibrator is working.

12. The sublimate consists mainly of the sulfide and of the carbamate of nonamethylenediamine. Contamination of the polymerizing powder should be prevented meticulously; otherwise discoloration occurs.

13. The checkers found that the white COS-salt turned yellow on standing in a tightly stoppered flask for several days before the polycondensation reaction was performed. The semipolymer resulting from the yellow salt was light tan and smelled of hydrogen sulfide. The yield was not affected, however.

14. When a higher molecular weight is preferred, the time of heating at 180° should be lengthened. The viscosity measurements may be made in 98% sulfuric acid.

3. Methods of Preparation

Polyureas can be synthesized generally by reacting alkylene diamines with CO_2 (5), with derivatives of carbon dioxide such as urea (6), phosgene (7), or with alkylene diisocyanates (8), alkylene diurethanes (9), or N,N′-carbonyl-imidazole (10).

4. References

1. Institute for Organic Chemistry TNO, Utrecht, P.O. Box 5009, The Netherlands.
2. Pioneering Research Laboratory, Textile Fibers Department, E.I. du Pont de Nemours & Co., Wilmington, Delaware 19898.
3. P. Klason, *J. prakt. Chem.,* (2) **36**, 64 (1887).
4. G. J. M. van der Kerk, H. G. J. Overmars, and G. M. van der Want, *Rec. trav. chim.,* 74, 1301 (1955).
5. G. D. Buckley and N. H. Ray, Brit. Pat. 619,275 (1949).
6. H. W. Arnold, U.S. Pat. 2,145,242 (1939).
7. E. L. Wittbecker, U.S. Pat. 2,816,879 (1957).
8. W. E. Hanford, U.S. Pat. 2,292,443 (1943).
9. Thüringische Zellwolle A. G. u. Zellwolle u. Kunstseide-Ring G.m.b.H., Belg. Pat. 449,368 (1943).
10. P. Beiersdorf & Co. A.G., Brit. Pat. 830,102 (1960).

Polyacrolein by Redox Polymerization

$$n\,CH_2{=}CH(CH{=}O) \xrightarrow{H_2O} \cdots{-}CH_2CH{-}CH_2{-}CH{-}\cdots$$

(structure: the two CH side groups of the polymer chain bridged by O; terminal HO and OH on the CH groups)

Submitted by R. C. Schulz (1) and H. Cherdron (2)
Checked by V. T. Kagiya and S. Morita (3)

Acrolein undergoes a redox polymerization (4) in the presence of water to form a polymer in which the aldehyde groups are hydrated or form hemiacetals. Several tetrahydropyran rings may be fused (5-7).

1. Procedure

In a 2-1., round-bottomed flask equipped with a stirrer, reflux condenser, dropping funnel, thermometer, and nitrogen inlet, 500 ml. of water is heated under reflux and in a nitrogen atmosphere for 30 min. After cooling to 20°, 4.75 g. of $K_2S_2O_8$ and 100 ml. of destabilized acrolein (*Caution! Acrolein is an extremely active lachrymator. All operations should be conducted in an efficient hood.*) (Note 1) are added. When the monomer is dissolved, a solution of 2.96 g. of $AgNO_3$ in 60 ml. of water is allowed to drop into the flask over a period of 10 min. accompanied by vigorous stirring; nitrogen is passed through the apparatus during this time. After several minutes the mixture becomes turbid and the polymer precipitates. The temperature should not rise above 20° (Note 2). After 2.5 hr. 500 ml. of water is added. The polymer is filtered with suction and washed twice with 500 ml. of water. To remove the silver salts, the filter cake is dispersed in a solution of 5 g. of $Na_2S_2O_3$ in 500 ml of water. After 1 hr. it is filtered with suction, washed several times, and finally dried under vacuum at 20°, then over $CaCl_2$, and lastly over concentrated H_2SO_4 (Note 3). The yield of polymer is 68 g. (80%).

2. Characterization

Polyacrolein prepared by redox polymerization is insoluble in all organic solvents at room temperature. The masked aldehyde groups of these polyacroleins are quite reactive (5-7) (Note 4), however. Numerous derivatives of the polymer with good solubility are available through chemical modifications. For many purposes it is advantageous to convert the polyacrolein into the water-soluble sodium bisulfite or sulfurous acid derivatives (8-10). Polyacroleins prepared by radical polymerizations have no softening points; they become discolored above 170° and sinter without melting at 220°.

For viscosity determination (4) 100 mg. of polyacrolein is mixed with 5 ml. of a 10% aqueous SO_2 solution (d = 1.0493) in a 10-ml. volumetric flask. This is allowed to stand for 24 hr. at room temperature and is then filled to the mark with 10% NaCl solution. Using this solution, the flow time is measured in an Ostwald viscometer (capillary diameter 0.43 mm.) by the usual method. The flow time of the solvent is determined with a 1:1 mixture of SO_2 solution and NaCl solution.

A generally valid viscosity-molecular weight relationship has not yet been derived. The osmotic molecular weight of several polyacrolein derivatives has been measured (11). With these values as a basis, the approximate molecular weight for specific viscosities measured at 10 g/1. in H_2SO_3/NaCl is as follows:

η_{sp}/c (dl./g.)	$\overline{M}$
0.01	5,600
0.05	56,000
0.10	140,000

The polyacrolein obtained from the polymerization described above has $\eta_{sp}/c = 0.3$ (dl./g.) and thus a molecular weight of approximately 28,000.

3. Notes

1. Immediately before use, acrolein is distilled under nitrogen through a 50-cm. packed column to remove the stabilizer, b.p. 52-53°/760 torr. The monomer should be at least 96% pure (12).

2. If the reaction mass becomes viscous, the flask must be repeatedly cooled with ice water.

3. Drying continues for several days. The total water content (bound and unbound) may be determined from the carbon analysis. Unbound water can be determined by the Karl Fischer method.

4. The checkers obtained a 97.5 mole % aldehyde group content by the phenylhydrazine method (13, 14).

4. Methods of Preparation

Polyacrolein with higher viscosity numbers can be obtained by redox polymerizations in aqueous emulsions (9,15). A polyacrolein-sulfurous acid solution is a particularly good emulsifier for this purpose.

The polymerization of acrolein can also be initiated by light (16) or γ-irradiation (17, 18). Initiation by azobisisobutyronitrile or peroxides in the presence of organic solvents is also possible (19,20). Ionic catalysts such as boron trifluoride, boron trifluoride etherate, sodium naphthalene, *n*-butyl lithium, sodium alcoholate, or phosphine also initiate the polymerization of acrolein. However, the structure and properties of these polymers differ fundamentally from those of polymers from radical polymerization (7).

For additional references on homo- and copolymerization of acrolein see ref. (21,22). Polymerization by electrochemical processes is described in ref. (23).

5. References

1. Organisch-Chemisches Institut der Universität, 65 Mainz, Germany.
2. Farbwerke Hoechst, Frankfurt-Hoechst, Germany.
3. Department of Hydrocarbon Chemistry, Faculty of Engineering, Kyoto University, Kyoto, Japan.
4. R. C. Schulz, H. Cherdron, and W. Kern, *Makromol. Chem.,* **24**, 141 (1957).
5. R. C. Schulz and W. Kern, *Makromol. Chem.,* **18/19**, 4 (1956).
6. R. C. Schulz, *Kunststoffe,* **47**, 303 (1957).
7. R. C. Schulz, *Angew. Chem. Intern. Ed. Engl.,* **3**, 416 (1964).
8. R. C. Schulz and I. Löflund, *Angew. Chem.,* **72**, 771 (1960).
9. H. Cherdron, *Kunststoffe,* **50**, 568 (1960).
10. T. L. Dawson and F. J. Welch, *J. Amer. Chem. Soc.,* **86**, 4791 (1964).
11. R. C. Schulz, E. Müller, and W. Kern, *Makromol. Chem.,* **30**, 39 (1959).
12. E. D. Peters, in C. W. Smith, Ed., *Acrolein,* John Wiley & Sons, Inc., New York, 1962, p. 240.
13. R. C. Schulz, R. Holländer, and W. Kern, *Makromol. Chem.,* **40**, 16 (1960).
14. R. C. Schulz and W. Passmann, *Makromol. Chem.,* **60**, 139 (1963).

15. H. Cherdron, R. C. Schulz, and W. Kern, *Makromol. Chem.*, **32**, 197 (1959).
16. F. E. Blacet, G. H. Fielding, and J. G. Roof, *J. Amer. Chem. Soc.*, **59**, 2375 (1937).
17. A. Henglein, W. Schnabel, and R. C. Schulz, *Makromol. Chem.*, **31**, 181 (1959).
18. Y. Toi and Y. Hachihama, *J. Chem. Soc. Japan*, Ind. Chem. Sect., **62**, 1924 (1959).
19. R. C. Schulz, S. Suzuki, H. Cherdron, and W. Kern, *Makromol. Chem.*, **53**, 145 (1962).
20. R. F. Fisher (Shell Development Co.), U.S. Pat. 3,079,357 (1963).
21. R. C. Schulz, "Acrolein Polymers," in *Encyclopedia of Polymer Science and Technology*, Interscience Publishers, New York, 1964, Vol. 1, p. 160.
22. R. C. Schulz, "Polymerization of Acrolein," in G. E. Ham, ed., *Vinyl Polymerization*, Vol. 1, Part 1, M. Dekker, New York, 1967, p. 403.
23. R. C. Schulz and W. Strobel, *Monatsh.* **99**, 1724 (1968); W. Strobel and R. C. Schulz, *Makromol. Chem.*, **133**, 303 (1970).

Polyacrolein by Anionic Polymerization

$$n\,\underset{\displaystyle HC{=}CH_2}{\overset{\displaystyle CH{=}O}{|}} \xrightarrow{NaCN} \cdots \left[\underset{\displaystyle HC{=}CH_2}{\overset{\displaystyle CH{-}O}{|}}\right]_n \cdots$$

Submitted by R. C. Schulz (1a) and G. Wegner (1a, b)
Checked by Y. T. Kagiya and S. Morita (2)

Anionic polymerization of acrolein at low temperature, excluding water carefully, yields a polymer the structural elements of which are linked together only by acetal bonds (3). The vinyl double bond is not induced to polymerize under these circumstances; no free carbonyl groups can be found in the polymer.

1. Procedure

A 250-ml. round-bottomed flask, equipped with a stirrer, dropping funnel, thermometer, nitrogen inlet, and a serum cap, is flushed with dry nitrogen and heated with the yellow flame of a Bunsen burner to remove the last traces of moisture. After cooling, the flask is charged, via the dropping funnel under nitrogen, with 20 ml. of acrolein (*Caution! Acrolein is an extremely active lachrymator. All operations should be conducted in an efficient hood.*) and 80 ml. of tetrahydrofuran (Notes 1 and 2).

The mixture is cooled to $-50°$ with vigorous stirring. Then 2.0 ml. of a solution of sodium cyanide (*Caution! NaCN is extremely toxic and it may be rapidly absorbed through the skin, particularly when dissolved in solvents such as dimethylformamide.*) in dimethylformamide (Note 3) is injected all at once through the serum cap. The temperature should not rise above $-50°$ during the polymerization. After 100 min. 2.0 ml. of methanol is injected to interrupt the polymerization. The polymer is precipitated by pouring the viscous solution into 1 l. of petroleum ether. It precipitates in white flocks that soon agglomerate and deposit at the bottom. The polymer is filtered, dissolved in tetrahydrofuran, and reprecipitated by pouring into a tenfold volume of petroleum ether. The polymer is filtered on a suction filter and dried over sulfuric acid. The yield of polymer is 9.1-10.5 g. (54-62%) (Note 4).

2. Notes

1. Commercial acrolein (at least 96%) is dried over calcium chloride, calcium hydride, or Molecular Sieve 4A and distilled under nitrogen using a 60-cm. packed column. One-third is separated as a light fraction, and one-third is left as a residue. Some hydroquinone is added to the main fraction with b.p. 52°. It is dried with calcium hydride until the evolution of hydrogen stops. The excess of calcium hydride is filtered, and acrolein is redistilled under nitrogen using a 60-cm. packed column. The fraction boiling at 52.7° is stabilized with hydroquinone, and a small quantity of finely powdered calcium hydride is added. The necessary quantity of acrolein is distilled under nitrogen from this stock immediately before each polymerization.

2. Tetrahydrofuran is adequately dried over sodium naphthalene. The necessary quantity of solvent is distilled under nitrogen immediately before each polymerization.

3. Sodium cyanide is finely powdered and dried over sulfuric acid under vacuum. (*Caution! Addition of NaCN to the acid will result in the evolution of HCN. Conduct all operations in an efficient hood.*) A saturated solution in anhydrous dimethylformamide is prepared (about 0.15 *N*). This solution (4 ml.) is diluted with dimethylformamide to 10 ml. The solution, prepared in this way and kept under nitrogen, maintains its full activity for 3 days.

4. The polymerization can be continued until the theoretical maximum conversion is reached. Side reactions occur, however, when the viscosity of the solution is too high, thus leading to cross-linking of the polymers. If a polymerization to a higher conversion is desired, it is advisable to increase the quantity of solvent. In this case, however, the reaction time also has to be prolonged.

3. Characterization

Polyacrolein, prepared at low temperature by anionic polymerization excluding water, is soluble in many organic solvents, e.g., in toluene, dioxane, tetrahydrofuran, acetone, chloroform, dimethylformamide, acetonitrile, nitrobenzene, pyridine, formic acid, and glacial acetic acid. It softens at 75° and forms a colorless melt at 140-150°. It also possesses a remarkable stability to acids and starts to decompose at an observable rate only above 150°. The double bond can undergo a multitude of reactions, e.g., addition of halogens and pseudohalogens, hydrogenation, crosslinking. The checkers obtained an 88.4 mole % vinyl double-bond content by hydrogenation in the presence of platinum dioxide in acetic acid. Viscosity measurements can be carried out in one of the solvents mentioned above.

Polyacrolein, prepared according to the procedure described above, gives, in dioxane at 25°, η_{sp}/c 0.2-0.4 dl./g. The degree of polymerization calculated from the reacted monomer/catalyst molar ratio is 540.

4. Methods of Preparation

Other catalysts which have been described for the anionic polymerization of acrolein include sodium naphthalene (4), trityl sodium (4, 5), butyl lithium (4), benzophenone potassium (4), sodium methoxide (4, 6), sodium amide (4), piperidine (4), ammonia (7), phosphine (8), and calcium zinc tetraethyl (9).

In many cases, however, partially or totally insoluble polymers are formed. The IR spectra as well as the chemical reactions show that the polymers obtained with these catalysts do not possess as uniform a structure as those prepared with sodium cyanide. They contain both vinyl groups and carbonyl groups as structural elements.

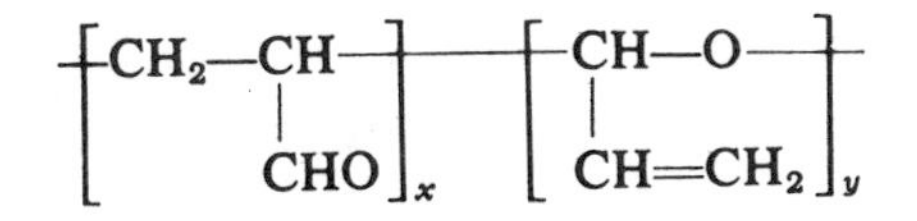

The ratio x/y depends on the experimental conditions. Sodium cyanide also initiates the polymerization of methacrolein or crotonaldehyde (3).

5. References

1a. Organisch-Chemisches Institut der Universität, Mainz, Germany.
1b. Current Address: Institute of Polymer Chemistry, The University of Freiburg, Freiburg i.Br., West Germany.
2. Department of Hydrocarbon Chemistry, Faculty of Engineering, Kyoto University, Kyoto, Japan.
3. R. C. Schulz, G. Wegner, and W. Kern, *J. Polym. Sci., C,* No. 16, 989 (1967); *Makromol. Chem.,* **100**, 208 (1967); **104**, 185 (1967).

4. R. C. Schulz and W. Passmann, *Makromol. Chem.*, **60**, 139 (1963).
5. R. C. Schulz, *Makromol. Chem.*, **17**, 62 (1955).
6. A. Zilkha, B. Feit, and M. Frankel, *Proc. Chem. Soc.*, 255 (1958).
7. R. Hank and H. Schilling, *Makromol. Chem.*, **76**, 134 (1964).
8. R. C. Schulz, *Chimia*, **19**, 143 (1965).
9. S. Inoue, T. Tsuruta, and J. Furukawa, *Makromol. Chem.*, **32**, 102 (1959).

Poly[2,2'-(m-phenylene)-5,5'-bibenzimidazole](PBI)

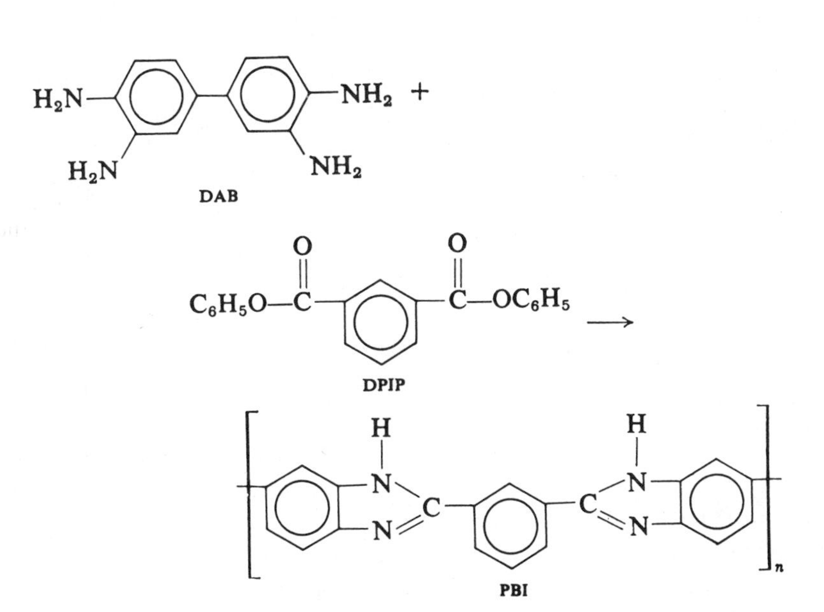

Submitted by A. B. Conciatori and E. C. Chenevey (1)
Checked by L. W. Frost and J. Gasper (2)

1. Procedure

A. Purification of 3,3'-Diaminobenzidine (DAB)

Caution! Benzidine has been found to be a carcinogen. Care with its derivatives should be observed because of possible similar activity.

A 5-l. flask equipped with stirrer, reflux condenser, and hemispherical mantle is charged with 3 l. of preboiled, deoxygenated water, 60 g. (0.28 mole) of

3,3′-diaminobenzidine (DAB), 20 g. of decolorizing charcoal, and 3 g. of filter paper pulp or other filter aid (Note 1). After refluxing for 20 min., the black suspension is filtered free of charcoal into a heated 5-1. flask, which is kept swept with nitrogen.

Final purification consists of percolating this boiling solution through a steam-jacketed charcoal-packed column into a nitrogen blanketed receiver (Note 2). On cooling, off-white crystals of DAB appear that are filtered off and dried in vacuum (Notes 3 and 4) to yield 21-30 g. of material melting at 176-178° (Note 5).

B. Polymerization

Polymerization is conducted in two stages. For the first stage a 1-1., three-necked flask is charged with 26.784 g. (0.125 mole) of DAB and 39.791 g. (0.125 mole) of diphenyl isophthalate (DPIP) (Notes 6 and 7). The flask is immersed in an oil bath and is equipped with a stirrer, Dean-Stark trap with condenser, and a nitrogen purge throughout the whole system. Degassing of the reactants and system is done by alternately evacuating with a vacuum pump and filling with nitrogen. A flow of nitrogen of about 100 ml./min. is begun and maintained throughout the reaction.

The reaction is stirred and heating is begun at a rate of about 2°/min. Reaction commences at about 215-225°. Phenol and water collect in the Dean-Stark trap. As the temperature increases and reaction proceeds, the yellow-brown mass becomes so stiff that stirring is impossible. The stirrer should be stopped when the temperature reaches 250-255° and about 15 ml. of condensate has been collected. After the stirrer is stopped, the mass foams and fills the flask about three-quarters full (Note 8). The polymer is heated to 290° and is held there for 1.5 hr. About 22 ml. of condensate is recovered (85-90% of theoretical).

On cooling, the yellow friable polymer is removed from the flask and is crushed (Notes 9 and 10). Because of the presence of phenol, more than the theoretical yield of prepolymer is obtained. The inherent viscosity of this prepolymer should be 0.2-0.3 dl./g. (0.4 g./dl. in 97% H_2SO_4 at 25°).

For the second stage of polymerization the prepolymer is charged into a flask (Note 11) and degassed in the same manner as in the first stage (Note 12). A nitrogen sweep of 60-120 ml./min. is used throughout the second stage. After immersion of the reactor in the heating bath (Note 13), the temperature is raised at a rate of about 1.5°/min. from 220° to 385° (Note 14). Polymerization is continued at 385° for 3 hr. After cooling and removal from the flask, a yellow-brown granulated polymer is found that should have an inherent viscosity of 0.8-1.2 dl./g. (0.4 g./dl. of 97% H_2SO_4 at 25°) (Note 15). Almost 100% solubility in concentrated H_2SO_4 at room temperature is evidence of negligible gel formation (Note 16). Almost complete solubility is also found in dimethylformamide and dimethylacetamide if heated in a suitable pressure vessel to 200-240°. Polymers of different inherent viscosities may be made by altering the second-stage heating cycle; however, temperatures of 400° cause formation of some gel.

2. Notes

1. DAB is produced by the American Aniline Products Division of Koppers Corporation. Small quantities are available from Aldrich Chemical Co. The checkers report that DAB from Burdick and Jackson Laboratories, Muskegon, Michigan, is suitable for use without further purification.

2. Production of white DAB requires the rigorous exclusion of oxygen from the system. For best results, construct a heated column (a West condenser heated with steam will suffice), fill the column with granular charcoal which has been deoxygenated by boiling in water, and connect the column through a stopcock to a receiver that can be evacuated. Ideally the receiver should be surrounded with ice to cool the DAB solution rapidly. Before passing the boiling DAB solution through the column, the receiver should be evacuated and refilled with nitrogen several times to remove oxygen. A nitrogen flow through the receiver should be maintained throughout the process. An increase in yield ensues if the column is washed with a little boiling water at the end.

3. Filtration may be conducted in air using a Buchner funnel with a rubber dam to squeeze the filter cake to minimize oxidation. Cooling to 10° allows recovery of most of the product.

4. Drying should be done in vacuum with a flush of nitrogen through the oven to help remove the large amounts of water present in the filter cake. Temperatures of 100-110° allow drying in several hours.

5. Melting points were measured with vacuum-sealed capillary tubes or on a differential scanning calorimeter.

6. DPIP of suitable purity to make polymer of high inherent viscosity can be obtained from Burdick & Jackson Laboratories, Muskegon, Michigan. This white material has a sharp melting point of 138-139°.

7. Because stoichiometry must be held to within 0.5% to obtain high inherent viscosity polymer, it is necessary to correct the charge for the volatiles present in

the reactants. A standard moisture balance is appropriate for this determination. It is preferable to weigh the reactants directly into the flask to eliminate the problem of transfer which is compounded by the buildup of electrostatic charges.

8. The amount of foam produced is influenced by the rate of heating and the point at which the stirrer is stopped. At temperatures of 270-280° the foam sets to a rigid porous structure and gradually becomes lighter yellow as reaction proceeds. It is desirable to allow the foam to form. Even if the foam plugs the flask neck, the foam becomes porous when the temperature increases. About 1 hr. is required to reach 290° from the time of initial melting.

9. Grinding in a Wiley mill (40 mesh) is preferable because static charges make the prepolymer difficult to handle when ground with a mortar and pestle.

10. If this prepolymer is to be kept, it should be degassed, capped under nitrogen, and shielded from light.

11. Although an ordinary flask can be used, a preferable reactor is a 100-ml. Ace Mini-Pot (Ace Glass Co., Vineland, New Jersey) equipped with a gas inlet stirrer and connected to a cold trap and bubbler tube.

12. Care must be used in filling the evacuated apparatus with nitrogen, as the fine prepolymer is easily blown around.

13. The heating bath may contain salt (such as HiTec, E. I. du Pont), Wood's metal, or 50/50 solder. It should be equipped with stirrer, thermoregulator, and thermocouple. It is satisfactory to immerse the cold reactor slowly over a 15-30 min. period in the molten bath at 200-230°.

14. Use of a thermocouple at these temperatures is necessary because the submitters found thermometers rapidly age and may be in error by 20°.

15. The inherent viscosity increases if the acid concentration increases above 98%. Inherent viscosities may also be measured in organic solvents such as DMF or DMAc; however, all the sample may not dissolve.

16. The checkers report that 6% of the polymer failed to dissolve in 97% sulfuric acid after soaking for 96 hr.

3. Methods of Preparation

DAB has also been purified by recrystallization from methanol (3). Polymerization has also been effected using similar procedures but employing high vacuum (3). An alternative polymerization using DAB or its hydrochloride and isophthalic acid or other derivatives in polyphosphoric acid has also been reported (4).

4. References

1. Celanese Research Co., Summit, New Jersey 07901.
2. Research and Development Center, Westinghouse Electric Corp., Churchill Boro, Pittsburgh, Pennsylvania 15235.
3. H. Vogel and C. S. Marvel, *J. Polym. Sci.,* **50**, 511 (1961).
4. Y. Iwakura, K. Uno, and Y. Imai, *J. Polym. Sci.,* **A2**, 2605 (1964).

Poly(ethylene succinamide)

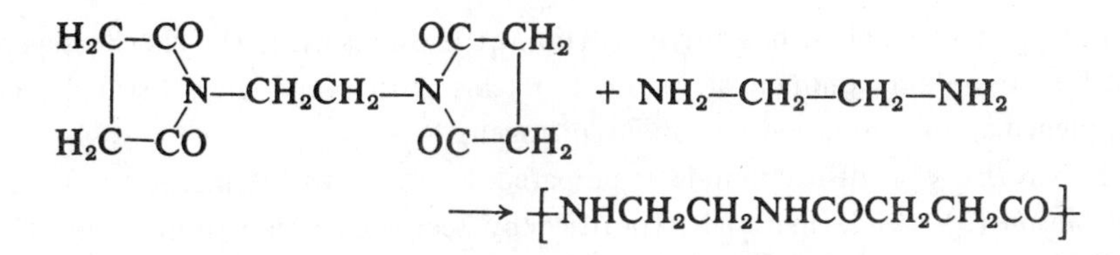

Submitted by V. T. Kagiyo, M. Izu, T. Matsuda, and K. Fukui (1)
Checked by R. M. Pierson and C. J. Suchma (2)

1. Procedure

Ethylenediamine (0.3 ml.) (Note 1) is placed in a weighed test tube of about 1.3 cm. diameter. The weight of ethylenediamine is exactly determined by reweighing the test tube. After addition of an equimolar quantity of N,N′-ethylene-disuccinimide (Note 2), the test tube is swept with nitrogen, sealed, and immersed in an oil bath at 200°. After 5 hr. the test tube is removed from the bath and allowed to cool to room temperature. After breaking the tube, the product is removed, ground in a mortar, and washed with water on a glass filter. The product and 120 ml. of water are placed in a 150-ml. flask and boiled for 1 hr. The polymer is separated by hot filtration and dried under vacuum at 60° for 48 hr. The yield of polymer is 0.81 g. (63%). *Analysis.* Calc.: C, 50.7%; N, 19.7%; H, 7.1%. Found: C, 49.4%; N, 20.1%; H, 7.3%.

2. Characterization

The polymer is soluble in formic acid, sulfuric acid, and trifluoroacetic acid, and insoluble in common organic solvents. The polymer does not show a clear melting point when heated in a sealed capillary under a nitrogen atmosphere, but decomposes at above 305°. For viscosity measurement the polymer (50 mg.)

is dissolved in 20 ml. of formic acid solution and the flow time is measured at 35° in an Ubbelohde viscometer. (The flow time of water at 30° is 90 sec.) The polymer has η_{sp}/c 0.21 (dl./g.). The infrared spectrum of the polymer measured on a KBr pellet shows the characteristic peaks of a secondary amide at 3310, 1642, and 1557 cm^{-1}. The shoulder at 1700 $cm.^{-1}$ is assignable to the succinimide group of the chain end.

3. Notes

1. Commercial anhydrous ethylenediamine (99.8%) is used. The checkers were unable to obtain commercial material of this purity, and they distilled 98% ethylenediamine from sodium under nitrogen.

2. N,N′-Ethylenedisuccinimide is prepared according to the method reported by Mason (3). Crude material is purified by recrystallization from water. The melting point is 251-253°. *Analysis.* Calc.: C, 53.6%; N, 12.5%. Found: C, 53.6%; N, 12.5%.

4. Methods of Preparation

This preparation is based on the paper by Kagiya, Izu, Matsuda, and Fukui (4). A similar method has been reported by Sambeth and Grundschober (5). These methods have been used for various polymers that contain succinamide units,–$NHCO(CH_2)_2CONH$–. It was shown by Kagiya and co-workers that the copolyamide prepared by these methods contained a crystalline portion of sequences having four amide linkages. Sambeth and Grundschober prepared a high molecular weight polyamide from bisglutarimide and diamine (5).

5. References

1. Department of Hydrocarbon Chemistry, Faculty of Engineering, Kyoto University, Kyoto, Japan.
2. The Goodyear Tire and Rubber Company, Akron, Ohio 44316.
3. A. T. Mason, *J. Chem. Soc.*, 55, 10 (1889).
4. T. Kagiya, M. Izu, T. Matsuda, and K. Fukui, *J. Polym. Sci.*, **A1**,5, 15 (1967).
5. J. Sambeth and F. Grundschober, paper presented at International Symposium on Macromolecular Chemistry, Tokyo, 1966.

cis-trans Isomerization of Polybutadienes and Polyisoprenes

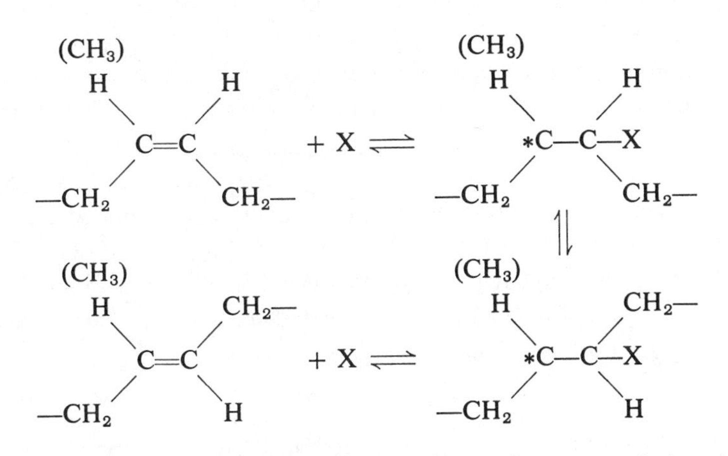

Where X = $C_6H_5S\cdot$ (part A) or SO_2 (part B). Other methods, including the case where X = Br•, are mentioned in part D.

A. Photosensitized Isomerization of Polybutadiene in Solution

Submitted by M. A. Golub (1) and D. B. Parkinson (2)
Checked by G. S. Trick and J. M. Ryan (3)

1. Procedure (Note 1)

Diphenyl disulfide (0.3 g.) is dissolved in 100 ml. of a benzene solution containing 1 g. of *cis*-1,4-polybutadiene. Aliquots of the sensitized solution are placed in several Pyrex® tubes that are flushed with nitrogen and stoppered

(Note 2). The tubes are placed 12 in. from a General Electric H85 mercury lamp and irradiated for 6 hr. (Note 3).

The benzene solution is poured into a beaker (Note 4), and 100 ml. of methanol is added slowly to precipitate the isomerized polymer. The precipitate is filtered, washed with methanol, and dried at 50° in a vacuum oven. The polymer structure is that of the equilibrium *cis-trans* ratio of 25/74 (Note 5).

2. Notes

1. This procedure is applicable to the *cis-trans* isomerization in solution of 1,4-polybutadiene and 1,4-polyisoprene as well as the isomerization of butadiene or isoprene segments that may be present as 1,4-polymerization units in copolymers.

2. An atmosphere of nitrogen is required to avoid extensive degradation of polymer that occurs when the ultraviolet irradiation is carried out in the presence of air.

3. The *cis-trans* ratios in the isomerizates, short of equilibrium, will depend on the placement of the tubes with respect to the ultraviolet source, on the intensity and spectral distribution of the source, and on the time of irradiation. Typical results for polybutadiene photoisomerization are shown in Table 1. The intrinsic viscosity of the isomerized polybutadiene is essentially unchanged from that of the starting polymer.

TABLE 1. STRUCTURE OF PHOTOISOMERIZED POLYBUTADIENE AS A FUNCTION OF REACTION TIME

Reaction Time, hr.	*trans*-1,4 Content, %
0	2
1	36
2	58
3	68
4	72
5	74
6	75

4. The photoisomerization is generally accompanied by the formation of some gelled polymer attached to the wall of the Pyrex tube. This is left in the tube when the benzene solution is decanted into a beaker.

5. The equilibrium *cis-trans* ratio of about 25/75 in polybutadiene is obtained starting from a nearly 100% *cis-* or 100% *trans*-polymer, whereas a nearly 100% *cis-* or *trans*-polyisoprene attains an equilibrium *cis-trans* ratio of about 40/60. The characterization of the microstructure is described in part C.

3. References

1. Ames Research Center, Moffett Field, California 94035.
2. Stanford Research Institute, Menlo Park, California 94025.
3. Research Division, Goodyear Tire & Rubber Co., Akron, Ohio 44316.

B. Bulk Isomerization of Polyisoprene with Sulfur Dioxide

Submitted by J. I. Cunneen (1a, b) and W. F. Watson (2)
Checked by M. A. Golub (3) and D. B. Parkinson (4)

1. Procedure (Note 1)

Natural rubber (RSSI, pale crepe, or highly purified rubber from the United States Rubber Co.) is masticated on a two-roll laboratory mill until its Mooney viscosity at 100° is 55-60. It is then formed into sheets of approximately 0.25 cm. thickness and 25 cm. width. Guttapercha is formed to the same dimensions with the mill rolls at 60°. Both polymers are rolled in glazed Holland paper.

The roll of polyisoprene (~400 g.) is placed in the reaction vessel, which is a glass cylinder 30 cm. long, 8 cm. dia., with a ground glass joint at one end, containing inlet and exit tubes for the passage of nitrogen or sulfur dioxide (*Hood!*) as required. The reaction vessel is immersed in an oil bath at 140 ± 1°, and the air is removed with a stream of nitrogen. The nitrogen is then replaced by sulfur dioxide (laboratory reagent grade) from a syphon. The sulfur dioxide is preheated by passing it through a copper coil immersed in an oil bath before it enters the reaction vessel. The flow of sulfur dioxide is rapid for the first few minutes to remove the nitrogen, but it is then reduced to a slow rate for the remainder of the experiment. A continuous flow of sulfur dioxide is not necessary because it functions as a catalyst and is not consumed. This procedure is used, however, to ensure that no air enters the apparatus. Isomerization occurs gradually, and the equilibrium composition of approximately 60% *trans* is reached in under 24 hr. (Note 2).

2. Properties

Configurational changes in the double bonds of natural rubber or gutta-percha cause marked changes in the strength, stress-strain (Note 3), crystallization (Note 4), and other physical properties of their vulcanizates (5,6). The effect of isomerization is strikingly obvious in the case of gutta-percha because this normally rigid polymer is converted into a rubber-like material (6).

Isomerized natural rubbers of increasing *trans* content have the following properties. Their processing behavior differs from that of normal natural rubber because far less breakdown occurs on cold milling (6). With pure gum vulcanizates (Note 5) tensile strength falls gradually at first but, when about 10% *trans* double bonds are present, a very rapid decrease occurs (5). The vulcanizate obtained from natural rubber containing about 6% *trans* double bonds is particularly interesting because, although its tensile strength and stress-strain properties are very similar to those of normal natural rubber, its rate of crystallization at low temperatures is several hundred times slower (5). This type of vulcanizate possesses distinct advantages over those made from normal natural rubber when used at low temperature ($-10°$ to $-40°$), because under these conditions the former would retain its elasticity for many years, whereas the latter would harden and lose its flexibility in a few days.

3. Notes

1. This procedure involves the treatment of the solid polymer with sulfur dioxide and is by far the most convenient method of obtaining the isomerized polymer in sufficient quantities (about 1 lb.) to enable the physical properties to be thoroughly evaluated. Larger amounts, up to 34 lb., have been made by treatment of natural rubber with butadiene sulfone in internal mixers (8).
2. A small increase occurs in intrinsic viscosity, gel content, and bulk viscosity, measured at 100° with a Wallace Rapid Plastimeter (9), as isomerization takes place. Changes in chemical unsaturation, determined by treatment with perbenzoic acid (10) are negligible (5) (Table 2).
3. Tensile strengths and stress-strain properties are measured on a Goodbrand testing machine by extending Type C British Standard 903 dumb bells at 600% per minute.
4. Rate of crystallization is measured by means of the relaxation of stress in a vulcanized test piece extended 150% and maintained at $-26°$ (at which temperature the rate is maximal). The time for reduction of stress by a factor

TABLE 2. PROPERTIES AND STRUCTURE OF ISOMERIZED POLYISOPRENES

Reaction time, hours	$[\eta]$, dl./g.[a]	Gel, %[b]	Bulk Viscosity	Unsaturation, %	*trans*-1,4 Content, %[c]
		A. *From Natural Rubber*			
0	3.64	0	44	95	0
0.5	3.60	0	46	97	5
1.5	3.50	0	43	97	9
3	3.31	0	46	98	21
5	3.54	0	47	97	28
8	4.24	6.0	68	97	39
24	4.54	8.8	80	98	56
		B. *From Gutta Percha*			
0	1.98	0	23	100	100
0.5	2.26	0	22	97	95
1.5	2.14	0	26	99	86
3	1.67	0	32	104	74
6	2.19	0	38	101	63
24	1.65	4.0	40		58

[a] Measured in benzene at 25°.
[b] Weight % of polymer insoluble in *n*-decane after 48 hr. of swelling at 25.0°.
[c] Determined by infrared spectroscopy (see part C).

of 2 has been shown to be an accurate measure of the time for half the primary crystallization to be achieved; stress half-life is thus an inverse measure of rate of crystallization (11).

5. Vulcanizates are prepared from the isomerized polyisoprenes by compounding on a two-roll laboratory mill according to the following recipe, heated in a press at 140° for 30 min.

Polyisoprene	100
Zinc oxide	5
Stearic acid	1
Sulfur	2.5
N-Cyclohexylbenzothiazole-2-sulfenamide	0.7
Phenyl-2-naphthylamine	1

4. References

1a. The Natural Rubber Producers' Research Association, Welwyn Garden City, Herts, England.
1b. Current Address: Universiti Sains Malaysia, Minden, Pulan Pinang, Malaysia.
2. Rubber and Plastics Research Association of Great Britain, Shawbury, Shrewsbury, Shropshire, England.
3. Ames Research Center, Moffett Field, California 94035.
4. Stanford Research Institute, Menlo Park, California 94025
5. J. I. Cunneen, *Rubber Chem. Techno.,* **33**, 445 (1960).
6. D. J. Elliott, *Trans. Inst. Rubber Ind.,* **40**, 180 (1965).
7. J. I. Cunneen and W. F. Watson, *J. Polym. Sci.,* **38**, 533 (1959).
8. J. I. Cunneen, P. M. Swift, and W. F. Watson, *Trans. Inst. Rubber Ind.,* **36**, 17 (1960).
9. H. W. Wallace and Co., Ltd., Croydon, England.
10. I. M. Kolthoff, T. S. Lee, and M. A. Mairs, *J. Polym. Sci.,* **2**, 199 (1947).
11. A. M. Gent, *Trans. Inst. Rubber Ind.,* **30**, 139, 144 (1954).

C. *Microstructure of Polybutadienes and Polyisoprenes (1)*

1. Infrared Analysis

In the case of polybutadiene or a butadiene-containing copolymer, the characterization of the isomerizate for *cis-trans* content is easily performed by means of infrared analysis of a thin film cast from a benzene solution of the polymer. The analysis involves measuring the absorbance (A) of the 13.6- and 10.4-μ bands (*cis-* and *trans-*, –CH=CH– units, respectively), and using the expression (2,3)

$$\%\ cis = \frac{100(4.4A_{13.6})}{4.4A_{13.6} + A_{10.4}}$$

This expression gives the percentage of the 1,4-polymerization units in the *cis* configuration, the *trans* content then being given by (100 - %*cis*). When 1,2-polymerization units (or $-CH{=}CH_2$ units) are present, and when the *cis-trans* content is to be expressed in terms of the total polybutadiene unsaturation, a more detailed infrared method may be used (4-6).

In the case of polyisoprene or an isoprene-containing copolymer, infrared analysis can be used to determine the *cis-trans* content (6), but a simpler and more accurate analysis is based on ^{1}H NMR spectroscopy. The infrared changes accompanying the isomerization of 1,4-polyisoprene are rather subtle, with the *cis* and *trans* $-C(CH_3){=}CH-$ units both having their main absorptions at around 12 μ (2). Analysis based on characteristic absorptions at 8.89 and 8.74 μ in Hevea and Balata, respectively, have been used, but they are not reliable for random distributions of *cis* and *trans* units (6). On the other hand, infrared is often preferred over NMR for determination of 1,2- and 3,4-polymerization units (or $-CH{=}CH_2$ and $-C(CH_3){=}CH-$ units, respectively), especially when these are present in small amounts (6).

Infrared examination reveals that, when diphenyl disulfide is used as photosensitizer, a small amount of $C_6H_5S\cdot$ radical becomes permanently attached to an isomerized polyisoprene backbone, but apparently not to an isomerized polybutadiene backbone.

The photolysis of diphenyl disulfide involves the reaction

$$C_6H_5S{-}SC_6H_5 \xrightarrow{h\nu} 2\ C_6H_5S\cdot$$

The incorporation of some $C_6H_5S\cdot$ into the polyisoprene backbone is believed to involve the reaction

$$C_6H_5S\cdot \; + \; \begin{matrix} CH_3 & & H \\ & \dot{C}{-}C{-}SC_6H_5 & \\ -CH_2 & & CH_2- \end{matrix} \rightarrow \begin{matrix} CH_3 & & H \\ C_6H_5S{-} & C{-}C{-}SC_6H_5 & \\ -CH_2 & & CH_2- \end{matrix}$$

This reaction presumably does not occur in polybutadiene. Isomerization of polyisoprene with sulfur dioxide yields a cleaner product, uncontaminated by adducts.

2. Nuclear Magnetic Resonance Analysis

In the case of polyisoprene or an isoprene-containing copolymer, the ^1H NMR spectrum of a 1 g./dl. benzene or carbon tetrachloride solution of isomerized polymer is typically obtained on a 60- or 100-MHz spectrometer at one-half to one-third the normal scan rate. The *cis-trans* ratio is calculated from the relative intensities of the well-resolved signals of the *cis*- and *trans*-methyl protons at 8.21 τ and 8.35 τ, respectively, in benzene solution, and at 8.33 τ and 8.40 τ, respectively, in carbon tetrachloride solution (2,7). The chemical shift values (τ) are given relative to tetramethylsilane.

In the case of polybutadienes the determination of the *cis-trans* ratio has not been possible by ordinary (i.e., undecoupled) ^1H NMR spectroscopy at 60- or 100-MHz. However, suitable decoupling of the methylene proton signals for polybutadiene at 100 MHz permits not only a determination of *cis-trans* content but also the distribution of diads (*cis-cis, cis-trans, trans-cis, trans-*

trans) (8). Moreover, decoupling the olefinic proton signals at 300 MHz allows a determination of the *cis-trans* triad sequence distribution as well as the *cis-trans* ratio (9).

The advent of proton-decoupled ^{13}C NMR spectroscopy has led to additional powerful methods for the analysis of both *cis-trans* content and sequence distributions in various polyisoprenes (10) and polybutadienes (11). Indeed, recent microstructural studies with the aid of decoupled ^{1}H or ^{13}C NMR spectroscopy have demonstrated that *cis-trans* isomerization of 1,4-polyisoprene or 1,4-polybutadiene results in random distributions of *cis* and *trans* units along the polymer chains (8-10).

3. References

1. M. A. Golub, Ames Research Center, Moffett Field, California 94035.
2. M. A. Golub, in J. P. Kennedy and E. G. M. Törnqvist, Eds., *Polymer Chemistry of Synthetic Elastomers,* Part II, Interscience Publishers, New York, 1969.
3. M. A. Golub, *J. Polym. Sci., Polym. Lett. Ed.,* **12**, 295 (1974).
4. R. R. Hampton, *Anal. Chem.,* **21**, 923 (1949).
5. R. S. Silas, J. Yates, and V. Thornton, *Anal. Chem.,* **31**, 529 (1959).
6. R. R. Hampton, *Rubber Chem. Technol.,* **45**, 546 (1972).
7. H. Y. Chen, *Rubber Chem. Technol.,* **41**, 47 (1968).
8. K. Hatada, Y. Tanaka, Y. Terawaki, and H. Okuda, *Polymer J.,* **5**, 327 (1973).
9. E. R. Santee, Jr., L. O. Malotky, and M. Morton, *Rubber Chem. Technol.,* **46**, 1156 (1973).
10. Y. Tanaka, H. Sato, and T. Seimiya, *Polymer J.,* **7**, 264 (1975).
11. See, for example, Y. Tanaka, H. Sato, M. Ogawa, K. Hatada, and Y. Terawaki, *J. Polym. Sci., Polym. Lett. Ed.,* **12**, 369 (1974); F. Conti, A. Segre, P. Pini, and L. Porri, *Polymer,* **15**, 5, 816 (1974); K. F. Elgert, G. Quack, and B. Stutzel, *Polymer,* **16**, 154 (1975).

D. *Methods of Preparation (1a, b)*

1. Methods of Isomerization

Polybutadienes can be isomerized in solution with ultraviolet light (2) or γ-rays (3) in the presence of suitable sensitizers including sulfides, disulfides, mercaptans, and organic bromine compounds such as allyl bromide. Unsensitized isomerization can occur on γ-irradiation of polybutadiene in solution or in the solid state (4), on ultraviolet irradiation of a polybutadiene film (5), or even thermally in the absence of a catalyst at temperatures below its decomposition temperature (6).

cis-trans Isomerization of polybutadiene also occurs in the course of vulcanization with sulfur at 140-160° (7-9), as well as under the influence of peroxides (7), nitrogen dioxide (10), sulfur dioxide (11), and selenium (12).

The equilibrium *cis-trans* ratio in the photosensitized isomerization is about 25/75, and in the direct photoisomerization it is about 30/70.

Natural rubber and gutta-percha can be isomerized by treatment with thiol acids, disulfides, butadiene sulfone, sulfur dioxide (13, 14), and elemental selenium (12), and photochemically with dibenzoyl disulfide, thiolbenzoic acid, and diphenyl disulfide as sensitizeıs (11). The reactions can be carried out in solution (11, 12, 15), in latex (15), or with thin sheets of either the raw or vulcanized polymers (11, 16-18) at 140° to yield equilibrium mixtures of similar isomeric composition (11-13), i.e., a *cis-trans* ratio of about 40/60. Polyisoprenes also undergo unsensitized photochemical (19) and radiation-induced (20) *cis-trans* isomerization.

With the organic sulfur compounds the isomerization probably occurs via a reversible step in a free-radical addition reaction (13, 16). Isomerization is much more rapid with the free thiol acid than with either acyl or other disulfides (11).

It has been shown that *cis-trans* changes are the only structural alterations occurring during the reaction between natural rubber and sulfur dioxide (21, 22) at 140°, and a "clean" isomerized polymer uncontaminated by adducts can be obtained only in this way. The isomerization presumably occurs via an "on-off" reaction at the double bond similar to that with the organic sulfur compounds, but free radicals are not inovlved because the rate of isomerization appears to be insensitive to free-radical catalysts and inhibitors (13).

For a survey of photochemical, radiation chemical and assorted catalytic methods for *cis-trans* isomerization of polymers containing C=C bonds, see ref. 23.

2. References

1a. J. I. Cunneen, The Natural Rubber Producers' Research Association, Welwyn Garden City, Herts, England.
1b. Current Address: Universiti Sains Malaysia, Minden, Pulao Pinang, Malaysia.
2. M. A. Golub, *J. Polym. Sci.*, **25**, 373 (1957).
3. M. A. Golub, *J. Amer. Chem. Soc.*, **80**, 1794 (1958); **81**, 54 (1959).
4. M. A. Golub, *J. Amer. Chem. Soc.*, **82**, 5093 (1960); *J. Phys. Chem.*, **69**, 2639 (1965).
5. M. A. Golub and C. L. Stephens, *J. Polym. Sci.*, **C16**, 765 (1967).
6. M. A. Golub, *J. Polym. Sci., Polym. Lett. Ed.*, **12**, 295 (1974).
7. W. A. Bishop, *J. Polym. Sci.*, **55**, 827 (1961).

8. J. J. Shipman and M. A. Golub, *J. Polym. Sci.,* **58**, 1063 (1962).
9. C. A. Caselli, T. Garlanda, M. Camia, and G. Manza, *Chim. Ind. (Milan),* **44**, 1203 (1962).
10. I. I. Ermakova, B. A. Dolgoplosk, and E. N. Kropacheva, *Dokl. Akad. Nauk SSSR,* **141**, 1361 (1961) [*C.A.,* **56**, 12717f (1962)].
11. J. I. Cunneen, G. M. C. Higgins, and W. F. Watson, *J. Polym. Sci.,* **40**, 1 (1959).
12. M. A. Golub, *J. Polym. Sci.,* **36**, 523 (1959).
13. J. I. Cunneen, *Rubber Chem. Technol.,* **33**, 445 (1960).
14. J. I. Cunneen and G. M. C. Higgins, *The Chemistry and Physics of Rubberlike Substances,* Maclaren, London, 1963, p. 25.
15. J. I. Cunneen and F. W. Shipley, *J. Polym. Sci.,* **36**, 77 (1959).
16. J. I. Cunneen, W. P. Fletcher, F. W. Shipley, and R. I. Wood, *Trans. Inst. Rubber Ind.,* **34**, 260 (1959).
17. J. I. Cunneen and W. F. Watson, *J. Polym. Sci.,* **38**, 521 (1959).
18. J. I. Cunneen and W. F. Watson, *J. Polym. Sci.,* **38**, 533 (1959).
19. M. A. Golub and C. L. Stephens, *J. Polym. Sci.,* **A-1, 6**, 763 (1968).
20. M. A. Golub and J. Danon, *Can. J. Chem.,* **43**, 2772 (1965).
21. J. I. Cunneen, G. M. C. Higgins, and R. A. Wilkes, *J. Polym. Sci.,* **A3**, 3503 (1965).
22. M. A. Golub, *J. Polym. Sci.,* **B4**, 227 (1966).
23. M. A. Golub, in J. Zabicky, Ed., *The Chemistry of Alkenes,* Vol. 2, Interscience Division, John Wiley and Sons, London-New York, 1970, pp. 449-460.

Poly(decamethyleneoxamide)

$$\begin{array}{l} CO_2(CH_2)_3CH_3 \\ | \\ CO_2(CH_2)_3CH_3 \end{array} + H_2N(CH_2)_{10}NH_2 \longrightarrow$$

$$H\left[HN(CH_2)_{10}NH\underset{\underset{O}{\|}}{C}-\underset{\underset{O}{\|}}{C}\right]_n O(CH_2)_3CH_3$$

Submitted by G. S. Stamatoff (1)
Checked by C. E. Hathaway (2)

1. Procedure

A mixture of 50 ml. of dry toluene (Note 1), 34.46 g. (0.20 mole) of pure decamethylenediamine (Note 2), and 1 ml. of a 0.5% solution of phosphorous acid in dry methanol is prepared in a 300-ml., three-necked flask fitted with a nitrogen inlet tube, a stirrer, and a drying tube. The stirred, nitrogen-blanketed mixture is warmed to 50° to effect complete solution, and 40.44 g. (0.20 mole) of pure di-*n*-butyl oxalate (Note 3) is added quickly. The last few drops are washed in with a few milliliters of dry toluene. Heat is evolved (Note 4), a white solid precipitates, and the mixture soon becomes too thick to stir. At this point the stirrer and drying tube are removed, the flask is placed in a liquid metal bath and is fitted with a distillation head and condenser. A 100 ml./min. stream of dry nitrogen is introduced just above the surface of the prepolymer. The bath is then brought to 260° as rapidly as distillation of the toluene and butanol will allow. Heating is continued at that temperature for 2 hr. more (Note 5). The flask and contents are then cooled to room temperature under nitrogen. A tough, lightly colored, opaque, solid polymer is obtained on breaking the flask. This polymer is soluble in a mixture of 7 parts of trichlorophenol and 10 parts of phenol by weight, and its inherent viscosity, ln $(\eta/\eta_0)/c$, determined at 25° in a 0.25% solution in this solvent is 1.25. This polymer melts sufficiently to be sticky at about 250°.

2. Notes

1. The toluene is dried over sodium.

2. Decamethylenediamine is available from Aldrich Chemical Co. It may also be made from sebacic acid according to directions found in the literature (3). Diamines of this type react with CO_2 at room temperature to form salts:

$$2\,RNH_2 + CO_2 \rightarrow [RNH_3^+][^-OC(=O)NHR]$$

Because such a salt gives some substituted urea and water on heating, it tends to consume diamine and limit molecular weight. It should therefore be stored in an atmosphere which is free from CO_2.

3. Di-*n*-butyl oxalate may be purchased from the Eastman Kodak Company or may be made by the method described for making diethyl oxalate in *Organic Syntheses* (4). The diester is freed from half-ester by shaking or stirring in a blender with 10% by weight of finely divided dry calcium hydroxide. The ester is separated by filtration while protecting it from water as well as possible. At this point the effectiveness of this treatment may be tested in the following manner. A 10 ml. portion of the ester is added to 100 ml. of distilled water containing a few drops of methyl orange in a glass-stoppered flask. It is immediately shaken vigorously for a few seconds. If there is an acid reaction with the indicator, the ester contains free acid and the treatment must be repeated. If not, the ester is fractionated through a 4-ft. column packed with glass helices and protected from contact with water in the atmosphere by drying tubes. The fraction boiling between 65.6° and 66.5°/0.45 torr is suitable for polymer preparation.

4. If larger preparations are attempted, provision must be made for removal of heat by cooling the flask or by condensation of refluxing toluene and butanol in a condenser of appropriate size.

5. Heating at temperatures higher than 260° results in lower molecular weight polymer, probably because of thermal degradation.

3. Methods of Preparation

Poly(decamethyleneoxamide) has been prepared by salt fusion (5) and by polymerization in the solid state (6). The first reference to the use of oxalate esters is in an early patent of Carothers (7). More recent patents on the use of diamines and oxalate esters describe techniques similar to those used here (8, 9).

4. References

1. Plastics Department, E. I. du Pont de Nemours and Co., Inc., Wilmington, Delaware 19898.
2. Monsanto Research Corp., Dayton Laboratory, Dayton, Ohio 45401.
3. E. C. Horning, *Org. Syn. Coll. Vol.,* 3, 229, 768 (1955).
4. H. Gilman, *Org. Syn. Coll. Vol.,* **1** (1st ed.), 56 (1932).
5. W. H. Carothers (E. I. du Pont de Nemours and Co., Inc.), U.S. Pat. 2,130,948 (September 20, 1938).
6. P. J. Flory (E. I. du Pont de Nemours and Co., Inc.), U.S. Pat. 2,172,374 (September 12, 1939).
7. W. H. Carothers (E. I. du Pont de Nemours and Co., Inc.), U.S. Pat. 2,158,064 (May 16, 1939).
8. S. J. Allen and J. G. N. Drewitt (Celanese Corp.), U.S. Pat. 2,558,031 (June 26, 1951).
9. G. S. Stamatoff (E. I. du Pont de Nemours and Co., Inc.), U.S. Pat. 2,704,282 (March 15, 1955).

Urea Resin

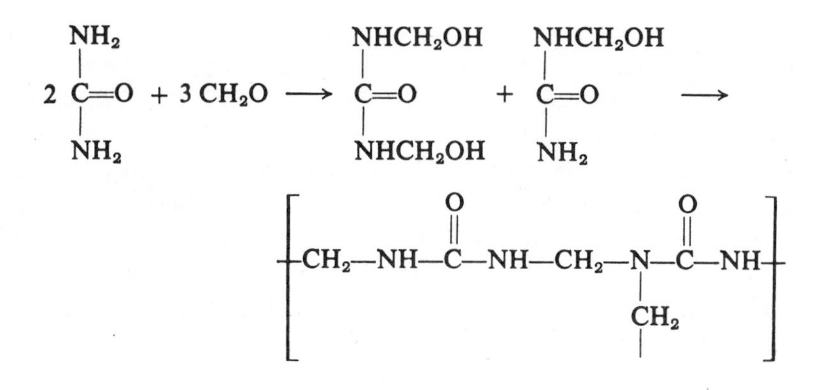

Submitted by L. Boldizar (1a, b)
Checked by H. Roth (2)

1. Procedure

Low methanol-37% formalin (Note 1) (2840 g., 35 moles), 2 g. of 80% triethanolamine, and 1133 g. (18.9 moles) of urea are charged with agitation into a 5-1., three-necked flask equipped with stirrer, reflux condenser, and thermometer. The urea is dissolved while the temperature of the solution is maintained between 20° and 25° (Note 2). When the urea is in solution, the pH is adjusted to 7.2-7.4 with triethanolamine or 5 *N* formic acid (Note 3). The reaction mixture is heated to reflux in 30-40 min., maintaining a uniform temperature rise, using a Glas-Col heating mantle, attached to a 1 KVA variable transformer, at 110 volts. The batch is refluxed between 98° and 102° with vigorous agitation until the reaction is completed. The time cycle is approximately 1-2 hr.

After 5 min. of refluxing, a sample is obtained and cooled to 25° for pH and viscosity tests (Note 4). The pH should be 6.4-6.8. The viscosity is checked every 10 min. and the readings are plotted against time until a value of 17-19 c.p. is anticipated by extrapolation. When the desired viscosity is reached, the heat is removed, the batch is cooled with a cold water or ice bath, and about 2 g. of triethanolamine is added to raise the pH to 7.5-7.7. The resin is concentrated under vacuum (Note 5) until a viscosity of 150-250 c.p. (Note 6) is reached. The pH should be 7.8-8.0 (Note 7).

2. Characterization

The resin-solids content of the syrup obtained by the described procedure is about 60% (Note 8), and the free formaldehyde in the syrup is between 2% and 4% (Note 9). The composition of the resin is $UF_{1.7}$. This means that, because of the equilibrium between the combined and unreacted formaldehyde, about 2.5 moles of the 35 moles of CH_2O charged is not reacted and is lost during concentration. Less than 50% of the combined formaldehyde is converted into methylene bridges during resinification. It can be assumed that the resin is a blend of mono- and dimethylol ureas, partially crosslinked, forming linear or branched units or both. These units are probably dimers and trimers; however, no molecular weight data are available.

3. Notes

1. The formalin should contain less than 2% methanol. If methanol-stabilized formalin (6-7% methanol) is used, the resinification is carried out at a somewhat lower pH level.
2. The urea is added slowly to maintain the desired temperature. The checker found that, if the total urea charge is added at one time, the temperature drops to 5-10°, because of the negative heat of solution, necessitating rapid external heating of the batch back to 20-25°. The resultant syrup is slightly opaque and, although it has the required pH and solids content, the viscosity is higher than desired.
3. If the temperature is allowed to rise above 25°, a rapid reaction starts with constantly changing pH. The pH is measured on a 25-50 g. sample with a glass electrode at 25°.
4. A Brookfield Model LVF viscometer with a No. 1 spindle is used at 60 r.p.m.
5. The temperature of the syrup should not exceed 50°. About 910 g. of condensate should be collected.

6. A Brookfield Model LVF viscometer with a No. 2 spindle is used at 60 r.p.m.

7. The equipment should be cleaned with hot water as soon as possible.

8. A 5-g. sample, weighed in a 3-in. diameter aluminum dish, is dried in an oven at 115° for 5 hr. to determine the solids content of the syrup.

9. The free formaldehyde is determined by the sodium sulfite method. A 5.0-g. sample of the resin is weighed into a 250-ml. beaker and dissolved in 100 ml. of distilled water. Thymolphthalein solution (10-15 drops) is added and the solution is neutralized with 0.5 *N* NaOH. Exactly 5.0 ml. of 0.5 *N* HCl is added from a buret with stirring. A saturated sodium sulfite solution (25 ml.) is added, the mixture is stirred for a few seconds and titrated with 0.5 *N* NaOH to the end point. A blank is run by diluting 25 ml. of saturated sodium sulfite solution with 100 ml. of distilled water. Thymolphthalein indicator (10 drops) is added and the solution is titrated to the end point with 0.5 *N* HCl.

% free formaldehyde =

$$\frac{[(5.0 - \text{blank}) \times 1\ N\ \text{HCl}] - [\text{ml. NaOH} \times 1\ N\ \text{NaOH}]}{\text{grams of sample}} \times 3.003$$

4. Methods of Preparation

The condensation of urea and formaldehyde has been widely investigated, and numerous review articles have been published (3-9).

The resin syrup as described above has a shelf life of at least 6 months. When properly catalyzed with acids or acid salts, it can be used in many binder applications, especially as wood adhesives.

5. References

1a. Plastics and Resins Division, American Cyanamid Co., Wallingford, Connecticut 06492.
1b. Current Address: Plastics Dept., American Cyanamid Co., Stamford, Connecticut 06904.
2. Central Research Laboratories, Inmont Corp., Clifton, New Jersey 07015.
3. A Einhorn, *Ann. Chem.,* **343**, 207 (1905).
4. A. Einhorn and A. Hamburger, *Ber. Deut. Chem. Ges.,* **41**, 24 (1908).
5. C. Ellis, *The Chemistry of Synthetic Resins,* Vol. 1, Reinhold Publishing Corp., New York, 1935, p. 613.
6. H. Kadowaki, *Bull. Chem. Soc. Japan,* **11**, 248 (1936).
7. J. I. deJong and J. deJonge, *Rec. trav. chim.,* **72**, 139, 1027 (1953).

8. H. P. Wohnsiedler, "Amino Resins and Plastics," in *Encyclopedia of Chemical Technology,* Vol. 2, 2nd ed., Interscience Publishers, New York, 1963, pp. 225-258.
9. W. R. Sorenson and T. W. Campbell, *Preparative Methods of Polymer Chemistry,* Interscience Publishers, New York, 1961, p. 300.

Melamine Resin

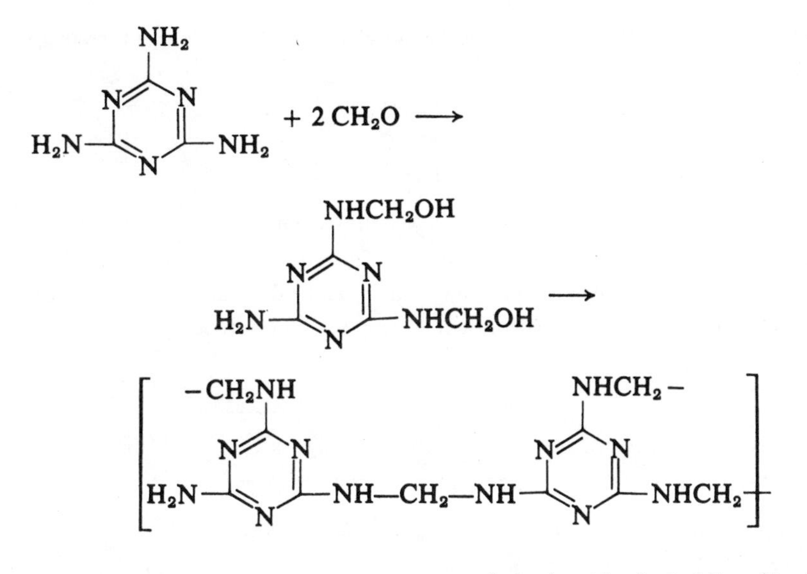

Submitted by L. Boldizar (1a, b)
Checked by H. Roth (2)

1. Procedure

Low methanol-37% formalin (Note 1) (2500 g., 30.9 moles) is charged into a 5-1. three-necked flask equipped with stirrer, reflux condenser, and thermometer. The pH (Note 2) of the formalin is adjusted to 8.0-8.2 with 2 *N* NaOH (about 3.8 ml. is required). Melamine (Note 3) (1890 g., 15 moles) is added with vigorous agitation. The slurry is heated to reflux in 30-40 min., maintaining a uniform temperature rise, with a Glas-Col heating mantle, attached to a 1 KVA variable transformer, at 110 volts. The batch is refluxed between 98° and 102° for 10 min. The heating mantle is removed and the resin syrup is cooled with a uniform cooling rate to 80-85° in 10-15 min. with a cold water bath. The resin is maintained at this temperature until the reaction is completed. The time cycle is approximately 1-2 hr.

The pH and degree of reaction are determined periodically by the "Hydrophobe Test" (Note 4). The pH is maintained at 8.8-9.2 at 25°. The test for Per cent Hydrophobe Solids is carried out at 10-min. intervals until an extrapolated value of 21-25% is obtained (Note 5). The syrup is cooled and the pH is adjusted to 9.2-9.4 by the addition of caustic solution (about 3 ml. of 2*N* NaOH is required). For optimum stability the syrup is stored at 40°. Improved room temperature stability is obtained by dilution to 50% resin solids (Note 6) using water or a mixture of water and alcohol (2B grade ethanol or isopropanol) (Note 7).

2. Characterization

The resin syrup can be dried in a vacuum oven at 50° and 25-in. mercury pressure in 20-24 hr. or by spray drying. The mole ratio of the dry product is 2:1 (formaldehyde/melamine). The resin is a blend of monomeric and dimeric methylolmelamines with the following composition (Note 8):

		Calculated			
		Number Average		Weight Average	
Hydrophobe Solids, %	Number Average Mol. Wt.	Monomer	Dimer	Monomer	Dimer
25	230	74	26	60	40

The triazine units are connected through methylene bridges formed by the elimination of water between methylol and $-NH_2$ or $-NH-$ groups. The formation of ether linkages between two methylol groups by splitting off water is practically non-existent under the mild alkaline and acidic conditions used during the processing and curing of this amino resin.

3. Notes

1. The formalin should contain less than 2% methanol.
2. The pH is measured with a glass electrode at 25°.
3. If recrystallized melamine is used in lieu of Melamine, Buffered (American

Cyanamid Company), the pH of the formalin should be adjusted to 8.4-8.6 before the addition of the melamine.

4. Syrup "Hydrophobe Test": 20 ± 0.2 g. of resin syrup is placed in a 250-ml. glass beaker. Distilled water is added in 1-ml. increments until a permanent white cloud is obtained at 30 ± 0.5° (the degree of cloudiness is not critical). Per cent Hydrophobe Solids (H.S.) at maximum dilution is calculated according to the equation

$$\% \text{ H.S.} = \frac{A \times B}{A + C}$$

where A = weight of syrup sample in grams; B = per cent theoretical solids level of the syrup = 64; C = milliliters of water required for a cloudy mix.

5. The value of % H.S. for each sample, taken at 10-min. intervals, is plotted against time. When the plot indicates that the next sample will give a % H.S. value of 21-25%, the reaction is continued to the indicated time, the syrup is cooled, and the pH is adjusted. The extrapolated % H.S. requires 31-40 ml. of water per 20 g. of resin syrup at 30°.

6. The theoretical amount of resin solids of the syrup is 64%.

7. The equipment should be cleaned with a 1:1 water-alcohol mixture as soon as possible.

8. The molecular weight is obtained using a Mechrolab Vapor Pressure Osmometer.

4. Methods of Preparation

The condensation of melamine and formaldehyde has been widely investigated (3-7). The resin syrup prepared according to the given procedure can be used to saturate pulp, etc., for the manufacture of molding compositions or of various types of papers for the laminating industry. The dry resin is a suitable binder in particle boards.

5. References

1a. Plastics and Resins Division, American Cyanamid Co., Wallingford, Connecticut 06492.
1b. Plastics Dept., American Cyanamid Co., Stamford, Connecticut 06904.
2. Central Research Laboratories, Inmont Corp., Clifton, New Jersey 07015.
3. A. Gams, G. Widmer, and W. Fisch, *Helv. Chim. Acta,* **24**, 302E (1941).
4. T. S. Hodgins, A. G. Hovey, S. Hewitt, W. R. Barrett, and C. Meeske, *Ind. Eng. Chem.,* **33**, 769 (1941).

5. K. Koeda, *J. Chem. Soc. Japan, Pure Chem. Sect.,* 75, 571 (1954).
6. R. Kreton and F. Hanousek, *Chem. Listy,* **48**, 1205 (1954).
7. H. P. Wohnsiedler, "Amino Resins and Plastics," in *Encyclopedia of Chemical Technology,* Vol. 2, 2nd ed., Interscience Publishers, New York, pp. 225-258.

Polythietane

$$n \; \text{(thietane)} \xrightarrow{BF_3:Et_2O} -\!\!\left[(CH_2)_3-S\right]\!\!-_n$$

Submitted by J. A. Empen and J. K. Stille (1)
Checked by S. K. Das (2a, b)

1. Procedure

A. Bulk

A 20-ml. flask with a side arm is flamed and equipped with a nitrogen inlet tube, a small Teflon® stirring bar, and a calcium chloride drying tube. The flask is charged with 5.0 g. (0.068 mole) of thietane (Note 1) and flushed with nitrogen for 10 min. The flask is externally cooled to 0° with an ice-water bath. The contents of the flask are rapidly stirred and 0.43 ml. (0.0034 mole) of boron trifluoride etherate (Note 2) is added gradually through the side arm in a counter current of nitrogen. The flask is sealed (Note 3) and the polymerization is allowed to proceed for 24 hr. at 10° (Note 4). The resulting polymeric mass is dispersed into 300 ml. of methanol, filtered, and dried under reduced pressure to give 2.7-3.2 g. (54-64%) of a rubbery polymer. After two reprecipitations (Note 5) this polymer has an inherent viscosity of 0.55-0.65 dl./g. (0.5 g./100 ml. of chloroform at 25°).

B. Solution

In a nitrogen-filled glove bag (Note 6) a 50-ml., rubber-capped, serum bottle is charged with 5.0 g. (0.068 mole) of thietane (Note 1), 0.31 g. (0.0034 mole) of epichlorohydrin (Note 7), and 25 ml. of methylene chloride (Note 8). The contents are shaken while 0.43 ml. (0.0034 mole) of boron trifluoride etherate (Note 2) is introduced into the solution with a calibrated hypodermic syringe. The bottle is capped and allowed to stand at room temperature for 72 hr. The polymerization is terminated by the addition of 1 ml. of methanol and the polymer is isolated by evaporation of the solvent. The yield of polymer is about 3.0 g. (60%) with an inherent viscosity (after two reprecipitations from chloroform into methanol) of 0.2-0.3 (0.5 g./100 ml. of chloroform at 25°). The polythietanes prepared by procedure A or by this procedure have glass transition temperatures (T_g) at −40 to −50° and soften at about 50-60°.

2. Notes

1. Commercial grade thietane (Eastman Organic Chemicals) was purified by first refluxing over calcium hydride followed by distillation from fresh calcium hydride under nitrogen. The purified monomer was stored in a glass-stoppered bottle over Molecular Sieves 4A (Linde).
2. A practical grade of boron trifluoride etherate (Matheson, Coleman and Bell) was distilled under nitrogen through a short Vigreux column before use, b.p. 126°.
3. The nitrogen inlet was replaced with a ground glass stopper and a pinch clamp was placed over the lead from the side arm.
4. If a higher temperature is employed, the polymerization becomes too rapid and hard to control. At 25° the polymerization was very rapid and an insoluble rubbery product was produced.
5. The polymer was dissolved in about 25 ml. of chloroform and reprecipitated into 250 ml. of methanol.
6. Purchased from I^2R of Cheltenham, Pennsylvania.
7. The epichlorohydrin was employed as a cocatalyst. The yields without a cocatalyst were considerably lower. Commercial grade epichlorohydrin (Eastman Organic Chemicals) was dried over anhydrous magnesium sulfate and distilled under nitrogen into a serum bottle containing molecular sieves, b.p. 117°.
8. Technical grade methylene chloride (Fisher Scientific Co.) was washed with concentrated sulfuric acid until the acid layer remained colorless. The organic layer was washed with water, dried by refluxing over phosphorus pentoxide, and finally distilled from fresh phosphorus pentoxide.

3. Methods of Preparation

Thietane has also been polymerized by Grignard reagents (3) and a variety of other catalytic systems (4). Substituted thietanes have also been reported to undergo cationic polymerization (4-6).

4. References

1. University of Iowa, Iowa City, Iowa 52240.
2a. Pittsburgh Plate Glass Company, Springdale, Pennsylvania 15144.
2b. Current Address: PPG Industries, Allison Park, Pennsylvania 15101.
3. C. C. Price and E. A. Blair, *J. Polym. Sci.,* **A1, 5**, 171 (1967).
4. J. K. Stille and J. A. Empen, *J. Polym. Sci.,* **A1, 5**, 273 (1967).
5. V. S. Foldi and W. Sweeney, *Makromol. Chem.,* 72, 208 (1964).
6. G. L. Brode, *Polymer Preprints,* **6** (2), 626 (1968).

Poly(cyclohexene sulfide)

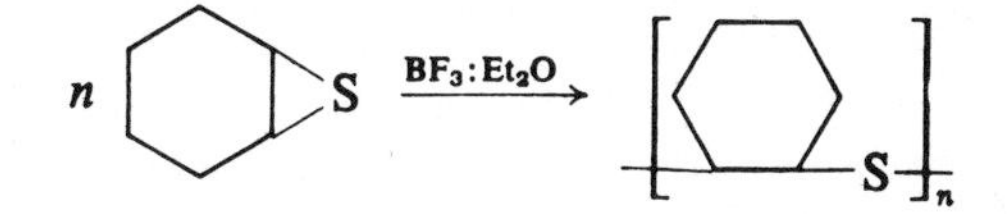

Submitted by J. A. Empen and J. K. Stille (1)
Checked by S. K. Das (2a, b)

1. Procedure

In a nitrogen-filled glove bag (Note 1) a previously flamed 50-ml. serum bottle is charged with 5 g. (0.044 mole) of cyclohexene sulfide (Note 2) and 25 ml. of anhydrous methylene chloride (Note 3). The bottle is capped and 0.28 ml. (0.0022 mole) of boron trifluoride etherate (Note 4) is added through the septum with a calibrated hypodermic syringe. The bottle is shaken occasionally during the addition and is then allowed to stand at room temperature for 48 hr.

The polymerization is terminated by the addition of 1 ml. of methanol, and the resulting polymer is isolated by evaporation of the solvent and drying under reduced pressure to give 4.5-5.0 g. (90-100%) of poly(cyclohexene sulfide). The polymer is dissolved in 25 ml. of chloroform and reprecipitated into cold methanol. After two reprecipitations the polymer is lyophyllized from benzene. The polymer prepared in this manner has an inherent viscosity of 0.2-0.3 dl./g. (0.5 g./100 ml. of chloroform at 25°). The poly(cyclohexene sulfide) has a glass transition temperature (T_g) of approximately $-17°$ and a softening point of 85-100° (Note 5).

2. Notes

1. The collapsible glove bag used in these polymerizations was purchased from I²R of Cheltenham, Pennsylvania.
2. Cyclohexene sulfide was prepared by the method of van Tamelen (3). The

monomer was heated under reflux over calcium hydride and finally distilled from fresh calcium hydride under nitrogen, b.p. 70°/19 torr. It was necessary to distill the monomer before each polymerization because it had a tendency to polymerize on standing.

3. Technical grade methylene chloride (Fisher Scientific Co.) was washed with concentrated sulfuric acid until the acid layer remained colorless. The organic layer was washed with water, dried by refluxing over phosphorus pentoxide and distilled from fresh phosphorus pentoxide. The solvent was stored over Molecular Sieves No. 4A (Linde).

4. A practical grade of boron trifluoride etherate (Matheson, Coleman and Bell) was distilled under nitrogen through a short Vigreux column before use, b.p. 126°.

5. The transition temperatures were obtained on a du Pont model 900 DTA. Lower molecular weight samples may have a T_g as low as −50°.

3. Methods of Preparation

Other catalysts have also proved successful for the polymerization of cyclohexene sulfide (4, 5). The bulk polymerization was too rapid and exothermic to allow sufficient control, however.

4. References

1. University of Iowa, Iowa City, Iowa 52240.
2a. Pittsburgh Plate Glass Company, Springdale, Pennsylvania 15144.
2b. Current Address: PPG Industries, Allison Park, Pennsylvania 15101.
3. E. E. van Tamelen, *Org. Syn. Coll. Vol.*, 4, 232 (1963).
4. R. Backsai, *J. Polym. Sci.*, A, 1, 2777 (1963).
5. J. K. Stille and J. A. Empen, *J. Polym. Sci.*, A1, 5, 273 (1967).

Anionic Polymerization of Formaldehyde

$$CH_2O \xrightarrow[\text{Heptane}]{(n-Bu)_3N} \left[CH_2O \right]_n$$

Submitted by R. N. MacDonald (1)
Checked by V. Jaacks (2a) and S. Iwabuchi (2a, b)

A. Tributylamine as Catalyst

1. Procedure

The apparatus shown in Fig. 1 is assembled in an efficient hood. A 500-ml., two-necked, round-bottomed pyrolysis flask (*A*) is connected to nitrogen inlet *B* and, with 1-in. dia. glass tubing, to trap *C* (Fig. 1) (Note 1). Trap *C* is filled with mineral oil (Note 2) about 1 in. deep so that the passage is just closed and the formaldehyde forms bubbles easily as it passes through the bottom of the trap. Trap *C* is connected to a series of three empty traps, *D-1, D-2,* and *D-3.* The empty traps are all U-shaped, 12 in. high, and made of 1-in. dia. tubing. Trap *D-3* is connected to reactor *E*, a 1-1., three-necked, round-bottomed, creased flask equipped with a mechanical stirrer, a thermometer, and connected by an exit tube (8 mm.) to a glass T-tube. The T-tube is connected to a nitrogen line *F* and a double mineral oil bubbler (Note 2). Bath *G* contains Dow Corning 550 Fluid as heat exchange fluid, bath *H*, a 1-gal. Dewar flask, contains acetone, and bath *J* contains water.

Polymerization. The apparatus is kept under a blanket of nitrogen through nitrogen inlets *B* and *F* as individual members are opened to admit reagents. In pyrolysis flask *A* is placed 50 g. (1.67 mole eq. of formaldehyde) of α-polyoxymethylene (Note 3). Traps *D-1, D-2,* and *D-3* are cooled to −15° with bath *H* and maintained at this temperature (Note 4) throughout the polymeri-

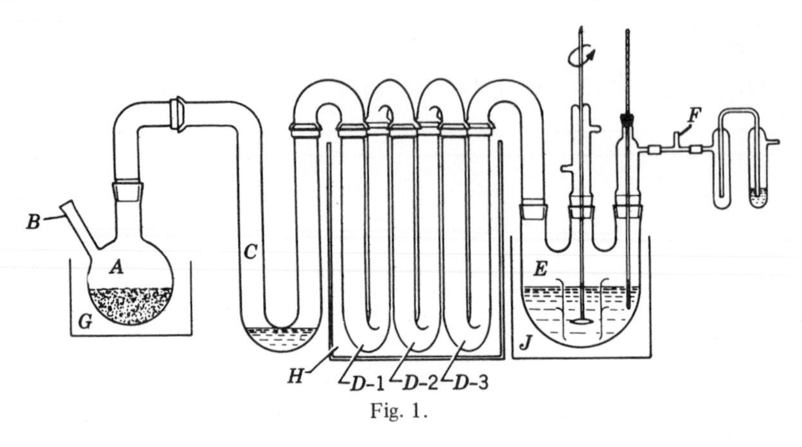

Fig. 1.

zation. *n*-Heptane (Note 5) is passed through a column of silica-gel (Note 6) directly into reactor *E* until 400 ml. is collected as determined by a calibrated mark on the wall of the reactor. Tri-*n*-butylamine (0.13 ml., 5.5 x 10^{-4} mole) and 0.1 g. (6 x 10^{-4} mole) of diphenylamine (Note 7) are added to the *n*-heptane. The nitrogen inlet *B* to the pyrolysis flask *A* is stoppered, and the nitrogen blanket is maintained with a nitrogen stream from *F*. As the reactor solution is stirred vigorously so that liquid splashes upon the walls, and bath *J* is held at 25°, bath *G*, preheated to 190°, is raised around flask *A*. This causes formaldehyde to be generated by pyrolysis as the bath is maintained at 190°. The formaldehyde passes through trap *C*, as evidenced by the bubbling, then through the three–15° traps, and into the top of reactor *E*. Here it is taken up by the vigorously splashing heptane solution. Care must be taken to have sufficient nitrogen pressure to prevent air from being sucked back through the bubbler. After an induction period of a few minutes the solution becomes cloudy from precipitating polyoxymethylene, and a gradual temperature rise of 3-5° is noted in the reactor. Pyrolysis is continued until about 80% of the α-polyoxymethylene is pyrolyzed; this requires 2-3 hr. The slurry of polyoxymethylene in the reactor is then filtered through a sintered glass funnel (Note 8). The solid polymer is washed thoroughly on the funnel with 1 1. of *n*-heptane in several portions, then slurried in an explosion-proof blender with 250 ml. of reagent acetone, and the acetone is decanted. This treatment in the blender is repeated eight times. The polymer is again collected on a sintered glass funnel and dried (Note 8). The yield of snow-white, solid polyoxymethylene is 24-30 g. (60-75%). The polymer melts on a hot stage microscope at 178° (onset of melting) and has η_{inh} 1-3 (Note 9).

2. Notes

1. All glassware should be thoroughly dried by heating for a minimum of 30 min. in a circulating air oven at 120° and assembled hot with a stream of nitrogen blanketing the apparatus as it cools. All connections are standard taper glass joints except that T-tube *F* is joined with Tygon tubing and the thermometer is inserted through a Tygon tubing sleeve.

2. The mineral oil used is Nujol (manufactured by Plough, Inc., New York). It is heated at 100° for 1 hr. as a stream of nitrogen is bubbled through it before use. This removes any traces of water or low-boiling substances. All ground-glass joints are lubricated with this dry mineral oil.

3. α-Polyoxymethylene is prepared conveniently in the laboratory by the following modification of Staudinger's procedure (4). In a 2-1. beaker, 670 ml. of water is heated to 90°, and 453 g. of paraformaldehyde (Note 10) is added as rapidly as bubbling and wetting permit. With the slurry held at 90° a 20% sodium hydroxide solution is added dropwise until the slurry changes from opaque white to opalescent. A few drops should suffice. The pH should be 7.0 at this point. The hot, cloudy solution is then filtered with suction through a sintered glass funnel. The suction flask should be cooled in an ice bath during the filtration. The clear filtrate is immediately transferred to a 2-1. creased flask and stirred vigorously at 40° as a solution of 6.1 g. of sodium hydroxide in 12.5 ml. of water is added dropwise over a 2-hr. period. Stirring is maintained at 40° for a total of 24 hr. reaction time. The slurry is filtered, washed on the filter until neutral with several 200-ml. portions of distilled water followed by 1 1. of reagent acetone, and dried in a vacuum oven overnight at 60°. α-Polyoxymethylene is obtained as 170 g. of snow-white, free-flowing solid.

4. The bath temperature is conveniently maintained at −15° with acetone cooled by addition of solid carbon dioxide from time to time.

5. The *n*-heptane is pure grade (99 mole-% minimum) manufactured by Phillips Petroleum Company, Special Products Division, Bartlesville, Oklahoma. Other aliphatic hydrocarbons in the boiling range of heptane, i.e., ligroin, b.p. 95-100°, may be used if properly purified to remove all ionic or chain transfer agents.

6. The silica-gel is refrigeration grade PA 100 manufactured by W. R. Grace, Co., Davison Chemical Div., Baltimore, Maryland. It is used directly from a freshly opened container. A column 12 in. high and 1 in. in diameter is used. The silica-gel column is purged by passing nitrogen upward before the *n*-heptane is admitted.

7. The tri-*n*-butylamine and diphenylamine are reagent grade chemicals obtained from Eastman Organic Chemicals, Distillation Products Industries, Rochester, New York.

8. Comparable washing can be achieved with a Soxhlet extractor. Distilled technical grade acetone may also be employed. During the filtrations the polymer should be protected from air as much as possible. Suction of air after no liquid is passing through the funnel is to be avoided. The apparently dry polymer still contains a considerable amount of acetone, which may be removed in a vacuum oven at 50° overnight.

9. The viscosity of a 0.5% solution of polyoxymethylene in *p*-chlorophenol containing 2% α-pinene as stabilizer is measured at 60° (3). The *p*-chlorophenol should be freshly distilled from sodium hydroxide under 60 torr pressure.

10. The paraformaldehyde used is the purified grade manufactured by Allied Chemical Corp., General Chemical Division, Morristown, New Jersey.

3. Methods of Preparation

High molecular weight polyoxymethylene may be obtained from pure formaldehyde with other initiators (4, 5) or from the cyclic oligomers such as trioxane with cationic initiators (6). Initiation of formaldehyde polymerization by radiation has also been reported to give high molecular weight polyoxymethylene (7).

4. Merits of the Preparation

Formaldehyde polymers prepared according to this method have very good thermal stability. Tough, bubble-free films which retain their toughness even after a week's aging at 105° in air, can be compression-molded.

5. References

1. Contribution No. 1164 from the Central Research and Development Department, E. I. du Pont de Nemours and Co., Wilmington, Delaware 19898.
2a. Organisch-Chemisches Institut der Universitat, Mainz, Germany.
2b. Current Address: Chiba University, Chiba, Japan.
3. R. N. MacDonald (E. I. du Pont de Nemours and Co.), U.S. Pat. 2,768,994 (1956); C. E. Schweitzer, R. N. MacDonald, and J. O. Punderson, *J. Appl. Polym. Sci.*, **1**, 158 (1959).
4. H. Staudinger, R. Signer, and O. Schweitzer, *Ber.*, **64**, 398 (1931).

5. J. Furukawa and T. Saegusa, *Polymerization of Aldehydes and Oxides,* Interscience Publishers, New York, 1963, p. 64.
6. A. K. Schneider (E. I. du Pont de Nemours and Co.) U.S. Pat. 2,795,571 (1957); D. E. Hudgin and F. M. Bernardinelli (Celanese Corporation of America), U.S. Pat. 2,989,506 (1963); W. Kern, *Chemiker-Ztg.,* 88, 623 (1964).
7. S. Okamura, S. Nakashio, K. Hayashi, and I. Sakurada, *Isotopes Radiation (Tokyo),* 3, 242 (1959); C. Chachaty, M. Magat, and L. Ter Minassian, *J. Polym. Sci.,* 48, 139 (1960).

B. Pyridine as Catalyst

$$CH_2O \xrightarrow{\text{Pyridine}} \left[CH_2{-}O \right]_n$$

Submitted by W. Kern and V. Jaacks (1)
Checked by I. H. Song and D. P. Wyman (2)

1. Procedure

A. Monomeric Formaldehyde

Anhydrous formaldehyde is generated in an efficient hood in the apparatus shown in Fig. 2 by pyrolysis of α-polyoxymethylene, a polycondensate obtained from formalin (3) (Note 1). The 500-ml., two-necked flask *A* is charged with 60 g. of α-polyoxymethylene and 60 g. of paraffin oil to serve as a heat transfer agent. Before attaching flask *A* to the rest of the apparatus it is temporarily equipped with a gas inlet and outlet, and nitrogen is passed over the surface of the slurry. The flask is heated in an oil bath until at ca. 130° rapid evolution of gaseous formaldehyde begins. Because the first formaldehyde contains the bulk of residual water from the α-polyoxymethylene, it is swept into the hood with the nitrogen.

After decomposition of about 10% of the α-polyoxymethylene one neck of flask *A* is closed with a well-greased stopper, and the flask is attached to the rest of the apparatus which has previously been flamed under reduced pressure and filled with dry nitrogen (Note 2). The apparatus consists of an empty glass tube (*T*) 2-3 m. long and 2-3 cm. in inner diameter. While passing through this tubing, the formaldehyde partially polymerizes to low molecular weight polyoxymethylene, thus binding water or methanol in the form of hydroxy and methoxy end groups, respectively (Note 3). The purified formaldehyde gas is then condensed in a cold trap (Note 4) containing 250 ml. of

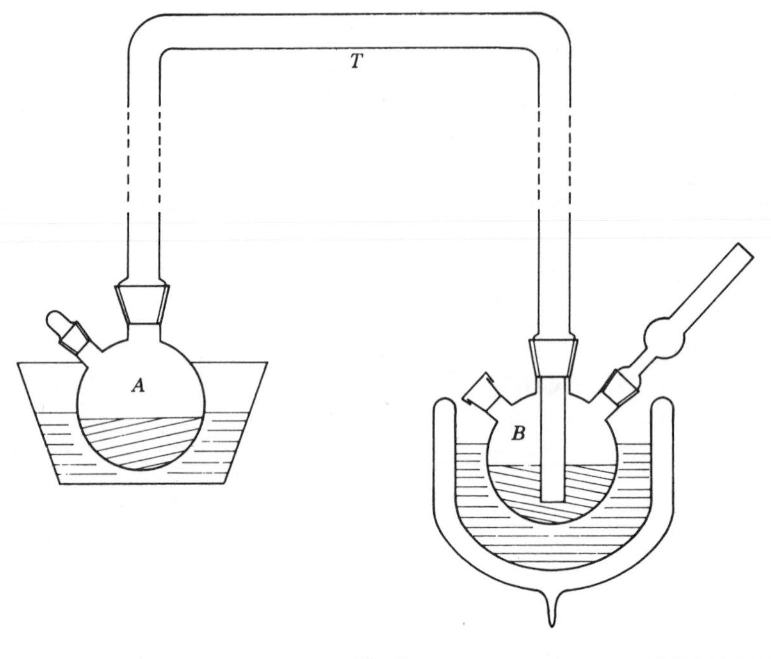

Fig. 2.

anhydrous ether, is distilled from sodium and cooled to −78°.

The pyrolysis of α-polyoxymethylene is continued at a fairly high rate while increasing bath temperatures to about 180°; all α-polyoxymethylene should be decomposed within 1 hr. (Note 5). About 40-45 g. of monomeric formaldehyde is absorbed in the ether, yielding an approximately 4 molar solution.

B. Polymerization

The gas inlet in flask *B* is replaced by an airtight mechanical stirrer (Note 6). Under vigorous stirring and efficient Dry Ice cooling, polymerization is initiated at −78° by cautiously injecting through the serum cap 2 mg. of pyridine dissolved in 5 ml. of anhydrous ether (Note 7). Polymerization, indicated by precipitation of white solid polyoxymethylene, should start before the pyridine addition is completed. If impurities in the monomer solution inhibit polymerization, more initiator should be used. The flask is kept at −78° for 45 min. after initiation of polymerization, during which time the contents solidify and stirring has to be interrupted. The ether and traces of unconverted formaldehyde are then removed under reduced pressure at slightly elevated temperature. The yield of dry powdery polymer is almost quantitative (ca. 40 g.).

2. Characterization

The white powdery polymer melts at 170-175°. Because of its high crystallinity, polyoxymethylene can be dissolved only at temperatures above 140° in solvents such as dimethylformamide, γ-butyrolactone, and benzyl alcohol (4). It may be dissolved at lower temperatures with gradual decomposition in slightly acidic solvents, e.g., in phenol at 90° and at room temperature in hexafluoroacetone sesquihydrate (5) buffered with 1 wt. % of triethylamine.

Viscosity measurements at 140° in dimethylformamide yield η_{sp}/c values between 1 and 2 dl./g., corresponding to molecular weights M_w between 100,000 and 200,000 (6).

Owing to unstable hemiacetalic hydroxyl end groups, the polymer depolymerizes almost quantitatively to formaldehyde when heated to 190° under nitrogen for a few hours. It can be stabilized by acetylation of the end groups with acetic anhydride in the same manner as described for polyoxymethylene prepared from trioxane (7).

3. Notes

1. α-Polyoxymethylene is prepared by acid-catalyzed polycondensation of formaldehyde in 35-60% aqueous solution (intermediate formation of methylene glycol $HO{-}CH_2{-}OH$) (3). It should be substantially dry. If no α-polyoxymethylene is available, paraformaldehyde can be used instead; it has a lower degree of polymerization and therefore contains more water chemically bound as hydroxyl end groups. If it contains appreciable amounts of acid or alkali, it should be washed neutral with water and dried under reduced pressure before use.

2. Careful exclusion of water during the entire process is necessary because water acts both as a retarder and as a chain transfer agent in ionic polymerization.

3. To prevent ground joints from being stuck together by polyformaldehyde they should be very well greased with petroleum jelly. The stopper on flask *A* acts as an emergency opening in case the apparatus should become plugged somewhere with polymer. The mobility should be frequently checked to avoid explosions of the apparatus caused by the high formaldehyde pressure.

4. A 500-ml., three-necked flask equipped with a serum-capped opening, a wide gas inlet tube, and a drying tube filled with fused powdered KOH is conveniently used. Diffusion of atmospheric moisture or of carbon dioxide from the cooling bath into the formaldehyde solution inhibits polymerization, because carbon dioxide is an effective inhibitor of anionic polymerizations.

5. If the formaldehyde is generated more slowly, too much of the monomer is lost by prepolymerization in the glass tube.

6. A sufficiently strong magnetic stirrer that can be operated from outside the Dewar vessel may be used instead.

7. Initiator should be added portionwise during 5 min. If the polymerization becomes too vigorous, the addition should be interrupted. When polymerizing batches of formaldehyde solution larger or more concentrated than indicated here, care must be taken to achieve good cooling and efficient transfer of the heat of polymerization to avoid explosive polymerization.

4. Methods of Preparation

Another method of preparing anhydrous formaldehyde utilizes trioxane as starting material. The trioxane is depolymerized at 220° on a phosphoric acid substrate (8).

Anhydrous formaldehyde can also be polymerized by cationic initiators. Molecular weights are usually not so high as in anionic polymerization, however, and acidic catalyst residues decrease the stability of the polymer. The cationic polymerization of trioxane is described in this volume (7).

5. References

1. Organisch-Chemisches Institut der Universität, 65 Mainz, Germany.
2. Borg Warner Corp., Marbon Chemical Div., Washington, West Virginia 26181.
3. J. F. Walker, *Formaldehyde,* 3rd ed., Reinhold Publishing Corp., New York, 1964, pp. 145-161.
4. R. C. Alsup, J. O. Punderson, and G. F. Leverett, *J. Appl. Polym. Sci.,* **1**, 185 (1959).
5. W. H. Stockmayer and Lock-Lim Chan, *J. Polym. Sci.,* **A2**, 4, 437 (1966).
6. L. Höhr, V. Jaacks, H. Cherdron, S. Iwabuchi, and W. Kern, *Makromol. Chem.,* **103**, 279 (1967).
7. W. Kern and V. Jaacks, this volume, p. 279.
8. A. Giefer, V. Jaacks, and W. Kern, *Makromol. Chem.,* 74, 39, 46 (1964).

Cationic Polymerization of 1,3,5-Trioxane

$$\text{(trioxane ring: } H_2C{-}O{-}CH_2{-}O{-}H_2C{-}O\text{)} \xrightarrow{BF_3 \cdot Et_2O} \left[O{-}CH_2{-}O{-}CH_2{-}O{-}CH_2 \right]_n$$

Submitted by V. Jaacks, (1a) S. Iwabuchi, (1a, b) and W. Kern (1a)
Checked by K. D. Kiss (2a) and I. Rosen (2a, b)

1. Procedure

To a flame-dried, 500-ml., round-bottomed flask equipped with a serum stopper are charged, under anhydrous conditions, in an efficient hood, 90 g. (1.0 mole) of pure anhydrous trioxane (Note 1) and 210 g. of ethylene dichloride, (*Caution! Many chlorinated hydrocarbons are toxic and should be used with adequate precaution.*) which has been refluxed and distilled over phosphorous pentoxide. Boron trifluoride etherate (70 mg., 0.5 mmole) in 7 ml. of ethylene dichloride is injected with a hypodermic syringe while the mixture is shaken vigorously (Note 2). The flask is then immersed in a bath at 45°.

After an induction period of about 2 min., during which small amounts of formaldehyde are formed, precipitation of insoluble polyoxymethylene begins. The reaction mixture should solidify within 10 min. (Note 3). After a total of 60 min. at 45° sufficient acetone is added to the polymer to form a slurry. The slurry is filtered and the polymer is thoroughly washed several times with acetone on a glass filter and sucked dry. To remove catalyst residues, which may degrade the polymer, as effectively as possible, it is boiled for 30 min. with 1 l. of ether containing 2 wt. % of tributylamine. Yield 40-50 g.

2. Characterization

The polyoxymethylene is a white powder of melting temperature 176-178° (in a sealed glass tube); it is soluble in phenol at 115°, in dimethylformamide or benzyl alcohol at 140° (3). Because of the unstable hemiacetalic hydroxyl end groups and unremoved catalyst residues, the polymer decomposes to formaldehyde almost entirely within 2 hr. at 190° even under pure nitrogen. Therefore it is recommended that molecular weights be determined by viscosity measurement after stabilizing the polymer by acetylation, otherwise degradation will interfere.

3. Stabilization by End Group Acetylation (4)

Polytrioxane (1 g.) is mixed with 100 ml. of dimethylformamide, 40 ml. of pure acetic anhydride, (*Hood!*) and 8 ml. of N,N-dimethylcyclohexylamine or a similar high-boiling tertiary amine (Note 4) in a 250-ml. flask equipped with a reflux condenser and a calcium chloride tube. The polymer is dissolved at about 145° by short and cautious heating over a free flame. The flask is then immersed in a bath at 140° for 30 min. The polymer, which precipitates on cooling, is filtered, washed with acetone, refluxed for 30 min. with 100 ml. of fresh acetone and dried. About 950 mg. of acetylated polymer is recovered that undergoes less than 10% decomposition during heating at 190° for 16 hr. under pure nitrogen (Note 4).

The stabilized polymer in a 1% solution in dimethylformamide at 140° has η_{inh} 0.50 dl./g., corresponding to a molecular weight $M_w \sim 50{,}000$ (5) (Note 5).

4. Notes

1. Trioxane (m.p. 64°) can be purified and dried by very careful distillation (b.p. 115°) through an efficient column (b.p. of azeotrope containing 30 wt. % of water, 91.3°.). Crystallization of trioxane during distillation and storage may cause partial polymerization and should therefore be avoided.

Trioxane can also be conveniently purified with chemical reagents, e.g., refluxing for 20 hr. over 5 wt. % of sodium or calcium hydride or with 10-20 wt. % of naphthalene-1,5-diisocyanate and 0.3% of dibutyl tin dilaurate or triethylenediamine as catalyst and subsequent fractional distillation.

In all cases, care should be taken to avoid adherence of water or methanol-containing condensates to the upper portion of the reflux condenser.

2. $SnCl_4$ may be used instead of $BF_3 \cdot Et_2O$.

3. The rate of polymerization depends greatly on the purity and dryness of the reagents and solvent. If polymerization is slow, more initiator should be added.

4. The checkers used tributylamine and obtained 875 mg. of acetylated polymer, η_{inh} 0.42 dl./g. The initial decomposition rate at 190° under pure nitrogen was 0.6%/min.

5. The molecular weights obtained are extremely dependent on the dryness of the trioxane solution; careful exclusion of water during the polymerization procedure is required.

5. Methods of Preparation

Polyoxymethylene of high molecular weight can also be made by anionic or by cationic polymerization of anhydrous formaldehyde (6-8). Purification, handling, and polymerization of trioxane, however, are much easier than that of anhydrous formaldehyde. Polycondensation of aqueous formaldehyde solutions yields polymers of comparatively low molecular weight (paraformaldehyde) (6).

6. References

1a. Organisch-Chemisches Institut der Universität, 65 Mainz, Germany.
1b. Current Address: Chiba University, Chiba, Japan.
2a. Research Department, Diamond Shamrock Co., T. R. Evans Research Center, Painesville, Ohio 44077.
2b. Current Address: Standard Oil Co., Cleveland, Ohio 44108.
3. R. G. Alsup, J. O. Punderson and G. F. Leverett, *J. Appl. Polym. Sci.,* **1**, 186 (1959).
4. H. Deibig, H. Höcker, V. Jaacks, and W. Kern, *Makromol. Chem.*, 99, 9 (1966).
5. L. Höhr, V. Jaacks, H. Cherdron, S. Iwabuchi, and W. Kern, *Makromol. Chem.,* **103**, 279 (1967).
6. J. F. Walker, *Formaldehyde,* 3rd ed., Reinhold Publishing Corp., New York, 1964, pp. 140-199.
7. W. Kern, H. Cherdron, and V. Jaacks, *Angew. Chem.*, 73, 177 (1961).
8. Z. Machácek, J. Mejzlik, and J. Pác, *J. Polym. Sci.,* **52**, 309 (1961).

Amorphous Polyacetaldehyde

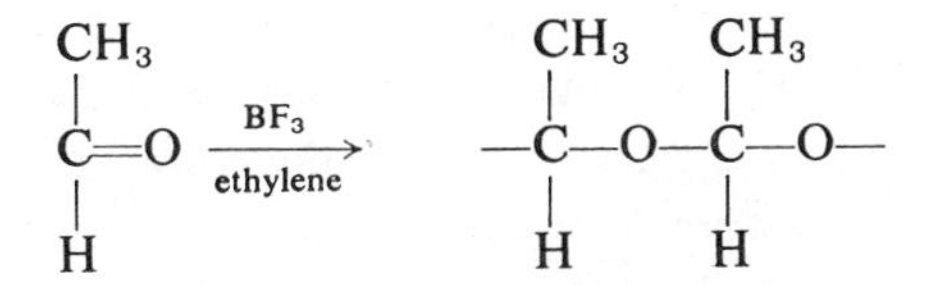

A. Boron Trifluoride as Catalyst

Submitted by O. Vogl (1a, b)

Checked by M. Goodman (2a, b) and G. C. C. Niu (2a, c)

1. Procedure (3)

Polymerization grade ethylene (250 ml.), b.p. −104°, is condensed with a liquid nitrogen bath (Note 1) in a 1-l., four-necked flask equipped with stirrer, thermometer (reading to −200°), gas inlet tube, and short condenser (Note 2). The condenser terminates in a Claisen adapter which is connected with a glass T to a dry nitrogen line and a double bubbler filled with mineral oil. The condensation of ethylene takes about 10-20 min. The gas inlet tube is replaced by an adapter with a serum stopper cap.

Stirring is started, and 39 g. (50 ml., 0.88 mole) of acetaldehyde (Note 3) is injected slowly to avoid excessive boiling of the ethylene using a cooled (polyethylene dry bag in refrigerator at 0°) hypodermic syringe. The internal temperature of the reaction mixture is kept between −110° and −120° by lowering and raising the liquid nitrogen bath. Care should be taken that no freezing of the reaction mixture occurs. Gaseous BF_3 (3 ml., 0.13 mmole) is now injected into the mixture with a hypodermic syringe (Note 4).

In 10-20 min. the stirrer stops because of the formation of polymer. The reaction is allowed to stand for an additional 30 min. while the ethylene is allowed to evaporate slowly. Excess ethylene is decanted with nitrogen blanketing, and 100 ml. of anhydrous pyridine is added. On stirring at room temperature, the polymer dissolves completely in the pyridine. If desired, the polymer may be isolated at this stage by pouring the pyridine solution into cold water.

For the end capping, 300 ml. of acetic anhydride is added to the pyridine solution of polymer, and the mixture is stirred under nitrogen for 1-2 hr. The brownish-green viscous solution is poured into 1 kg. of ice and 1 l. of water, thereby precipitating the polymer. The polymer is then kneaded by hand in several changes of water and ice to destroy the acetic anhydride completely. (Rubber gloves should be worn during this operation.) The washing is complete when the wash water is colorless. Dry weight of the polymer 34-37 g. (87-94%), η_{inh} = 1-2 dl./g. (Notes 5 and 6).

2. Notes

1. Instead of using a liquid nitrogen bath directly, which is often difficult to control, to cool the polymerization flask, it was found advantageous to use other low-temperature baths. A methylcyclohexane (m.p. $-126°$) bath can be used at $-120°$ with a liquid nitrogen cooling bath. A pentane bath can be held at $-120°$ by bubbling liquid nitrogen slowly through the pentane. The latter bath poses a considerable fire hazard, however, and must be used in a ventilated hood.

2. To obtain high molecular weights all operations must be carried out under rigorously dry conditions. The reaction flask should be flamed, the remaining parts of the apparatus and the syringes should be assembled while hot and then cooled in a stream of dry nitrogen.

3. Pure acetaldehyde may be obtained by decomposition of paraldehyde (4) or by careful fractional distillation of acetaldehyde (3). For the preparation of elastomeric polyacetaldehyde even of very high molecular weight, the latter procedure is adequate. Acetaldehyde, obtained from Eastman in a sealed vial, was stirred for 1 hr. at 0° with $Na_2CO_3 \cdot H_2O$, followed by stirring with $MgSO_4$ for 30 min. It was distilled in a low-temperature still from an antioxidant [0.1% of AgeRite White (Note 7)] in a nitrogen atmosphere, the first 20% of the distillate being discarded. The receiver was kept at 0°, and the purified sample was used immediately. The distillation apparatus was rinsed with soap solution or dilute NaOH solution after each distillation to eliminate acid buildup in the distillation column. Acidic cleaning solutions, particularly for the Dry Ice condenser should be avoided. Acidity on the glass surface causes rapid polymer formation and plugging of the condenser.

4. BF_3 etherate may be used instead of gaseous BF_3. The etherate brings about polymerization much faster than does gaseous BF_3. Introduce the corresponding amount of catalyst (16 ml. of BF_3 etherate) to the reaction mixture with a hypodermic syringe.

5. The inherent viscosity of polyacetaldehyde is measured in 0.5% butanone solution. Polymers having up to η_{inh} 5 dl./g. have been obtained by the method described in this preparation.

6. To obtain a stable polymer all impurities must be removed from the rubbery polymer. This is done by dissolving the crude capped polymer in 10-20 times its volume of peroxide-free ether and washing it as follows. Pyridine is removed first by shaking the ethereal solution in separatory funnel with water containing 1% of acetic acid, followed by distilled water. Residual acetic acid and initiator are then removed by shaking several times with 1% sodium carbonate solution, followed by repeated washings with distilled water. It is sometimes necessary to break the emulsions by adding sodium chloride. The polymer solution is finally dried over magnesium sulfate and filtered. Antioxidant and thermal stabilizer are added, and the solvent is evaporated under reduced pressure (oil pump vacuum) at room temperature. Removal of the last traces of solvent requires good vacuum.

7. This antioxidant is *sym*-di-β-naphthyl-*p*-phenylenediamine obtained from R. T. Vanderbilt Co., Inc., New York, New York.

3. Methods of Preparation

Elastomeric polyacetaldehyde has also been prepared by using a number of other Lewis and protic acids as initiators (3). Aluminum alkyls, aluminum alkoxides, zinc alkyls, and phosphines also give polyacetaldehyde in good yield. The latter polymers, however, contain varying degrees of crystalline isotactic polyacetaldehyde. Acidic alumina and other metal oxides have been shown to give amorphous polyacetaldehyde (5).

4. Merits of the Preparation

Although most methods of acetaldehyde polymerization give good yields of relatively low molecular weight amorphous polymer often accompanied by some paraldehyde formation, the method described gives high yields of very high molecular weight polymer free of paraldehyde. This method is also very suitable for a larger-scale preparation.

5. References

1a. Central Research Department, Experimental Station, E. I. du Pont de Nemours and Company, Wilmington, Delaware 19898.
1b. Current Address: Institute of Polymer Science, University of Massachusetts, Amherst, Massachusetts 01002.
2a. Polytechnic Institute of Brooklyn, Brooklyn, New York 11201.
2b. Current Address: University of California/San Diego, LaJolla, California 92037.
2c. Current Address: Soflens Div., Bausch & Lomb Corp., Rochester, New York 14602.
3. O. Vogl, *J. Polym. Sci.,* A2, 4591 (1964).
4. M. Letort and J. Petry, *J. Chim. Phys.,* 48, 594 (1951).
5. J. Furukawa and T. Saegusa, *Polymerization of Aldehydes and Oxides,* Interscience Publishers, New York, 1963.

B. Fluorisil-Phosphoric Acid as Catalyst

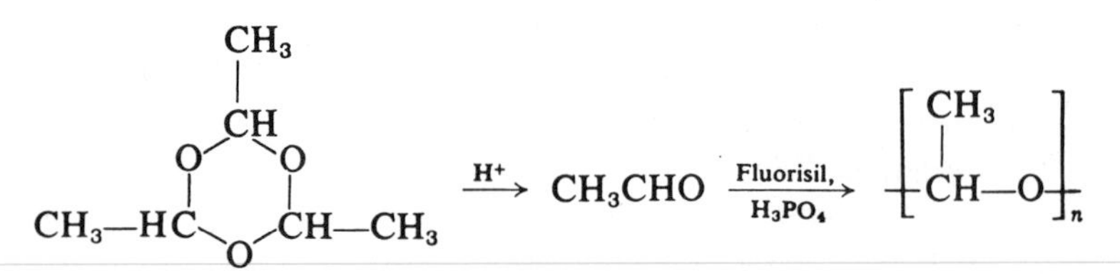

Submitted by J. Brandrup (1a, b) and M. Goodman (1a, c)
Checked by O. Vogl (2a, b) and L. Hammond (2a)

1. Procedure

A 50 ml., three-necked flask, a fractionating column (Note 1) 15-20 cm. long, a condenser, and a 25 ml., two-necked receiver are assembled hot (Note 2) under nitrogen (Note 3). The system is closed by a bubble trap filled with mineral oil. Fluorisil (1 g.) is placed in the receiver after disconnecting the bubble trap from the receiver. A 5% solution of 100% phosphoric acid (Note 4) in absolute ether (1 ml.) is introduced the same way (Note 5). Predried paraldehyde (25 ml.) (Note 6) is placed in the distilling flask under nitrogen, and dry cupric sulfate (1 g.) (Note 7) is added in a similar manner. After all additions have been made, the condenser is cooled to approximately $-60°$ to $-70°$ and the receiver is placed in a Dewar flask filled with Dry Ice-acetone ($-78°$) (Note 8).

Distillation (oil bath) of the paraldehyde commences. The decomposition of paraldehyde is regulated so that all acetaldehyde is cooled to at least $-50°$

before contacting the catalyst (Note 9). Distillation occurs over about a 1 hr. period.

After nearly all monomer is distilled, the receiver is disconnected and quickly stoppered. The receiver is kept at −78° for 1-2 days. The molecular weight increases with increasing reaction time. Very high molecular weight material forms from the glassy, clear solid (at −78°). The polymer is collected after 2-3 days. At the end of the reaction time, 20 ml. of pyridine is added to the solution (Note 10) and the flask is warmed. The polymer is filtered from the Fluorisil and precipitated by decantation into water with agitation. It is advisable to redissolve the polymer in methanol and to add a methanolic solution of a stabilizer. One such additive is composed of Ultramid 1C and β-naphthylamine (Note 11). The stabilized polymer is precipitated by decantation in water and dried under reduced pressure at room temperature. Rubbery, amorphous polyacetaldehyde is obtained that has a reduced viscosity of 0.16 dl./g. in a 0.1% dimethylformamide solution, yield 8-13 g. (40-70%).

The polymer is unstable if not properly handled (avoid contact with acid). It should be stored in a cool place.

2. Notes

1. Many types of fractionating columns such as Vigreux or packed columns can be used.

2. All glassware must be washed with alkali to remove any trace of acid to avoid undesired polymerization. The glassware is dried at 120-150° overnight and is then immediately assembled under dry nitrogen starting with the three-necked flask to ensure dryness.

3. Oxygen-free, dry nitrogen is obtained by passing commercial prepurified nitrogen through concentrated sulfuric acid and sodium hydroxide.

4. Phosphoric acid (85%) is treated in ether with the appropriate amount of P_2O_5 to bring it to 100%.

5. Fluorisil from the Floridin Company, Tallahassee, Florida, was heated at 450-500° for 24 hr., weighed hot, and immediately introduced into the flask. Any other protonic acid may be used in place of the phosphoric acid (3).

6. Paraldehyde from Baker and Adamson is stored over calcium hydride. Before use it is decanted into a flask containing aluminum triisobutyl (1-3% by volume based on paraldehyde). The paraldehyde is then distilled immediately before use to remove any trace of water (b.p. 121° at 760 mm.). *Caution! Aluminum alkyls may be pyrophoric.*

7. Copper sulfate is dried shortly before use by heating over an open flame until the entire mass is white.

8. Methanol is circulated by a Cole-Parker pump No. 7000 through a copper spiral kept at $-78°$ with Dry Ice-acetone.

9. If the distillation occurs too rapidly, the monomer is not cooled sufficiently and paraldehyde is formed on contact with the catalyst.

10. Amorphous polyacetaldehyde is very sensitive to acid degradation. The polymer can be degraded in a very short time during the workup if the catalyst is not destroyed. Any basic substance (aniline, ammonia, amines) can be used in place of pyridine.

11. Ultramid 1C, BASF, Germany, is a methanol-soluble amide which forms a synergistic stabilizer with β-naphthylamine for aldehydes (4).

3. References

1a. Polymer Research Institute and Department of Chemistry, Polytechnic Institute of Brooklyn, Brooklyn, New York 11201.

1b. Current Address: Hoechst, Frankfurt am Main, West Germany.

1c. Current Address: University of California/San Diego, La Jolla California 92037.

2a. Central Research Department, E. I. du Pont de Nemours and Co., Experimental Station, Wilmington, Delaware 19898.

2b. Current Address: Institute of Polymer Science University of Massachusetts, Amherst, Massachusetts 01002.

3. M. Letort, *Compt. Rend.,* **240**, 86 (1955); **241**, 651, 1765 (1955).

4. K. Weissermel and W. Schmieder, *Makromol. Chem.,* **51**, 39 (1962).

Poly(ethylene oxide)

$$H_2C\underset{O}{\overset{}{—}}CH_2 \xrightarrow[\substack{C.\ Ca(NH_2)_2—C_2H_4O \\ D.\ KOH}]{\substack{A.\ R_2Zn—CH_3OH \\ B.\ FeCl_3—C_3H_7O\ \text{complex}}} \left[CH_2—CH_2—O\right]_n$$

Submitted by F. E. Bailey, Jr., and H. G. France (1)
Checked by C. C. Price, R. Spector, and Y. Atarashi (2)

A. $R_2Zn—CH_3OH$ as Catalyst

1. Procedure

Under a nitrogen atmosphere 15 ml. of dry toluene and 15 g. of ethylene oxide are charged to a suitable Pyrex® pressure tube (Note 1) followed by addition of 1 ml. of a toluene solution of methanol-modified dibutyl zinc (*Caution! Dibutyl zinc is pyrophoric and highly reactive.*) (Note 2). The tube is capped with a rubber policeman and placed in a Dry Ice-acetone bath. The tube is flame-sealed (Note 3) and then warmed slowly to about 50° and placed in a constant-temperature bath (Note 4) at 90° for about 24 hr. After this time the tube is cooled (Note 5) and opened, and the solid polymer is dissolved in 300 ml. of benzene. The polymer is precipitated by slowly adding the benzene solution to 2 1. of heptane which is rapidly agitated with a suitable stirrer. The white, tough, solid polymer is recovered from the heptane by filtration and is dried under vacuum at 40° for 12 hr. There is obtained a 50-70% conversion to polymer having a reduced viscosity in acetonitrile of 1.8-3.3 dl./g. (Note 6).

2. Notes

1. A suitable pressure tube can be prepared from a 9-in. piece of 22-25 mm. Pyrex® tubing that is sealed on one end. To the other end is sealed a 3-in. piece of 8-mm. tubing. The monomer, diluent, and catalyst are charged to the polymerization tube with a thistle tube that passes through the 8-mm. neck and extends to the bottom of the tube. *Caution! Because these tubes are under pressure during polymerization, they should be properly annealed, and extreme care should be exercised in handling them.*

2. Dibutylzinc can be prepared as described in *Organic Syntheses* (3). The methanol modification is conducted by dissolving 10.5 g. of dibutylzinc in 141 ml. of dry toluene and slowly adding 0.96 g. of methanol. This solution should be stored at temperatures below 0° to retard degradation of the catalyst. Diethylzinc can also be used as the catalyst. The use of different catalysts and different degrees of modification of catalyst alters the yield and properties of the polymer product.

3. The polymerization tube is cooled in Dry Ice-acetone for a period of time sufficient to ensure good sealing because of the partial vacuum obtained in the tube. This vacuum can be determined by observing the decrease in pressure on the rubber policeman used to cap the tube.

4. Any suitable bath that can be maintained at 90° and that allows gentle rocking or rotating of the polymerization tubes can be used.

5. The tube should be cooled to Dry Ice-acetone temperature to reduce the pressure in the tube prior to opening.

6. Reduced viscosity is determined at a concentration of 0.2 g./100 ml. of acetonitrile at 30°. The molecular weight can be controlled by changes in catalyst concentration, monomer/solvent ratio, and the ratio of modifier in the catalyst preparation.

B. $FeCl_3—C_3H_7O$ Complex as Catalyst

1. Procedure

A Pyrex® polymerization tube is charged in a manner similar to that in part A, except that 0.15 g. of ferric chloride-propylene oxide (Note 1) complex is used as catalyst. The tube is sealed and placed in a constant-temperature bath at 90° for 24 hr., after which time the polymer is recovered as in part A.

There is obtained a 70-100% yield of polymer having a reduced viscosity in acetonitrile of 1.2-1.6 dl./g. (Note 2).

2. Notes

1. The ferric chloride-propylene oxide complex catalyst is prepared as described in the literature (4).

2. Variations in catalyst preparation and changes in the monomer/solvent ratio alter both the yield and the viscosity of the polymer product.

C. $Ca(NH_2)_2$—C_2H_4O as Catalyst

1. Procedure

Under a nitrogen atmosphere, poly(ethylene oxide) can be prepared at room temperature using an ethylene oxide-modified calcium amide catalyst (Note 1). A 2-1., four-necked flask equipped with stirrer, thermometer, a gas inlet tube which extends to the bottom of the flask, and a reflux condenser is charged with 1 1. of dry heptane and a quantity of catalyst suspension (Note 1) equivalent to 1 g. of contained calcium. This heptane/catalyst suspension is stirred rapidly, and ethylene oxide is bubbled into the reaction mixture at sufficient rate to give a "blow-off" of not less than 40 bubbles/min. The temperature increases from 25° to 45° in about 10-15 min. and is maintained by cooling with an ice bath at a temperature below 50° (Note 2) for about 7 hr. The resulting finely divided polymer is recovered by filtration and dried under vacuum at 40° for 12 hr. A white granular polymer (about 70-80 g.) having a reduced viscosity in acetonitrile of about 20-30 (Note 3) is obtained.

2. Notes

1. Ethylene oxide-modified calcium amide is prepared by a procedure similar to that described in U.S. Pat. 2,971,988. Calcium (5 g.) is slowly added to 200 ml. of liquid ammonia in a 1-1. glass flask, in an efficient hood, exposure of the contents to the atmosphere being avoided. In this procedure it is best to use a distilled granular calcium metal rather than calcium turnings or wire, because the presence of calcium oxide reduces the activity of the catalyst. The

ammonia should contain less than 0.1% water, because the presence of water also diminishes the catalyst activity. If desired, the calcium-ammonia solution can be filtered to remove the insoluble calcium oxide present. (*Caution! Calcium hexammoniate, which is formed on evaporation of the excess ammonia, is very pyrophoric.*) Ethylene oxide (3 g.) dissolved in 50 ml. of liquid ammonia is slowly added to the calcium-ammonia solution over a 15-30 min. period. The ammonia solution turns from dark blue on addition of calcium to a dark grey color after the addition of the ethylene oxide. The reaction mixture is stirred and the excess ammonia is allowed to evaporate. After about 4-6 hr. 50 ml. of dry heptane is slowly added to displace the last 50 ml. of excess ammonia, and the catalyst is obtained as a fine greyish white suspension in heptane. On titration for calcium this suspension contains approximately 1 g. of contained calcium per 10 ml. of suspension. It is desirable to charge a glass jar containing small glass beads with the catalyst suspension and place it on a roller for about ½-1 hr. to obtain the catalyst as a very fine suspension.

2. The temperature on addition of ethylene oxide should not be allowed to exceed 50° because the polymer will tend to agglomerate and decrease the catalyst activity. The temperature can be controlled by raising an ice bath under the reaction flask and maintaining the temperature between 45° and 50° during the initial polymerization stage. The checkers found that some heating was needed in the latter stages to maintain the temperature above 40°.

3. Polymers prepared by this method have extremely high molecular weight, and measurement of the reduced viscosity is difficult. The polymer dissolves very readily in refluxing acetonitrile. The presence of moisture converts the catalyst residue into insoluble salts that appear as haze and seeds in the solution. The solution must be filtered before viscosity determinations are made, or samples of the polymer should be dissolved, filtered, and precipitated to remove the catalyst residue.

D. KOH as Catalyst

1. Procedure

Anhydrous (Note 1) diethylene glycol (118 g.) is added to a 2-1., three-necked flask equipped with reflux condenser, stirrer, thermometer, and gas inlet tube which extends below the surface of the liquid. Freshly fused potassium hydroxide (2.5 g.) is added to the diethylene glycol and the mixture is stirred and heated to 100°, at which point all the potassium hydroxide is dissolved. The heat source is removed and a cold water bath is placed in position beneath the flask. Dry ethylene oxide gas is fed into the flask. An extremely exothermic reaction occurs and the temperature of the reaction is controlled at 100-120° with the water bath and the regulation of the ethylene oxide feed. The reaction is allowed to continue for about 8-9 hr., or until no further absorption of ethylene oxide is observed (Note 2). The hot reaction mixture is poured from the flask into a 2-l. beaker and allowed to cool to room temperature. A tan waxy solid (about 700-900 g.) that has a freezing point range of about 25-30°, indicating a molecular weight of about 600-700 (Note 3), is obtained. When heated to 250° at 2-3 torr this material shows no appreciable loss in weight, and no volatiles are distilled.

2. Notes

1. Diethylene glycol is dried for 16 hr. over calcium hydride and then distilled at 129° at 10 torr. The reactants should be kept as nearly anhydrous as possible in order to obtain the required molecular weight ranges (see Note 3).
2. A bubbler connected to the "blow-off" side of the reflux condenser can be used to determine when the ethylene oxide ceases to react. The temperature of the reaction decreases as the absorption of ethylene oxide gas slows, and external heat is supplied to maintain the temperature at 100-120°. The checkers found no decrease in reactivity in 9 hr. and they interrupted the polymerization at that time.
3. Trace quantities of water present in the reaction decrease the molecular weight and yield of the product. Catalyst/diethylene glycol ratio also affects the molecular weight. Higher molecular weight polymers, up to about M_w6000, can be prepared by dividing the obtained product in half, adding more potassium hydroxide (2 g.) and continuing the ethylene oxide feed. The temperature should be increased to about 130-140°. Work has continued on various systems for polymerizing ethylene oxide (5-10).

E. References

1. Research and Development Department, Union Carbide Corp., Chemicals and Plastics, South Charleston, West Virginia 25303.
2. Department of Chemistry, University of Pennsylvania, Philadelphia, Pennsylvania 19104.

3. C. R. Noller, *Org. Syn.,* Coll. Vo. **2**, 184 (1943).
4. C. C. Price, and M. Osgan, *J. Amer. Chem. Soc.,* 78, 4787 (1956).
5. C. C. Prcie and D. D. Carmelite, *J. Amer. Chem. Soc.,* 88, 4039 (1966).
6. C. C. Price and R. Spector, *J. Amer. Chem. Soc.,* 88, 4171 (1966).
7. C. E. H. Bawn, A Ledwith, and N. McFarlane, *Polymer;* **10**, 653 (1969).
8. A. A. Solonvanov and K. S. Kozanskii, *Vysokomol. Soyed.* **A12**, 2114 (1970).
9. K. S. Kazanskii, A. A. Solovyanov, and S. G. Entelio, *European Polym. J.* 7, 1421 (1971).
10. S. Nenna and J. E. Figurello, *European Polym. J.* **11**, 511 (1975).

Polypromellitamic Acids and Polypyromellitimides

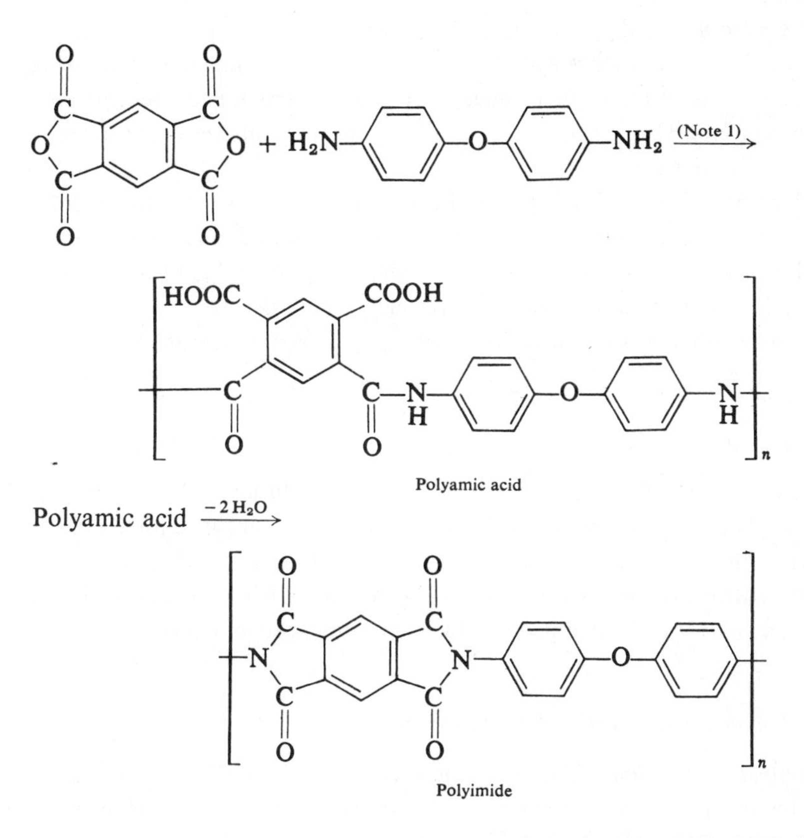

Submitted by C. E. Sroog (1)
Checked by P. M. Hergenrother (2a, b)

1. Procedure

A. Polyamic Acid

A 500-ml. flask fitted with mercury seal stirrer, nitrogen inlet, drying tube, and stopper is carefully flamed to remove traces of water and is allowed to cool under a stream of dry nitrogen in a dry box. In the dry box 10.0 g. (0.05 mole) of bis(4-aminophenyl) ether (Note 2) is added to the flask through a dried powder funnel, and residual traces are washed in with 160 g. of dry dimethylacetamide (Note 3). Pyromellitic dianhydride (10.90 g., 0.05 mole) (Note 4) is added to the flask through a second, dried powder funnel over a period of 2-3 min. with vigorous agitation (Note 5). Residual dianhydride is washed in with 28 g. of dry dimethylacetamide. The powder funnel is replaced by the stopper, and the mixture is stirred for 1 hr. A small surge in temperature to 40° occurs as the dianhydride is first added, but the mixture rapidly returns to room temperature. Cooling may be necessary to prevent an initial temperature surge (Note 6).

This procedure yields a 10% solution that can be stirred without difficulty. Sometimes the mixture becomes extremely viscous, and dilution to 5-7% solids may be required for efficient stirring. Polyamic acids so prepared exhibit η_{inh} 1.5-3.0 dl./g. at 0.5% in dimethylacetamide at 30°.

The polymer solutions may be stored in dry, sealed bottles at −15°.

B. Polyamic Acid Films

Thin layers (10-25 mils) of polyamic acid solutions are drawn onto dry glass plates with a doctor knife and dried for 20 min. in a forced draft oven (with nitrogen bleed) at 80°. The resulting colorless to pale yellow films are only partly dry and, after cooling, can be peeled from the plates, clamped to frames, and dried further under reduced pressure at room temperature.

C. Polyimide: Conversion from Polyamic Acid

Thermal Conversion. The cyclization of polyamic acid to polyimide can be followed spectrophotometrically by disappearance of the N–H bands (3.08 μ), and appearance of imide bands (5.63, 13.85 μ). Films of the polyamic acid that have been dried to a solids level of 65-75% are clamped to frames, carefully heated in a forced draft oven to 300° over a period of 45-60 min. (Note 7), and then heated at 300° for an additional hour. Further heating at

300° causes essentially no increase in the imide infrared absorption at 5.63 and 13.85 μ. The product is a tough, lemon-yellow film with outstanding thermal stability.

Chemical Conversion (6). Films of the polyamic-acid, after casting on glass plates, are placed into a bath of equal volumes of acetic anhydride and pyridine. After a few minutes a tough, yellowish film can be stripped from the plate. This gel film is then immersed overnight in acetic anhydride-pyridine: benzene (10:10:30) mixture. The next day it is removed, clamped to stainless steel frames and heated gradually to 300° and left for 1 hr.

2. Notes

1. The diamine may also be *p*-phenylene, *m*-phenylene, *p,p'*-biphenylene, 4,4'-diphenylene sulfide, sulfone, methylene, or isopropylidene.
2. Bis(4-aminophenyl ether) obtained from Du Pont, Organic Chemicals Department, was purified by simple vacuum drying at 50° at $\leqq$1 torr for 8 hr. m.p. 193-194°.
3. Dimethylacetamide, Du Pont technical grade material, was distilled from phosphorus pentoxide at 30 torr.
4. Pyromellitic dianhydride, technical grade material fo the Du Pont Explosives Department was sublimed through silica-gel at 220-260° and 0.25-1 torr.
5. A red-orange color may be observed at the solid-liquid interface that rapidly lightens to a lemon yellow as the pyromellitic dianhydride dissolves and reacts with the diamine.
6. Best results are obtained at temperatures of 15-75°; above 75° a decrease in the molecular weight of the polyamic acid becomes marked. Above 100°, cyclization to imide is appreciable, causing eventual precipitation of polyimide as well as a lowered molecular weight of the polyamic acid. Above 150°, the cyclization is so rapid that polyimide sometimes precipitates before all the dianhydride can be added.
7. It is necessary to heat the film initially at low temperature to avoid blistering and cracking, during cyclization and water removal. Once this initial heating period is completed, the film can be heated rapidly to 300°.

3. Methods of Preparation

The patent literature (3-5) describes the preparation of meltable polyimides by melt fusion of salt from diamines and pyromellitic acid or diamine and the diacid-diester.

4. References

1. Film Department, E. I. du Pont de Nemours and Co., Wilmington, Delaware 19898.

2a. Narmco Research and Development Division, Whittaker Corp., San Diego, California 92101.

2b. Current Address: NASA Langley Research Center, Hampton, Virginia 23665.

3. W. M. Edwards and I. M. Robinson (E. I. du Pont de Nemours and Co.), U.S. Pat. 2,710,853 (1955).
4. W. M. Edwards and I. M. Robinson (E. I. du Pont de Nemours and Co.), U.S. Pat. 2,880,230 (1959).
5. W. M. Edwards, I. M. Robinson, and E. N. Squire (E. I. du Pont de Nemours and Co.), U.S. Pat. 2,867,609 (1959).
6. A. L. Endrey (E. I. de Nemours and Co.), U.S. Pat. 3,179,630 (1965).

Alternating Copolyhydrazide of Terephthalic and Isophthalic Acids

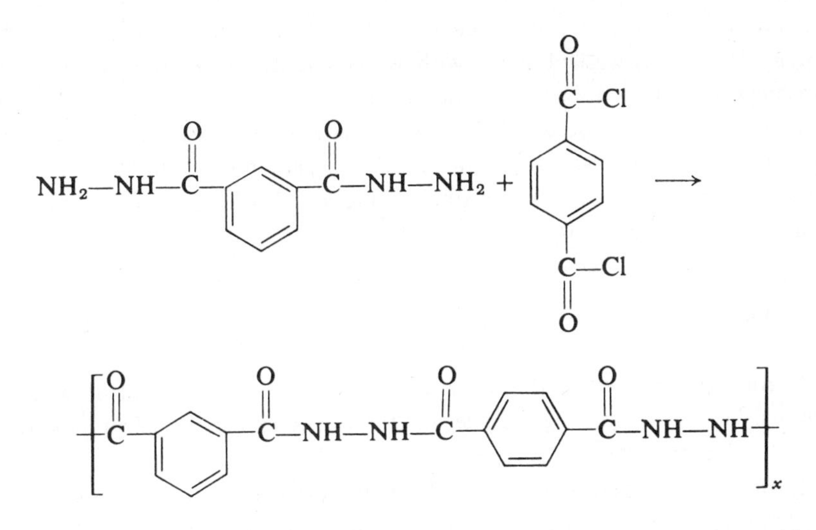

Submitted by A. H. Frazer (1a, b) and T. A. Reed (1a)
Checked by C. L. Schilling, Jr., and J. E. Mulvaney (2)

1. Procedure

In a nitrogen-filled (Note 1) dry box, 250 ml. of hexamethyl phosphoramide (*Caution! Hexamethylphosphoramide has been found to be a potent carcinogen in test animals. Extreme care in the use and disposal of this solvent should be exercised.*) (Note 2) and 19.10 g. (0.0985 moles) of isophthalic dihydrazide (Note 3) are placed in a 500-ml. resin kettle fitted with nitrogen bleed, drying tube, air-driven shear stirrer, and stoppered openings for addition of dry solids (Note 4). The sealed system is transferred to a hood for connection to a nitrogen line and attachment of the shear stirrer to an air motor. The hydrazide is dissolved by stirring at a water bath temperature of 50°. When solution is completed, the hot water bath is replaced with an ice bath, and the system is cooled to a solution temperature of 5°.

On addition of the first portion of terephthaloyl chloride (*Caution! Lachrymator.*) (10 g.) (Note 5) the solution becomes syrupy. After additional cooling for 1 hr., a second portion (10 g.) of terephthaloyl chloride is added (total 20.0 g., 0.0985 mole). The solution rapidly becomes thick and excessively difficult to stir. After 2 hr. more of cooling, the ice bath is removed and slow stirring is continued for 1 hr. The mixture is then allowed to stand unstirred, under nitrogen, overnight.

The viscous polymer solution is poured into distilled water and stirred rapidly in an explosion-proof blender. It is collected on a sintered glass filter after coagulation, washed twice with water and finally with methanol. The washings are accomplished by returning the collected polymer to the blender and agitating at high speed with the appropriate solvent. After drying overnight at 100° in a vacuum oven under flowing nitrogen at a pressure of 20-30 torr, the product (32 g., 100%) is a white solid having η_{inh} 1.4-1.6 dl./g. (Note 6) and a polymer melt temperature of 390°.

2. Notes

1. All nitrogen used in this work was dried by passing through two drying towers, the first containing silica-gel and the second containing calcium hydride. After drying it contained 0.6 p.p.m. water.

2. Hexamethyl phosphoramide (*Caution! Carcinogen*) was obtained from Eastman Chemical Products, Inc. It was further purified by distillation from calcium hydride at reduced pressure (5-8 torr) through a 10-in. Vigreux column. The first 10% of the distillate was discarded, and the distilled solvent was stored in sealed bottles in a nitrogen-filled dry box.

3. Isophthalic dihydrazide was made by the method of Sorenson and Campbell (3). After one recrystallization from water-methanol, it was dried for 4 days under flowing nitrogen in a vacuum oven (20-30 torr) at 120°.

4. All glassware used in this work was dried overnight at 150°. It was removed from the oven while still hot and allowed to cool in a nitrogen-filled dry box before use.

5. Terephthaloyl chloride was obtained from Hooker Chemical Company. It was distilled at reduced pressure through a 10-in. Vigreux column, recrystallized from dry toluene, and, after overnight evacuation to remove the last traces of toluene, it was sublimed in vacuum.

6. Inherent viscosities were measured in dimethyl sulfoxide at a concentration of 0.5% and a temperature of 30°. Viscosities as high as 2-3 dl./g. may be obtained. Polymer having η_{inh} 2.15 dl./g. has a molecular weight of 173,600 by light scattering.

3. Methods of Preparation

This method is a general method for the preparation of aliphatic, aromatic, aliphatic-aromatic, aromatic-oxalic, and aromatic-heterocyclic polyhydrazides (4-6). Polyhydrazides have also been prepared by melt (7, 8) and high-temperature solution polymerization (9).

4. References

1a. Pioneering Research Division, Textile Fibers Department, E. I. du Pont de Nemours and Co., Wilmington, Delaware 19898.
1b. Current Address: Central Research & Development Dept. E. I. du Pont de Nemours Co., Wilmington, Delaware 19898.
2. Department of Chemistry, The University of Arizona, Tucson, Arizona 85721.
3. W. R. Sorenson and T. W. Campbell, *Preparative Methods of Polymer Chemistry,* Interscience Publishers, New York, 1961, p. 103.
4. A. H. Frazer and F. T. Wallenberger, *J. Polym. Sci.,* **A2**, 1137 (1964).
5. A. H. Frazer and F. T. Wallenberger, *J. Polym. Sci.,* **A2**, 1147 (1964).
6. A. H. Frazer (E. I. du Pont de Nemours and Co.), U.S. Pat. 3,130,182 (April 21, 1964).
7. J. W. Fisher, *Chem. Ind. (London),* 244 (1952).
8. J. W. Fisher and E. E. Wheatly, U.S. Pat. 2,512,633 (June 27, 1950).
9. J. B. McFarlane and L. L. Miller, U.S. Pat. 2,615,862 (October 28, 1952).

Poly[1,3-phenylene-2,5-(1,3,4-oxadiazole)-1,4-phenylene-2,5-(1,3,4-oxadiazole)]

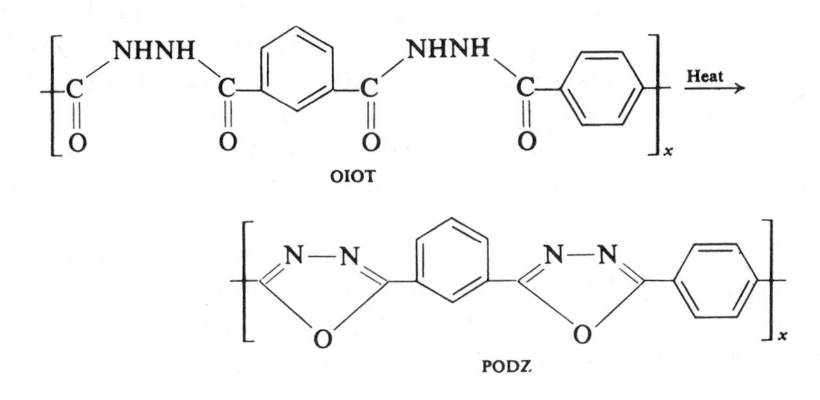

Submitted by A. H. Frazer (1a, b) and T. A. Reed (1a)
Checked by C. L. Schilling, Jr., and J. E. Mulvaney (2)

1. Procedure

A 1-g. sample (Note 1) of the alternating copolyhydrazide of terephthalic and isophthalic acids (3) is placed in a beaker with an insulated bottom (Note 2) and heated under flowing nitrogen (Note 3) at 290° for 48 hr. A quantitative yield (0.8-0.9 g.) (Note 4) of yellowish tan material (Note 5) is obtained. It emits a strong blue-white luminescence under long-wavelength ultraviolet light, has an oxygen analysis of 11.4% (Note 6) and η_{inh} 0.6 (Note 7). It is no longer soluble in common organic or aprotic solvents, and it has a polymer melt temperature in excess of 400°. Infrared spectra of the polymer show little or no amine or carbonyl absorption bands, but have strong bands corresponding to "C–H" and aromatic absorption (Note 8).

2. Notes

1. The sample may be in the form of powder, filament, or film.
2. An insulated bottom was used to prevent overheating by radiation from the exposed heating coils of the furnace.
3. The nitrogen used was preheated before entering the beaker by passing it through stainless steel coils inside the furnace.
4. Because polyhydrazides absorb water readily and retain it strongly, yields based on weight differences are unreliable.
5. The color varies depending on the form of the polymer. Films are almost colorless, fibers and powder are a light golden tan or yellow.
6. The theoretical oxygen content for this polymer is 11.1%. Because of combustion difficulties involved in completely burning this thermally stable material, however, the analytical value obtained for the oxygen content may vary.
7. Inherent viscosity was measured in 100% sulfuric acid at a concentration of 0.5% and a temperature of 30°. Nothing is known about the relation of molecular weight and inherent viscosity in the case of this polymer. The inherent viscosity of fibers prepared in a similar manner has been as high as 0.70 dl./g. (H_2SO_4).
8. The "N–H" band at 3.06 μ and the "C=O" band at 6.03 μ are both extremely strong in films of the copolyhydrazide. During the cyclodehydration of films (0.0004 in. thick) they rapidly decrease in intensity. After 24 hr. of heating, these bands have essentially disappeared.

3. Methods of Preparation

In addition to this preparation of poly(1,3,4-oxadiazoles) (4-7), these polymers have also been prepared by the reaction of dicarboxylic acid dihydrazides with bisiminoesters (8) and the reaction of dicarboxylic acid chlorides with bis-tetrazoles (9). The preparation of the isomeric poly-(1,3,4-oxadiazoles) has been reported via the cyclization of poly(acylamide oximes) (10).

4. References

1a. Pioneering Research Division, Textile Fibers Department, E. I. du Pont de Nemours and Co., Wilmington, Delaware 19898.
1b. Current Address: Central Research and Development Department, E. I. du Pont de Nemours & Co., Wilmington, Delaware 19898.
2. Department of Chemistry, The University of Arizona, Tucson, Arizona 85721.

3. A. H. Frazer and F. T. Wallenberger, *J. Polym. Sci.,* **A2**, 1147 (1964).
4. A. H. Frazer, W. Sweeny, and F. T. Wallenberger, *J. Polym. Sci.,* **A2**, 1157 (1964).
5. A. H. Frazer and F. T. Wallenberger, *J. Polym. Sci.,* **A2**, 1171 (1964).
6. A. H. Frazer and F. T. Wallenberger, *J. Polym. Sci.,* **A2**, 1181 (1964).
7. E. I. du Pont de Nemours & Co., Brit. Pat. 884,973 (Dec. 20, 1961).
8. J. Kranz, H. Pohlemann, F. Schauder, and H. Weidinger, Ger. Pat. 1,154,626 (Feb. 24, 1962).
9. E. J. Abshire and C. S. Marvel, *Makromol. Chem.,* **44-46**, 388 (1961).
10. D. C. Blomstom, U.S. Pat. 3,044,994 (July 17, 1962).

Poly[2,5-bis(p-phenylene-1,3,4-oxadiazole) isophthalamide]

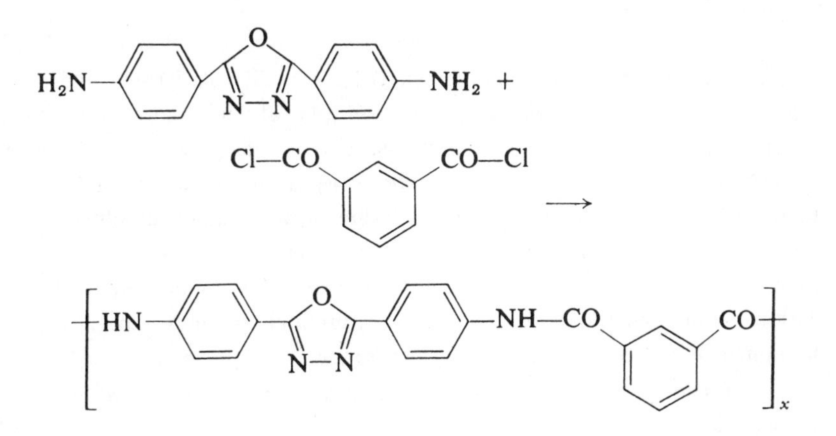

Submitted by J. Preston and W. B. Black (1)
Checked by J. K. Stille (2a) and J. Higgins (2a, b)

Several heteroaromatic diamines can be polymerized via solution polycondensation with simple or heteroaromatic diacid chlorides to yield wholly aromatic ordered amide-heterocycle copolymers (3).

1. Procedure

To a three-necked resin flask fitted with a heavy-duty stirrer with an efficient stirrer blade and a nitrogen bleed are added 7.56 g. (0.03 mole) of 2,5-bis(*p*-aminophenyl)-1,3,4-oxadiazole (Note 1) and 60-ml. of N,N-dimethylacetamide (Note 2) containing 3% dissolved lithium chloride. The solution (Note 3) is cooled to $-30°$ and 6.09 g. (0.03 mole) isophthaloyl chloride (Note 4) is added all at once. The reaction mixture is stirred rapidly for 30 min., the cooling bath is removed, and the temperature of the solution is maintained at 0° for 30 min. The temperature of the solution is then allowed to rise to room temperature, and the solution is stirred for at least 1 hr.

The ordered oxadiazole-amide copolymer is isolated (Note 5) by pouring the clear, viscous solution into rapidly stirred water in a blender jar (Note 6). The tough, cream-colored polymer is washed with water (two 100-ml. washes) and is then dried in a vacuum oven at 60°. The yield of polymer is 96-100%; the inherent viscosity of the polymer is 1.65 dl./g. determined at 30° on a 0.5% solution in N,N-dimethylacetamide (containing 5% dissolved lithium chloride) (Note 7). The polymer does not melt below 400°.

2. Notes

1. Syntheses for 2,5-bis(*p*-aminophenyl)-1,3,4-oxadiazole have been described elsewhere (4, 5); catalytic hydrogenation (4) of the dinitro intermediate prepared using oleum (5) is a particularly convenient synthetic route. The diamine so prepared may need further purification to be suitable for polymerization. This is done by heating to boiling a dilute aqueous solution of the diamine with an equivalent of hydrochloric acid and filtering. The filtrate is made basic with sodium hydroxide and the diamine is collected. Recrystallization from 2-methoxyethanol yields a product of m.p. 260-262°. This monomer should be dried thoroughly before use.

2. The solvent should be of high purity and free of water, which can be reduced to a suitably low level by storing the solvent over Linde molecular sieves (Type 5A, $^1/_{16}$ inch pellets). The lithium chloride should be dried in an oven at 150° before addition to the solvent. As a precaution the freshly prepared solution containing dissolved lithium chloride is stored for several hours over molecular sieves before use to remove any water which might have been absorbed by the hygroscopic lithium chloride. Another solvent that can be used in this polymerization is N-methylpyrrolidone.

3. The mixture first can be warmed if necessary to effect solution.

4. Isophthaloyl chloride, m.p. 42-43°, should be recrystallized from dry hexane (50 ml. of hexane for 100 g. of acid chloride before use) (6). To effect crystallization, the solution should not be cooled below 21°. If the starting material is impure, it is advisable to distill it under reduced pressure (b.p. 123-125°/5 torr) before recrystallization.

5. The resulting viscous solution can be fabricated into fibers or films directly without prior isolation of the polymer. If desired, the hydrochloric acid in the

polymer solution can be neutralized before fabrication of the polymer by the addition of 2.52 g. (0.06 mole) of lithium hydroxide monohydrate.

6. Care should be taken to add the polymer solution very slowly to the water or the tough polymer will form lumps and stall the blender. Alternatively, the very viscous solution may be thinned with additional solvent before precipitation.

7. Determined with a Cannon-Fenske Series 100 viscometer.

3. References

1. Chemstrand Research Center, Inc., Durham, North Carolina 27709.
2a. Department of Chemistry, University of Iowa, Iowa City, Iowa 52240.
2b. Current Address: Cosden Oil Co., Big Spring, Texas 79720.
3. J. Preston and W. B. Black, *J. Polym. Sci.*, **B4**, 267 (1966).
4. J. Preston, *J. Heterocyclic Chem.*, **2**, 441 (1965).
5. A. E. Siegrist, E. Moergeli and K. Hoelzele, U.S. Pat. 2,765,304 (October 2, 1956); [*C.A.*, **51**, 12,983 (1957)].
6. W. R. Sorenson and T. W. Campbell, *Preparative Methods of Polymer Chemistry*, Interscience, Publishers, New York, 1961, p. 120.

Poly(terephthaloyl trans-2,5-dimethylpiperazine) by Low-Temperature Solution Polycondensation

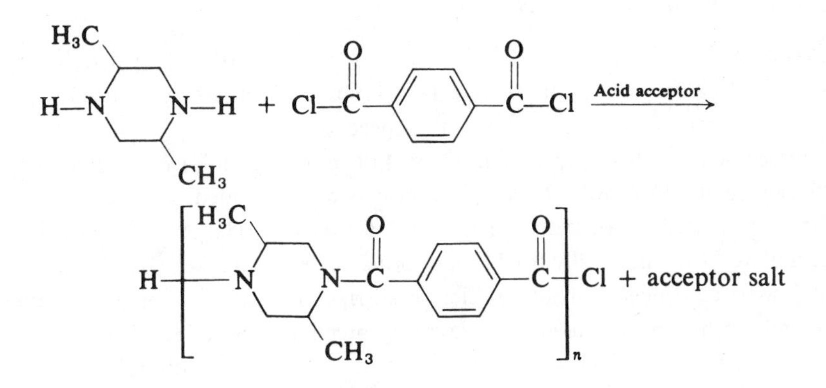

Submitted by P. W. Morgan and S. L. Kwolek (1)
Checked by H. Pikas and E. V. Gouinlock, Jr. (2)

1. Procedure

A. Tertiary Amine as Acid Acceptor

trans-2,5-Dimethylpiperazine (2.28 g., 0.02 mole) (Note 1) and 5.6 ml. of triethylamine (Notes 1 and 2) are dissolved in 100 ml. of chloroform (*Caution! Many chlorinated hydrocarbons are quite toxic and inhalation of vapors and skin contact should be minimized. Polymer preparation and isolation should be conducted in an efficient hood. Such solvents should be disposed of in a manner compatible with safety and environmental considerations.*) (Note 3) in a 500-ml. Erlenmeyer flask. To this solution is added, with swirling, a solution of 4.06 g. (0.02 mole) of terephthaloyl chloride (Note 1) in 80 ml. of chloroform. More chloroform (20 ml.) is used to rinse in at once the last of the terephthaloyl chloride solution. The mixture remains clear, but the temperature rises quickly from 25° to 42° and there is an increase in solution viscosity. After 5 min. (Note 4) the product is coagulated by pouring the solution into 200-400 ml. of hexane (*Caution! Stirring devices should be non-sparking or shielded as suggested for home blenders in ref 7., p. 489.*) with stirring. A fibrous precipitate of polymer and crystals of triethylamine hydrochloride are obtained. The precipitates are collected on a fritted glass filter (Note 5) and washed thoroughly with water and then with acetone. After drying (Note 6) an 85-92% yield of polymer is obtained which has η_{inh} 2-3 dl./g. (*m*-cresol).

B. Excess Diamine as Acid Acceptor

trans-2,5-Dimethylpiperazine (4.56 g., 0.04 mole) (Note 1) is dissolved in 100 ml. of chloroform (Note 3) in a 1-qt. home blender (Note 7) fitted with a funnel for reagent addition. While the speed of the stirrer is simultaneously increased with a rheostat, a solution of 4.06 g. (0.02 mole) of terephthaloyl chloride (Note 1) in 80 ml. of chloroform is rapidly added. More chloroform (20 ml.) is used to rinse in the last of the acid chloride. A thick gelatinous precipitate of diamine dihydrochloride forms at once, but the mixture becomes fluid with continued stirring. Moderate stirring is continued for 5 min. then 350 ml. of hexane is added. The resulting precipitate is collected on a fritted glass filter (Note 5) and washed well with water and then with acetone. After drying (Note 6) the granular product (92-98% yield) has η_{inh} between 2 and 3 dl./g. (*m*-cresol).

2. Characterization

The polymer is soluble in formic acid, chloroform combined with methanol or formic acid, and *m*-cresol. Isolated polymer is not soluble in chloroform, although it forms a metastable or supersaturated solution when first prepared in this solvent (Note 8).

The inherent viscosity $[\eta_{inh} = \ln \eta_{rel}/c]$ of a dilute solution is determined in *m*-cresol at 30° with 0.5 g. of polymer per 100 ml. of solution.

The polymer does not melt on a hot metal block or bar at 400°. A polymer having an intrinsic viscosity of 3.1 (Note 9) had 14 moles of carboxyl and 33 moles of amine end groups per 10^6g., which corresponds to a number average molecular weight ($\overline{M}_n$) of 42,600. For a series of polymer preparations with intrinsic viscosities between 0.2 and 10 the relation of $[\eta]$ to $\overline{M}_n$ was found to be $[\eta] = 2.96 \times 10^{-4}\overline{M}_n^{0.87}$. A limited number of values for weight-average molecular weights ($\overline{M}_w$), determined by the light-scattering method, yielded a linear plot fitting the equation

$$[\eta] = 1.69 \times 10^{-4}\overline{M}_w^{0.85},$$

within the range of intrinsic viscosities between 0.5 and 11. dl./g.

3. Notes

1. The required chemicals, which are commercially available, are purified in the following manner: *trans*-2,5-dimethylpiperazine is recrystallized from acetone or methyl ethyl ketone, followed by sublimation. The melting point of pure diamine is 118° (sealed tube). Terephthaloyl chloride is recrystallized from dry hexane; alternatively, it may be sublimed at reduced pressure. The melting point of pure acid chloride is 81.2°. Triethylamine is distilled through a spinning-band column or a 10-in. helices-packed column from potassium hydroxide pellets in order to remove such interfering impurities as water and primary or secondary amines. The acid chloride is stored in a dry box or desiccator. Incomplete solubility of the acid chloride indicates hydrolysis to acid and a need for repurification. On the other hand, ready solubility is not a reliable indication of purity.

2. Other tertiary amines which give high molecular weight polymer in chloroform or dichloromethane systems include tri-*n*-propylamine, tri-*n*-butylamine, N-ethylpiperidine, N-allylpiperidine, N-ethylmorpholine, N-methylmorpholine, and N-allylmorpholine. A detailed discussion of the relationship between acceptor base strength and solvent selection and the degree of polymerization appears in the literature (3).

3. Chloroform is A.C.S. reagent grade freed of alcohol stabilizer as follows. It is washed three times with equal volumes of water, dried with calcium chloride, and stored over a mixture of equal parts of potassium carbonate and calcium chloride in an amber bottle. Overnight drying is necessary for good results. Chloroform washed and well-dried in the preliminary step is stable for several weeks. Decomposing chloroform yields hydrogen chloride and phosgene, which may be detected by moist litmus paper in the vapors or by the immediate development of turbidity on the addition of ethylenediamine.

4. Polymer of high molecular weight can be obtained in high yield in 15 sec. If the polymerization is allowed to continue, the average molecular weight of the product increases slowly for several minutes.

5. Paper filters contaminate the polymer with cellulose fiber.

6. The polymer is conveniently dried overnight in a vacuum oven at 80-100°.

7. A home blender, or equally efficient stirring device, should be used for polymerization systems that yield precipitates of polymer quickly. The blender is not required for the chloroform system described in procedure B.

In preparation for use, the blender jar and stirrer bearing are dried. The bearing is sealed and lubricated with an inert substance, such as "Celvacene" medium vacuum grease (Distillation Prodcuts Industries). The top of the jar is covered with aluminum foil, and over this is placed the plastic cap. A wide-mouthed funnel for reagent addition is inserted through a ¾-in. hole in the center of the cap and foil. The motor housing is blanketed with nitrogen as a safety precaution.

8. Several other solvents form metastable or supersaturated solutions of polymer of varying stability. The stabilities of these solutions decrease in the order chloroform, dichloromethane, 1,1,2-trichloroethane, and 1,2-dichlorethane. In the last liquid, formation of a translucent, pasty precipitate occurs at once.

9. The intrinsic viscosity was measured in *sym*-tetrachloroethane-phenol (40/60 by weight) at 30°.

4. Methods of Preparation

Poly(terephthaloyl *trans*-2,5-dimethylpiperazine) has been prepared in high molecular weight by the interfacial polymerization process (5, 6). Other variations of the low-temperature solution preparation of poly(terephthaloyl *trans*-2,5-dimethylpiperazine) are given in ref. 4 and 7; application of the solution method to many related polymers is described.

5. Applications

Fibers have been prepared by dry spinning (8). Films of this polymer have been evaluated as reverse osmosis membranes (9) and related copolymers have been studied for medical implant applications (10).

6. References

1. Pioneering Research Division, Textile Fibers Department, E. I. du Pont de Nemours & Co., Wilmington, Delaware 19898.
2. Central Research and Development Department, Hooker Chemical Corp., Niagara Falls, New York 14302.
3. P. W. Morgan and S. L. Kwolek, *J. Polym. Sci.,* **A2**, 209 (1964).
4. P. W. Morgan and S. L. Kwolek, *J. Polym. Sci.,* **A2**, 181 (1964).
5. E. L. Wittbecker, U.S. Pat. 2,913,433 (Nov. 17, 1959); E. I. du Pont de Nemours & Co., Brit. Pat. 785,214 (Oct. 23, 1957).
6. M. Katz, *J. Polym. Sci.,* **40**, 337 (1959).
7. P. W. Morgan, *Condensation Polymers by Interfacial and Solution Methods,* Interscience Publishers, 1965.
8. W. W. Moseley and R. G. Parrish, U.S. Pat. 3,318,849 (May 9, 1967).
9. L. Credali, P. Parrini, E. Leonelli, and A. Chiolle, *Chim. Ind.* (Milan), **56**, 19 (1974).
10. D. S. Bruck, *Polymer,* **10**, 939 (1969).

Poly(hexamethyleneadipamide) by Melt Polymerization

$$H_2N\text{—}(CH_2)_6\text{—}NH_2 + HO\text{—}\overset{\overset{\displaystyle O}{\|}}{C}\text{—}(CH_2)_4\text{—}\overset{\overset{\displaystyle O}{\|}}{C}\text{—}OH \longrightarrow$$

$$(H_3\overset{+}{N}\text{—}(CH_2)_6\text{—}\overset{+}{N}H_3)(^{-}O\text{—}\overset{\overset{\displaystyle O}{\|}}{C}\text{—}(CH_2)_4\text{—}\overset{\overset{\displaystyle O}{\|}}{C}\text{—}O^{-}) \longrightarrow$$

$$\left[HN\text{—}(CH_2)_6\text{—}NH\text{—}\overset{\overset{\displaystyle O}{\|}}{C}\text{—}(CH_2)_4\text{—}\overset{\overset{\displaystyle O}{\|}}{C} \right]_x + H_2O$$

Submitted by P. E. Beck and E. E. Magat (1)
Checked by S. K. Das (2a, b)

1. Procedure

A. Hexamethylenediamine-Adipic Acid Salt

In a 500-ml. Erlenmeyer flask 29.2 g. (0.2 mole) of adipic acid (Note 1) is dissolved in 250 ml. of warm ethanol, and the solution is cooled to room temperature. A solution of 23.43 g. (0.202 mole) (Note 2) of hexamethylenediamine (Note 1) in 50 ml. of ethanol is added to the adipic acid solution with gentle stirring (Note 3). Immediate precipitation of the salt takes place. The pH of the salt is adjusted to 7.6 (Note 4). The solvent is evaporated, and the salt is air-dried to constant weight. A quantitative yield of white, crystalline salt is obtained, m.p. 198-200°. The pH of the dry salt (1% solution in water) is 7.6 (Note 4).

B. Poly(hexamethyleneadipamide)

A glass polymer tube (Note 5) is charged with 32.0 g. of hexamethylenediamine-adipic acid salt (Note 6). A constriction is made in the upper half of the neck of the tube with a glass-blowing torch. The tube is connected to a three-way stopcock; the other two inlets are connected to a vacuum pump and a source of low pressure nitrogen (about 1 atm.), respectively. The polymer tube is placed on a 30° incline to the horizontal and purged of air by alternately evacuating and filling the tube with nitrogen (4-5 cycles). The tube is then sealed at the constriction with a torch while under vacuum (Note 7). The sealed polymer tube is clamped just below the neck in a vapor bath (Note 8). The polymer tube is heated in a vapor bath (naphthalene) at 218° for 2 hr. (Note 9). (*Caution! See Note 10.*) When the heating cycle is completed, the polymer tube is allowed to cool to room temperature. The tube is then opened about 1 in. from the tip after first scoring with a file. The rough edges are fire-polished.

A 60° bend is put in the neck of the polymer tube. The polymer tube is clamped in an upright position in a vapor bath. The open end is connected to a trap by pressure tubing (Note 11). The trap is connected with a three-way stopcock to a vacuum pump and a 1-atm. source of nitrogen. The tube is purged of air by alternately evacuating and flushing with nitrogen. The polymer tube is heated under nitrogen in the vapor bath (dimethylphthalate) at 283° for 2 hr. (Note 12). The polymer is then cooled to room temperature under nitrogen.

The polymer is obtained by breaking the glass tube, which is wrapped in a towel, with a hammer (Note 13). The yield is quantitative except for mechanical losses.

The tough, white, opaque polymer has a crystalline melt temperature of 265°. The inherent viscosity (η_{inh}), determined at 25° in 0.5% solution in *m*-cresol is 1.0-1.4 dl./g. and the number average molecular weight ($\overline{M}_n$) (Note 14) is 18,700-20,000 (Note 15). End group analysis is the following: carboxyl ends, 40-51; amine ends, 56-61 (Note 16). The polymer is soluble in formic acid.

Fibers and films may be obtained by melt methods or from formic acid solution.

2. Notes

1. All ingredients must be of the highest purity: adipic acid, m.p. 152°, N.E. 72.1 ± 0.2; hexamethylenediamine, m.p. 41-42°, b.p. 98-100° (20 torr), N.E. 58.1 ± 0.2.

2. A small excess (1%) of diamine is added to ensure that the salt is rich in diamine. This is desirable because the diamine is the most volatile component.

3. Addition of the diamine solution is made quantitative by repeatedly rinsing the container holding the diamine with ethanol. About 100 ml. of ethanol is used for this purpose.

4. The pH of the salt is determined on a 1% solution of the salt in water using a pH meter. This is accomplished by withdrawing from 1-2 g. of wet salt from the slurry and dissolving it in 100 ml. of distilled water and then determining the pH. A pH of 7.6 ± 0.2 is acceptable, although an imbalance on the basic side is more desirable. (See Note 2). Salt imbalance may be corrected by the addition of a small amount of the indicated component to the slurry and stirring well. The pH is then redetermined. A salt with a low pH gives a polymer containing a high carboxyl end content; alternatively, a salt with a high pH produces a polymer with high amine end content.

5. Heavy-walled polymer tubes, purchased from Labglass, Inc., Vineland, New Jersey have the following dimensions: 12-in long body with 38 mm. O.D.; 12-in. long neck with 13 mm. O.D.

6. The salt is conveniently introduced into the polymer tube through a powder funnel connected to the neck of the tube by a short length of rubber tubing.

7. A thick, well-annealed heavy seal is necessary to withstand pressures that develop during heating.

8. The vapor bath consists of a 50 × 400 mm. test tube about one-quarter filled with a suitable liquid. In use, the upper walls of the test tube act as an air condenser to prevent loss of vapors. The vapor bath is heated with a mantle or other electrical heating device. For safety considerations a steel tube vapor bath of these dimensions is recommended and heating should be conducted in a hood behind a shield.

9. The polymer tube is adjusted so that the bottom of the tube is not immersed in the boiling liquid, and so that the vapor rises to about 1 in above the solid in the polymer tube. The top of the vapor bath may be wrapped with aluminum foil to prevent vapors from escaping.

10. *Caution! This operation is extremely hazardous because there is a possibility that the sealed tube may shatter when heated because of the pressures that develop.* This may be caused by a poor seal, defects in the glass, or overheating. *The vapor bath containing the sealed tube must be shielded from all sides but left open at the top. It is recommended that this phase of the procedure be performed in a hood. The hand and arm used to manipulate the tube should be protected with leather, asbestos, or equivalent protective gloves.*

11. The trap consists of a 50-ml., round-bottomed flask with an adapter. A small vacuum trap suffices, however.

12. Higher molecular weight polymer can be obtained by applying vacuum for the last 15 min. of the heating cycle.

13. The polymer pulls away from the glass on cooling.

14. The number average molecular weight is calculated from the equation

$$\overline{M}_n = \frac{2 \times 10^6}{\text{amine ends} + \text{carboxyl ends}}$$

15. Lower molecular weight polymer can be obtained by the addition of acetic acid to the salt in the sealed tube cycle. For example, 1 mole % of acetic acid gives polymer having $\overline{M}_n$ 12,000.

16. Amine and carboxyl ends are reported as ends per 10^6 g. of polymer and are determined by pH and conductometric titrations, respectively (3, 4).

3. Methods of Preparation

Poly(hexamethyleneadipamide) has been prepared from hexamethylenediamine and adipoyl chloride by interfacial polymerization (5) and low-temperature solution polycondensation (6).

4. References

1. Carothers Research Laboratory, Textile Fibers Department, E. I. du Pont de Nemours & Co., Wilmington, Delaware 19898.
2a. Coatings and Resins Division, Pittsburgh Plate Glass Co., Springdale, Pennsylvania 15144.
2b. Current Address: PPG Industries, Allison Park, Pennsylvania 15101.
3. J. E. Waltz and G. B. Taylor, *Anal. Chem.*, **19**, 448 (1947).
4. G. B. Taylor, *J. Amer. Chem. Soc.*, **69**, 635 (1947).
5. R. G. Beaman, P. W. Morgan, E. L. Wittbecker, and E. E. Magat, *J. Polym. Sci.*, **40**, 329 (1959).
6. S. L. Kwolek and P. W. Morgan, *J. Polym. Sci.*, **A2**, 2693 (1964).

4-Nylon

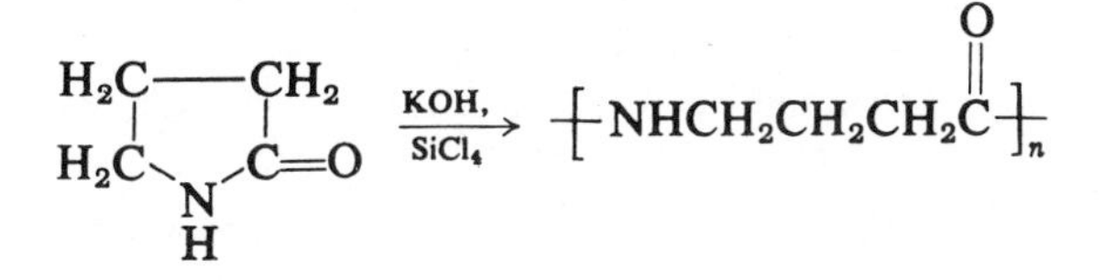

Submitted by R. W. Wynn, S. A. Glickman, and M. E. Chiddix (1)
Checked by J. M. Hoyt and K. Koch (2)

1. Procedure

A 250 ml., three-necked, round-bottomed flask equipped with stirrer, thermometer, and Claisen head suitable for vacuum distillation is charged with 120 g. of freshly distilled 2-pyrrolidone (Note 1). The charge is heated under nitrogen to 80° with a Glas-Col mantle. Flake potassium hydroxide (97%) (3.4 g.) is added (Note 2). The water formed together with about 20 ml. of monomer is rapidly distilled from the flask at 1 torr (Note 3). The hot solution is rapidly transferred to an 8-oz. polyethylene bottle previously purged with nitrogen. While the solution is still hot, 0.5 g. of silicon tetrachloride is added (Notes 4 and 5). The bottle is capped, agitated by hand, and allowed to cool to room temperature. After 10 min. and at a temperature of about 50° polymerization is indicated by precipitation of solid polymer. After 24 hr. at room temperature the mixture is very hard. It is broken with a hammer and the bottle is cut open for its removal (Note 6). The lumps are then blended with 150 ml. of water containing 0.1% formic acid in a blender (Note 7). The powdered product is filtered and washed on the filter with 150 ml. of 0.1% formic acid followed by three 100-ml. washings with distilled water. It is finally washed with alcohol and dried at 3 torr at 70°. There is obtained 73.0 g. of white polymer (Note 8) (74% conversion based on monomer). The polymer has η_{rel} 2.38 dl./g. (Note 9).

2. Notes

1. The 2-pyrrolidone was distilled from commercial 2-pyrrolidone (GAF Corp). A fraction boiling at 77.0-78.5°/0.7-0.8 torr was used for the polymerization.

2. The submitters used potassium hydroxide, technical flake, from J. T. Baker Chemical Company. Analysis showed 97% KOH. This grade varies between 90% and 98% and can be used as is. The checkers successfully used both flake potassium hydroxide, technical grade, and pellet potassium hydroxide, A.C.S. grade, 85% minimum assay. Sodium hydroxide (3.4 g.), 97% minimum assay, flake or pellet, can also be used, but a larger amount of silicon tetrachloride (1.0 g.) is required for reproducible results.

3. This distillation requires about 15 min.

4. These operations should be performed with a minimum exposure to the atmosphere.

5. The silicon tetrachloride was used as obtained from Matheson, Coleman, and Bell without purification. The use of silicon tetrachloride as an initiator for the polymerization is described by Taber (7).

6. The checkers recommend cutting the larger polymer lumps with a hammer and a small wood-chisel (½ in.) in a suitable metal dish.

7. The checkers found that the blending operation puts considerable stress on the blender; the glass blender jar broke in one experiment. The jar should be taped to protect the operator from possible injury. Sodium hydroxide produces polymers which are easier to grind than the polymers made with potassium hydroxide.

8. In all cases, light brown polymers were obtained by the checkers.

9. The relative viscosity was determined on a 0.5% solution in *m*-cresol.

3. Methods of Preparation

The preparation of 4-nylon from 2-pyrrolidone containing an alkali metal salt of 2-pyrrolidone as the catalyst has been described (4). Initiators that may be used for the polymerization include N-acyl-2-pyrrolidone (5), N-substituted secondary amides (6), and Group IV metal halides (7). The alkali metal salt may be prepared through the use of the alkali metal (4), its hydroxide (4), its hydride (3), or its amide (3).

4. References

1. Dyestuff and Chemical Division, General Aniline & Film Corp., Easton, Pennsylvania 18042.
2. U. S. Industrial Chemicals Company, Research Department, Cincinnati, Ohio 45202.
3. J. M. Butler, R. M. Hedrick, and E. H. Mattus, U.S. Pat. 3,028,369 (Apr. 3, 1962).
4. W. O. Ney, W. R. Nummy, and C. E. Barnes, U.S. Pat 2,638,463 (May 12, (1953).
5. C. E. Barnes, W. O. Ney, and W. R. Nummy, U.S. Pat. 2,809,958 (Oct. 15, 1957).
6. S. A. Glickman and E. S. Miller, U.S. Pat 3,016,366 (Jan. 9, 1962).
7. D. Taber (General Aniline & Film Corp.), Brit. Pat. 850,160 (Sept. 28, 1960).

Poly[methylene bis(4-phenylcarbodiimide)]

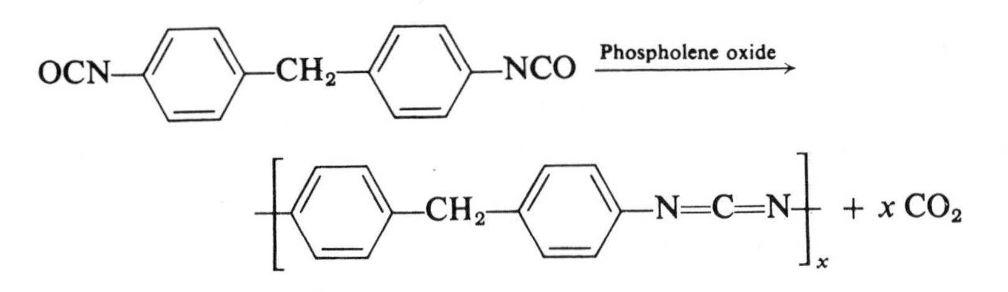

Submitted by T. W. Campbell (1) and V. S. Foldi (2)
Checked by S. H. Metzger, Jr. (3)

1. Procedure

In a 500 ml., three-necked flask equipped with stirrer, condenser, and nitrogen inlet are placed 150 ml. of benzene, 50 ml. of chlorobenzene (Note 1), 25 g. of methylene bis(4-phenylisocyanate) (*Caution!*) (Note 2), and 0.04 g. of 1-ethyl-3-methyl-3-phospholene oxide (Note 3). The mixture is gradually heated to reflux (Note 4) with a stream of nitrogen passing over the stirred mixture; an increasingly viscous solution is formed (Note 5) with copious evolution of carbon dioxide until the mass in the flask finally forms a stiff gel.

The gel can be treated to give a coarse, fibrous polymer by repeated washings in methanol in an explosion-proof blender and drying in a vacuum oven at 100°. The yield of polymer is 95-100%. The polymer is insoluble in all common organic solvents. Clear, very tough films can be pressed from it in a Carver press at 300°.

2. Notes

1. The polymerization can also be performed in other solvents or solvent mixtures such as toluene, xylene, chlorobenzene, *o*-dichlorobenzene and mixtures thereof. If the reaction is run in xylene, for example, the solution becomes

milky, then a second liquid phase begins to separate. This liquid phase becomes more and more viscous and eventually yields high molecular weight fibrous material (4). Solvents should be dry to avoid decomposition.

2. Methylene bis(4-phenylisocyanate) is available from Mobay Chemical Company. It may be purified by distillation (b.p. 148-150°/0.12) or recrystallization from hexane (5).

3. The catalyst can be prepared according to procedure described in ref. 6.

4. The checker used a nitrogen bleed that passed into the liquid phase, and he noted serious foaming as the mixture became viscous. The submitters did not encounter this problem with this procedure, if heating was not too rapid; however, caution is dictated.

5. This solution can be used for casting of films, but is metastable and it gels on standing.

3. References

1. Benger Laboratory, Textile Fibers Department, E. I. du Pont de Nemours & Company, Waynesboro, Virginia 22980.
2. Pioneering Research Division, Textile Fibers Department, E. I. du Pont de Nemours & Company, Wilmington, Delaware 19898.
3. Mobay Chemical Company, New Martinsville, West Virginia 26155.
4. T. W. Campbell and K. C. Smeltz, *J. Org. Chem.*, **28**, 2069 (1963).
5. V. S. Foldi, T. W. Campbell, and D. J. Lyman, *Macromol. Syn.*, **1**, 63 (1963); Coll. Vol. I, p. 69.
6. W. B. McCormack, *Org. Syn.*, **43**, 73 (1964).

Poly-ε-caprolactone

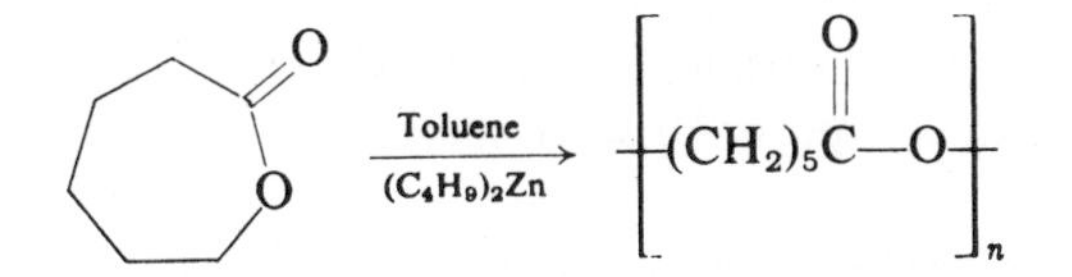

Submitted by E. F. Cox, F. Hostettler, and R. R. Kiser (1)
Checked by M. Goodman (2a, b) and J. Brandrup (2a, c)

1. Procedure

A 1-1., four-necked flask is fitted with a mechanical stirrer, a condenser protected with a drying tube, a thermometer, and a gas inlet. The flask is swept with a slow stream of dry nitrogen and dried by flaming. To the flask are added 500 ml. of anhydrous toluene (Note 1) and 114 g. (108 ml., 1.0 mole) of refined ε-caprolactone (Note 2). The solution is heated to 80° under a nitrogen atmosphere, and 0.20 ml. di-*n*-butylzinc (*Caution! Pyrophoric.*) (Note 3) is added using a syringe. The solution becomes viscous in a few minutes and the polymerization is essentially complete in 15-20 min. but stirring is normally continued for 1 hr. at 80° to ensure completion (Note 4). The polymer is precipitated by pouring a fine stream of the viscous toluene solution into a petroleum ether-Dry Ice mixture at -80° while stirring vigorously. The polymer is dried for 16 hr. in a vacuum oven at 40° and 1 torr. The product, obtained in nearly quantitative yield, is a white, fluffy, fibrous solid having a reduced viscosity in chloroform of 1.0-1.5 dl./g. (0.2% conc., 30°) and a polymer melt temperature around 60°. It may be melt-pressed to films at this temperature, and drawable fibers may be pulled from the polymer melt. The product is soluble in most aromatic and halogenated hydrocarbon solvents, and films may be cast from the solutions.

2. Notes

1. The toluene is analytical reagent grade dried over sodium wire.

2. Best results are obtained from ϵ-caprolactone that is distilled over a high-boiling diisocyanate such as diphenylmethane-4,4′-diisocyanate to remove hydroxyl and carboxylic acid impurities such as hydroxycaproic acid. About 10 g. of diisocyanate is used per liter of monomer. Caprolactone purified in this manner and stored under nitrogen may be used for polymerizations after standing several months. Caprolactone is commercially available from Union Carbide Corporation, Chemicals Division.

3. Directions for preparing di-*n*-butylzinc are given in *Org. Syn. Coll. Vol.*, **2**, 184 (1943). Diethyl- and dibutylzinc are commercially available from Stauffer Chemical Co., Anderson Chemical Division. Other initiators such as butyllithium, dimethylcadmium, or trialkyl aluminums compounds are equally effective.

4. The checkers used diethylzinc and noted an induction period and a slow increase in viscosity. After a 3.5 hour reaction period the yield was 74%.

3. Methods of Preparation

Cyclic esters have been polymerized to molecular weights of 100,000 or higher using both anionic (3, 4) and cationic (5) initiators. The polymerization may be performed in solution as described above or in bulk. The bulk polymerization is quite exothermic and tends to give higher molecular weight polymer, but removal of the tough solid from the reaction flask is difficult. The solution method is preferred for laboratory preparation.

4. References

1. Research and Development Department, Union Carbide Corporation, Chemicals Division, South Charleston, West Virginia 25303.

2a. Department of Chemistry, Polytechnic Institute of Brooklyn, Brooklyn, New York 11201.

2b. Current Address: University of California,/San Diego, LaJolla, California 92037.

2c. Current Address: Hoechst, Frankfurt am Main, West Germany.

3. E. F. Cox and F. Hostettler (Union Carbide Corporation), U.S. Pats. 3,021,309 through 3,021,317 (1962) [*C. A.*, 57, 12,719 (1962)].

4. H. Cherdron, H. Ohse, and F. Korte, *Makromol. Chem.*, **56**, 179 (1962).

5. H. Cherdron, H. Ohse, and F. Korte, *Makromol. Chem.*, **56**, 187 (1962).

Oil-modified Alkyd Resin

Formulation	
Safflower oil, non-break	1220.0
Glycerin, 99%	144.0
Pentaerythritol, pure or mono (A)	162.0
Litharge (technical PbO)	0.43
Pentaerythritol, pure or mono (B)	121.0
Diethylene glycol	24.2
Phthalic anhydride	770.0
	2441.63
Theoretical water loss	93.63
Theoretical yield, solid resin	2348.0
Xylene	70
Mineral spirits	2278
Theoretical total yield	4698.0

Submitted by R. B. Graver (1)
Checked by H. Roth (2)

1. Procedure

A 5-l. flask fitted with a mechanical stirrer, a thermometer, a reflux condenser and Dean-Stark trap, and an inert gas inlet tube, is charged with safflower oil, glycerin, and pentaerythritol (A). It is heated to 150-160° with agitation under an inert gas blanket (CO_2 or N_2), with an electric heating mantle controlled with a rheostat. At 150-160° litharge is added through the opening used for the thermometer fitting. Heating is continued until the temperature reaches 230-235°, which is maintained for 1 hr. A sample taken at this time should form a clear solution when mixed 1:1 by volume with absolute methanol and warmed slightly. If it is not clear in 1 hr., the temperature is increased to 240° and is maintained for 1 hr. more.

After cooling to 150°, pentaerythritol (B), diethylene glycol, and phthalic anhydride are added. The water trap is filled with xylene. The solution is reheated to 105-110°. Use of the inert gas blanket is continued. After a temperature of 105° is reached, a sample is taken every 30 min. and the acid value (Note 1) and the viscosity at 50% concentration in mineral spirits (Note 2) are measured. When the acid value is 6-10 and viscosity is 9-17 stokes at 50%, the reaction is cooled to 180° and mineral spirits is added. The reaction is then cooled to 100° and is filtered or strained.

The final resin should have these approximate characteristics:

% NV (Note 3)	50%
Viscosity	9-17 stokes
Acid value	6-10
Color (Note 4)	4-5

2. *Notes*

1. The acid value is the milligrams of KOH required to neutralize 1 g. of resin. It is determined as follows. Approximately 1 g. of resin is dissolved in 50 ml. of benzene solvent that contains phenolphthalen indicator and that has been titrated to a pink color. The solution is titrated to the pink end point with approximately 0.1 N alcoholic KOH solution. The acid value is calculated as follows:

$$\text{Acid value} = \text{A.V.} = \frac{(\text{ml. KOH}) \times N \times 56.1}{(\text{sample wt. in g.})}$$

2. Viscosity is measured directly in stokes by using Gardner-Holdt viscosity tubes at 25° (See Gardner and Sward, *Paint Testing Manual,* 12th ed., Gardner Laboratories, Inc, Bethesda, Maryland, 1962, pp. 171-173).

3. Per cent NV or % non-volatile resin content can be determined by weighing a 0.5 g. sample into an aluminum cup 2 in. in diameter and having ¼-in. sides, diluting with 1-5 ml. of benzene and evaporating on a 150° hot plate for 15 min.

$$\% \text{NV} = \frac{(\text{final wt. sample and cup}) - (\text{wt. cup})}{(\text{initial wt. sample and cup}) - (\text{wt. cup})}$$

4. Color can be determined in Gardner units by use of comparator as described on p. 23 of *Paint Testing Manual* (Note 2).

3. References

1. Polymer Division, Archer Daniels Midland Co., Minneapolis, Minnesota 55401.
2. Organic Chemistry Department, Central Research Laboratories, Inmont Corp., Clifton, New Jersey 07015.

Alfin Polybutadiene with Molecular Weight Control (1)

$$CH_2{=}CH{-}CH{=}CH_2 \xrightarrow[\text{Alfin catalyst (2)}]{\text{Hexane}} \begin{array}{l}\text{Polybutadiene} \\ >65\%\ trans\text{-}1,4 \\ <10\%\ cis\text{-}1,4 \\ 25\%\ \text{vinyl}\end{array}$$

Submitted by H. Greenberg and L. Grinninger (3)
Checked by E. E. Bostick (4a,b) and R. C. Gueldner (4a)

1. Procedure

A. Alfin Catalyst

Commercial hexane (11.), 100 ml. of *n*-butyl chloride, and 100 ml. of isopropanol are dried by allowing each to stand over one-tenth of its volume of fresh No. 3A molecular sieve (Note 1) overnight.

Dry hexane (700 ml.) is charged into a 1-1., three-necked Pyrex flask equipped with a Dry Ice-methanol condenser, a stirrer (200 r.p.m.), thermometer, and nitrogen blanket. Using a hypodermic needle, 46.0 g. (0.8 g-atom of Na) of a 40% dispersion of sodium in petrolatum-mineral spirits (Note 2) is transferred to the flask containing the hexane. *n*-Butyl chloride (37.9 g., 0.403 mole) is added dropwise to the stirred suspension of sodium in hexane at such a rate that the temperature does not rise above 56°. This takes approximately 2 hr. After the reaction subsides it is stirred for 20 min. *Caution! Lack of immediate rise in temperature after the first 2 ml. of butyl chloride has been added indicates the need for overcoming an induction period.* The addition of 0.25 ml. of ispropanol (hypodermic syringe) generally causes the reaction to proceed in about another 15 min. If reaction fails to start, a second increment of alcohol is added. If the reaction still "holds up," it is discarded and the reagents are purified further.

To the sodium butyl suspension 12.0 g. (0.2 mole) of isopropanol is added over 15-20 min. maintaining the temperature preferably between 20° and 40° but not over 56°. Propylene gas (Note 3) is introduced until there is a steady reflux from the Dry Ice-methanol condenser with the temperature of the slurry at 20-25°. The suspension is stirred for 1 hr. after charging the propylene. It is then allowed to stand overnight.

The catalyst is now ready for use. It may be siphoned (Note 4) into a clean, dry, 1-1. flask. The flask should preferably be closed with a tightly wired serum cap (Note 5) for subsequent withdrawal of portions of the catalyst slurry using a hypodermic syringe.

B. Polymerization

Dry hexane (150 ml.) is added to a 16-oz. beverage bottle which has been dried in an oven at 110° for several hours, purged with nitrogen, and stoppered. Commercial 75% 1,4-dihydronaphthalene (0.8 g.; 0.6 g. of 100% material) is added as molecular weight moderator (Note 6) with a small hypodermic syringe. The moderator is introduced by loosening the stopper slightly to insert the needle between the stopper and bottle. Butadiene (Note 7) [30 g. (±2 g.) (Note 8)] is condensed into the hexane by cooling to −20° in a Dry Ice-methanol bath, introducing the butadiene through a small two-holed rubber stopper, and carrying the inlet tube halfway into the bottle. The outlet tube is attached to a nitrogen blanket. Two rough volume calibration marks on the beverage bottle should have been established to read 150 ml. for the hexane and for the 30 g. of condensed butadiene, respectively. The weight of the butadiene is obtained accurately by weighing the bottle and contents before and after condensing the butadiene. The charging stopper is removed and quickly replaced with a metal "Slip Seal" (Note 9). The contents of the bottle must not be allowed to warm from the −20° cooling bath temperature before adding the catalyst.

The alfin catalyst is shaken and 30 ml. of catalyst is transferred into the butadiene-hexane solution. It is capped securely and is shaken immediately to disperse the catalyst.

Polymerization starts at once, and the liberated reaction heat causes a temperature rise from −20° to about 40° in about 10 min. Agitation is continued until the contents of the bottle become almost solid. The bottle is allowed to stand at room temperature for 2 hr. There will then be a vacuum in the bottle and the polymer will be a semi-solid "plug".

The beverage bottle is opened and, with small amounts of absolute ethanol and a suitable metal wire hook, the polymer mass is transferred into a 1000-ml. explosion-proof blender. The polymer is precipitated by adding sufficient anhydrous ethanol (about 500 ml.) containing antioxidant (1 g. of phenyl β-naphthylamine per 500 ml.) (Note 10) slowly to give a "crumb" slurry. The slurry is filtered and the "crumb" is washed twice with 500-700 ml. portions of ethanol in the blender to further reduce the hexane level. The polymer crumb is washed twice as above with water to remove traces of residual alkali. It is then washed twice with acetone containing 1% antioxidant to displace the water and dried in a vacuum desiccator to constant weight. Yield 30 g. (100%). Microstructure by infrared analysis is 66.6% *trans,* 7.8% *cis,* 25.6% vinyl; intrinsic viscosity (in toluene at 25°) 2.3 dl./g.; molecular weight (5), 229,000. Without the addition of 1,4-dihydronaphthalene the intrinsic viscosity would be about 15 dl./g., which is equivalent to a molecular weight of over 5 million. The constants used in the calculation for molecular-weight in the formula $[\eta] = KM^a$ are $K = 0.0011$, $a = 0.62$.

2. Notes

1. Molecular sieves can be obtained from Linde Division, Union Carbide Corporation, 61 E. Park Drive, Tonawanda, New York.

2. Pint samples of 40% sodium dispersions in petrolatum-mineral spirits suitable for this preparation can be purchased from the Gray Chemical Co., Gloucester, Massachusetts. This sodium dispersion is a fluid, not a paste as might be expected. In handling sodium, standard precautions such as eye protection are used. Care must also be taken to avoid bringing sodium and water together accidentally.

3. Lecture cylinder propylene (Matheson) is sufficiently pure as it comes from the cylinder. The amount to cause slow reflux at room temperature in the catalyst preparation reaction mixture is a considerable excess, but this is not detrimental.

4. Care should be exercised in handling alfin catalyst slurries. They are mildly pyrophoric and can ignite the hexane solvent if spilled in the air.

5. A good type of serum cap for this use may be purchased from Harshaw Scientific Co., catalog No. 5300.

6. Dihydronaphthalene-1,4, 75% pure, may be purchased from Gallard Schlesinger Chemical Mfg. Corp., Garden City, New York.

7. "Polymerization grade" butadiene (Matheson) needs no further purification for this experiment.

8. *Caution! Under no conditions should one attempt to polymerize more than 30 g. of butadiene in less than 150 ml. of hexane in a beverage bottle.* Of course, with pure butadiene monomer the bottle would explode from the heat and pressure of the reaction on adding alfin catalyst.

9. The resealing beverage bottle closure readily available at hardware stores under the trade name "Slip-Seal" is recommended. These simple devices have a thumbscrew that exerts positive pressure against a rubber gasket on top of the beverage bottle.

10. If infrared measurements or other tests, in which antioxidants can interfere, are to be made on the polymer, it is best to omit addition of antioxidant. Instead, store the gum under a high vacuum or under an inert gas atmosphere.

3. *References*

1. V. L. Hansley and H. Greenberg (National Distillers and Chemical Corp.), U.S. Pat. 3,067,187 (Dec. 4, 1962); V. L. Hansley and H. Greenberg, *Rubber Age,* **94**, 87 (1963).
2. A. A. Morton, et. al., *J. Amer. Chem. Soc.*, **68**, 93 (1946); **69**, 160, 161, 172, 950, 167 (1947); **70**, 3132 (1948); **71**, 481, 487 (1949); *Ind. Eng. Chem.*, **42**, 1488 (1950).
3. Research Division, U.S. Industrial Chemicals Co., Division of National Distillers and Chemical Corporation, 1275 Section Road, Cincinnati, Ohio 45237.

4a. Research Laboratory, General Electric Co., P.O. Box 1088, Schenectady, New York 12345.

4b. Current Address: General Electric Co., Mt. Vernon, Illinois 62864.

5. H. Mark and A. V. Tobolsky, *Physical Chemistry of High Polymeric Systems,* Interscience Publishers, New York, High Polymer Series, Vol. 2, p. 290.

Poly-p-xylylene

$$Cr_2(SO_4)_3 \cdot 15\,H_2O + Zn \xrightarrow[80°C.]{H_2O,\ CH_3OH} 2\,CrSO_4 + ZnSO_4$$

$$n\,ClCH_2\text{-}C_6H_4\text{-}CH_2Cl + 2n\,Cr^{2+} \longrightarrow \left[\text{-}CH_2\text{-}C_6H_4\text{-}CH_2\text{-}\right]_n + 2n\,CrCl^{2+}$$

Submitted by M. Hoyt and D. R. Fitch (1)
Checked by Y. P. Castille (2)

1. Procedure

Caution! Some individuals have developed dermatitis when preparing poly-p-xylylene by this general method (Note 1).

A. Apparatus

An aqueous methanolic solution of chromous sulfate is prepared in the 1-l. upper reaction flask of Fig. 1 which is modified with a 6-mm stopcock for bottom "draw-off" and is heated with an inverted upper half of a Glas-Col spherical heating mantle. The solution flows by gravity through filter *A* (Note 2) into the 2-l. lower reaction flask, where it reacts at reflux with α,α'-dichloro-*p*-xylene added in *p*-dioxane from the 250-ml. pressure-equalizing addition funnel. The lower flask should be equipped with an efficient bulb condenser. Asco Teflon glands (Note 3) provide convenient stirrer bearings. Glass or stainless steel stirrer shafts and blades may be used. The 1-l. filter flask *B* serves as a ballast to prevent "suck-back" of air. An oil bubbler *C* to indicate nitrogen flow and a mercury pressure release valve *D* are included.

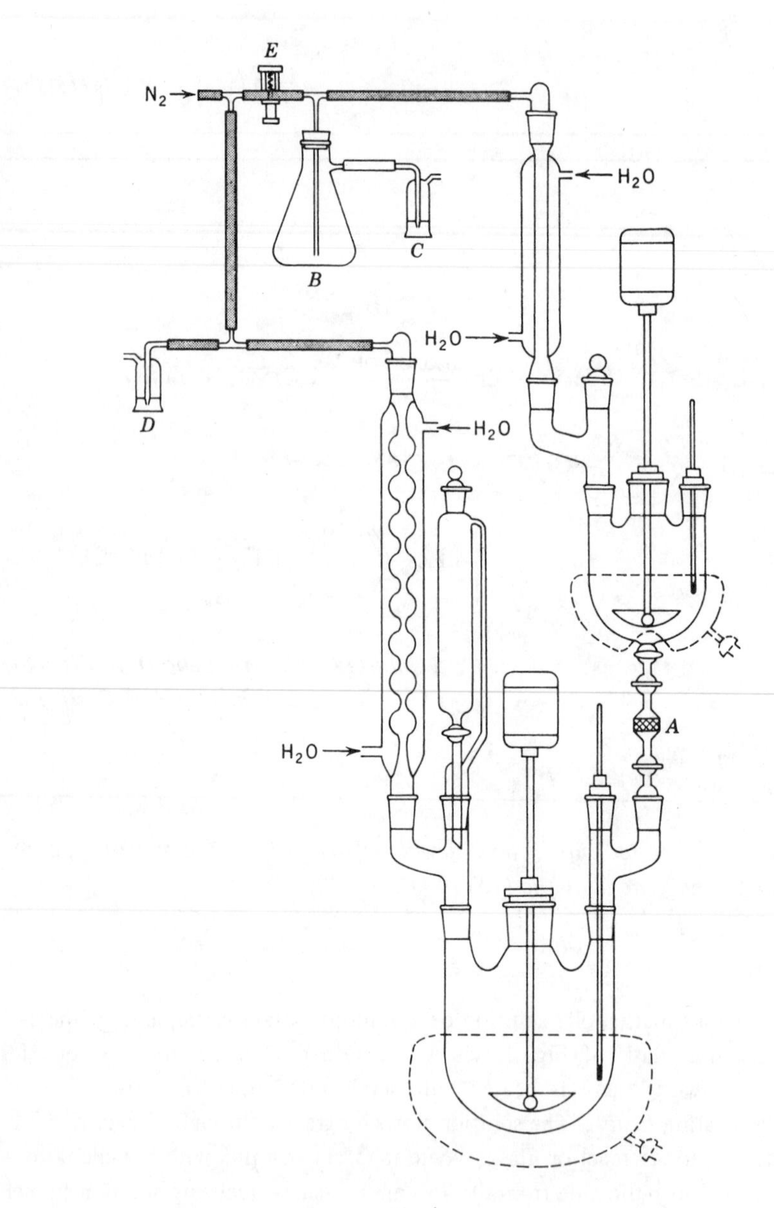

Fig. 1.

B. *Nitrogen Purge*

Because chromous ion solutions are rapidly oxidized by air, the entire system must be flushed with nitrogen before starting and must be kept under nitrogen throughout the reaction. Pinch clamp *E* is closed and nitrogen is passed through the two reaction flasks and the oil bubbler *C* for about 20 min. Care is taken to flush the addition funnel. The pinch clamp *E* is then removed and the nitrogen stream is adjusted so that a moderate flow passes through the oil bubbler *C*, thus blanketing the entire system.

C. *Chromous Sulfate*

Into the upper flask are introduced 116 g. (0.175 mole) of chromium sulfate pentadecahydrate (Note 4), 10.0 g. (0.153 g. atom) of 20-mesh granular zinc (Note 5), 425 ml. of deionized water, and 250 ml. of methanol (Note 6). The contents of the flask are stirred and heated to reflux (80°) as rapidly as possible. Refluxing and stirring are continued until only a few particles of zinc are visible; this usually requires 1-1.5 hr. (Note 7). The blue-green solution contains 270-295 mmoles of chromous sulfate (Note 8). It is filtered into the lower flask and rapidly reheated to reflux. The filtration requires 10-20 min. (Note 9).

D. *Polymer Preparation*

A solution of 17.5 g. (0.100 mole) of α,α′-dichloro-*p*-xylene (Note 10) in 150 ml. of dioxane (Note 11) is added from the pressure-equalizing funnel to the refluxing, filtered chromous sulfate solution as rapidly as possible, with good stirring. The addition requires 20-30 sec. After 30 min. of refluxing and stirring, the reaction is complete. About 90-100 mmoles excess chromous sulfate remains unreacted (Note 8). The flask is cooled at once with a bucket of cold water.

E. *Polymer Isolation (Caution! Hood. See Note 1.)*

Without delay the polymer is collected on a tared, fritted disk funnel (Note 12). The filter cake is promptly washed with ten successive 100-ml. portions of deionized water. The cake is dispersed well with a spatula after each addition of water and is then sucked dry with a water aspirator. Immediately thereafter it is washed with ten 50-ml. portions of acetone (Note 13) and dried in vacuum at 50°. Yield 8.8-9.1 g. (85-88%) of white, friable solid (Note 14); ash content 0.05-1.6% (Note 15).

2. Characterization

Inherent viscosity is conveniently determined in a Cannon-Fenske viscometer (Note 16) heated in a benzophenone vapor bath (305°). Arochlor 1242 (Note 17) is used as solvent, and an inhibitor (Note 18) must be added (1%). The polymer sample, solvent, and inhibitor are placed in the viscometer, and nitrogen is slowly bubbled through the mixture for about 20 min. A nitrogen atmosphere is maintained in the viscometer throughout the determination. The viscometer is placed in the vapor bath, and a stopwatch is started to show cumulative time in the bath. The polymer dissolves completely in about 3-7 min. Successive viscosity determinations are started as soon as solution is complete and are repeated as rapidly as possible over a period of about 20-30 min. noting the cumulative bath time at the start of each determination. A plot is made of the decreasing inherent viscosities against cumulative bath time. A smooth, downward trend should be observed. Extrapolation to zero bath time gives inherent viscosities ranging from 0.83 to 1.24 dl./g. (0.40 g./100 ml.).

The infrared absorption spectrum of the polymer is in accord with a published spectrum for poly-*p*-xylylene (3). An x-ray diagram of one of the polymers showed *d*-spacings of 5.3, 4.4, and 3.9 Å, indicating the presence of both the α- and the β-forms of poly-*p*-xylylene (4,5). When placed on a temperature gradient bar (modified Dennis bar) (6) at temperatures above 400°, the polymer melts with decomposition.

3. Notes

1. Workers of the Research Division of the U.S. Industrial Chemicals Co. developed a skin irritation resembling sunburn during the initial study of this reaction. Careless handling of α,α'-dichloro-*p*-xylene may have been responsible, but the evidence is not conclusive. The submitters repeatedly used the procedure described here without such difficulty. The apparatus shown in Fig. 1 was attached to a rack in a well-ventilated laboratory. All operations with α,α'-dichloro-*p*-xylene, however, such as recrystallization and solution makeup and also polymer isolation, were confined to a hood.

2. The detachable filter was fabricated from a reduced-end sealing tube, Corning No. 39580, containing a 40-mm. dia. coarse-porosity glass frit, and was fitted with 18/9 ball and socket joints.

3. Asco glands are available from Matheson Scientific, Inc.

4. Chromium sulfate, reagent, crystals, $Cr_2(SO_4)_3 \cdot 15\ H_2O$, from Matheson, Coleman and Bell, was used.

5. Zinc metal, A.C.S., 20-mesh, from Matheson, Coleman and Bell was preferred. Mossy zinc, A.C.S., from the same source serves equally well. Zinc dust is not recommended because it appears to react more slowly.

6. A.C.S. grade methanol from Fisher Scientific Co. was employed.

7. The checker found that a 3-hr. reflux period was needed. Even then, 0.23- and 0.71-g. residues of unreacted zinc were recovered in duplicate experiments. The length of tubing leading to the bottom stopcock of the upper flask should be minimized to reduce trapping of zinc.

8. As indicated by Castro (24), chromous sulfate can be determined by removing 1-ml. samples with a syringe and injecting them into 25-ml. portions of 0.25 *M* aqueous ferric chloride solution under nitrogen. A drop of Ferroin indicator (*o*-phenanthroline-ferrous sulfate complex, 0.025 M, Fisher Scientific Co.) is added and the solution is titrated to the green end point with standard ceric ammonium sulfate solution. The titrations are unnecessary if essentially all the zinc has been consumed and the system is well-blanketed with nitrogen.

9. The checker observed slower filtration (40-60 min.)

10. α,α'-Dichloro-*p*-xylene, m.p. 98.5-99.5°, from Diamond Alkali Co. was used as received.

11. *p*-Dioxane, a purified grade from Fisher Scientific Co., was used as received. Futher purification did not lead to improved results. The *p*-dioxane solution of α,α'-dichloro-*p*-xylene was sparged for about 15 min. with nitrogen before addition (*Caution! Hood*).

12. A 350 ml., coarse-porosity, sintered glass funnel is recommended.

13. After about eight acetone washes, the last wash should not show a cloud on dilution with water. If it does, acetone washing should be continued until no cloud is observed on dilution.

14. The checker obtained 9.1 and 9.3 g. (88% and 89%).

15. The submitters found 0.05-0.24% ash (3 polymers); the checker found 1.6% (1 polymer). The need for prompt and thorough washing of the polymer at the end of the synthesis is emphasized.

16. Poly-*p*-xylylene is soluble only at elevated temperatures (~300°.) in solvents such as the Arochlors and benzyl benzoate. Schaefgen (15) has shown that it degrades rapidly in solution at these temperatures. With adequate precautions, however, reasonably reproducible inherent viscosities can be measured.

The submitters used a special viscometer similar to that described by Schaefgen (15). The solvent was distilled and stored under nitrogen, and the samples were carefully degassed. The checker, however, obtained comparable inherent viscosities with a Cannon-Fenske viscometer and undistilled Arochlor

1242 containing 1% antioxidant 2246 (Note 18). His method seems more convenient and is recommended.

The accompanying tabulation shows comparative viscosity determinations made in both laboratories.

Schaefgen (15) used the term "inherent viscosity at zero time" to denote the viscosity when complete solution was achieved. The submitters and the checker

	Inherent Viscosity[a]	
Polymer No.	Results of Submitters, Viscometer of ref (15)[b]	Results of Checker, Cannon-Fenske Viscometer[c]
Submitters'		
No. 1	0.94	0.83
No. 2	1.23	1.19
No. 3	1.31, 1.38	1.13
Checker's		
No. 1	–	1.12
No. 2	–	1.24

[a] Extrapolated to zero bath time, 0.40 g./100 ml. at 305°.
[b] No inhibitor used.
[c] Antioxidant 2246 used (1%).

adopted the alternative practice of reporting inherent viscosity extrapolated to zero bath time. The extrapolated values are a little (<10%) higher than the first measured values, having been obtained after 5-10 min. of bath time.

A mixture containing 0.50 g. of polymer per 100 ml. of Arochlor 1242 at room temperature contains 0.40 g./100 ml. at 305° because the solvent density falls from 1.37 g./cc. at 25° to 1.10 g./cc. at 300° (25).

17. Arochlor 1242, Monsanto Chemical Co., is a chlorinated biphenyl fraction. The checker found that distilled solvent and technical grade solvent sparged with nitrogen give the same viscosity if inhibitor is used.

18. Antioxidant 2246, 2,2′-methylene-bis-(4-methyl-6-*t*-butylphenol) was obtained from American Cyanamid Co.

4. Methods of Preparation

The method described here has been reported in a preliminary communication (7). It provides a convenient laboratory synthesis of soluble, high molecular weight poly-*p*-xylylene from commercially available starting materials.

Soluble poly-*p*-xylylene of comparable molecular weight can also be prepared by the Hofmann degradation of *p*-methylbenzylammonium salts (6, 8, 9), but the starting materials are somewhat less readily available.

Poly-*p*-xylylene is also obtained by the fast-flow, vacuum pyrolysis of *p*-xylene first reported by Szwarc (10) and later studied by other workers (11-14). Elaborate equipment and extreme temperatures (800-1000°) are required. The very reactive intermediate, *p*-xylylene or *p*-quinodimethan, is formed in the pyrolysis and polymerizes as it condenses on cool surfaces. The pyrolysis polymer appears to be more or less crosslinked (4, 11-13, 15). Quenching the pyrolysis stream in cold (−78°) solvents affords fairly stable solutions of monomeric *p*-xylylene (14). When the *p*-xylylene solutions are allowed to polymerize to completion at low temperatures, a relatively tractable, soluble polymer is produced (4).

Recently the pyrolysis method was modified to yield initially the cyclic dimer, di-*p*-xylylene (16, 17). Thermolysis of the latter in vacuum at 600° to regenerate *p*-xylylene permits thin films of the polymer to be deposited on various substrates. Elaborate equipment and careful control are again required, but the method provides commercially useful forms of the polymer (Parylene polymers of Union Carbide Corp.).

The literature contains many references to the conversion of xylylene dihalides to xylylene polymers with such coupling agents as sodium (18-20), lithium (21), various transition metal powders (22), and phenyllithium (21).

The chemistry of *p*-xylylene, its analogs, and its polymers has been reviewed (23).

5. References

1. Research Division, U.S. Industrial Chemicals Co., Cincinnati, Ohio 45202.
2. Pioneering Research Division, Experimental Station, E. I. du Pont de Nemours and Co., Wilmington, Delaware 19898.
3. R. A. Nyquist, *Infrared Spectra of Plastics and Resins,* 2nd ed., The Dow Chemical Co., Midland, Michigan, May 3, 1961, Spectrum No. 124.
4. L. A. Errede and R. S. Gregorian, *J. Polym. Sci.,* **60**, 21 (1963).
5. C. J. Brown and A. C. Farthing, *J. Chem. Soc.*, 3270 (1953).
6. W. R. Sorenson and T. W. Campbell, *Preparative Methods of Polymer Chemistry,* Interscience Publishers, New York, 1961, pp. 49-50.
7. J. M. Hoyt, K. Koch, C. A. Sprang, S. Stregevsky, and C. E. Frank, *Polymer Preprints,* **5**, No. 2, 680 (1964).
8. F. S. Fawcett (E. I. du Pont de Nemours and Co.), U.S. Pat. 2,757,146 (July 31, 1956) [*C.A.* **50**, 14268g (1956)].
9. T. E. Young (E. I. du Pont de Nemours and Co.), Brit. Pat. 807,196 (Jan. 7, 1959) [*C.A.* **53**, 8711h (1959)]; U.S. Pat. 2,999,820 (Sept. 13, 1961), described in ref. 6, pp. 275 ff.

10. M. Szwarc, *J. Chem. Phys.*, **16**, 128 (1948).
11. L. A. Auspos, C. W. Burnam, L. A. R. Hall, J. K. Hubbard, W. Kirk, Jr., J. R. Schaefgen, and S. B. Speck, *J. Polym. Sci.*, **15**, 19 (1955).
12. M. H. Kaufmann, H. F. Mark, and R. B. Mesrobian, *J. Polym. Sci.*, **13**, 3 (1954).
13. R. S. Corley, H. C. Haas, M. W. Kane, and D. I. Livingston, *J. Polym. Sci.*, **13**, 137 (1954).
14. L. A. Errede and B. R. Landrum, *J. Am. Chem. Soc.*, 79, 4952 (1957).
15. J. R. Schaefgen, *J. Polym. Sci.*, **41**, 133 (1959).
16. W. F. Gorham, *J. Polym. Sci.*, **A1, 4,** 3027 (1966).
17. *Chem. Eng. News,* **43**, No. 38, 68 (1965).
18. R. A. Jacobson, *J. Amer. Chem. Soc.*, **54**, 1513 (1932).
19. A. A. Vansheidt, E. B. Mel'nikova, M. E. Krakovyak, L. V. Kukhareva, and G. A. Gladovskii, *J. Polym. Sci.*, **B1**, 339 (1963).
20. F. R. Dammont, *J. Polym. Sci.*, **Part B, 1,** 339 (1963).
21. J. Golden, *J. Chem. Soc.*, 1604 (1961).
22. K. Sisido, N. Kusano, R. Noyori, Y. Nozaki, M. Simosaka, and H. Nozaki, *J. Polym. Sci.*, **A1**, 2101 (1963).
23. L. A. Errede and M. Szwarc, *Quart. Rev.*, **12**, 301 (1958).
24. C. E. Castro, *J. Amer. Chem. Soc.*, **83**, 3262 (1961).
25. Arochlor Plasticizers, *Tech. Bull.* PL-306, Monsanto Chemical Co., 1960, p. 43.

Polyphenylene Sulfide (1)

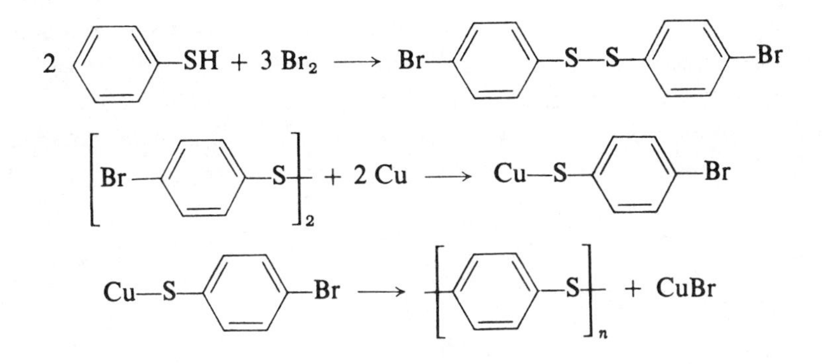

Submitted by C. E. Handlovits (2)
Checked by R. A. Fouty (3a,b) and A. W. Anderson (3a)

1. Procedure

A. Bis-(p-bromophenyl)disulfide

Bis-(*p*-bromophenyl)disulfide may be prepared quite easily in yields greater than 95% by the reaction of bromine with thiophenol in the absence of solvents (4). Bromine (65.5 g., 0.410 mole) is placed in a 250-ml., three-necked flask fitted with a stirrer, a condenser, and a dropping funnel. The flask and contents are cooled to 10° and 15.1 g. (0.173 mole) of thiophenol is added dropwise over a 20-30 min. period to keep the temperature at 10-15°. After the addition is completed, the reaction mixture is allowed to warm to room temperature; after 2-2½ hr. the contents are heated to 50-55° and the excess bromine is removed by distillation. The residual solid is washed with a dilute (10%) aqueous solution of sodium bisulfite to remove residual bromine, then dissolved in boiling ethanol. The solution is cooled and crystallization is effected. The yield is 26.3 g. (99.4%) of bis-(*p*-bromophenyl)disulfide, light cream to white plates, m. p. 91-92°.

B. Copper p-Bromothiophenoxide

A solution of 20 g. (0.053 mole) of bis-(*p*-bromophenyl)disulfide in 250 ml. of *n*-hexanol is placed in a 500-ml., three-necked flask equipped with a stirrer, a gas inlet, and a condenser to which is attached a gas outlet. The flask and contents are flushed with an inert gas (nitrogen or argon), and a flow of inert gas through the apparatus is maintained throughout the reaction. Copper metal, purified electrolytic dust, from Fisher Scientific Company (5.4 g., 0.084 mole) (Note 1) and 5.25 ml. (0.1 mole) of pyridine are added to the reaction vessel. The reactants are heated to a gentle reflux with stirring for 4-6 hr. The reaction must be carried out until the copper metal is completely consumed. The copper salt precipitates as a bright yellow solid, which is filtered, washed with methylene chloride and then acetone, and dried to give a nearly quantitative yield of copper *p*-bromothiophenoxide. The material should have m.p. 260-275°.

C. Polymerization of Copper p-Bromothiophenoxide

Solid State Polymerization. For solid state polymerization 10 g. (0.0397 mole) of copper *p*-bromothiophenoxide is placed in a 75-ml., heavy-walled, Pyrex® ampoule. The ampoule and monomer are evacuated and flushed at least three times with an inert gas (argon or purified nitrogen) and sealed with a torch at atmospheric pressure. The ampoule is heated at 200° for 5 days (Note 2), and the necessary safety precautions are taken to ensure that the glass ampoule will be contained if it ruptures while being heated.

After cooling to room temperature, the ampoule is cooled to Dry Ice temperature, opened, and the solids are ground if necessary. The solids are mixed with 200 ml. of concentrated HCl for 1 hr, filtered, and water-washed, and the concentrated HCl wash is repeated twice more. The light grey to dark grey solid phenylene sulfide polymer is oven-dried. Yield 95-100%.

Use of a Liquid Heating Medium. When freshly distilled pyridine is used as the liquid, the procedure is exactly the same as for solid state polymerization, except that a 10-g. (0.0397-mole) charge of copper *p*-bromothiophenoxide and 20 ml. of pyridine are placed in a 125-ml. ampoule and the reaction is carried out for only 24 hr. at 200°.

The use of triethylene glycol as the liquid heating medium is carried out by placing 50 g. (0.199 mole) of copper *p*-bromothiophenoxide and 500 ml. of distilled triethylene glycol in a 1-1., three-necked flask equipped with a stirrer, controlled heating mantle, condenser, and gas inlet. The reactor and reactants are flushed continuously with an inert gas (nitrogen or argon) and heated to 200°. When the temperature reaches ~70-75°, 15.7 g. (0.199 mole) of distilled pyridine is added. The reaction is heated with stirring for 7 days, cooled, filtered, water-washed, and extracted 2-3 times by mixing each time with 1 l. of concentrated HCl, filtering and water-washing. The phenylene sulfide polymer is oven-dried to give 20-21 g. of material having a softening point of 265-275°.

2. Characterization

On a Fisher-Johns melting point apparatus the polymer softens at ~250-280°. Crystalline melting point determinations with a Kofler hot stage on a polarizing microscope gave a value of 280-287°.

The polymer can be fractionated into three molecular weight ranges very easily. First it is extracted with boiling toluene, and the toluene soluble fraction is precipitated by slow addition, with vigorous stirring, to cold methanol. Generally little or none of the polymer is soluble in boiling toluene. The remainder is dissolved in boiling diphenyl ether (<2% solution) filtered hot, and the diphenyl ether-soluble fraction is precipitated in methanol, filtered, washed with acetone, and dried. Approximately 60-100% of the polymer is obtained as material soluble in diphenyl ether, 0-10% is soluble in toluene, and the remainder is insoluble in diphenyl ether.

An estimate of the molecular weight can be obtained from a bromine end group analysis. The bromine content should be ~1.0%. The copper analysis should be ~0.1%. The end group analysis can be correlated with solution viscosity in diphenyl ether at 250°, or the melt viscosity of the phenylene sulfide polymer at 303°.

3. Notes

1. Freshly prepared copper metal may be used. This is prepared by the method of Vogel (5). Care must be exercised in maintaining a liquid layer over the copper metal at all times. The water is removed by decanting and is replaced with *n*-hexanol.

2. Temperatures as high as 250-265° can be used without appreciably affecting the polymer. The time can be extended to any reasonable period, 20-30 days for example. Both the temperature and the length of the polymerization affect the molecular weight of the polymer, as would be expected with a condensation polymerization.

4. References

1. R. W. Lenz, C. E. Handlovits, and H. A. Smith, *J. Polym. Sci.,* 58, 351 (1962); J. T. Edmonds, Jr. and H. W. Hill, *Adv. in Chem.* **140**, 174 (1975); U.S. Pat. 3,354,129, Nov. 21, 1967.
2. Plastics Department Research Laboratory, The Dow Chemical Company, Midland, Michigan 48640.

3a. Plastics Department , E. I. du Pont de Nemours and Company, Experimental Station, Wilmington, Delaware 19898.

3b. Current Address: E. I. du Pont de Nemours and Company, Lincolnwood, Çhicago, Illinois 60646.

4. E. Bourgeois and A. Abraham. *Rec. trav. chim.,* **30**, 407 (1911).
5. A. I. Vogel, *Textbook of Practical Organic Chemistry,* 2nd ed., John Wiley & Sons, New York, p. 188 (1951).

Expandable Polystyrene Beads

Submitted by A. R. Ingram, (1a,b) E. Lyle, (1a) and E. M. Fettes (1a,c)
Checked by R. Westphal (2)

1. Procedure

A. One-Step Process

To each of one or more 12-oz. bottles (Note 1) are added 5.0 g. of a 2% solution of poly(vinyl alcohol) (Note 2) and 135 ml. of distilled or deionized water. After swirling the bottle to dilute the poly(vinyl alcohol), 60 g. (66 ml.) of styrene is added (Note 3); it contains 0.18 g. of benzoyl peroxide. (*Caution! Benzoyl peroxide is a strong oxidant and should be handled with care*), 0.09 g. of *t*-butyl perbenzoate (*Caution! Peroxidic materials are strong oxidants.*) (3), and 5.7 g. (9 ml.) of *n*-pentane (Note 4). The bottles are sealed with crown caps having an aluminum liner (Note 5).

The sealed bottles are inserted in a bottle polymerizer (Note 6) and rotated end over end at a speed of 180 r.p.m. during a schedule consisting of 1 hr. to 90°, 12 hr. at 90°, 1.2 hr. to 115°, 7 hr. at 115°, and cooling to less than 30°.

The bottles are opened in a hood to allow excess pentane to escape, and the contents are poured onto a Buchner funnel containing four thicknesses of cheese cloth or onto a 60 mesh stainless steel sieve (Note 7 and 8). The beads are washed with water until the filtrate is clear. The beads are spread in evaporating dishes or shallow trays and permitted to dry in a fume hood or circulating oven at a temperature below 35°. As soon as the beads are free-flowing, they are stored in glass jars with tightly fitting screw caps.

B. Two-step Process

Polymerization. In a 2-1. resin kettle (Note 9) 0.03 g. of Ultrawet K (Note 10) is dissolved in 500 g. of distilled water, 3.0 g. of finely divided tricalcium phosphate is added, then stirred to wet the powder thoroughly (Note 11) (4). The kettle is fitted with an 18-in. × ¾-in. air-cooled reflux condenser, a thermometer in a thermometer well, which also serves as a baffle (Note 12), a Pyrex® stopper, and a stainless steel, flat blade, T-impeller agitator (Note 13), which is ½ in. from the bottom of the kettle. Heat is supplied by a heating mantle regulated with a variable transformer. In a separate beaker 1.5 g. of benzoyl peroxide (*Caution! Peroxides are strong oxidants. Benzoylperoxide should be handled with care.*) and 0.25 g. of *t*-butyl perbenzoate (*Caution! Peroxidic materials are strong oxidants.*) (3) are dissolved in 500 g. of styrene (Note 3) and the solution is added to the resin kettle. With the agitator revolving at 400 r.p.m. the suspension is heated according to the schedule 0.8-1.0 hr. to 85°, 0.3-0.5 hr. to 90° (Note 14), 7 hr. at 90 ± 1°, and cooling to less than 30°.

A magnetic stirrer is used in the open kettle bottom to maintain a uniform suspension while 160 g. of slurry is transferred to each of four 12-oz. bottles (Note 1). The bottles are capped (Note 5), inserted in a bottle polymerizer (Note 6) and rotated end over end at a speed of 180 r.p.m. according to the schedule 1-1.5 hr. to 115°, 4 hr. at 115°, and cooling to less than 30°.

While the preparation of the beads could be carried out with 160-g. charges in 12-oz. bottles, it is instructive to perform the suspension polymerization in kettles so that the progress of polymerization and bead formation can be observed periodically.

Impregnation. Into each bottle from the preceding polymerization step are charged 40 g. of distilled water containing 0.0048 g. of dissolved Ultrawet K (Note 10), and 0.32 g. of tricalcium phosphate (Note 11). The bottle is shaken vigorously to disperse the phosphate. With a buret, 13 ml. of *n*-pentane (Note 4) is added rapidly, the bottle is immediately sealed with a new cap and shaken well (Note 15). The bottles are then returned to the polymerizer and rotated end over end according to the schedule 1 hr. to 90°, 10 hr. at 90°, then cooling to less than 30°.

Each bottle is opened carefully in a hood to allow excess pentane to escape, and the contents are poured into a 1-1. beaker. Approximately 7 ml. of concentrated hydrochloric acid is slowly added to the slurry (final pH below 2) to dissolve the tricalcium phosphate. The temperature of the slurry should be kept below 40° by the addition of water if needed. The acidified slurry is allowed to stand for 3 min. before washing.

The beads are recovered by filtration through a 60-mesh stainless steel sieve or a coarse, fritted glass funnel. (Filters made from cotton or brass ahould be avoided because of attack by hydrochloric acid.) The beads are washed under cold running water for 10-15 min. while the bead layer is agitated by the force of the water. Drying and storage are carried out in the same manner as in the one-step process.

2. Characterization

For best foam properties the pentane content should be between approximately 5.5% and 7%. It can be determined by heating a 1-g. sample, weighed to ±0.0001 g., in a 2-in. dia. aluminum foil weighing dish in a vented oven at 150° for 2 hr., followed by cooling in a desiccator (Note 16). The percent weight loss is commonly called the volatile content.

Bead size distribution can be measured by passing a weighed sample of the beads through a series of sieves of increasingly finer mesh (Note 17). Sizes may also be determined by comparison with standards or by microscopic (3-10X) examination against a ruler or grid.

The molecular weight range satisfactory for expandable beads is indicated by intrinsic viscosities of 0.5-1.0 in toluene at 30°. (Note 18).

Residual styrene monomer content should be less than 0.3% because excessive monomer causes shrinkage in molded foams (see also Note 16). Monomer can be determined by ultraviolet spectroscopy (5).

3. Expansion of Beads

The pentane-impregnated beads from either the one-step or the two-step procedure expand on being heated. A simple device for this operation is a household double boiler with the bottom of the upper pan replaced by a 100-mesh stainless steel screen (Note 19). Steam is generated by boiling water in the lower pan. If this is not available, the beads may be expanded directly in boiling water. In this procedure the water in the container (pan or liter beaker) is at least 2 in. deep, and the top of the container is at least 2 in. above the water level. The expanded beads are removed from the boiling water by a perforated scoop such as a tea strainer. The effect of time on expansion can be seen if the beads are permitted to remain in contact with the water or steam for different lengths of time such as 2, 3, and 5 min.

4. Notes

1. Twelve-ounce citrate of magnesia bottles without stoppers, Preiser

Scientific, Inc. Charleston, W. Va., catalog number 10-4698 were used.

2. The 2% solution of poly(vinyl alcohol) is prepared by adding 2 g. of Vinol 540 (Air Products & Chemicals Co.) to 98 g. of water with agitation. The solution is heated to 90° and maintained there for 1 hr. with stirring. The solution is cooled before use. Erratic results in obtaining desired sizes of beads can occur if the solution is used after about 3 days.

3. The styrene should be free of polymer and inhibitor. The presence of polymer can be detected by adding a few drops of the styrene to 100 ml. of methanol and to determine if clouding occurs. The monomer can be separated from polymer by vacuum distillation. Inhibitor is readily removed by passing the monomer through a column of activated alumina or by washing the monomer with aqueous sodium hydroxide.

4. *n*-Pentane, technical grade, 95%, Phillips Petroleum Co.

5. Beer bottle caps with cork lining and aluminum foil spots, American General Supply Co., 112 Smithfield St., Pittsburgh, Pennsylvania 15222. An industrial type of capper is recommended; a suitable model is available from Crown Cork and Seal Co., Baltimore, Maryland, Model AXA.

6. Information about the bottle polymerizer may be obtained from Research Appliance Co., Route 8, Allison Park, Pennsylvania 15101.

7. The finely divided solid present in the bottles along with the beads is to be discarded. It will pass through the cheesecloth or sieve but will be retained by rapid filter paper.

8. Stainless steel screen, 60-mesh, Fisher Scientific Co., catalog number 4-881-10, sieve No. 60.

9. Resin kettle with cover, Corning Glass Works, catalog number 6947.

10. Ultrawet K, Arko Chemical Co., or other source of sodium dodecylbenzene sulfonate.

11. "Tricalcium phosphate," NF grade, hydroxyapatite, $3Ca_3(PO_4)_2 \cdot Ca(OH)_2$, Stauffer Chemical Co., or Monsanto Co.

12. Thermowell, Kontes Glass Co., catalog number 20100, or other ½ in. O.D. × 5½ in. Pyrex®, with joint to fit kettle top.

13. Agitator blades 3½ in. dia. × ¾ in., blades tilted to 45° angle to provide a "lifting" action. A Teflon® bearing is used, 1½ in. O.D. × 6 in. with 34/45 standard taper inner joint.

14. The polymerization reaction is exothermic, therefore the final temperature increase should be very gradual.

15. To show the effect of pentane content on the characteristics of the foam, the amounts of pentane can be varied from 5 to 9 ml. in the other three bottles, if desired.

16. Occluded water or excessive residual monomer interferes with this determination. It is important that the beads be free of surface water.

17. Number 10, 20, and 40 sieves of 3-in. dia., Dual Manufacturing Co., Chicago, Illinois.

18. The sample weight for the viscostiy determination must be corrected for the volatile content.

19. A photograph of this equipment can be found in ref. 6 or ref. 7.

5. References

1a. Koppers Company, Inc., Monroeville, Pennsylvania 15146.
1b. Current Address: Arco Polymer Inc., Monroeville, Pensylvania 15146. The assistance of H. A. Wright in the preparation of the revised procedure is thankfully noted.
1c. Current Address: Nothern Petrochemical Co., Des Plaines, Illinois 60018.
2. Foster Grant Co., Inc., Leominster, Massachusetts 01453.
3. G. F. D'Alelio (Koppers Co., Inc.), U.S. Pat. 2,692,260, October 19, 1954 [*C. A.* **49**, 2118e (1955)].
4. J. M. Grim (Koppers Co., Inc.), U. S. Pat. 2,673,194, March 23, 1954 [*C. A.* **48**, 11839i (1954)].
5. J. J. McGovern, J. M. Grim, and W. C. Teach, *Anal. Chem.*, **20**, 312 (1948).
6. A. R. Ingram and H. A. Wright, *Mod. Plastics,* **41**, No. 3, 152 (1963).
7. A. R. Ingram, *J. Cellular Plastics,* **1**, 69 (1965).

Water-Solubule Sulfonated Polystyrene

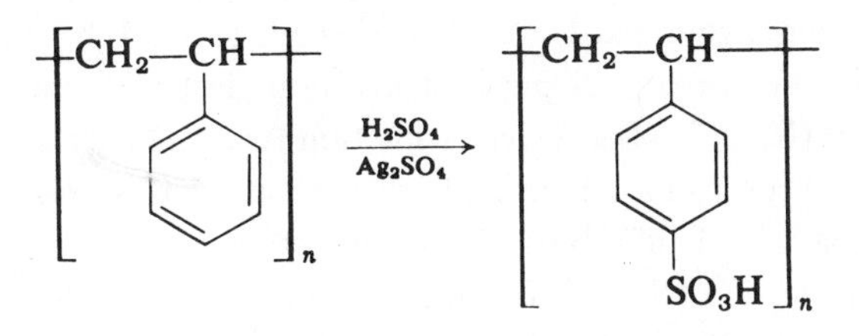

Submitted by A. R. Mukherjee and C. R. Raha (1)
Checked by W. C. Bauman (2)

1. Procedure

Styrene is polymerized in the presence of benzoyl peroxide (*Caution! Benzoyl peroxide is a strong oxidant and should be handled with care.*) under a nitrogen atmosphere (Note 1) and the polymer is processed by the technique of Overberger (3). A 500 ml., three-necked flask equipped with a variable speed stirrer, a thermometer, and a calcium chloride tube, is charged with 50 ml. of concentrated sulfuric acid (sp. gr. 1.84) and 10 mg. of silver sulfate; the flask is kept in a constant level water bath at 98-99°. Polystyrene (5 g.) in the form of 40-mesh powder (Note 2) is added to the hot liquid mass in ten increments (0.5 g. each) at regular intervals of 15 min. For each addition the stirring is stopped, the thermometer is instantly removed, and the polymer is inserted with a suitable spoon. More sulfuric acid (10 ml.) is added in two equal portions through the same opening, one after 1 hr. and the second after 2.5 hr. to wash any polystyrene powder adhering to the sides into the flask. The reaction is allowed to continue for 4 hr. (Note 3); the product is a light brown, transparent, viscous liquid containing a very small amount of unreacted polystyrene powder (Note 4). It is then carefully filtered through glass wool, previously washed with boiling distilled water to warm it, into a 250-ml. beaker containing 160 ml. of cold (approximately 10°) distilled water (4). The flask is rinsed with 5 ml. of sulfuric acid and the rinsed liquid is also added to the beaker. The contents of the beaker are stirred thoroughly to a clear, homogeneous, brown liquid. The beaker is kept overnight in a refrigerator at 5-10°, and sulfonated polystyrene separates (Note 5) as a thick, brown, gel-like mass. The supernatant liquid is decanted and the solid mass is immediately washed with 10 ml. of cold (about 10°) distilled water to remove as much as possible of the adhering mineral acid. The entire mass is transferred to a bag made of cellophane, 15 cm. X 15 cm., with about 10 ml. of water, the mouth of the bag is tied with a cord, and the bag is suspended from a glass rod in 150 ml. of water in a 1-1. beaker and dialyzed (Note 6) (4). The water is changed at half-hour intervals; the total time for dialysis is 2 hr. (Note 7). The aqueous solution (ca. 150 ml.) of the polymer in the bag is passed through a column of Dowex 3 ion exchange resin powder (J. T. Baker Chemical Co., Phillipsburg, New Jersey) for removal of any trace of remaining sulfuric acid (Note 8) (4-6). The liquid is allowed to drop at the rate of 80-90 drops/min. the column is washed with another 50 ml. of distilled water to elute any additional material. Further washing of the column can elute only negligible amounts of the material. The entire solution (about 200 ml.) is then evaporated on a water bath in a 500-ml. capacity, flat-bottomed porcelain basin to a semi-solid consistency. The semisolid mass is dried to constant weight in a vacuum desiccator. Yield 6.61 g.

2. Characterization

Sulfonated polystyrene prepared by this technique is a sheet-like, hard, light brown solid, completely soluble in water, methanol, ethanol, and propanol. It is insoluble in benzene, carbon tetrachloride chloroform and methyl ethyl ketone. The aqueous solution above a concentration of 1% is quite viscous and shows typical polyelectrolytic behavior during viscosity measurements, i.e., η_{sp}/c value increases with decrease in concentration of the solution, resulting in a parabolic η_{sp}/c against c (c = concentration) plot. In replicate experiments the nature of this curve remains unaltered, though absolute values of η_{sp}/c change considerably. This result shows that, because of the heterogeneity of the medium, the extent of sulfonation is different in different samples. The titer value of this resin is 47.2 ml. of 0.1 N caustic soda solution/g. of the resin, corresponding to

78 SO_3H groups/100 styrene units, excluding possible degradation, if any, in the polymer chain.

3. Notes

1. In a 100-ml., glass-stoppered, conical flask, fitted with inlet and outlet tubes for nitrogen flushing, are placed 25.00 ml. (22.68 g.) of distilled styrene and 0.25 g. of benzoyl peroxide. The monomer is polymerized under a nitrogen atmosphere by heating the flask at 50-55° for 22 hr. The polymer weighs 21.4 g. (94.3% yield); $[\eta] = 0.41$ at 35° in benzene solution.

2. The polymer flakes, dried in a vacuum desiccator, are ground to powder in a mortar with a pestle and sieved in a 40-mesh B.S. sieve.

3. As the molecular weight of the polystyrene sample increases, more time is required for dissolution. Polystyrene samples with molecular weights up to 250,000 could be sulfonated by this method. Also, the product was para-sulfonated, predominantly but perhaps not exclusively (7).

4. The end point of the reaction is indicated when a drop of the brown viscous product, drawn in a warm pipet, given a complete solution on vigorous shaking with water.

5. The beaker is placed in a porcelain or glass basin containing water as the outside contacting agent. This procedure gives better and more complete separation of the product.

6. If the cellophane bag cracks, it is wrapped in cellophane. The time for dialysis is then approximately doubled.

7. During dialysis, both the sulfonated polymer and sulfuric acid come out through the pores of the membrane, the latter definitely at a much faster rate. After 2 hr. no more sulfuric acid comes out, as evidenced by a barium chloride test which indicates its removal from the system (barium chloride gives a precipitate with both polystyrene sulfonic acid and sulfuric acid but, whereas the precipitate of the former is completely soluble in a 6 *N* HCl solution of 2.5% barium chloride, the precipitate of the latter remains completely insoluble in such a solution, even on boiling).

8. For preparation of such a column the weakly basic anion exchange resin (20-50 mesh) is washed with 0.05% caustic soda solution, packed to 15 cm. height in a column with a porous porcelain bed, and washed alkali-free with 1 l. of distilled water. Equivalent results can be obtained through the use of the

weak-base resin Amberlite IR-45 (Rohm and Haas) or the strong-base resin Dowex 1, as suggested by the checker.

4. Methods of Preparation

The use of concentrated sulfuric acid as the sulfonating agent and the solvent (4, 5, 7-10) has been described. Sulfuric acid of 95% strength has been employed (11), and concentrated sulfuric acid in a halohydrocarbon solvent has similarly been used (12). Sulfur trioxide alone or its complexes have been used as the sulfonating agent (13-19). The use of chlorosulfonic acid as the sulfonating agent has also been reported (20).

5. References

1. Indian Association for the Cultivation of Science, Jadavpur, Calcutta-32, India.
2. The Dow Chemical Co., Chemicals Department, Midland, Michigan 48640.
3. C. G. Overberger, *Macromol. Syn.*, **1**, 4 (1963); Coll. Vol. I, p. 5.
4. H. P. Gregor, H. Jacobson, R. C. Shair, and D. M. Wetstone, *J. Phys. Chem.*, **61**, 141 (1957).
5. M. H. Waxham, B. R. Sundheim, and H. P. Gregor, *J. Phys. Chem.*, **57**, 969 (1953).
6. H. H. Roth, U.S. Pat. 2,789,944 (1957) [*C. A.*, **51**, 17235 (1957)].
7. R. Hart and R. Janssen, *Makromol. Chem.*, **43**, 242 (1961).
8. G. B. Bachman, H. Hellman, K. R. Robinson, R. W. Finholt, E. J. Kahler, L. J. Filar, L. V. Heisey, L. L. Lewis, and D. D. Micucci, *J. Org. Chem.*, **12**, 108 (1947).
9. R. F. Prini and A. E. Lagos, *J. Polym. Sci.*, **A2**, 2917 (1964).
10. I. H. Spinner, J. A. Ciric, and W. F. Graydon, *Can. J. Chem.*, **32**, 143 (1954).
11. A. Takahashi, S. Shimoyama, and I. Kagawa, Koygo Kagaku Zasshi, **61**, 1617 (1958) [*C. A.*, **56**, 74961 (1962)].
12. H. H. Roth, E. R. Cowherd, W. C. Bauman, and R. M. Wiley, Brit. Pat. 960,502 (1964) [*C. A.*, **61**, 7194b (1964)].
13. M. Baer, U.S. Pat. 2,533,210 (1950) [*C. A.*, **45**, 3651g (1951)].
14. H. H. Roth and H. B. Smith, U.S. Pat. 2,663,700 (1953) [*C. A.*, 48, 6161h (1954)].
15. J. Fantl, U.S. Pat. 2,718,514 (1955) [*C. A.*, **50**, 1367h (1956)].
16. D. D. Reynolds and J. A. Cathcart, U.S. Pat. 2,725,368 (1955) [*C. A.*, **50**, 9786c (1956)].
17. R. Signer, A. Demagistri, and C. H. Muller, *Makromol. Chem.*, **18/19**, 139 (1956).
18. A. S. Teot, U.S. Pat. 2,763,634 (1956) [*C. A.*, **51**, 6218b (1957)].
19. A. F. Turbak, U.S. Pat, 3,072,618 (1963) [*C. A.*, 58, 13851c (1963)].
20. A Colin-Russ, Brit. Pat. 825, 422 (1959) [*C. A.*, **54**, 9365b (1960)].

Cellulose Acetate

$$C_6H_7O_2(OH)_3 + (CH_3CO)_2O \xrightarrow{H_2SO_4} C_6H_7O_2(OCOCH_3)_3$$

Submitted by J. E. Kiefer (1)
Checked by A. J. Rosenthal (2)

1. Procedure

Wood pulp (Note 1) is activated by soaking it in water for 16 hr. The pulp is pressed in a Büchner funnel to remove most of the water. The remaining water is removed by displacement with glacial acetic acid on the Büchner funnel, and the final mixture is pressed until a product containing 1 part of cellulose to 2 parts of acid is obtained. In a 1-1. flask equipped with a mechanical stirrer capable of mixing viscous solutions is placed 30 g. of activated wood pulp wet with 60 g. of glacial acetic acid. A solution of 2.1 g. of 98% sulfuric acid in 390 g. of glacial acetic acid is added, and the mass is mixed at 25-30° for 30 min. Acetic anhydride (*Lachrymator!*) (150 g.) is added at the rate of 5 ml./min. while mixing. The reaction mixture is maintained between 25° and 30° during the addition. The mass is then warmed to 35° and mixed until a clear, fiber-free, viscous solution is obtained. The reaction normally requires about 2 hr. (Note 2).

A solution consisting of 4 g. of magnesium acetate dissolved in 40 g. of water and 40 g. of glacial acetic acid is added to the reaction mixture at the rate of 5 ml./min. (Note 3). The cellulose acetate is precipitated by pouring it into water. The product is isolated by filtration and washed with water until the water washings are neutral. It is then washed with 21. of water containing 3 g. of magnesium acetate, isolated by filtration, and dried at 60°. The white, flocculent cellulose triacetate (270 g.) contains 43.5-44.5% acetyl and is soluble in dichloromethane, chloroform, and formic acid (Note 4). The inherent viscosity of the regenerated cellulose, 0.1% in cupriethylene diamine, is 1.63 dl./g.

2. Notes

1. Special acetate cellulose (P-IB) wood pulp obtained from Rayonier, Inc., was used. Other purified grades of wood pulp, cotton linters, filter paper, or a regenerated cellulose (viscose) can be acetylated by this procedure, however.

2. The reaction rate varies considerably with the type of cellulose being acetylated. Regenertated cellulose is much more reactive than is filter paper, wood pulp, or cotton linters. For example, a regenerated cellulose can be acetylated in 40-60 min. under these conditions, whereas cotton linters or a high molecular weight wood pulp requires 2-4 hr. The completion of the reaction is best determined by microscopic examination of a drop of the solution pressed between two glass slides. Prolonging the reaction at this point causes a reduction in the molecular weight of the product.

3. The temperature may rapidly rise about 25° as a result of hydrolysis of excess acetic anhydride.

4. Cellulose triacetate, normally called primary acetate, is insoluble in acetone. An acetone-soluble ester can be obtained by hydrolysis of the primary acetate to a secondary acetate (3). The secondary acetates contain 35-42% acetyl. Because of their solubility characteristics and compatibility with plasticizers, the secondary acetates are preferred in most commercial applications.

3. Methods of Preparation

Cellulose has been acetylated with acetic anhydride using the following catalysts: sulfuric acid (4), zinc chloride (5), perchloric acid (6), methanesulfonic acid (7), aromatic sulfonic acids (8), sulfur dioxide (9), pyridine (10), and basic salts (11). Acetyl chloride (12) and ketene (13) have also been used as the acetylating agents for preparing cellulose acetate.

4. Merits of the Preparation

There are several advantages to using sulfuric acid as the acetylation catalyst. Mainly the sulfuric acid combines with the cellulose during the acetylation. The combined sulfate groups enhance the solubility of cellulose acetate in acetic acid (the common esterification diluent) and a more uniform acetylation results. Also, sulfuric acid is a good catalyst for hydrolysis. Consequently an acetone-soluble product can be obtained by adding water after the esterification is completed. The hydrolysis can then be carried out without isolating the cellulose triacetate.

5. References

1. Research Laboratories, Tennessee Estman Co., Division of Eastman Kodak Co., Kingsport, Tennessee 37662.
2. Research Laboratories, Celanese Corp., Summit, New Jersey 07901.
3. H. Bates, J. W. Fisher, and J. R. Smith, U.S. Pat. 2,775,585 (1956).
4. G. W. Miles, Brit. Pat. 19,330 (1905).
5. K. Hess and G. Schultze, *Ann. Chem.*, **455**, 81 (1927).
6. C. J. Malm, U.S. Pat. 1,645,915 (1927) [*C. A.*, **22**, 164 (1928).
7. Soc. des usines Chemiques Rhône-Poulenc, Fr. Pat. 705,546 [*C. A.*, **23**, 5287 (1929)].
8. H. S. Mark, A. D. Little, and W. H. Walker, U.S. Pat. 709,922 (1902).
9. K. Murata, U.S. Pat. 2,903,481 (1959). [*C. A.*, **54**, 894d (1960)].
10. K. Hess and N. Ljubitsch, *Ber.*, **61B**, 1460 (1928) [*C.A.*, **22**, 4793 (1928)].
11. R. C. Blume, U.S. Pat. 2,632,006 (1953) [*C. A.*, **47**, 5682h (1953)].
12. C. J. Malm, J. W. Mench, D. L. Kendall, and G. D. Hiatt, *Ind. Eng. Chem.*, **43**, 684 (1951).
13. Ketoid Co., Brit. Pat. 237,591 (1924) [*C. A.*, **20**, 1522 (1926).

Cellulose Nitrate

$$C_6H_7O_2(OH)_2 + 2\ HNO_3 \rightarrow C_6H_7O_2(OH)(ONO_2)_2 + 2\ H_2O$$

Submitted by G. P. Touey (1)
Checked by A. J. Rosenthal (2)

1. Procedure

An 88% sulfuric acid solution is prepared by adding 92 g. of reagent grade (96%) acid to 8 ml. of water in a 250-ml. beaker. The beaker is cooled to 10° in an ice bath and 60 g. of reagent grade (70%) nitric acid is added (*Caution!*) (Note 1).

The beaker is removed from the ice bath, and the solution is allowed to warm to room temperature. A 2.5-g. sample of finely shredded (¼ × ¼ in.) filter paper (Note 2) is added while the mixture is gently stirred with a glass stirring rod. The Beaker is placed in a water bath maintained at 36-38° and the reaction is allowed to proceed at this temperature for 45 min. During this time the mixture is stirred occasionally (Note 3).

The acid mixture is drained (Note 4), and the cellulose nitrate is immediately added to 21. of rapidly agitated distilled water. The product is filtered and washed in another 21. of distilled water. The washed and filtered cellulose nitrate is boiled in 11. of 50% acetic acid solution for 1 hr., then boiled in 500 ml. of 2% sodium bicarbonate solution for 10 min. (Note 5). After being washed free of bicarbonate, the filtered but undried ester is stored in 100 ml. of ethyl alcohol.

2. Notes

1. This solution should be prepared in a hood, because some fuming occurs.

The final mixture contains 21% nitric acid, 61.6% sulfuric acid, and 17.4% water. With this amount of water (16-19%) in the mixed solution, the resulting cellulose nitrate is soluble in an ethyl alcohol-ether mixture and contains 11.5-12.5% nitrogen. This nitrogen content corresponds to an average degree of substitution of slightly over two nitrate groups per anhydroglucose unit. If a product with a higher nitrogen content is desired, a mixed acid solution containing less water is used. For example, a mixture containing 25% nitric acid, 62% sulfuric acid, and 13% water yields a product containing 13.2-13.6% nitrogen under the reaction conditions stipulated in this procedure. A completely nitrated product would contain 14.2% nitrogen. However, it is more desirable that samples prepared in the laboratory contain 13% nitrogen or less, because more completely esterified products are difficult to stabilize.

2. A good grade of ashless filter paper is used because it is a pure form of cellulose and is readily available in a laboratory. However, pure grades of commercial wood pulp or cotton linters can be used. Because the amount of cellulose needed is small (2.5-3.0 g.), the ratio of mixed acid solution to cellulose can be maintained at 50:1 without employing a large volume of the acid. The use of a high ratio of mixed acid to cellulose also produces a more uniformly esterified product. In addition, it is a precaution not to prepare large quantities of cellulose nitrate in a laboratory because of its incendiary nature.

3. Because the esterification takes place rapidly, the temperature range of the water bath can be controlled by the occasional addition of hot water. The mixture should be agitated and the temperature adjusted about every 5 min. At temperatures of 25° or below the reaction rate is slow. Temperatures above 40° result in excessive degradation of the cellulose.

4. The liquid is not filtered because the material might be excessively exposed to the air, and oxidative degradation could occur. Also, exposure of the filtered but unwashed product at this stage could cause some hydrolysis and thus lead to a product insoluble in organic solvents.

5. This is a convenient laboratory procedure for stabilizing cellulose nitrate. It is an important part of the procedure, because a trace of nitric acid or sulfuric acid in the product leads to instability. Boiling in the acetic acid solution also hydrolyzes any combined sulfuric acid which may be present as acid sulfate groups:

$$\text{Cellulose–OH} + H_2SO_4 \rightarrow \text{Cellulose-OSO}_3\text{H} + H_2O$$

$$\text{Cellulose-OSO}_3\text{H} + \text{H}_2\text{O} \xrightarrow{\text{Acid catalyst}} \text{Cellulose-OH} + \text{H}_2\text{SO}_4$$

Boiling with sodium bicarbonate solution neutralizes any unhydrolyzed sulfate groups to the more stable salt form:

$$\text{Cellulose-OSO}_3\text{H} + \text{NaHCO}_3 \rightarrow \text{Cellulose-OSO}_3\text{Na} + \text{H}_2\text{O} + \text{CO}_2$$

Commercially, the stabilization of cellulose nitrate is an involved process requiring prolonged heating of the ester in water under pressure, followed by boiling it in water containing a trace of calcium carbonate. Samples of cellulose nitrate prepared in the laboratory should not be stored in the dry state for extended periods because of possible incomplete stabilization. For prolonged storage the ester should be dampened with water or alcohol and stored in a closed container.

One simple procedure for destroying laboratory samples of cellulose nitrate is to boil (*Caution! Hood.*) the ester in a 5% sodium hydrosulfide solution. After about 1 hr. of boiling, the product should be completely denitrated. It can then be removed by filtration and discarded without creating a fire hazard. Destroying the ester by burning is dangerous, because the product burns extremely rapidly.

3. Methods of Preparation

Cellulose nitrate has been well described in a number of textbooks (3-8). However, there are very few references that give specific instructions for its prepartion in the laboratory. One excellent reference (4) deals with three laboratory procedures for making highly nitrated (13% N or higher) cellulose using the sulfuric acid-nitric acid mixture. Another method (9) utilizes a mixture of nitric acid, phosphoric acid, and phosphorus pentoxide to produce a substantially undegraded product.

4. Merits of the Preparation

This laboratory procedure for making cellulose nitrate was chosen because of its simplicity. In addition, the procedure employs a nitrating mixture which is used commercially. The laboratory product obtained using this mixture has a nitrogen content similar to that of the industrial cellulose nitrates.

5. References

1. Research Laboratories, Tennessee Eastman Co., Division of Eastman Kodak Co., Kingsport, Tennessee 37662.
2. Research Laboratories, Celanese Corp., Summit, New Jersey 07901.
3. A Steyermark, *Quantitative Organic Microanalysis,* The Blakiston Co., New York, 1951, pp. 56-81.
4. C. Doree, *The Method of Cellulose Chemistry,* Chapman and Hall Ltd., London, 1947, pp. 233-250.
5. E. F. Heuser, *Chemistry of Cellulose,* John Wiley and Sons, New York, 1944, pp. 172-222.
6. J. T. Marsh and F. C. Wood, *An Introduction to the Chemistry of Cellulose,* D. Van Nostrand Co., Princeton, N.J., 1939, pp. 183-191.
7. E. Ott, H. M. Spurlin, and M. W. Grafflin, *Cellulose and Cellulose Derivatives,* Part 2, Interscience Publishers, New York, 1954, pp, 713-762.
8. E. C. Worden, *Technology of Cellulose Esters,* Vol. 1, Part 3, D. Van Nostrand Co., Princeton, N.J., 1921, pp. 1933-2271.
9. W. J. Alexander and R. L. Mitchell, *Anal. Chem.,* **21**, 1497 (1949).

Polymerization of Dimethylketane

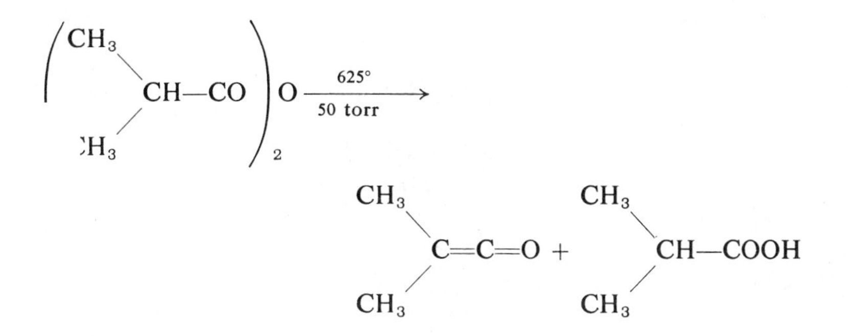

Submitted by G. F. Pregaglia and M. Binaghi (1)
Checked by C. G. Overberger (2a, b) and P. S. Yuen (2a, c)

1. Procedure (Note 1)

Caution! In the presence of air, dimethylketene yields highly explosive peroxides; These are dangerous also in the form of semi-invisible films on glassware. Vapors are harmful and irritating. All treatments must be carried out in an efficient hood with suitable shielding. All the parts that have come into contact with dimethylketene must be thoroughly washed with methanol soon after use and the washes should be destroyed.

Pyrolysis of the anhydride is carried out in the appartus shown in Figs. 1 and 2. The pyrolysis oven consists of a copper tube having a length of 100 cm. and an inner diameter of 1.5 cm. (Note 2). The copper tube is covered with a stainless steel tube. Thermocouples for temperature control are placed on the steel tube. The oven is heated with electric resistances having a power of about 2 kilowatts.

A nitrogen atomsphere is attained in the apparatus by repeated evacuation and refilling with purified nitrogen (Note 3). All stopcocks of the apparatus are

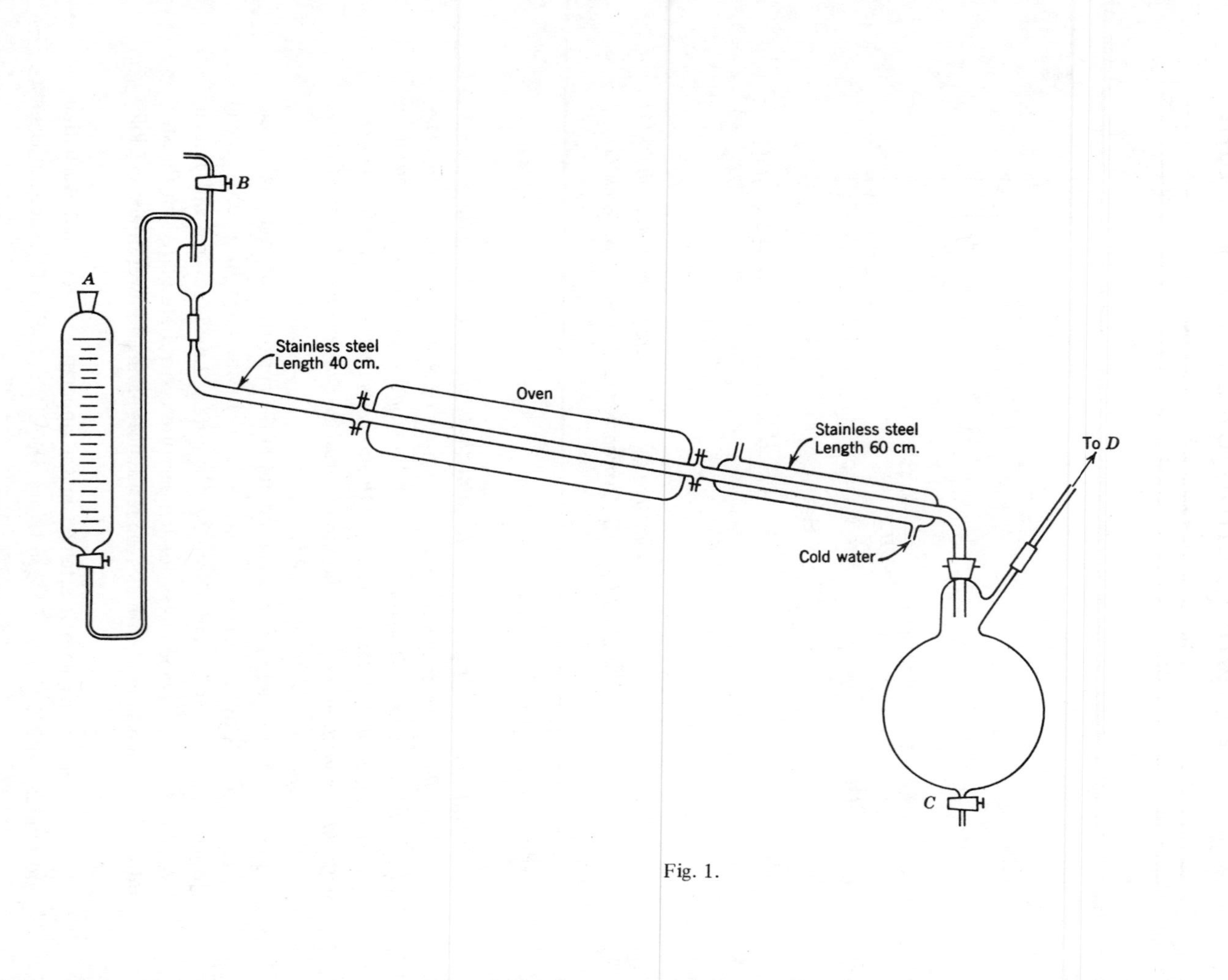
B
A
Stainless steel
Length 40 cm.
Oven
Stainless steel
Length 60 cm.
To D
Cold water
C

Fig. 1.

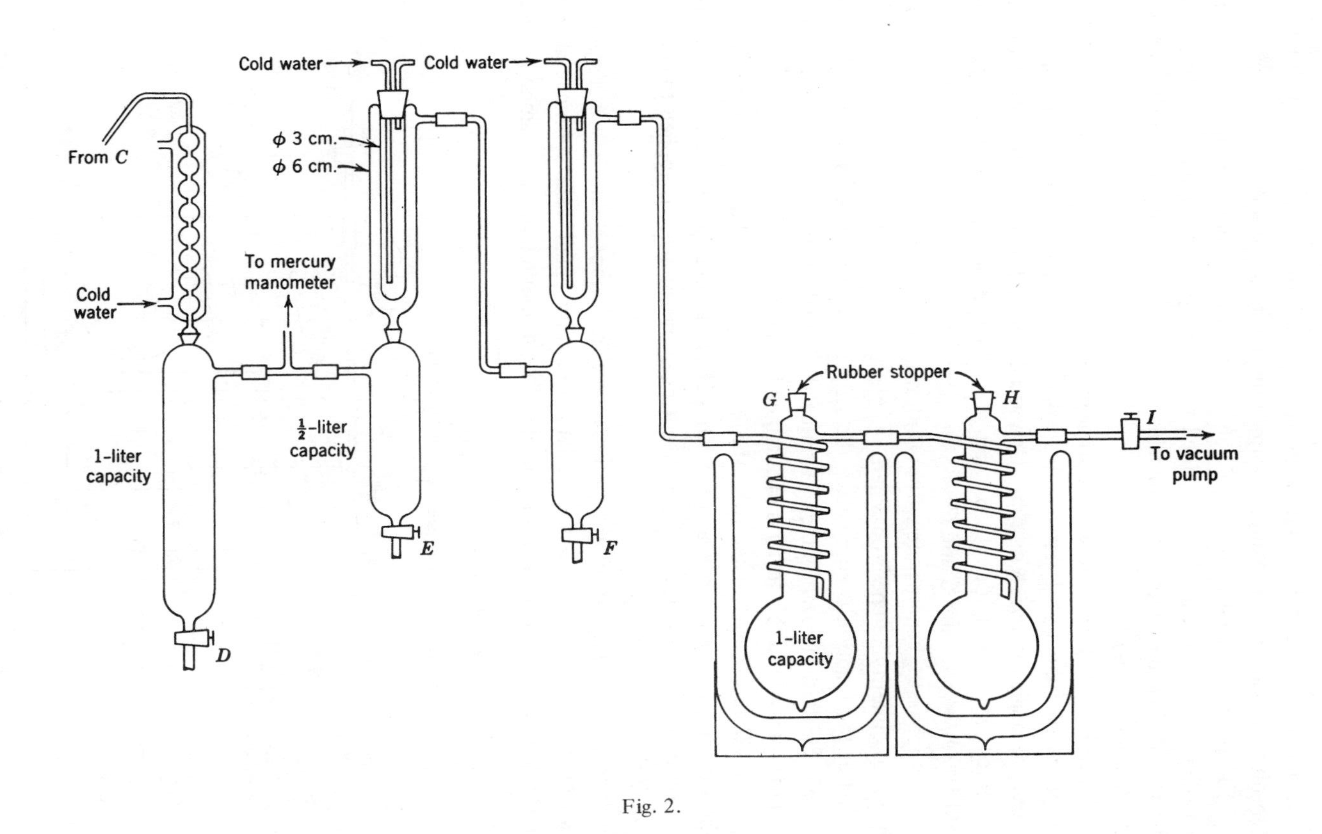

Cold water
Cold water
From C
ϕ 3 cm.
ϕ 6 cm.
To mercury manometer
Cold water
1-liter capacity
½-liter capacity
D
E
F
Rubber stopper
G
H
I
To vacuum pump
1-liter capacity

Fig. 2.

closed except stopcock *I*. The internal pressure is regulated to 50-100 torr. (absolute pressure) by a vacuum pump connected to stopcock *I*.

The oven is heated to 600-650°, and cold water is circulated in the coolers. Distilled isobutyric anhydride is added to the graduated dropping funnel *A*, and traps *G* and *H* are cooled by immersion in Dewar flasks containing Dry Ice and methanol (Note 4).

When both temperature and internal pressure of the oven become stable, the stopcock of funnel *A* is regulated so that 270 g./hr. of isobutyric anhydride passes through the pyrolysis tube. Unreacted anhydride and isobutyric acid are collected in vessels *C, D, E,* and *F,* while dimethylketene is collected (90-100 g./hr.) in traps *G* and *H*.

The apparatus can operate without interruption for about 6 hr.; approximately 1 l. of dimethylketene is obtained.

After the reaction is completed, the stopcock of funnel *A* and stopcock *I* are closed. Purified nitrogen is then introduced through stopcock *B*. A slight nitrogen overpressure (a few torr.) is maintained in the apparatus.

Traps *G* and *H* are isolated from the remaining part of the apparatus by closing the rubber joint between *F* and *G* with a Hoffman clamp. The vacuum pump is removed from stopcock *I*, and a slight nitrogen overpressure is applied to traps *G* and *H* through this stopcock.

It is now possible to draw dimethylketene from traps *G* and *H* with a siphon, as shown in Fig. 3. To allow complete siphoning, small wells are made at the bottom of traps *G* and *H*. Immediately after siphoning, both the siphon tubes

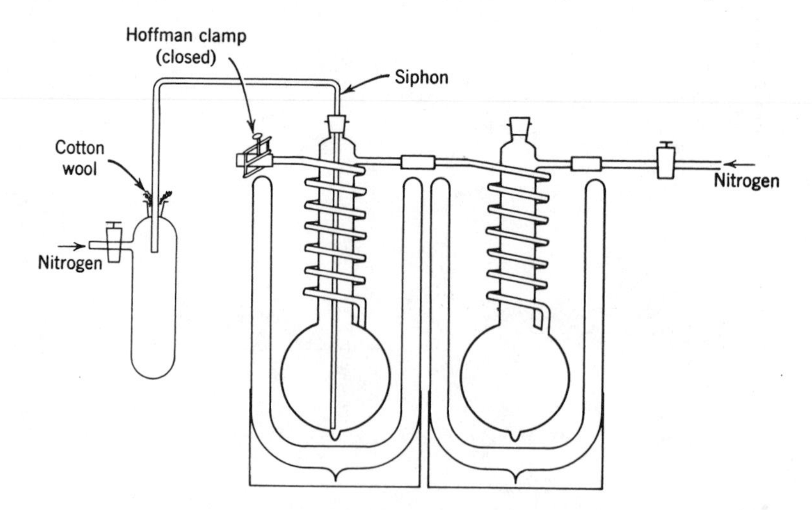

Fig. 3.

and the traps are washed with methanol to prevent the formation of peroxides.

Dimethylketene obtained by this method is already 98-99% pure in the crude state. It is important to avoid the presence of sulfur-containing compounds during the pyrolysis.

Crude dimethylketene can be easily stored over a period of some weeks under nitrogen atmosphere at $-78°$ in 100-250 ml. graduated test tubes fitted with a ground glass joint at the top and a side-arm stopcock (Note 5). To prevent the formation of peroxides it is necessary to add a trialkyl aluminum (*Caution! Trialkyl aluminum compounds are generally pyrophoric and react violently with water.*) (e.g., triethylaluminum or triisobutyl aluminum) to dimethylketene (3-5 ml. of heptane solution with 10% by weight of alkyl aluminum per 100 ml. of dimethylketene) (Note 6). The trialkyl aluminum solution must be slowly added to dimethylketene which must be maintained at $-78°$, and poured along the test tube walls to avoid violent reactions. The presence of trialkyl aluminum also allows freeing the monomer of impurities, such as isobutyric anhydride, which might modify the course of the polymerization (Note 7).

Dimethylketene must be distilled shortly before use. This operation can be suitably conducted under reduced pressure (18-20 torr) with the device shown in Figs. 4 and 5.

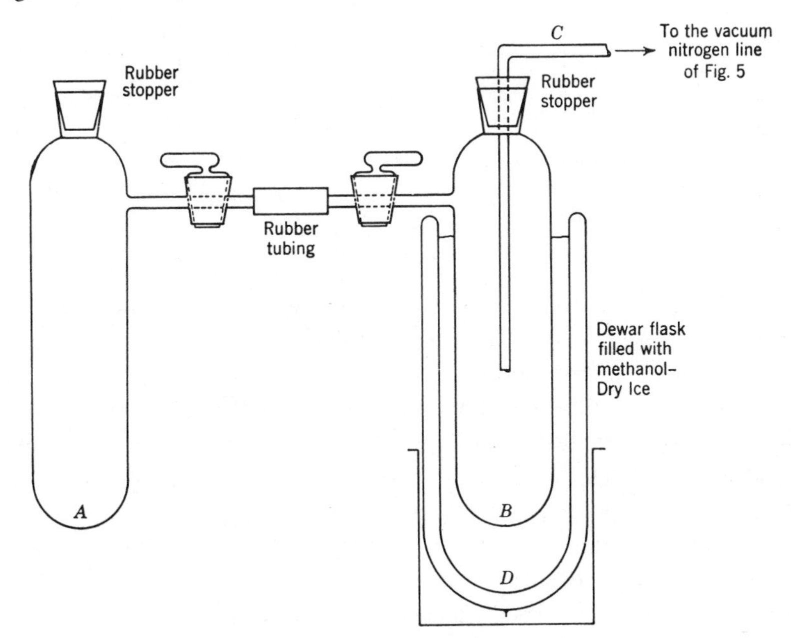

Fig. 4.

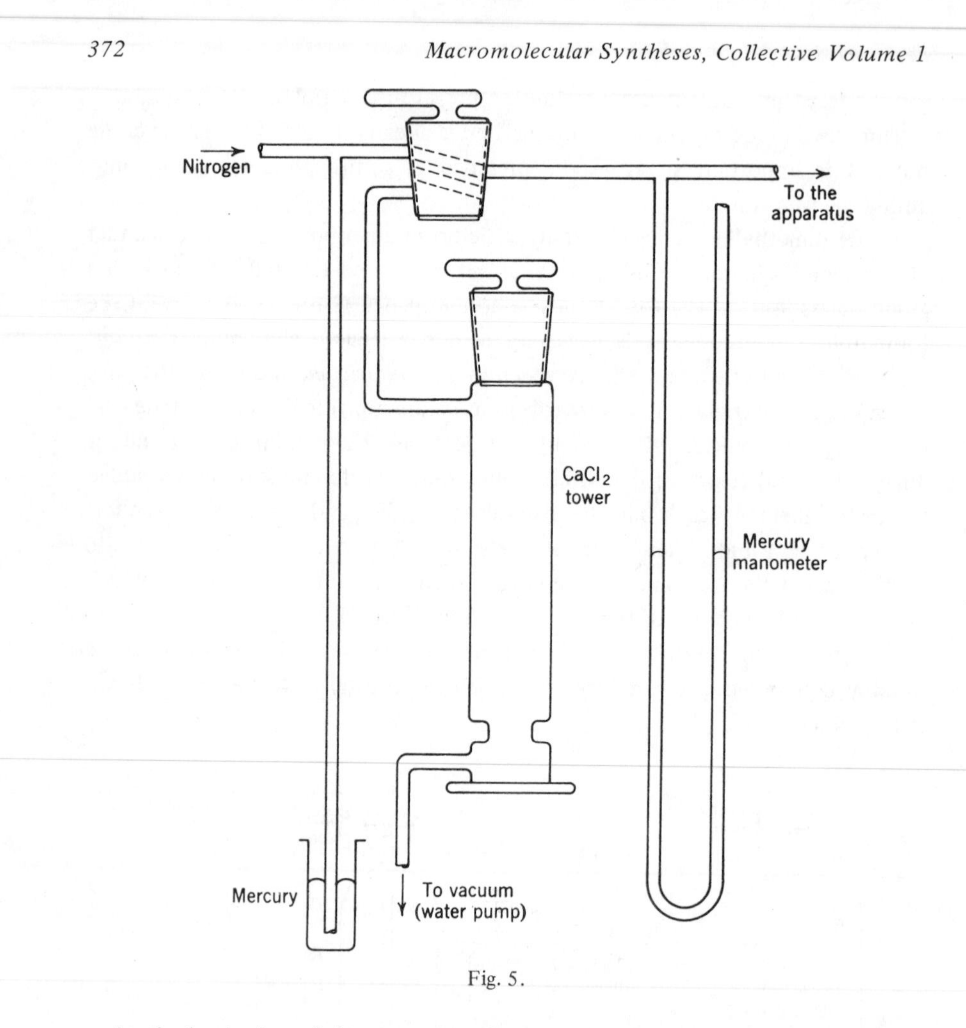

Fig. 5.

At the beginning of the operation the test tube *A* which contains dimethylketene with alkyl aluminum is immersed in the Dewar flask *D*, while tube *B* is maintained at room temperature.

The stopcock of tube *A* is closed and that of tube *B* open; vacuum and nitrogen are repeatedly applied through the inner tube *C*. Finally tube *B* is evaculated (18-20 torr), test tube *A* is removed from the Dewar flask, replaced with test tube *B*, and the stopcock of test tube *A* is opened.

To prevent bumping of the liquid dimethylketene, test tube *A* must be kept horizontal during the distillation. As soon as the temperature of the liquid in *A* has reached about $-30°$, dimethylketene begins to condense in *B*.

After distillation of the requried amount of dimethylketene, nitrogen is introduced from the tube *C*, the stopcocks of test tubes *A* and *B* are closed and the rubber tubing is removed. Test tube *A* is then immersed in a bath at $-78°$. Nitrogen is introduced from the stopcock of test tube *B*; the stopper with the inserted tube *C* is removed and replaced with an ordinary rubber stopper. As soon as tube *C* has been removed, it must be thoroughly washed with methanol to prevent formation of peroxides.

Distilled dimethylketene should be used within 24 hr. It must be stored at $-78°$ until use and withdrawn under nitrogen with a pipet connected to a syringe, or with a siphon. Siphons and pipets must be thoroughly washed with methanol immediately after use to prevent formation of peroxides.

The residual distilled dimethylketene can be decomposed by dilution with 5-10 parts of toluene and subsequent addition of methanol, or can be stored for a longer time by addition of trialkyl aluminum in amounts similar to those reported above.

2. Notes

1. This preparation is essentially the one reported earlier (3).

2. A much shorter pyrolysis tube (45 cm.) can be used; however, it results in a slightly decreased yield, i.e., 64% instead of 74-82%.

3. The nitrogen should not contain more than 10 p.p.m. of oxygen or water.

4. For traps *G* and *H*, liquid nitrogen for cooling is preferable to Dry Ice and methanol.

5. To seal test tubes, rubber stoppers are preferred to the ground glass ones.

6. Care must be taken to adhere to this concentration of trialkyl aluminum. Excess of trialkyl aluminum causes vigorous reaction and polymerization during distillation.

7. Dimethylketene obtained by methods other than the pyrolysis of isobutyric anhydride generally contains a greater amount of impurities which impair the polymerization. They can be removed, however, by repeated distillation of dimethylketene over trialkyl aluminum by the same methods and with the same amounts of aluminum alkyl mentioned above. In general, two or three distillations are sufficient in the case of dimethylketene obtained by pyrolysis of dimethylmalonic anhydride and of 2,2,4,4-tetramethyl-1,3-cyclobutanedione. The dimethylketene obtained from α-bromoisobutyryl bromide, according to the method reported in *Organic Syntheses* (4), is not recommended for the polymerization because complete elimination of ethyl acetate used as solvent is very difficult.

3. Methods of Preparation

Dimethylketene has been prepared by the treatment of α-bromoisobutyryl bromide with zinc (4, 5) and by the pyrolyses of isobutyrylphthalimide (6), dimethylmalonic anhydride at 180° (7), α-carbomethoxy-α,β-dimethyl-β-butyrolactone (8), the dimer of dimethylketene, 2,2,4,4-tetramethyl-1,3-cyclobutanedione (9), isobutyric acid and anhydride at about 700° (3), and 3-diazobutanone at 180° (10).

4. References

1. Direzione Centrale delle Ricerche, Montecatini Edison S.p.A., Milano, Italy.
2a. Polytechnic Institute of Brooklyn, Brooklyn, New York 11201.
2b. Current Address: University of Michigan, Ann Arbor, Michigan 48104.
2c. Current Address: Stauffer Chemical Co., Eastern Research Center, Dobbs Ferry, New York 10522.
3. M. Mugno and M. Bornengo, *Chim Ind. (Milan)*, **45**, 1216 (1963); **46**, 5 (1964).
4. C. W. Smith and D. G. Norton, *Org. Syn. Coll. Vol.*, **4**, 348 (1963).
5. H. Staudinger and H. W. Klever, *Ber.*, **39**, 968 (1906).
6. C. D. Hurd and M. F. Dull, *J. Amer. Chem. Soc.*, **54**, 2432 (1932).
7. H. Staudinger, *Helv. Chim. Acta*, **8**, 306 (1925).
8. E. Ott, *Ann.*, **401**, 159 (1913).
9. W. E. Hanford and J. C. Sauer, *Org. Reactions*, **3**, 136 (1946).
10. W. Ried and P. Junker, *Angew. Chem. Int. Ed. Eng.*, **6**, 631 (1967).

B. *Polymerization of Dimethylketene*

Submitted by G. F. Pregaglia and M. Binaghi (1)
Checked by C. G. Overberger (2a, b) and H. Mukamal (2a, c)

Polydimethylketene (Polyacetalic Structure)

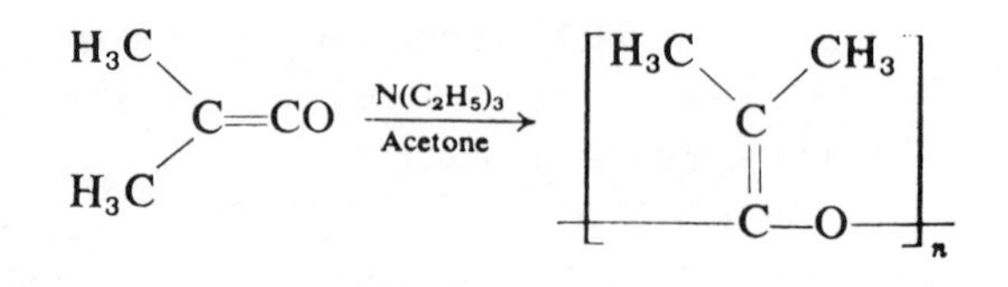

1. Procedure

Caution! See p. 367.

A 100-ml., four-necked flask is fitted with a mechanical stirrer, a low temperatures thermometer and a tube connected to a vacuum nitrogen line, constructed as shown in Fig. 4.

A nitrogen atmosphere is attained in the apparatus by repeated evacuation with vacuum release to purified nitrogen (Note 1). The apparatus is kept under a slight nitrogen overpressure. The flask is charged wtih 200 ml. of anhydrous acetone (Note 2) and cooled externally with a Dry Ice-methanol bath so that the inner temperature is lowered to about $-70°$. Freshly distilled dimethylketene (20 ml.) (Note 3) is added from a pipet connected to a syringe. The stirring of the solution is started soon after; then 0.4 ml. of triethylamine (Note 4) is added. The solution becomes cloudy almost immediately with a decrease of the yellow color. After 30 min. the polymerization is stopped by addition of 50 ml. of methanol (Note 5), and the contents of the flask are transferred to a beaker with the aid of 500 ml. of methanol. The suspension is allowed to settle for several hours. The upper clear liquid is siphoned, and the very finely divided residual suspension is centrifuged (Note 6). The sticky solid that separates is suspended in 100 ml. of methanol and recentrifuged. This procedure is repeated three times; finally the solid is dried under vacuum at room temperature (Note 7). After drying, 11-12 g. of a finely divided white polymer is obtained (69-81% yield).

2. Characterization

High molecular weight polydimethylketene with acetalic structure is partially soluble in several organic solvents (e.g., benzene, chloroform, cyclohexanone). It is completely soluble in boiling diphenyl ether. In a sealed capillary tube it does not decompose until a temperature of 290°. Between 290° and 310° it melts with gas release.

The polymer has little stability in air and degrades slowly to a light yellow viscous liquid.

The powder x-ray diffraction pattern shows a highly crystalline structure. The infrared absorption spectrum is obtained from a Nujol mull. It shows a characteristic absorption band at 5.85 μ (3).

3. Notes

1. Prepurified or lamp grade nitrogen containing not more than 10 p.p.m. of oxygen or water should be used.

2. Pure reagent grade acetone, dried over anhydrous calcium sulfate and distilled under nitrogen, is used.

3. See part A.

4. Pure reagent grade triethylamine, dried over anhydrous calcium sulfate and distilled under nitrogen is used.

5. If the reaction mixture is still strongly yellow, the addition of methanol causes a vigorous reaction because of the presence of large amounts of unreacted dimethylketene. In this case, methanol must be added very slowly from a dropping funnel. The presence of dimethylketene 30 min. after the start of the reaction indicates that the products used, and in particular dimethylketene, are not sufficiently pure. Dimethylketene can be further purified by redistilling it over a trialkyl aluminum compound.

6. Because the solid is very finely divided, it is very difficult to separate it by filtration.

7. This operation is preferably carried out by applying vacuum directly to the vessel in which the polymer has been centrifuged. This prevents losses on transferring the solid caused by its sticky nature.

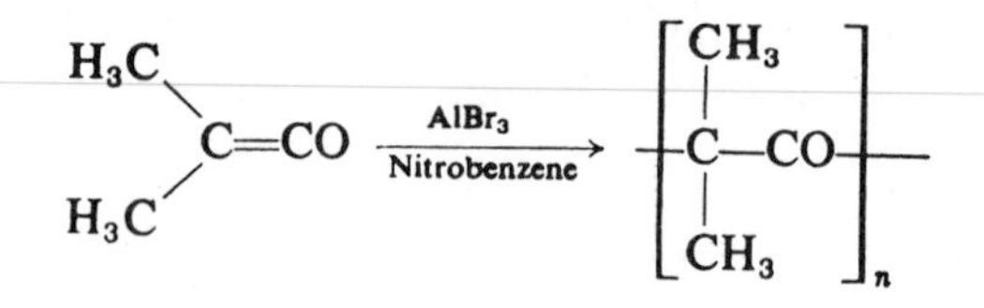

1. Procedure

Caution! See p. 367.

A 100-ml. four-necked flask is equipped with a 10-ml. dropping funnel with pressure-equalizing tube, mechanical stirrer, and tubing connected to a vacuum-nitrogen line, constructed as shown in Fig. 4. A nitrogen atmosphere is attained in the apparatus by repeated evacuation with vacuum release to purified nitrogen (Note 1). The apparatus is maintained under a slight nitrogen overpressure. Anhydrous carbon tetrachloride (20 ml.), 13 ml. of anhydrous nitrobenzene (Note 2), and 20 ml. of freshly distilled dimethylketene (Note 3) are introduced into the flask. Nitrobenzene (7 ml.) and 3 ml. of a nitrobenzene solution of $AlBr_3$ (Note 4) are introduced in the dropping funnel. The flask is cooled externally with a bath at $-30°$ and stirring is started. The contents of the dropping funnel are dropped into the flask over a period of about 15 min. and stirring is continued for a further 75 min.

The reaction is stopped by the addition of 20 ml. of methanol (Note 5) and the contents of the flask are transferred to a beaker containing 250 ml. of methanol. Concentrated hydrochloric acid (5 ml.) is added and the colored polymer lumps are crushed with the aid of a stirrer. The polymer is allowed to stand for a few hours, then it is filtered, washed with methanol, and resuspended in 250 ml. of methanol containing 5 ml. of concentrated hydrochloric acid. This procedure is repeated until the polymer is perfectly white; it is filtered and dried under vacuum at room temperature.

After drying, 9-10 g. of fiber-forming white polymer is obtained (56-63% yield).

2. Characterization

Polydimethylketene having a ketonic structure is soluble in hot nitrobenzene, acetophenone, anisole, and cyclohexanone. Its intrinsic viscosity in nitrobenzene at 135° is 0.8-0.9; its melting point in a capillary tube is 195-205°. The polymer may be pressed into film between aluminum foils at 200-220°. The powder x-ray diffraction pattern shows a highly crystalline structure (4). The infrared absorption spectrum is obtained from a Nujol mull. It shows a characteristic band at 5.99 μ (5).

3. Notes

1. See Note 1 on p. 375.
2. Reagent grade carbon tetrachloride that was dried over anhydrous calcium sulfate and distilled under nitrogen is used. Reagent grade nitrobenzene is dried over calcium chloride, distilled under reduced pressure, and maintained under a nitrogen atmosphere.
3. See part A.
4. A solution containing 350 g. of $AlBr_3$ per liter of solvent is used. It is prepared from anhydrous nitrobenzene (Note 2) and reagent grade anhydrous aluminum bromide distilled under nitrogen.
5. See Note 5 on p. 376.

4. Methods of Preparation

Polydiphenylketene have a polyketonic structure has been obtained by polymerization with $PbCl_4$ as catalyst (7).

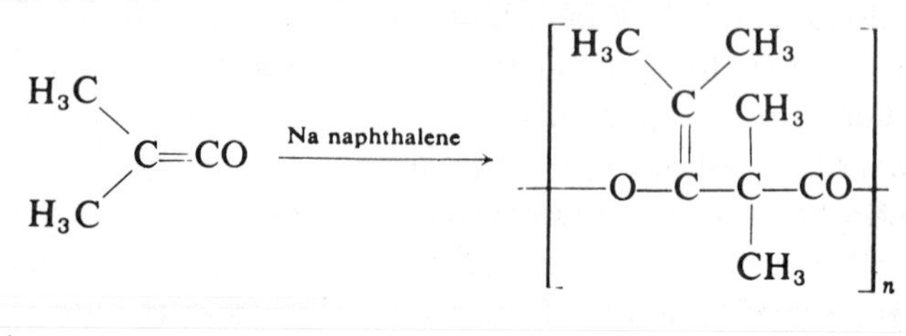

1. Procedure

Caution! See p. 367.

A 250-ml., four-necked flask is equipped with a mechanical stirrer, a reflux condenser, and a 50 ml. dropping funnel with pressure equalizing tube. The top of the reflux condenser is connected to a vacuum-nitrogen line constructed as shown in Fig. 4. A nitrogen atmosphere is attained in the apparatus by repeated evacuation with vacuum release to purified nitrogen (Note 1). The system is maintained under a slight nitrogen overpressure.

Anhydrous ether (120 ml.) (Note 2) and 1 ml. of a 0.1 *M* solution of sodium naphthalene in tetrahydrofuran (Note 3) are introduced into the flask at room temperature. Anhydrous ether (30 ml.) and 20 ml. of dimethylketene (Note 4) are introduced into the dropping funnel. Stirring is immediately started and the dimethylketene solution is added at the rate of about 5 ml./minute. A few minutes after the addition starts, the mixture begins to cloud and the solvent begins to reflux. Occasional external cooling with a cold water bath may be necessary to moderate the boiling rate. After dimethylketene has been added, stirring is continued for 2 hr., then the reaction is stopped by the addition of 20 ml. of methanol (Note 5). The contents of the flask are transferred to a beaker with the aid of 50 ml. of methanol. The polymer is repeatedly washed with methanol, filtered, and dried under vacuum at room temperature. A finely divided white polymer (14-15 g.) is obtained 87-94% yield).

2. Characterization

Polydimethylketene having a polyester structure is soluble in different hot solvents, such as toluene, tetralin, anisole, cyclohexanone. Its intrinsic viscosity in tetralin at 135° is 0.3-0.4. Its melting point in a capillary tube is 195-220°. The polymer may be pressed into a film between aluminum foils at 150-170°. The powder x-ray diffraction pattern shows a highly crystalline structure. The infrared absorption spectrum is obtained from a film cast on an NaCl disk from a chloroform solution. It shows a characteristic absorption band at 5.75 μ (6).

3. Notes

1. See Note 1 on p. 375.
2. Reagent ether is refluxed over lithium aluminum hydride during 10 hr. and then distilled twice under nitrogen.
3. The sodium naphthalene solution is prepared from tetrahydrofuran distilled twice over lithium aluminum hydride under a nitrogen atmosphere. The amounts of clean sodium and reagent grade naphthalene necessary to obtain a 0.1 M solution are added to tetrahydrofuran. The mixture is left at room temperature until the sodium completely dissolves.
4. See part A.
5. See Note 5 on p. 376.

4. Methods of Preparation

Polymethylphenylketene and polydiphenylketene having a polyester structure have been obtained by polymerization with anionic catalysts (7, 8, 9).

5. References.

1. Direzione Centrale dell Ricerche, Montecatini Edison S.p.A., Milano, Italy.

2a. Polytechnic Institute of Brooklyn, Brooklyn, New york 11201.

2b. Current Address: Department of Chemistry, University of Michigan, Ann Arbor, Michigan 48104.

2c. Current Address: E. I. Du Pont de Nemours & Co., Richmond, Virginia 23221.

3. G. F. Pregaglia, M. Binaghi, and M. Cambini, *Makromol. Chem.*, **67**, 10 (1963).
4. I. W. Bassi, P. Ganis, and P. A. Temussi, *J. Polym. Sci.*, **C16**, (5) 2867 (1967).
5. G. F. Pregaglia, M. Peraldo, and M. Binaghi, *Gazz. chim Ital.*, **92**, 488 (1962).
6. G. Natta, G. Mazzanti, G. F. Pregaglia, and M. Binaghi, *Mackromol Chem.*, **44**, 537 (1961).
7. Sh. Nodzhimutdinov, E. P. Chernova, and V. A. Kargin, Spektrosk. Polim, Sb. Dokl. Vses. Simp. 108 (1965).
8. S. Nunmoto and Y. Yamashita, Kogyo Kagaku Zasshi, **71**, 2067 (1968).
9. T. Tsunetsugu, K. Azimoto, T. Fueno, and J. Fuzukawa, *Makromol. Chem.*, **112**, 210 (1968).

Urethane Block Copolymer from Polyoxymethylene and Poly(propylene adipate)

$$\mathrm{HOCH_2CH_2{+}CH_2O{+}_x CH_2CH_2OH} \;+\; \mathrm{HOCH_2\underset{\displaystyle CH_3}{\underset{|}{C}}HO{+}\overset{\displaystyle O}{\overset{\|}{C}}(CH_2)_4\overset{\displaystyle O}{\overset{\|}{C}}OCH_2\underset{\displaystyle CH_3}{\underset{|}{C}}HO{+}_y H}$$

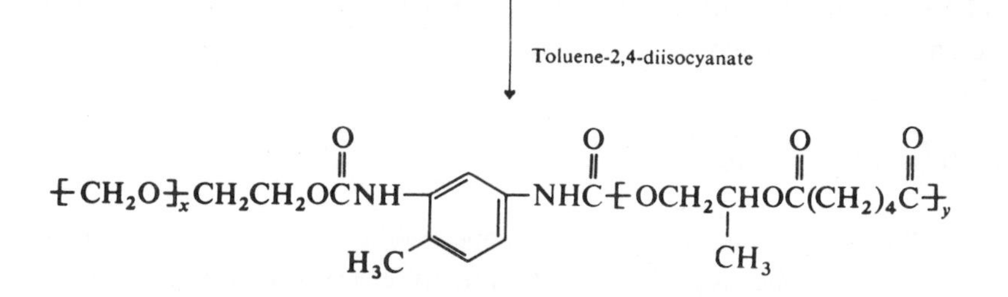

Submitted by T. W. Brooks, C. Bledsoe, and J. Rodriguez (1)
Checked by G. Statton and T. Ryan (2)

1. Procedure

A. Copolymerization of Trioxane and Dioxolane (Note 1).

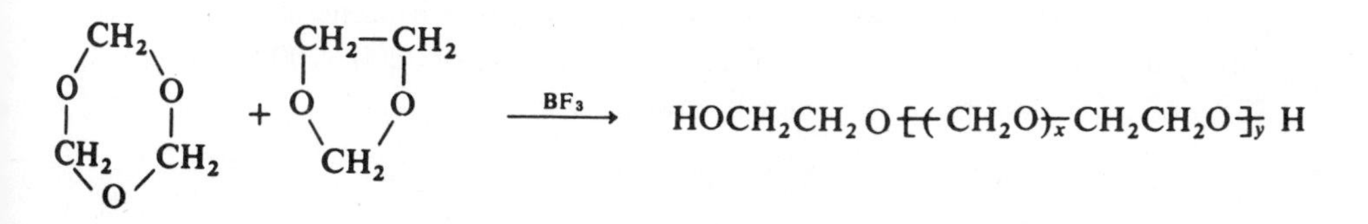

A 1-qt. brown bottle is fitted with a two-holed rubber stopper into which are fixed two short lengths of glass tubing. One tube is fitted with a rubber septum and the other is connected to a nitrogen source and low-pressure release valve which keep the contents of the bottle under a blanket of dry nitrogen. The flamed or oven-dried bottle is then charged with 210 ml. of dry cyclohexane (Note 2), 51.3 g. (0.57 mole) of trioxane (Note 3), and 2.22 g. (0.03 mole) of dioxolane (Note 4). The bottle is then lowered into an oil bath at 55° and the contents are stirred magnetically under a blanket of nitrogen. When all the trioxane has dissolved, 0.30 g. of freshly distilled boron trifluoride etherate is injected through the rubber septum with a small hypodermic syringe. The polymerization is allowed to proceed at 55° for 20 hr., during which time copolymer precipitates as a fine white solid. At the end of this period the reaction mixture is cooled to room temperature and to it is added 100 ml. of a 5% solution of triethanolamine in methanol. The polymer is then collected by filtration and washed with three 50-ml. portions of methanol. The polymer is slurried in 250 ml. of 1% aqueous triethanolamine and the mixture heated at 100° for 1 hr. (Note 5). The slurry is then filtered and the polymer is washed thoroughly with methanol and vacuum dried at 40° for about 8 hr. The yield of copolymer, with an inherent viscosity of 0.3-0.5 dl./g. at 60° in *p*-chlorophenol containing 2% α-pinene at a copolymer concentration of 0.5 g./dl., is 26-32 g. (50-60%).

B. Poly(propylene adipate)

A 500-ml., three-necked flask, equipped with a stirrer, a nitrogen inlet extending below the level of the reaction mixture, a straight distilling head, and a condenser, is charged with 292 g. (2.0 moles) of adipic acid, 159.5 g. (2.1 moles) of propylene glycol (Note 6), and 0.5 g. of zinc acetylacetaonate. This mixture is stirred and heated in an oil bath at 140° with a slow stream of nitrogen passing through the melt until all of the water has been removed by distillation and the acid number of the pot material is 1 or less. At this point the temperature is raised to 200° and the pressure gradually reduced to about 1 torr. Glycol distills rapidly at first and then more slowly as interesterification proceeds. Interesterification is allowed to continue until the viscous syrupy polyester has an inherent viscosity of 0.13-0.16 dl./g. in benzene at 30°. This corresponds to a number average molecular weight range of 3000-3700 (Note 7). The product is a syrupy, light-yellow-to-brown mass.

C. Poly(oxymethylene-propylene adipate) Block Copolymer

This procedure is designed to give a urethane-type block copolymer that is approximately 30 mole % in oxymethylene units and 65 mole % in propylene adipate units. Block copolymers with higher precentages of oxymethylene units and correspondingly lower precentages of propylene adipate units may be prepared in the same way by making appropriate adjustments in the relative amounts of prepolymers and diisocyanate used.

A 50-ml., semimicro resin kettle, equipped with a stirrer, a condenser, a nitrogen-inlet tube, and a rubber serum cap, is charged with 1.5 g. of poly-oxymethylene (number average molecular weight, 6000) (Note 8) and 9.3 g. of poly(propylene adipate) (number average molecular weight 3000). The kettle is then heated in an oil bath at 180° and the molten prepolymer mixture is stirred under nitrogen for 30 min. With a dry syringe or a micropipet 0.52 g. of toluene-2,4-diisocyanate is added to the stirred melt all at once. This viscous mass is then stirred for 2-4 hr. during which time chain-coupling is evidenced by a pronounced viscosity increase that may cause the reaction mass to climb the stirrer. At this point the reaction mixture is cooled to room temperature, and the rubbery product is cut away from the stirrer and the kettle with a knife. The product is weighed to determine the extent of weight loss during reaction (Note 9). The crude block copolymer is then cut into small pieces and extracted with hot acetone for 24 hr. in a Soxhlet extractor. After the extraction residue is vacuum-dried at 60° for 8-12 hr. it is again weighed to determine the amount of material extracted. From the material balance, by taking into account the weight lost during reaction and extraction and by neglecting the urethane units, a close approximation to the final composition of the block copolymer can be calculated (Note 10). The weight lost during reaction is generally 2-10% of the total charge, and the weight lost during extraction may vary from 15 to 70% of the total charge, depending on the relative amounts of prepolymers in the charge and, presumably, on the accuracy of the prepolymer molecular weight determinations.

2. Notes

1. Because of the inherent thermal instability of polyoxymethylene homopolymer, it is necessary to use the more stable formaldehyde-ethylene oxide

copolymer in the diisocyanate end-linking reaction that gives the block copolymer.

2. Technical-grade cyclohexane dried by distillation from calcium hydride is suitable.

3. Commercial trioxane should be purified by recrystallizing twice from methylene chloride. (*Caution! Halogenated solvents may be toxic and should be used with care.*)

4. Eastman 1,3-dioxolane, b.p. 74-75°, may be used as received.

5. This treatment removes polyoxymethylene end segments to leave the polymer terminated predominantly by $-CH_2CH_2OH$ groups. The weight loss resulting from this treatment is about 5%.

6. U.S.P. 1,2-propylene glycol, b.p. 189°, is purified for use first by reaction with sodium metal (1 g./100 ml.), then refluxed for 1 hr. under nitrogen, and finally distilled under nitrogen.

7. Because valid molecular weight data on poly(propylene adipate) are difficult to obtain by the usual end-group titration methods, it is preferable to determine the number average molecular weight with a commercial vapor-pressure osmometer. For users of this procedure who may not have such an instrument readily available to them, the inherent viscosity range indicated in the procedure gives an approximation of the molecular weight desired for the subsequent block copolymerization. Experience has shown this approximation to be adequate for determining the appropriate quantities of reagents used in the end-linking reaction.

8. Number average molecular weight of polyoxymethylene may be calculated from the inherent viscosity by the expression $\eta_{inh} = 4.13 \times 10^{-4} M_w^{0.724}$. This calculation gives the weight average molecular weight, which, because the polymer has the most probable molecular weight distribution, may be divided by two to give the number average molecular weight (3).

9. Weight loss during the dissocyanate end-linking reaction is attributable to decomposition of the polyoxymethylene. The extent of weight loss will vary from about 2-10%, depending on the amount of polyoxymethylene employed.

10. The quantitites used in this procedure should give block copolymer that is approximately 30 mole % in polyoxymethylene and 65 mole % in poly(propylene adipate), calculated on the basis of the repeating unit formula weights of the respective prepolymers, i.e., $-CH_2O-$ (30) and $-CO(CH_2)_4\ CO_2CH(CH_3)$-$CH_2O-$ (186).

3. Characterization

Solubility. The block copolymer is soluble in dimethylformamide, dimethylsulfoxide, formic acid, *p*-chlorophenol, and hexafluoroacetone sesquihydrate.

Viscosity Measurements. The inherent viscostiy of the block copolymers should be 1.2-1.8 dl./g. determined at 60° in *p*-chlorophenol containing 2% alpha-pinene, for a concentration of 0.5 g. in 100 ml. of solvent.

Thermal Behavior. Differential thermal analysis of the block copolymers reveals endotherms at -29-32°, 152-159°, and 177-200°, which correspond, respectively, to the glass transition temperature of poly(propylene adipate), the crystalline melting temperature of polyoxymethylene, and thermal decomposition. Films may be melt-pressed between sheets of Teflon at 170°, and test specimens may be compression-molded at the same temperature in molds lined with Teflon film or some other mold release agent.

4. Methods of Preparation

A large number of patents exist on the subject of polymerization or copolymerization of formaldehyde and its cyclic oligomer, trioxane. Much of this work is reviewed in a monograph by Furakawa and Saegusa (4). The boron trifluoride-initiated copolymerization of trioxane with alkylene oxides or cyclic acetals, such as a dioxolane, was first utilized as a means of preparing thermally stable polyoxymethylenes by Walling, Brown, and Bartz (5). This is the basis for the thermal stability of the commercial acetal resin Celcon, marketed by Celanese Corporation of America. This process is probably the most convenient for the preparation of small quantities of such copolymers on a laboratory scale. The procedure for copolymerizing trioxane and dioxolane described above is essentially a modified version of that reported by Inoue (6).

Preparation of urethane elastomers from low molecular weight polyoxymethylenes and polyesters such as poly(ethylene adipate) have been reported (7). Oxymethylene block copolymers that incorporate vinyl polymers (8, 9, 10), oelfin oxide polymers (8, 11, 12), aldehyde polymers (8), alkyl isocyanate polymers (8), and polyamides (13) have also been reported.

5. References

1. PCR, Inc. [formerly Peninsular ChemResearch, Inc. (subsidiary of Calgon Corporation)], Gainesville, Florida 32501.
2. ARCO Chemical Company, Glenolden, Pennsylvania 19036.
3. H. J. Wagner and K. F. Wissbrun, *Makromol. Chem.*, **81**, 14 (1965).
4. J. Furakawa and T. Saegusa, *Polymerization of Aldehydes and Oxides*, Wiley-Interscience, New York, 1963.
5. C. T. Walling, F. Brown, and K. W. Bartz (Celanese Corporation of America), U.S. Patent 3,027,352 (March 22, 1962) [*C. A.*, **57**, 4887c (1962)].
6. M. Inoue, *J. Appl. Polym. Sci.*, **8**, 2225 (1964).
7. Farben-fabriken Bayer A.-G., Netherlands Appl. 6,512,142 (March 18, 1966) [*C. A.*, **65**, 5642a (1966)].
8. E. J. Kirkland and W. J. Roberts (Celanese Corporation of America), Belg. Patent 636,370 (December 16, 1963) [*C. A.*, **63**, 6590f (1965)].
9. K. Noro, H. Kawazura, T. Moriyama, and S. Yoshioka, *Makromol. Chem.*, **83**, 35 (1965).
10. T. J. Dolce, D. F. Stuetz, and W. J. Roberts (Celanese Corporation of America), U.S. Patent 3,281,499 (October 25, 1966) [*C. A.*, **61**, 7136c (1964)].
11. Farbwerke Hoechst A.-G., Neth. Appl. 6,408,971 (February 9, 1965) [*C. A.*, **63**, 3076g (1965)].
12. E. I. du Pont de Nemours & Company, Brit. Patent 807,589 (January 21, 1959) [*C. A.*, **55**, 2199d (1961)].
13. K. D. Kiss (Diamond Alkali Company), U.S. Patent 3,299,005 (January 17, 1967) [*C. A.*, **66**, 55966n (1967)].

Poly(1,3-cyclohexadiene)

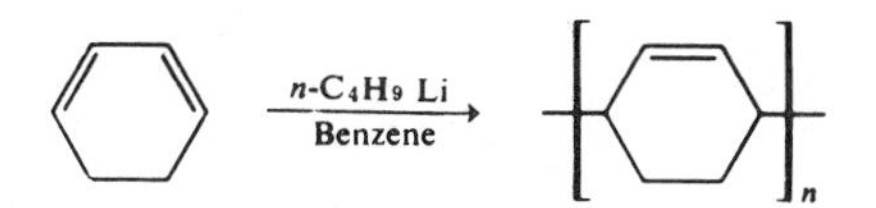

Submitted by P. E. Cassidy (1) and C. S. Marvel (2)
Checked by H. J. Harwood, E. Thall and C. A. Soman (3)

1. Procedure

Into a capped, dry, 25-ml. syringe bottle, which has been flushed with purified nitrogen (Note 1), are injected 5.0 ml. of dry benzene (Note 2) and 4.4 g. (0.055 mole) of 1,3-cyclohexadiene (Note 3). With a syringe, 1.0 ml. of a 1.5 *M* *n*-butyllithium solution (Note 4) is added (Note 5) and the mixture is shaken intermittently by hand. After 17 hr. the reaction mixture is light yellow and viscous.

After 120 hr. the viscous yellow solution is poured slowly into 250 ml. of rapidly stirred methanol containing a trace of *N*-phenyl-2-naphthylamine anti-oxidant. The white fibrous precipitate is then washed with methanol in a blender and dried *in vacuo* to yield 4.1 g. (93%) of product with an inherent viscosity of 0.29 in benzene at 30° (Note 6).

2. Notes

1. The purification train utilizes Fieser's solution, sulfuric acid, and sodium hydroxide pellets.
2. The benzene is distilled from sodium and stored over sodium.
3. The preparation of 1,3-cyclohexadiene has been described (4). In this preparation monomer of 98% purity from Chemical Samples Company,

Columbus, Ohio, is used. The diene disproportionates readily to cyclohexene and benzene and forms peroxides on contact with air and is therefore stored under nitrogen in the cold. Distillation of 1,3-cyclohexadiene on a spinning band column just prior to use is recommended.

4. Commercial *n*-butyllithium in hexane is used.

5. In large-scale preparations the addition of the catalyst in one portion causes an exothermic reaction that results in lower yields.

6. An inherent viscosity of 0.20 dl./g. corresponds to a molecular weight of approximately 17,000.

3. Methods of Preparation

A polymer of 1,3-cyclohexadiene has also been prepared by the use of Ziegler catalysts and cationic initiators (6, 7, 8). The product has been brominated and dehydrobrominated to yield poly-*p*-phenylene (5-8).

Dawans and LeFebvre (8) have characterized poly-1,3-cyclohexadiene by nmr and infrared spectral analyses and propose that the polymerization is terminated by chain transfer and disproportionation to give cyclohexadienyl and phenyl end groups. They also report evidence of the presence of 1,2- and 1,4-repeating units.

4. References

1. Department of Chemistry, Southwest Texas State University, San Marcos, Texas 78666.
2. Department of Chemistry, University of Arizona, Tucson, Arizona 85721.
3. Institute of Polymer Science, The University of Akron, Akron, Ohio 44304.
4. J. P. Schaefer and L. Endres, *Organic Syntheses,* **47**, 31 (1967); Coll. Vol. **V**, 285 (1973).
5. P. E. Cassidy, C. S. Marvel, and S. Ray, *J. Polym. Sci.,* **A3**, 1553 (1965).
6. D. A. Frey, M. Hasegawa, and C. S. Marvel, *J. Polym. Sci.,* **A1**, 2057 (1963).
7. C. S. Marvel and G. E. Hartzell, *J. Amer. Chem. Soc.,* **81**, 448 (1959).
8. F. Dawans and G. LeFebvre, *J. Polym. Sci.,* **A2**, 3277 (1964).

Anionic Polymerization of α-Methylstyrene: α,ω-Poly(α-methylstyrene)dicarboxyllic Acid

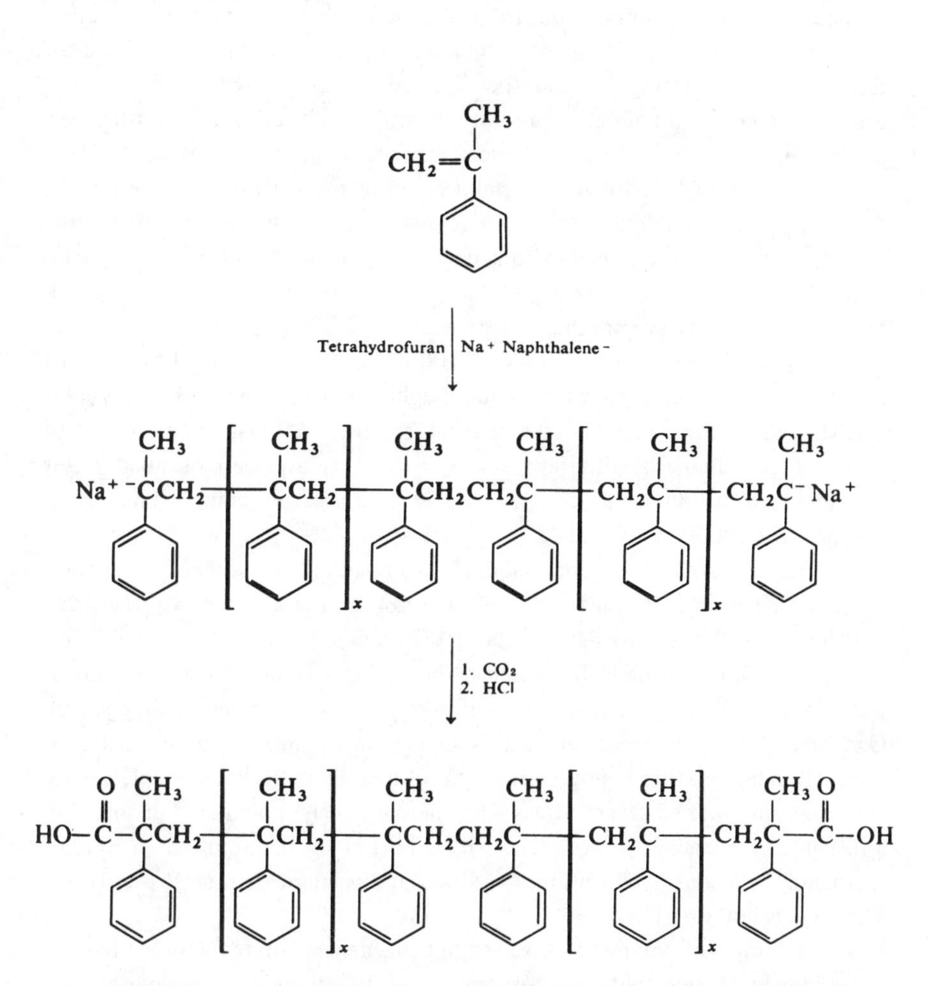

Submitted by L. C. Anand (a), A. Takahashi (a, b), and N. G. Gaylord (1a)
Checked by J. C. Falk, G. Matula, and J. E. Burroughs (2)

1. Procedure

A 5-1., three-necked flask, equipped with a mechanical stirrer, a gas inlet and outlet, and a serum stopper, is flushed with purified argon (Note 1) while still hot, after being oven-dried at 120°. After the flask has cooled to room temperature, 3 1. of purified tetrahydrofuran (Note 2) and 600 ml. of α-methylstyrene (Note 3) are added to the flask under an argon atmosphere. When the reaction mixture has cooled to -30° to -40° and the flow of argon has stopped, a solution of sodium naphthalene in tetrahydrofuran (Note 4) is added with a hypodermic syringe through the serum stopper until a red color persists in the reaction mixture. An additional 170 ml. of sodium naphthalene solution is quickly added through the serum stopper, and the mixture is stirred slowly while argon is passed through the gas outlet to prevent the entry of adventitious impurities.

The dark red solution becomes viscous and equilibrium conversion at -30 to -40° is established in 1 hr. At this point 5 ml. of the reaction mixture is withdrawn with a hypodermic syringe, precipitated in methanol, and dried; the η_{sp}/c vs. c is then determined in toluene at 25°. The one-point value for a concentration of 0.2 g. polymer in 100 ml. toluene was 0.20 dl./g., which corresponds to a viscosity average molecular weight of 50,000 (Note 5).

To obtain a molecular weight of 70,000-80,000 an additional 500 ml. of α-methylstyrene is titrated with sodium naphthalene solution until a red color persists and is then added to the reaction mixture. After 45 min. at -30 to -40° a 5-ml. sample is withdrawn, and the viscosity average molecular weight is determined in toluene at 25°; $[\eta]$ of 0.275 dl./g. corresponds to a molecular weight of 72,000 (Notes 6 and 7).

To prepare a carboxyl-terminated polymer the serum stopper is removed from the reaction flask under a stream of argon that is directed into the flask. A cylinder of carbon dioxide (Note 8) is inverted and the valve is inserted into the open neck of the flask. Powdered carbon dioxide is introduced as rapidly as possible until the red color of the reaction solution has completely disappeared. One liter of 0.2 N hydrochloric acid is added and the mixture is stirred for 30 min. After the mixture is poured into 6 1. of methanol, the solid is collected by filtration and washed several times with methanol. The polymer is dissolved in 1500 ml. of benzene, reprecipitated by addition of the solution to 6 1. of methanol, collected by filtration, and dried in a vacuum oven at 40° at 1-torr. The yield is 900 g. (91%).

A 5-g. sample of polymer is dissolved in benzene and titrated with methanolic potassium hydroxide. With the assumption of difuctionality a number average molecular weight of 70,000 is calculated. Also with the assumption of a molecular weight of 72,000, as determined by viscosity, the carboxyl content is calculated to be 96% (Note 9).

2. Notes

1. Research-grade argon is purified by successive passage through two columns of Drierite, two columns of Linde 4A molecular sieves, and two columns of sodium naphthalene in tetrahydrofuran.

2. Tetrahydrofuran is distilled from lithium aluminum hydride under an argon atmosphere, stored over pieces of sodium with stirring, and then transferred to the reaction flask by applying a positive pressure of argon.

3. α-Methylstyrene is distilled over pieces of sodium under an argon atmosphere, stirred for 30 min. over fresh pieces of sodium, and then transferred to the reaction flask, which contains tetrahydrofuran, by applying a positive pressure of argon. During the transfer operation the α-methylstyrene is passed through dry glass wool to remove pieces of sodium and other foreign matter.

4. The sodium naphthalene solution is prepared by the addition of 5 g. of resublimed naphthalene to 125 ml. of distilled tetrahydrofuran and the addition of 500 mg. of sodium. The flask is capped with a serum stopper and stirred with a magnetic stirrer for 2 to 3 hr. at room temperature. A 2-ml. sample of the green solution is withdrawn with a hypodermic syringe and added to a mixture of 20 ml. of methanol and 5 ml. of distilled water. Titration with a standardized 0.1 *N* HCl solution with phenolphthalein indicated that the sodium naphthalene solution was 0.235 *M*.

5. The 5-ml. sample is precipitated in 200 ml. of methanol; the solid is collected by filtration and washed with methanol several times. It is then redissolved in 5 ml. of benzene for ease of handling. The benzene solution is evaporated in a rotating vacuum evaporator at 1 torr at 40-50° for 20 min. A 0.2-g. portion of the dried polymer is dissolved in 100 ml. of toluene, and the relative viscosity is determined at 25°. Because the intrinsic viscosity plot is essentially parallel to the axis, a one-point determination of η_{sp}/c vs. c can be used in conjunction with a published plot (3) to give the molecular weight.

6. A third sample withdrawn 15 min. later had an $[\eta]$ 0.275, which indicates that equilibrium conversion is obtained in 45 min.

7. To obtain a nontelechelic poly(α-methylstyrene) the polymer may be isolated at this point by the addition of dilute hydrochloric acid and methanol, analogous to the isolation procedure utilized after carboxylation reaction.

8. Instrument-grade carbon dioxide of 99.99% purity is used. The addition of solid carbon dioxide is made as rapidly as possible to avoid side reactions

and condensation of water. Neither gaseous carbon dioxide nor Dry Ice are free of such undesirable effects.

9. A 5-g. sample of polymer is dissolved in 150 ml. of distilled benzene, and the solution is titrated with 0.02 *N* methanolic potassium hydroxide solution with phenolphthalein as indicator. The number average molecular weight is obtained from the following equation with the assumption of difunctionality:

$$\text{molecular weight} = \frac{y}{x \cdot N} \times 2 \times 1000$$

where y = weight of sample in grams
x = milliliters of potassium hydroxide solution.
N = normality of the potassium hydroxide solution

The functionality may be determined from the following equation by use of the molecular weight obtained from the viscosity determination:

$$\text{percent carboxyl} = \frac{y}{x \cdot N} \times \frac{2 \times 1000}{\text{molecular weight}} \times 100$$

3. Methods of Preparation

Nonterminating or "living" polymers of α-methylstyrene have been prepared with catalysts based on sodium-naphthalene (3, 4, 5, 6) and other sodium-aromatic hydrocarbon complexes (3), as well as sodium metal in tetrahydrofuran (5). Styrene (7, 8), butadiene and isoprene (9), "living," and block copolymers (10) have been prepared in a similar manner. The polymers may be terminated not only with carbon dioxide (11) but also with ethylene oxide and carbon disulfide (10).

4. References

1a. Gaylord Research Institute Inc., Newark, New Jersey 07105.
1b. Current Address: Hooker Chemical Co., Niagara Falls, New York 14302.
2. Roy C. Ingersoll Research Center, Borg-Warner Corporation, Des Plaines, Illinois 60018.
3. A. A. Tobolsky, A. Rembaum, and A. Eisenberg, *J. Polym. Sci.*, **45**, 347 (1960).
4. D. J. Worsfold and S. Bywater, *J. Polym. Sci.*, **26**, 299 (1957).
5. H. W. McCormick, *J. Polym. Sci.*, **25**, 488 (1957); **41**, 327 (1959).
6. A. F. Siriani, D. J. Worsfold, and S. Bywater, *Trans. Faraday Soc.*, **55**, 2124 (1959).
7. M. Szwarc. M. Levy, and R. Milkovich, *J. Amer. Chem. Soc.*, **78**, 2656 (1956).

8. F. Wenger, *J. Amer. Chem. Soc.*, **82**, 4281 (1960).
9. M. Szwarc, H. Brody, M. Ladscki, and R. Milkovich, *J. Polym. Sci.*, **25**, 221 (1957).
10. M. Szwarc, *Adv. Polym. Sci.*, **2**, 275 (1960).
11. H. Brody, D. H. Richards, and M. Szwarc, *Chem. Ind. (London)*, No. **45**, 1473 (1958).

Glycidyl Vinyl Ether
[1-(Vinyloxy)-2,3-epoxypropane]

$$H_2C\overset{O}{\diagup\!\diagdown}CHCH_2OH + (CH_2{=}CH\text{—})_2O \xrightarrow[\text{90–100° Autogenous Pressure}]{(CH_3CO_2\text{—})_2Hg} H_2C\overset{O}{\diagup\!\diagdown}CHCH_2OCH{=}CH_2 + CH_3CHO$$

Submitted by W. H. Gumprecht and W. J. Borecki (1)
Checked by J. A. Empen and J. E. Mulvaney (2)

1. Procedure

Caution! Vinyl ethers react violently with acidic materials. Care should be taken to maintain neutral or slightly alkaline conditions during the reaction and product isolation.

In a 3-1. pressure vessel (Note 1), cooled externally with ice water (Note 2), are sealed under nitrogen a cold solution of 127 g. (0.4 mole) of reagent-grade mercuric acetate in 297 g. (4.0 moles) of glycidol (Notes 3) and 1120 g. (16.0 moles) of vinyl ether (Note 4). The vessel is heated rapidly with agitation to 90-100° and maintained at this temperature for 8 hr. The vessel is cooled to below room temperature with ice water, vented, in an efficient hood and discharged.

The solution (Note 5) is distilled from a 500-ml. flask (Note 6) without fractionation at a pressure less than 10 torr until the temperature in the distillation flask reaches 90°; the distillate is collected in a 3-1. receiver (Note 7) cooled in a Dry Ice bath. The residue (Note 8) can be discarded. The cold distillate is washed with three 400-ml. portions of an ice-cold 20% sodium chloride solution and then with two 300-ml. portions of ice water. The organic layer is dried over molecular sieves (Note 9), and 2 g. of *N*-phenyl-1-naphthylamine is added (Note 10). The drying agent is removed by gravity filtration through a coarse-grade sintered-glass funnel and washed with 175 ml. of *p-t*-butyltoluene (Note 11).

The combined filtrates are distilled through a 45-cm., vacuum-jacketed column filled with Heli-pak packing (Note 12) from a 500-ml. distillation flask (Note 6). Initially, vinyl ether is collected to a b.p. of 35° (Note 13). The product, collected at 132-133°, n_D^{25} 1.4329 (Note 14), is a colorless liquid weighing 280-310 g. (70-77%) (Note 15).

2. Notes

1. A stainless steel, water-jacketed autoclave equipped with a mechanical stirrer was used.
2. Cooling of the vessel is advisable to prevent excessive losses of vinyl ether.
3. Glycidol is available from several suppliers. A supply obtained from the Ott Chemical Company was used in developing the synthesis. Because it decomposes slowly, even at room temperature, glycidol, b.p. 49-51° (7 torr), should be distilled before use.
4. Vinyl ether is sold under the trade-name of "Vinethene" by Merck, Sharp & Dohme. As supplied, it contains *N*-phenyl-1-naphthylamine and alcohol as stabilizers. Vinyl ether, b.p. 28-29°, should therefore be distilled through a Vigreux column just before use.
5. A small amount of flocculent precipitate is usually present in the solution. It need not be removed, because it remains in the residue from the first distillation.
6. Because of the large volume of vinyl ether that must be carried through both distillations, it is advisable, in the interest of obtaining an optimum yield, to use a small distillation flask and add the solutions to it with an addition funnel. (The checkers found it more convenient to use a 1-1. flask to prevent flooding after removal of the low-boiling material.)
7. To obtain maximum recovery of vinyl ether the receiver should be fitted with a Dry Ice-cooled condenser at its outlet.
8. The residue, a viscous syrup, consists of mercuric salts and glycidyl polymers.

9. Type 3A molecular sieves, availalbe from the Linde Company, was used in the form of $^{1}/_{16}$ in. pellets.

10. *N*-Phenyl-1-naphthylamine acts both as an antioxidant to maintain alkalinity. It can be used as it is supplied by any of several vendors, such as Matheson, Coleman and Bell.

11. In addition to washing product from the molecular sieves, *p-t*-butyltoluene, b.p. 190-191°, acts as a distillation "chaser," It can be used as it is obtained from any of several suppliers, such as Eastman Organic Chemicals.

12. This kind of packing is supplied by Podbielniak, Inc. It was chosen because of its efficiency and low liquid retention The particular packing used was made of "Hastelloy 'B'" in the dimensions of 0.092 × 0.175 × 0.175 in.

13. The recovered vinyl ether, which contains small amounts of acetaldehyde, can be recycled through the synthesis after redistillation. It should be stabilized with a small amount of *N*-phenyl-1-naphthylamine as soon as it is collected.

14. Glycidyl vinyl ether obtained in this way has a purity of 98-99% by vapor phase chromatography (polypropylene glycol on firebrick at 120°).

15. Redistillation of combined small fractions, b.p. 120-132° and 133-136°, affords an additional 16-24 g. (4-6%) of product.

3. Methods of Preparation

Glycidyl vinyl ether has been prepared by dehydrochlorination of glycidyl 2-chloroethyl ether (3). The method described here is a modification of the vinyl *trans*-etherification reaction (4). The preparation of glycidyl vinyl ether from glycidol and vinyl ethers using mercuric acetate as catalyst, as well as polymerization to homo- and co-polymer through the vinyl or epoxy groups, has been described (7-10). Co-polymers with pendent epoxy groups have been reported to be obtained from glycidyl vinyl ether (11).

4. Merits of the Preparation

Because of the sensitivity of the epoxide ring to nucleophiles, glycidyl vinyl ether cannot be obtained by the base-catalyzed vinylation of glycidol with acetylene (5), a method used for the preparation of most of the vinyl ethers of commerce. The same deterrent severely restricts the yield in the dehydrochlorination of glycidyl 2-chloroethyl ether with sodium hydroxide (3).

Two factors limit the yield of glycidyl vinyl either to about 18% by the standard vinyl *trans*-etherification method (4). Again, the sensitivity of epoxides, in this case to by-product alcohols, is a problem. In addition, the necessity of shifting the equilibrium prevailing in the process requires a slow distillation of the product at a reaction temperature near 160°. The elevated temperatures and prolonged contact permit competing reactions to destroy the epoxide ring to a considerable extent.

The method described here, though a vinyl *trans*-etherification, does not yield a by-product alcohol or involve shifting an equilibrium. It should find application in vinylating alcohols containing other reactive groups and in the preparation of vinyl ethers from high-boiling or solid alcohols.

Glycidyl vinyl ether can be copolymerized with many vinyl monomers, such as tetrafluoroethylene and other fluorinated olefins (6), maleic anhydride, vinyl chloride, vinylidene chloride, and ethylene, to yield polymers that can be cross-linked by the epoxide ring.

5. References

1. Contribution No. 395 from the Organic Chemicals Department, E. I. du Pont de Nemours & Company, Wilmington, Delaware 19899.
2. Department of Chemistry, The University of Arizona, Tucson, Arizona 85721.
3. W. Kawai and S. Tsutsumi, *J. Chem. Soc. Japan, Pure Chem. Sect.*, **80**, 88 (1959) [*C. A.*, **55**, 4466a (1961)].
4. W. H. Watanabe and L. E. Conlon, *J. Amer. Chem. Soc.*, 79, 2828 (1957).
5. J. W. Copenhaver and M. H. Bigelow, *Acetylene and Carbon Monoxide Chemistry*, Reinhold, New York, 1949, Chapter II.
6. I. Pascal (E. I. du Pont de Nemours & Company), Brit. Patent 948,998 (Feb 5, 1964) [*C. A.*, **60**, 13408h (1964)].; U.S. Pat. 3,132,121 (May 5, 1964).
7. S. W. Tinsley and E. A. Rick (Union Carbide Corp.) U.S. Pat. 3,203,939 (Aug. 31, 1965).
8. H. Sorkin (Air Reduction Co.) U.S. Pat. 3,414,634 (Dec. 3, 1968).
9. G. B. Payne (Shell Oil Co.,) Can. Pat. 820,339 (Aug. 12, 1969).
10. T. I. Wang and G. M. Nakaguchi (Atlantic Richfield Co.) U.S. Pat. 3,699,131 (Oct. 17, 1972).
11. B. Vecellio (Pirelli Soc. per Azioni) Ger. Pat. 1,520,584 (April 10, 1969).

Poly(2,5-distyrylpyrazine)

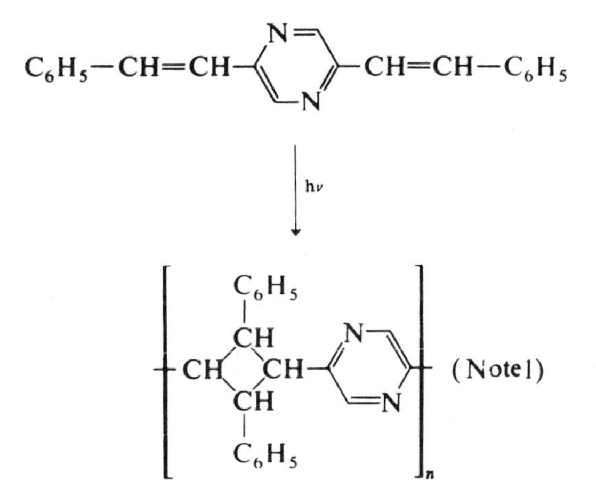

Submitted by M. Hasegawa and Y. Suzuki (1)
Checked by Z. G. Garlund and J. Laverty (2)

1. Procedure

A. With Dispersant

Finely crystalline 2,5-distyrylpyrazine (0.7 g.) (3) is dispersed in 125 ml. of cyclohexane (Note 2) and, with stirring, irradiated with a 100-watt high-pressure mercury lamp located at the center of the reaction flask (Ushio Electric Inc.) for 2 hr. at room temperature. The sparkling yellow of the starting compound gradually changes into the white of the polymer during radiation. After the product is collected by filtration, it is washed with acetone to remove the monomer and dried *in vacuo*. The reduced viscosity of poly(2,5-distyrylpyrazine) thus obtained is 3.1 dl./g. (0.35 g. of polymer in 100 ml. of trifluoroacetic acid, 30°) and the yield is nearly quantitative (Note 3). The polymer, which melts at 339-343° with considerable decomposition (Note 4), is highly crystalline (4) and is insoluble in most common polymer solvents, such as *m*-cresol, formic acid, and dimethylformamide. However, it does dissolve in concentrated sulfuric, dichloroacetic, and trifluoroacetic acids.

B. *Without Dispersant*

Fine crystals of 2,3-distyrylpyrazine in a rotary flask (Pyrex®) are exposed to sunlight for 10 hr. The product is treated in the manner already described. The reduced viscosity is 1.3 dl./g. (0.31 g. of polymer in 100 ml. of trifluoroacetic acid, 30°) and the yield is 80%.

C. *In Solution*

A tetrahydrofuran solution of the monomer (1.41×10^{-2} M) yielded 60% of oligomers ($\overline{M}_n$(VPO) 900) when irradiated with filtered light (Note 5) for 48 hr. at room temperature (9).

2. Notes

1. The *trans*-head-to-tail configuration of the poly(2,5-distyrylpyrazine) has been established by crystallographic studies (5, 6).
2. Cyclohexane, a dispersant, is employed to provide homogeneous irradiation to the monomer surface.
3. The reduced viscosity and the yield of poly(2,5-distyrylpyrazine) are varied by controlling the irradiation time (7).
4. 2,5-Distyrylpyrazine is recovered nearly quantitatively by heating the polymer *in vacuo* at a temperature higher than its melting point (8).
5. The filtered light ($\lambda > 380$nm) was produced with a cut-off filter (Corning 3-75) on a 500 W Xenon lamp.

3. References

1. Research Institute for Polymers and Textiles, 4 Sawatari Kanagawa–ku, Yokohama, Japan.
2. Research Laboratories, General Motors Technical Center, Warren, Michigan 48090.

3. R. Frank, *Chem. Ber.*, 38, 3727 (1905).
4. M. Iguchi, H. Nakanishi, and M. Hasegawa, *J. Polym. Sci.*, **A1**, **6**, 1055 (1968).
5. Y. Sasada, H. Shimanouchi, H. Nakanishi, and M. Hasegawa, *Bull. Chem. Soc. Japan*, **44**, 1262 (1971).
6. H. Nakanishi, M. Hasegawa, and Y. Sasada, *J. Polym. Sci.*, **A2**, 10, 1537 (1972).
7. H. Nakanishi, Y. Suzuki, F. Suzuki, and M. Hasegawa, *J. Polym. Sci.*, **A1**, 7, 753 (1969).
8. M. Hasegawa and Y. Suzuki, *J. Polym. Sci.*, **B,5**, 813 (1967).
9. J. Suzuki, T. Tamaki, and M. Hasegawa, *Bull. Chem. Soc. Japan*, 47, 210 (1974).

Polymerization of α-Phenylethyl Isocyanide

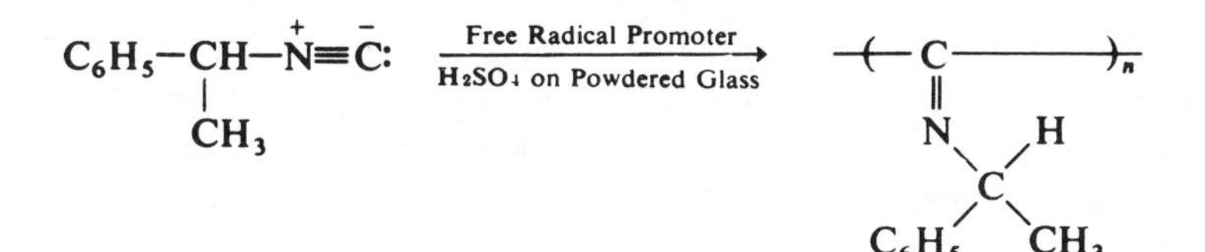

Submitted by F. Millich, R. G. Sinclair, II, and G. K. Baker (1) (2)
Checked by O. Vogl (3)

1. Procedure

A. Powdered Glass Catalyst

Twenty grams of powdered glass (Note 1), approximately 200 mesh, is washed with benzene, dried, soaked overnight in a 0.1 *M* sodium dichromate solution in conc. sulfuric acid, washed clear with water, then washed successively on a fritted-glass funnel with 50-ml. portions of a 5 *N* sodium hydroxide solution, distilled water, and 6 *N* sulfuric acid. The last of the excess acidic wash fluid is thoroughly removed on the filter under a blanket of dry nitrogen by the application of suction for 60 min. The moist, acid-coated ground glass is transferred to a single-necked, round-bottomed flask that is attached to a rotary evaporator. The flask is rotated at a 45° angle in an oil bath maintained at 90° for 16 to 20 hr. at 0.4 torr. Some droplets of acidic viscous condensate that form around the neck of the flask may be removed by wiping with several sheets of absorptive tissue. Titration indicates that a coating of 0.015-0.02 g. of sulfuric acid/g. of ground glass catalyst is present. The preparation, which may be stored in this flask after stoppering, retains its initial activity even after storing for 6 months. The powdered glass preparation is white and has some tendency to form lumps; it is neither cakey nor a free-flowing powder.

B. Polymerization

Caution! This preparation should be carried out in a well-ventilated hood, and provision should be made to avoid contact of isocyanide with the skin (Note 2).

In a 100-ml., round-bottom flask 2.0 g. of fresh, acid powdered glass is covered with 10 ml. of *dry n*-heptane (previously treated with metallic sodium and distilled). A magnetic stirrer provides brisk agitation, and 3.0 g. of freshly distilled α-phenylethyl isocyanide (Note 3) is added to the reaction flask. An orange color often occurs immediately at active catalyst sites. After a vertical condenser is fitted into the flask, oxygen (Note 4), which is first passed through anhydrous calcium chloride and then a flask of *n*-heptane, is introduced into the reaction mixture with a fine-tipped capillary at a rate of 10 ml. per min. The reaction flask is heated 20 to 40 hr. at 50° with stirring and continuous introduction of oxygen. The powdered glass becomes coated with red-brown curds of polymer, and the solution turns yellow to red-brown.

When the *n*-heptane mixture is decanted through coarse filter paper, the powdered glass-polymer matrix remains in the flask. Dissolution of the polymer (Note 5) requires treating the matrix with 25 to 50 ml. of benzene to produce a solution of such viscosity that it will flow through filter paper. Gravity filtration, must be used. The filter paper is washed with benzene until free of nearly all color; this may require cutting the filter paper, immersion of the pieces in fresh benzene, and finally filtration of the mixture through fresh filter paper. The combined benzene phase is reduced in volume at room temperature to approximately 25-50 ml. This solution is added dropwise to a tenfold excess of vigorously stirred methanol to precipitate the polymer. The precipitate is collected on a Büchner funnel, air-dried partly, dissolved in approximately 30 ml. of benzene, and recollected by lyophilization of the solution at 0.5-1.0 torr. A yield of 60-70% is commonly experienced after 40 hr.; under the best conditions the yield is quantitative (Note 6).

The fluffy solid appears yellow. The polymer is soluble in a variety of solvents, including aromatic hydrocarbons, tetrahydrofuran, and haloalkanes, from which clear, brittle films may be formed by evaporation. The characterization of polyisocyanides has recently been reviewed (4). The preparation described above most frequently produced (±20%) number average molecular weights (determined by dynamic membrane osmometry) of 50,000, and weight average molecular weights (determined by light scattering) of 80,000 (Note 6) at the temperature of polymerization of 50°.

Elemental analysis reveals that 0.5-1.8% oxygen by weight is variously incorporated into the polymer. Possibly for this reason these polymer samples show a decrease in molecular weight on exposure to light or the application of heat (5, 6). Samples heated on a melting point hot stage show resinification and weight loss at temperatures greater than 200° (7).

2. Notes

1. The powdered glass may be obtained from Fisher Scientific Company.
2. Isocyanides, especially the more volatile compounds, should be treated with caution. As does carbon monoxide, isocyanides complex with metals and metal ions. However, the following commentary has been published (8a): "With a few exceptions, isonitriles exhibit no appreciable toxicity to mammals. As has been found in the toxicological laboratories of Farbenfabriken Bayer A.-G., Elberfeld, Germany, oral and subcutaneous doses of 500-5000 mg./kg. of most of the isonitriles can be tolerated by mice, yet there are exceptions like 1,4-diisocyanobutane, which is extremely toxic ($LD_{50,mice}$ (10 mg./kg.)." Nearly all lethal doses of isocyanides known for a variety of animals are described in early literature. Thus one finds reported in 1896 that the lethal doses of ethyl isocyanide, applied subcutaneously to mouse, rabbit, and pigeon are 74,110 and 99 mg/kg. body weight (9). The sparse literature on the subject has recently been reviewed (10), wherein the statement is made: "As far as we know, no such data are available for human beings." The high values given above are more commonly experienced with some of the non-volatile isocyanides. Nausea, headaches, interference with the digestive system, and also symptoms akin to those of jaundice have been cited by humans exposed to isocyanides (10). The patent literature also cites pronounced biocidal properties of multifunctional isocyanides and of isocyanide metal complexes (10). Although most of these volatile compounds have intense and vile odors, it has been our experience that continued exposure to isocyanide vapors tends to dull olfactory sensitivity. Ethyl isocyanide has been known to explode (11). Glassware and gas streams can be purged of isocyanides with mineral acids and water.
3. The monomer is conveniently prepared by dehydration of *N*-(α-phenylethyl)formamide in an 87% yield with phosgene. Careful distillation will free the monomer of basic substances that would adversely affect the acid catalyst. Alternatively, the formamide may be treated with chlorodimethyl formiminium chloride prepared *in situ* from thionyl chlroide and N,N- dimethylformamide (12). The advantages of this method are that it avoids the use of

phosgene and the product is much more easily freed by distillation of contaminants detrimental to the catalyst. Other synthetic methods for the preparation of isocyanides have been reviewed recently (8b). Some physical characteristics of the monomer are colorless liquid, b.p. 54° (0.6 torr.); n_D^{25} = 1.5041; d^{32} = 0.952 g./ml. The infrared spectrum shows no amide absorption bands, but has a strong characteristic band at 2150 cm.$^{-1}$ (4.7 μ); the ultraviolet spectrum possesses an ϵ_{max} of 200 at 257 nm. in tetrahydrofuran. A preparation of the formamide is available in the literature (13). Some additional physical characteristics for the formamide are b.p. 189-91° (26 torr); m.p. 40 ± 2°; the ultraviolet spectrum shows an ϵ_{max} of 353 at 257 nm. The monomer, stored in amber bottles at room temperature in the presence of adventitious amounts of oxygen, is subjected to polymerization over the course of several months. Pretreatment of storage containers to eliminate acidic and basic contaminants and the exclusion of oxygen will extend the shelf life of the monomer.

4. Oxygen from a compressed gas cylinder (indrustrial grade, 99.9% pure, Puritan Compressed Gas Corporation) was employed. Occasionally a brisk stream of air was substituted for oxygen without producing a significant difference.

5. When completely insoluble linear poly (isocyanides) are synthesized by this method the glass-polymer matrix is agitated in a cylinder with carbon tetrachloride, or other suitable, dense liquid in which the powdered glass sinks and the polymer floats.

6. The preparation of the catalyst, the amount of oxygen, the degree of agitation, the polymerization temperature, the reaction time, and other conditions (6) profoundly affect both the rate of monomer conversion and the molecular weight. Thus the number average molecular weights may range from 25,000 to 200,000, with corresonding intrinsic viscosities, respectively, of 0.4-2.0 dl./g. in toluene at 30° (14), the highest $\overline{M}_n$ having been obtained in a polymerization that ran 19 days at 27°, without dibenzoyl peroxide or the introduction oxygen and that used a coating of 0.004 g. of sulfuric acid/g. of ground glass (15). A slower rate will reduce the apparent yield for reactions terminated at 20 hr. duration, the yields ranging upward from 25%.

2. Methods of Preparation

A merit of the method described is that by-product formation and charring of the monomer is avoided and the conversion can be quantitative. The slow rate of polymerization promises to allow kinetic rate studies. The solubility of the polymer has permitted the only reported solution characterization of polyisocyanides. Optically active α-phenylethyl isocyanide has been polymerized to yield polymer of very high specific molar rotation (15). Heretofore no catalytic method had been available for the controlled polymerization of bulky aliphatic isocyanides to yield a linear polymer of high molecular weight. One modification of the technique described above which the present authors have found convenient is to dissolve 0.05 g. of dibenzoyl peroxide in the monomer before addition of the latter to the reaction flask. The presence of oxygen in the air above the reaction mixture is still desirable to ensure high yields, but the process of bubbling a stream of oxygen through the mixture may be eliminated when the preparation must be left unattended, or when higher molecular weight and greater thermal stability is desired. A further modification of this alternate procedure permits the substitution of a 1% sodium bisulfate solution in place of the 6 *N* sulfuric acid solution used to apply the acidic coating to the neutral ground glass (15). A recent modification obviates the use of ground glass (16). In this procedure a fine dispersion of sulfuric acid droplets is generated in heptane in a high-speed blender and an aliquot is quickly transferred to a heptane solution of the monomer. However, polymodal molecular weight distributions have been observed by gel permeation chromatography with this catalyst.

The authors (6) and others (17-21) have also found that Lewis acids, at reduced temperatures, catalyze rapid polymerization. Ziegler catalysts (18, 19) and metal carbonyls (13) have also beeen investigated. This subject has been reviewed (4, 22). Recently, the use of nickel salts and chelates as catalysts has been described (23). Sulfuric acid has been used exclusively for the preparation of the only other known soluble homopolymers, i.e., poly(β-phenylethylisocyanide) (16) and poly(methyl α-isocyanopropionate) (24).

4. References

1. University of Missouri at Kansas City, Polymer Chemistry Division, Chemistry Department, Kansas City, Missouri 64110; taken, in part, from the doctoral theses of R. G. Sinclair and G. K. Baker.
2. Acknowledgment is made to the donors of the Petroleum Research Fund, administered by the American Chemical Society, for grant support of the research from which this synthesis resulted.
3. Department of Polymer Science and Engineering, University of Massachusetts, Amherst, Massachusetts 01002.
4. F. Millich, *Adv. Polym. Sci.*, **19**, 117 (1975).

5. F. Millich and R. G. Sinclair, *J. Polym. Sci.*, **C22**, 33 (1968) and earlier publications cited therein.
6. F. Millich and R. G. Sinclair, *J. Polym. Sci.*, **A6**, 1417 (1968).
7. F. Millich, *Chem. Rev.*, **72**, 101 (1972).
8. I. Ugi, Ed., *Isonitrile Chemistry*, Academic, New York, 1971; (a) p. 2; (b) Chapter 2.
9. H. Wedekind: Thesis, Kiel (1896).
10. R. W. Stephany: Ph.D. Dissertation, Utrecht (1972).
11. P. Lemoult, *Compt. Rend.*, **143**, 902 (1906).
12. H. M. Walborsky and G. E. Niznik, *J. Org. Chem.*, **37**, 187 (1972).
13. M. L. Moore, *Organic Reactions*, **5**, 316 (1949).
14. For a discussion of intrinsic viscosities of fractionated poly (α-phenylethyl isocyanide), see F. Millich, E. W. Hellmuth, and S. Y. Huang, *J. Polym. Sci.*, Polym. Chem. Ed. **13**, 2143 (1975).
15. F. Millich and G. K. Baker, *Macromolecules*, **2**, 122 (1969).
16. F. Millich and G.-M. Wang, VI-A.C.S.-Midwest Chemistry Conference, Lincoln, Nebraska, Oct. 28-30, 1970, Abstract No. 617.
17. R. W. Stackman, *J. Macromol. Sci.-Chem.*, **A2(2)**, 225 (1968).
18. T. Saegusa, N. Taka-ishi, and H. Fujii, *Polymer Letters*, **5**, 779 (1967).
19. Y. Yamamoto and N. Hagihara, *Nippon Kagaku Zasshi*, **89**, 78 (1968).
20. Y. Yamamoto, T. Takizawa, and N. Hagihara, *Nippon Kagaku Zasshi.*, **87**, 1355 (1966).
21. S. Iwatsuki, K. Ito, and Y. Yamashita, *Kogyo Kagaku Zasshi*, **70**, 1822 (1967).
22. F. Millich, in H. F. Mark, N. G. Gaylord, and N. M. Bikales, Eds., "*Encyclopedia of Polymer Science and Technology*, Vol. 15, Wiley-Interscience, New York, 1971, p. 395.
23. R. J. M. Nolte, R. W. Stephany, and W. Drenth, *Rec. trav. Chim.*, **92**, 83 (1973).
24. F. Millich and J. Chenvanij, unpublished results.

Block Copolymerization of Methyl Methacrylate with Poly(ethylene oxide)

$$-\!\!\left(CH_2CH_2OCH_2CH_2O\right)\!\!- \xrightarrow{\text{Stirring}} \left\{\begin{matrix} \sim CH_2CH_2\cdot + \cdot OCH_2CH_2\sim \\ \text{or} \sim CH_2\cdot + \cdot CH_2OCH_2CH_2\sim \end{matrix}\right\}$$

$$\xrightarrow[\text{Methacrylate}]{\text{Methyl}} \text{Block copolymer}$$

Submitted by Y. Minoura (1) and A. Nakano (2)
Checked by K. Goto and H. Fujiwara (3)

1. Procedure

Poly(ethylene oxide) (8.0 g.) (Note 1) is dissolved (Note 2) in 200 ml. of methyl methacrylate (Note 3) by heating (Note 4) to about 45° in a 300 ml. round-bottomed flask (Note 5). After the system becomes homogenous, a stopcock is connected to the flask and the solution is degassed twice (note 6). At the last stage of degassing the flask is filled with an inert gas (Note 7) and the degassed solution is decanted from the flask to a 250 ml. container of a mixer also filled with inert gas. The container is attached tightly (Note 8) to the cover of the mixer (Note 9) and then placed in a water bath at 20° (Note 10).

After inert gas has bubbled through the solution for about 10 min. the solution is stirred at a speed of 30,000 r.p.m. for 5 min. (Note 11) under a steady flow of gas. As the temperature of the solution rises to about 50° (no higher), the stirring is stopped for about 15 min. to cool the solution by standing (Note 11). The stirring-cooling process is repeated twelve times and requires about 60 min. of stirring. At the end of this time the container is removed from the apparatus, and the solution is poured slowly, with rapid stirring, into 10 1. of ethyl ether to precipitate the polymer. The polymer is separated on a sintered-glass filter, washed with ether, and dried to yield 10.24 g. of polymer (1.2%

conversion of the methyl methacrylate) (Note 12). The intrinsic viscosity of the product (benzene, 30°) is 1.7, dl./g. whereas that of the poly(ethylene oxide) is 4.2 dl./g. Because all of the polymer obtained is soluble in hot methanol, no homopolymer of methyl methacrylate is present. Therefore the block efficiency is close to 100% (Note 13).

2. Notes

1. Poly(ethylene oxide) supplied by Seitetsu Kagaku Company, Ltd. or Union Carbide Chemical Company, Ltd. can be used. In this experiment PEO-5N (molecular weight. 1.0×10^6; degree of polymerization, 22,700), made by Seitetsu Kagaku, is used. The higher the molecular weight, the better for block copolymerization because degradation occurs more easily. The relationship between the molecular weight and intrinsic viscosity is as follows:

$$[\eta] = 6.4 \times 10^{-5} \cdot M^{0.82} \text{ (aqueous solution, at } 35°) \tag{4}$$

$$[\eta] = 6.1 \times 10^{-4} \cdot M^{0.64} \text{ (benzene solution, at } 30°) \tag{5}$$

2. The system must be stirred mechanically to dissolve the poly(ethylene oxide).

3. Commercially availble methyl methacrylate is distilled with steam and then redistilled under vacuum.

4. Heating is required to dissolve poly(ethylene oxide). To prevent homopolymerization of methyl methacrylate the temperature is maintained below 50°.

5. A round-bottomed and mechanically strong flask is used for degassing.

6. Degassing is carried out by the usual method. After the solution is frozen in a Dry Ice-methanol bath, the flask is evacuated and allowed to thaw twice.

7. Nitrogen may be used, but a denser inert gas, such as argon, is preferrred to minimize contact with air during transfer.

8. A packing made of Teflon is used.

9. A suitable high-speed mixer is supplied by Tokushukika Kogyo Company, Ltd. Its parts, such as the container, turbine, an stator, are made of stainless steel. The container has a diameter of 7 cm. and a depth of 9 cm.

10. The mechanical parts of the mixer become quite hot. If the stirring is prolonged, the mixer, especially the Bakelite gear, will be destroyed.

11. The time required depends on the temperature of the bath. If the water used is below 20°, however, the polymer precipitates on the container wall.

12. The amount of monomer polymerized is estimated from the difference between the weight of the polymer precipitated from a constant weight of solution before and after stirring. In another experiment, in which 1 g. of *N*-phenyl-2-naphthylamine is added as an inhibitor into a mixture of 200 ml. of methyl methacrylate and 8 g. of poly(ethylene oxide), little increase in weight is found, as shown below.

Stirring time (min.):	Weight of polymer precipitated from 10 g. of solution (g.)
0	0.40
10	0.40
20	0.41
40	0.41
60	0.41
80	0.41

Therefore the increase in weight from 8 g. of starting poly(ethylene oxide) to 10.24 g. of product represents the amount of methyl methacrylate incorporated as block copolymer.

13. The checkers report that under the conditions described here pure methyl methacrylate does not homopolymerize.

3. References

1. Department of Chemistry, Faculty of Engineering, Osaka City University, Osaka, 558, Japan.
2. Wireless Research Laboratory, Matsushita Electric Industrial Company, Ltd., Osaka, Japan.
3. Department of Applied Chemistry, Osaka Institute of Technology, Osaka, 535, Japan.
4. F. E. Bailey, Jr., and R. W. Callard, *J. Appl. Polym. Sci.,* **1**, 56 (1959).
5. Y. Minoura, T. Kasuya, S. Kawamura, and A. Nakano, *J. Polym. Sci.,* **A2, 5,** 125 (1967).

Poly(phenolphthalein terephthalate)

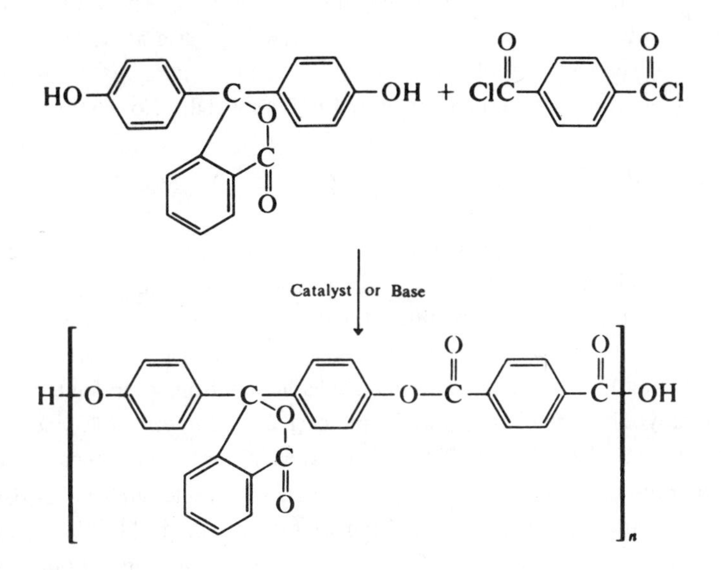

Submitted by P. W. Morgan (1)
Checked by E. Pearce (2a, b) and L. T. C. Lee (2a)

1. Procedure

A. Interfacial Method (3)

Phenolphthalein (0.01 mole; 3.18 g.) (Note 1), 0.80 g. of sodium hydroxide (Note 2), and 1.0 g. of tetraethylammonium chloride (Note 3) are dissolved in 100 ml. of water in a 1-quart blender jar (Note 4). The top of the jar is covered with aluminum foil and a corner is temporarily raised to permit introduction of the acid chloride. The blender is started at an intermediate speed controlled by a variable transformer. A solution of 2.03 g. (0.01 mole) of terephthaloyl chloride (*Lachrymator!*) (Note 5) in 30 ml. of distilled 1,2-dichloroethane (*Caution! Many halogenated solvents are toxic and should be used with care.*) contained in a 50-ml. Erlenmeyer flask is introduced into the blender jar rapidly and smoothly as the stirring speed is simultaneously increased to near maximum. The top is closed at once and stirring is continued for a minimum of 5 min. The initial blood-red color fades rapidly and should reduce to a tinge of pink. Hexane (250 ml.) is added to precipitate the polymer (Note 6), which is collected on a medium-poroisty sintered-glass funnel and washed repeatedly with distilled water by stirring vigorously in the blender and filtering. The washed granular polymer is dried in a vacuum oven in a shallow glass dish at 60-80° for 15-20 hr. (Note 7). The yield of polymer is about 95% and the inherent viscosity is 0.9-1.5 dl./g. (*sym*-tetrachloroethane-phenol, 40/60 by weight; Note 8).

B. Solution Method at Low Temperature (3)

Phenolphthalein (3.18 g.) is suspended in 25 ml. of 1,2-dichloroethane (Note 9) in a 250-ml. Erlenmeyer flask with a ground-glass joint and on which is mounted a 60-ml. graduated, stoppered addition funnel with a pressure-equalizing by-pass (Note 10). After a magnetic stirring bar is placed in the liquid, 2.80 ml. (0.02 mole) of distilled triethylamine is added from a graduated pipette. Terephthaloyl chloride (2.03 g.) in 25 ml. of 1,2-dichloroethane is placed in the addition funnel. The flask and funnel are then mounted in ice-water bath (a 125 × 65-mm. crystallizing dish) on a magnetic stirrer. Medium-speed stirring is started and, after 10 min. the addition of terephthaloyl chloride solution is begun at a rate that produces total addition in about 40 min. About 5 ml. of solvent is then added to the funnel to dissolve the acid chloride residue and this is added slowly over about 5 min. During the reaction the mixture in the flask remains colorless and become syrupy. There is, however, a precipitate of triethylamine hydrochloride. After the mixture is allowed to stand at room temperature for 1 hr. it is then diluted with 50 ml. of solvent and the dilute mixture is poured into 250 ml. of hexane in a beaker. The product is collected on a filter and then shredded and washed as in Procedure A. The yield is 4.34 g. The range of η_{inh} to be expected is 0.8 to 1.5 dl./g.

C. Solution Method at Elevated Temperature (3)

A three-necked, 250-ml., round-bottomed flask is equipped with a mechanical stirrer, a water condenser, and a 60-ml. graduated addition funnel with a pressure-equalizing by-pass. A source of dry nitrogen is connected to the top of the addition funnel, and 3.98 g. (0.0125 mole) of phenolphthalein, 0.20 g. of triethylamine hydrochloride (Note 11), and 25 ml. of distilled *o*-dichlorobenzene are placed in the flask. A solution of 2.564 g. (0.0125 mole) of terephthaloyl choride in 25 ml. of *o*-dichlorobenzene is placed in the addition funnel. After a slow flow of nitrogen is started and the stirrer is set at a moderate speed, the flask is heated with an oil bath (Note 12) to gentle reflux and the acid chloride solution is added over a period of about 30 min. The heating and stirring are continued until evolution of hydrogen chloride ceases (about 4 hr.). Some dichlorobenzene may be lost in the gas stream and should be replaced if the loss is excessive. At the end of reaction the flask attachments are disconnected and the syrupy polymer solution, allowed to cool, forms a soft gel which is dispersed in 250 ml. of hexane. The polymeric precipitate is collected by filtration and washed successively with ethanol, 50% aqueous ethanol, and water. Drying is done as in Procedure A. The yield is about 90% and the inherent viscosity is 1.0 dl./g. or above. The polymer may have a light yellow coloration not shown by the low-temperature preparations.

2. Characterization

The polymer from the low-temperature procedures is essentially colorless and forms bright, transparent, flexible films when solutions in chloroform, dichloroethane, 1,1,2-trichloroethane, or tetrahydrofuran are spread on glass plates. Thin films (0.1 mil) are used to obtain infrared spectra. There are two absorptions attributed to the carbonyl group at 1779 and 1739 cm.$^{-1}$ and no carboxyl or aliphatic OH bands. Solutions (1% by weight) of polymer (Procedure B with η_{inh} of 1.45 dl./g., Note 13) are made in reagent-grade chloroform and tetrahydrofuran at about 27°. When an equal volume, or even less, of the second solvent is added to each of the solutions or portions of the two solutions are mixed, a polymer precipitate forms. This demonstrates an antagonistic solubility effect (3, 4).

The polymer does not melt up to 400° on short-time contact with a metal surface (5). A film (ca. 8 × 2 × 0.001 in.) cast (5) from a 10% solution in tetrahydrofuran is air-dried on the glass plate, soaked in distilled water for 6 hr. and dried at 80° under vacuum. Strips about $^1/_8$ in. wide are cut with a pair of razor blades mounted side by side. A film strip sticks to a brass block at 318-320° and breaks under light load at 340° (6) when heating and contact are begun at about 300°. Thermogravimetric analysis under nitrogen with heating at 20°/min. shows the beginning of rapid weight loss at about 470°. The tensile strength/elongation of 1/8 -in. film strips measured on an "Instron" tester at 70°F. and 65% RH at 1-in. jaw separation and elongation rate of 100%/min. is about 4800 p.s.i./5.5% (average of three breaks).

3. Notes

1. The phenolphthalein is Baker and Adamson Reagent Grade sold by the General Chemicals Division of Allied Chemical Corporation.

2. The sodium hydroxide is conveniently handled and dispensed by preparing a standardized solution (ca. 0.20 g./ml.). The solution is stored in a bottle equipped with an automatic buret and protected from the air with a soda-lime tube. The bottle joint should have a Teflon® coating and the stopcock should be of Teflon®.

3. The tetraethylammonium chloride is Eastman White Label No. 3022. Polymer with lower molecular weight is obtained when the quaternary ammonium salt is omitted. Such salts, in wide variety, are effective polymerization accelerators (4, p. 349). The amount to be used to obtain maximum molecular weight in the polymer must be determined experimentally for each polymer-salt-solvent-stirring combination.

4. The Waring Blender is preferred because of the quality of mixing and minimum difficulty with leakage in the seals. If the stirring shaft does leak, it should be cleaned and lubricated with Celvacene medium-vacuum grease (Distillation Products Industries). The jar should be glass in four-leaf clover shape.

5. The 1,2-dichloroethane is Eastman White Label No. 132. Terephthaloyl chloride is purified by crystallization from dry hexane; in an alternate or sequential step it may be sublimed at reduced pressure. The melting point of pure acid chloride is 81.2°.

6. Acetone can be used as a precipitant but may reduce the yield by 5%. There is a fire hazard in the use of actone or hexane with an unsealed electric motor, but the blender motor can be sealed and blanketed with nitrogen as shown in Reference 4, p. 489. Alternatively, the precipitation can be done by pouring the blender contents into hexane in a beaker and stirring manually. After filtration the product is shredded in water in the blender.

7. To avoid condensation of moisture on the walls of the oven and to accelerate drying a slow bleed of air or nitrogen through the oven is desirable.

8. The inherent viscosity ($\eta_{inh} = [2.3 \log \eta_{rel.}]/c$) is determined at a concentration (c) of 0.5 g./100 ml. of solution and a temperature of 30°. This 30° temperature is convenient when ambient temperature is above 25°. Other

solvents may be suitable for viscosity determination. Suggested solvents are *sym*-tetrachloroethane, chloroform, cyclohexanone, 1,1,2-trichloroethane, and *m*-cresol. The viscosity numbers from the several solvents will differ.

9. Some gain in polymer viscosity may be attained by distillation of the solvent.

10. The stopcock is lubricated with Dow-Corning high-vacuum silicone grease.

11. The triethylamine hydrochloride is synthesized by passing anhydrous hydrogen chloride (*Hood!*) into a solution of pure triethylamine in dry ether. The salt is collect and dried under vacuum. It can be generated in the polymerization mixture by adding triethylamine (0.2 ml.) to *o*-dichlorobenzene with the phenolphthalem and passing in a small amount of hydrogen chloride or by introducing about 2 ml. of the acid chloride solution. Either step should be done *before* heating is started.

12. The level of the oil bath should be kept somewhat below the level of the stirred reaction mixture to minimize evaporation and caking of forming polymer on the walls of the flask.

13. Thermal and tensile properties may vary considerably with inherent viscosity, being lower in magnitude as the inherent viscosity is lower.

4. Other Preparative Variations

Other solvents for the low-temperature solution method have been examined (7). High-temperature polymerization of aromatic polyesters has been carried out without catalysts at temperatures up to 330° (8, 9) and with a variety of catalysts (10, 11).

5. References

1. Pioneering Research Division, Textile Fibers Department, E. I. du Pont de Nemours & Company, Wilmington, Delaware 19899.
2a. Allied Chemical Corporation, Morristown, New Jersey 07960.
2b. Current Address: Polytechnic Institute of New York, Brooklyn, New York, 11201.
3. P. W. Morgan, *J. Polym. Sci.*, **A2**, 437 (1964).
4. P. W. Morgan, *Condensation Polymers by Interfacial and Solution Methods*, Wiley-Interscience, New York, 1965, pp. 340, 342, 497, 507.
5. W. R. Sorenson and T. W. Campbell, *Preparative Methods of Polymer Chemistry*, 2nd ed., Wiley-Interscience, New York, 1968.
6. R. G. Beaman and F. B. Cramer, *J. Polym. Sci.*, **21**, 223 (1956).
7. V. V. Korshak, S. V. Vinogradova, and V. A. Vasnev, *Vysokomol. Soedin., Ser. A*, **10**,

1329 (1968) [*C. A.*, **69**, 36504f (1968)]; V. V. Korshak, V. A. Vasnev, S. V. Vinogradova, and T. I. Mitaishvili, *Vysokomol. Soedin., Ser. A.*, **10**, 2182 (1968) [*C. A.*, **70**, 12032v (1969)]; V. V. Korshak, S. V. Vinogradova, V. A. Vansev, and T. I. Mitaishvili, *Vysokomol. Soedin., Ser. A*, **11**, 81 (1969) [*C. A.*, **70**, 68813t (1969)].

8. F. F. Holub and S. W. Kantor (General Electric Company), U.S. Patent 3,160,603 (Dec. 8, 1964) [*C. A.*, **62**, 9257g (1965)].
9. V. V. Korshak, S. V. Vinogradova, P. M. Valetskii, and Yu. V. Mironov, *Vysokomol. Soedin.*, **3**, 66 (1961) [*C. A.*, **55**, 26506g (1961)]; V. V. Korshak, S. V. Vinogradova, G. L. Slonimskii, S. N. Salazkin, and A. A. Askadskii, *Vysokomol. Soedin.*, **8**, 548 (1966) [*C. A.*, **64**, 19789h (1966)]; V. V. Korshak, S. V. Vinogradova, P. M. Valetskii, and A. N. Bashakov, *Doklady Chem. USSR*, **174**, 516 (1967) [*C. A.*, **68**, 59909k (1968)].
10. M. Matzner and R. Barclay, *J. Appl. Polym. Sci.*, **9**, 3321 (1965).
11. R. M. Ismail, *Angew. Makromol. Chem.*, **8**, 99 (1969).

Block Copolymers of Styrene and Ethylene Oxide

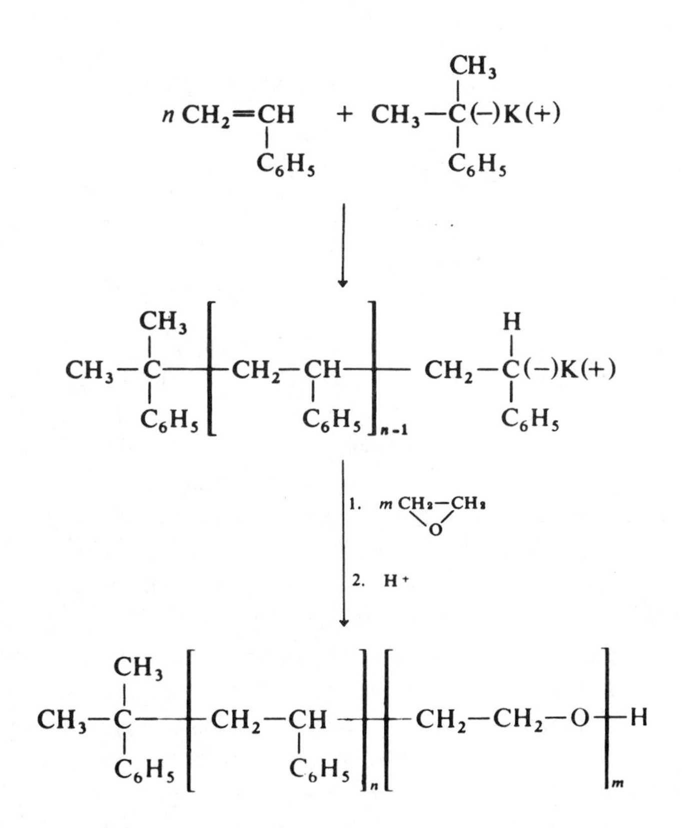

Submitted by J. J. O'Malley (1a, 1b,) and R. H. Marchessault (1a, c)
Checked by D. J. Worsfold (2)

1. Procedure

Syntehsis of polystyrylpotassium in tetrahydrofuran solution is carried out in the apparatus shown in Fig. 1. After the apparatus is carefully dried and evacuat-

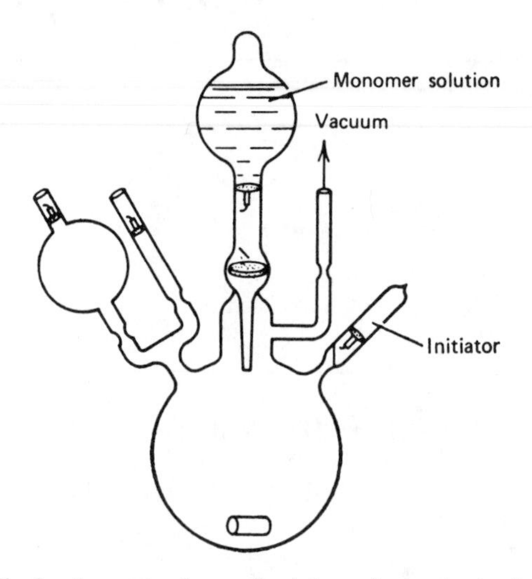

Fig. 1. Apparatus for synthesizing polystyrylpotassium.

ed to 10^{-5}-10^{-6} torr, it is removed from the vacuum line by sealing the constriction. A red cumylpotassium initiator solution (4.0 ml., 0.095 *N*) (Note 1) in tetrahydrofuran (Note 2) is added to the reaction flask *via* the breakseal which is broken with a glass-covered iron nail. Any initiator solution remaining in the ampoule can be washed into the flask by cooling the ampoule and condensing solvent in it. After stirring is started, the monomer bulb containing 7.6 g. (0.073 mole) of styrene (Note 3) in 38 ml. of tetrahydrofuran is sprayed into the initiator solution at 0° through a 0.5-mm. bore capillary tube. The nonterminating polymerization is completed almost instantaneously and the red polystyrylpotassium solution is transferred into ampoules, which are sealed from the apparatus at the constrictions and stored in the refrigerator until used (Note 4). The molecular weight of polystyrene prepared in this manner is 20,000 (5).

The polystyrene-poly(oxyethylene) block copolymer is prepared by adding purified ethylene oxide (Note 5) to a tetrahydrofuran solution of polystyrylpotassium, Ampoules of polystyrylpotassium (17 ml., 7.6×10^{-3} *N* in active end groups) and ethylene oxide (2.5 g., 0.057 mole) are attached to the apparatus shown in Fig. 2. After the apparatus has been dried and evacuated on the

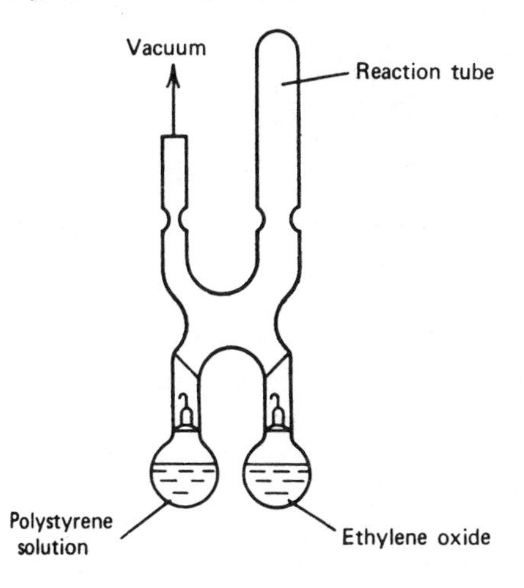

Fig. 2. Apparatus for synthesizing block copolymers.

vacuum line, it is removed by sealing the constriction. The polymer solution is then emptied into the thick-walled Pyrex tube (Note 6) and the tube is cooled to -78°. After the addition of the ethylene oxide the reaction tube is sealed from the apparatus and the contents shaken for a few minutes while the tube warms to ambient temperature. During this warming period the characteristic red color of the polystyrylpotassium disappears, an indication that copolymerization has started. To complete the ethylene oxide polymerization the tube is kept at 70° in a sand bath for 4 days (Note 7). At the end of this period the polymer solution (Note 8) is neutralized with acetic acid and the contents are isolated by precipitation in 10 volumes of heptane followed by centrifugation and decantation.

Copolymers prepared by the procedure described above contain varying amounts of both homopolymers because of transfer reactions; (Note 4) hence the polymeric product should be purified to remove unwanted materials. Polystyrene homopolymer is eliminated by dissolving the precipitate in benzene and titrating the solution with diethyl ether to precipitate the block copolymer and any poly(oxyethylene) homopolymer present. After the nonsolvent mixture has been decanted, the precipitate is dissolved in a little benzene and the solvent is distilled in a rotary evaporator to form a thin polymeric film on the walls of the round-bottomed flask. Repeated washing of this film with distilled water will remove salts and any homo-poly(oxyethylene) present (Note 10). The pure copolymer is dried by lyophilization of a 1% benzene solution. The yield of pure block copolymer is usually about 80%.

2. Characterization

The block copolymer contains 45% polystyrene and has a number average molecular weight of 46,000. The calculated molecular weight, based on the molecular weight of the homo-polystyrene (20,000) and the copolymer composition, is 44,000. The intrinsic viscosity of the copolymer in toluene at 35.0° is 0.33 dl./g.

The semicrystalline block copolymer exhibits a melting temperature of 59° [the poly(oxyethylene) block] and a glass transition temperature of 90° [the polystyrene block]—both measurements by differential scanning calorimetry.

3. Notes

1. Cumylpotassium is prepared by Ziegler's method (3) from methyl cumyl ether and sodium-potassium alloy. A course fritted-glass filter should be used to remove the insoluble sodium methoxide formed in this reaction. Finer filters may be used thereafter if deemed necessary. The concentration of cumylpotassium is determined by the addition of a known volume of the solution to distilled water, followed by titration with 0.01 *N* hydrochloric acid to the phenolphthalein end point.

2. Tetrahydrofuran is purified by heating under reflux for 24 hr. over potassium metal and collecting the middle fraction of the distillate at 66°. Sodium metal, benzophenone, and a stirring bar are then introduced into the distilled solvent which is degassed on the vacuum line. The solvent will gradually take on the purple color of the sodium benzophenone dianion when the tetrahydrofuran is dry. All solutions are prepared by distilling tetrahydrofuran from this reservoir.

3. Styrene is purified by two distillations on the vacuum line from calcium hydride. Residence time on calcium hydride between distillations is 24 hr. The pure monomer should be stored in a refrigerator and used as soon as possible.

4. Proton abstraction reactions, reported (4) to occur at room temperature, result in the formation of a 1,3-diphenylallyl anion at the end of the polymer chain.

5. Commercially pure (99.7%) ethylene oxide is purified by distilling three times at -40° over fresh calcium hydride under high vacuum. Because of its low boiling point (10.7°), ethylene oxide must be kept cold during these operations.

6. A Carius combustion tube is convenient for this purpose.

7. Pressure buildup caused by the ethylene oxide will occur and necessary precautions should be taken.

8. At high ethylene oxide concentrations the block copolymer solution may exist as an amber-colored gel at room temperature or below (6, 7).

9. The presence of homopolymer may be detected by GPC or density gradient ultracentrifugation.

10. Some block copolymer will be lost in this step, especially if the copolymer is rich in ethylene oxide. In the latter case other solvent-nonsolvent systems may be preferred, but the final choice is governed by the chain length of the respective sequences.

4. Methods of Preparation

The block copolymer described above is type A-B. Copolymer structures of the type A-B-A have been prepared (8, 9) with a difunctional anionic initiator, where B corresponds to polystyrene and A to poly(oxyethylene). Anionic polystyrene teminated with phosgene to form an acid chloride end group has been condensed with poly(oxyethylene) to synthesize copolymer structures of type B-A-B (10). If the dianionic polystyrene is used in the latter scheme, then $A{-}(B{-}A)_n$ structures should result. A comparable synthesis of styrene-isoprene block copolymers has been described (11).

5. References

1a. Chemistry Department, State University of New York, College of Forestry, Syracuse, New York, 13210.
1b. Current Address: Xerox Corpostion, Research Laboatories, Webster, New York 14580.
1c. Current Address: University of Montreal, Montreal Quebec, Canada.
2. Applied Chemistry Division, National Research Council. Ottawa, Canada.
3. K. Ziegler and H. Dislich, *Ber.*, **90**, 1107 (1957).
4. G. Spach, M. Levy, and M. Szwarc, *J. Chem. Soc.*, 355 (1962).
5. R. Waach, A. Rembaum, J. D. Coombes, and M. Szwarc, *J. Amer. Chem. Soc.*, 79, 2026 (1957).
6. A. Skoulios, G. Finaz, and J. Parrod, *Compt. Rend.*, **251**, 739 (1960).
7. J. J. O'Malley, Ph.D. Dissertation, State University of New York, College of Forestry, 1967.
8. D. H. Richards and M. Szwarc, *Trans. Faraday Soc.*, **56**, 1644 (1959).
9. M. Baer, *J. Polym. Sci.*, **A2**, 417 (1964).
10. G. Finaz, P. Rempp. and J. Parrod, *Bull. Soc. Chim., France*, 262 (1962).
11. J. Prud'Homme, J. E. L. Roovers, and S. Bywater, *European Polym. J.*, 8, 901 (1972).

Preparation of Isotactic Poly(1-butene)

$$\underset{\substack{|\\ CH_2CH_3}}{CH_2{=}CH} \xrightarrow[TiCl_3]{Et_2AlCl} -(CH_2-\underset{\substack{|\\ CH_2CH_3}}{CH})_n-$$

Submitted by W. Kern, (1a) H. Schnecko, (1a, 1b) W. Lintz, and L. Kollar (1a)
Checked by M. Goodman (2a, b) and T. Furuyama (2a, c)

1. Procedure

A. Apparatus (Fig. 1)

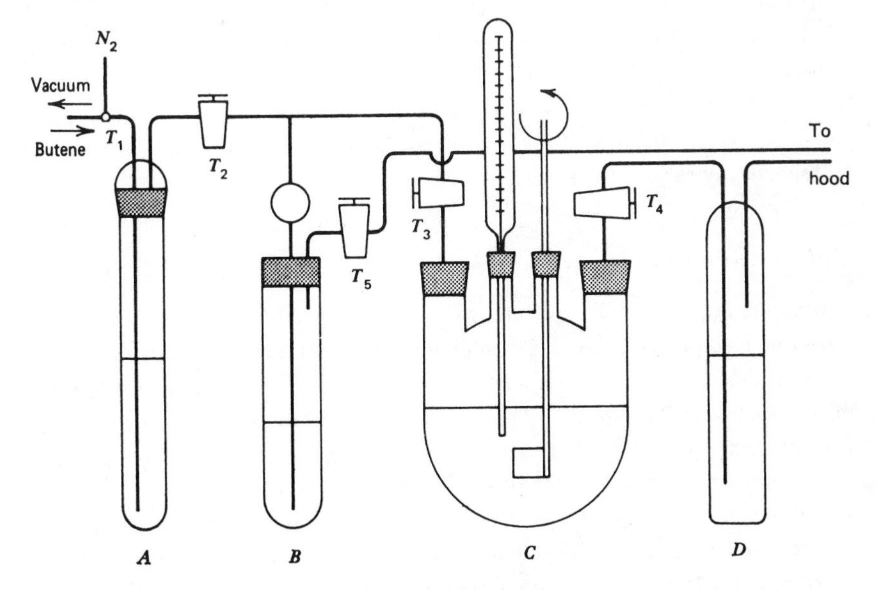

Fig. 1. Polymerization apparatus.

A 1-l., four-necked, round-bottomed flask *C* (Note 1), fitted with a stirrer, a thermometer, a gas inlet (ca. 0.8 mm. O.D.), and an outlet, is attached to a pressure control vessel *B*. The outlet at *B* is connected by tubing to an efficient hood and the inlet, to a gas wash bottle *A*.

A three-way stopcock T_1 at the beginning of the line allows evacuation of the system or filling with dry purified nitrogen. With stopcocks T_2 and T_3 the pressure control vessel can be closed; T_4 and T_5 permit closing of the whole system. Another gas flow control vessel *D* is behind flask *C*.

B. Preparation of Reaction System

The cleaned and dried (Note 1) empty system is closed by stopcocks T_4 and T_5, followed by three consecutive evacuations and fillings with nitrogen via stopcock T_1. Stopcocks T_4 and T_5 are then opened and with nitrogen passing through the system, vessels *A, B, D* and flask *C* are half filled with dry ligroin (Note 2), ca. 50 ml. for vessels *A, B, D,* respectively, and ca. 500 ml. for flask *C*. To the wash bottle *A*, 5 ml. of diethylaluminum chloride is added to make a wash solution for the monomer. At stopcock T_1 the vacuum pump is replaced by a line from a tank of 1-butene (Note 3), and the monomer is passed through the whole system to saturate the ligroin; simultaneously, flask C is heated to 60° and stirring (600-800 r.p.m.) is started. After stopcock T_4 (excess monomer escapes via vessel *B*) is closed, 20 ml. of diethylaluminum chloride (160 mmoles) and 25 ml. of a ligroin suspension containing 2.5 g. of titanium trichloride (16.2 mmoles) (*Caution!*) (Note 4) are added through the thermometer neck. The order of catalyst addition should not be reversed, because the excess of the aluminum compound serves partly to remove any solvent impurities.

C. Polymerization

Monomer uptake will start immediately after catalyst addition and can be monitored at pressure vessel *B*; preferably, the 1-butene flow should be regulated so that only a minimum amount escapes from *B*. Depending on the purity of the components, there will be a more or less sharp rate increase in the first 30 min., thereafter the rate will be constant or drop slightly; the polymerization is quenched after 1.5 hr. by addition of 30 ml. of *n*-butanol which decolorizes the brown mixture and stops further monomer uptake; the reaction mixture is allowed to cool slowly to room temperature without stirring and left for several hours or overnight. Then another 300 ml. of an *n*-butanol-methanol mixture (2:1) is added and the precipitated polymer is collected by suction-filtration, washed thoroughly with methanol, dilute hydrochloric acid, and water, and dried in vacuo at 50°; 55-65 g. of polymer is obtained. Poly(1-butene) is a white crystalline material with a melting range of 110-130°, depending on the steric purity (Note 5). It is partly soluble in carbon tetrachloride; for viscosity measurements aromatic solvents are used at elevated temperatures; e.g., xylene or tetralin above 80°. An $[\eta] - \overline{M}_w$ relationship (3)

$$[\eta] = 9.49 \times 10^{-5} \cdot \overline{M}_w^{0.73}$$

exists for poly(1-butene) at 115° that can also be applied at 135° to a first approximation (4). For stabilization an antioxidant (e.g., 0.2% by weight of the polymer) of *N*-phenyl-2-naphthylamine is added. Intrinsic viscosities (135°, decalin) are 1.5-1.8 dl./g. Films for infrared and x-ray spectra may be obtained from carbon tetrachloride solutions or by pressing the bulk polymer at 100°.

2. Notes

1. A three-necked flask is sufficient, if one ground joint is adapted for gas inlet and outlet. All glass parts should be well cleaned and flamed.
2. A high-boiling ligroin fraction (e.g., 100-130°) is preferred; it should be distilled in a nitrogen atmosphere over sodium.
3. The purity of the monomer should be in the range of 95%; a certain amount of saturated gaseous hydrocarbons will not be detrimental, but polar impurities (e.g., carbonyl- or hydroxyl-containing compounds) may result in a decreased polymerization activity.
4. Organoaluminum compounds react vigorously with mositure and air and can be handled only in an inert gas atmosphere (nitrogen, argon). For this reason water should not be used as a thermostat liquid. STAUFFER titanium trichloride-AA (supplied by the Stauffer Chemical Company, Weston, Michigan, or 380 Madison Avenue, New York, New York 10017) is a particularly active catalyst component. Because it is easily hydrolyzed, it is advisable to prepare a suspension in ligroin under nitrogen or argon. It can be used several times and is easily handled. About 10 g. of titanium trichloride is placed in a dry, nitrogen-purged 250-ml., three-necked flask fitted with an efficient stirrer and gas inlet and outlet. Dry purified ligroin (100 ml.) is then added. Before use this mixture is stirred rapidly to produce a uniform suspension.

An accurate determination of active Ti^{3+} can be made by titration with Ce^{4+} solution. For this purpose 5 ml. of the suspension is added to 50 ml.

of 1 *N* sulfuric acid in an Erlenmeyer flask which is kept under a carbon dioxide atmosphere by the addition of small pieces of solid carbon dioxide. This two-phase system is titrated with aqueous 0.1 *N* ceric sulfate solution with 2-3 drops of a diphenylamine solution in sulfuric acid (430 mg./50 ml. of conc. sulfuric acid) as an indicator. The colorless mixture becomes blue with excess Ce^{4+}.

5. X-ray powder spectra are indicative of a highly crystalline structure. Poly(1-butene), however, shows polymorphism, three modifications being possible. The individual spectrum depends on the method of isolation as well as the age, history, and treatment of the sample (6).

3. Methods of Preparation

Poly(1-butene) may also be prepared with the catalyst systems titanium trichloride-triisobutylaluminum (7) or titanium tetrachloride-triethyl aluminum. (8). Because titanium tetrachloride is a liquid, it permits easier handling and more accurate dosage. The polymers, however, are less isotactic than those prepared with titanium trichloride-diethylaluminum chloride. In all cases the rate of polymerization can be followed more accurately by using gas burets to monitor the monomer supply (5). For a description of the industrial process developed by Chemische Werke Hüls, Germany to manufacture poly (1-butene), see ref. (9).

4. References

1a. Organisch-Chemisches Institut, Universität, Mainz, Germany.
1b. Current Address: Dunlop AG, 6450 Hanau, West Germany.
2. Department of Chemistry, Polytechnic Institute of Brooklyn, Brooklyn, New York 11201.
2b. Current Address: The University of California/San Diego, La Jolla, California 92037.
2c. Current Address: Asahi Industries America, New York, N.Y. 10001.
3. S. S. Stivala, R. J. Valles, and D. W. Lewi, *J. Appl. Polym. Sci.,* 7, 97 (1963).
4. W. Lintz, unpublished results.
5. H. Schnecko, M. Reinmöller, K. Weirauch, W. Lintz, and W. Kern, *Makromol. Chem.,* 69, 105 (1963).
6. F. Danusso, G. Gianotti, and G. Polizotti, *Makromol. Chem.,* **80**, 13 (1965); C. Ceacintov, R. B. Miles, and H. Schuurmans, *J. Polym. Sci.,* **C14**, 283 (1966).
7. M. H. Jones and M. P. Thorne, *Can. J. Chem.,* **40**, 1510 (1962).
8. A. I. Medalia, A. Orzechowsky, J. A. Trinchera, and J. P. Morley, *J. Polym. Sci.,* **41**, 241 (1959).
9. M. D. Rosenzweig, *Chem. Eng.* 8, 56 (1973).

Poly[2,2'-(m-phenylenedioxyl)diethanol -bisphenol A carbonate]

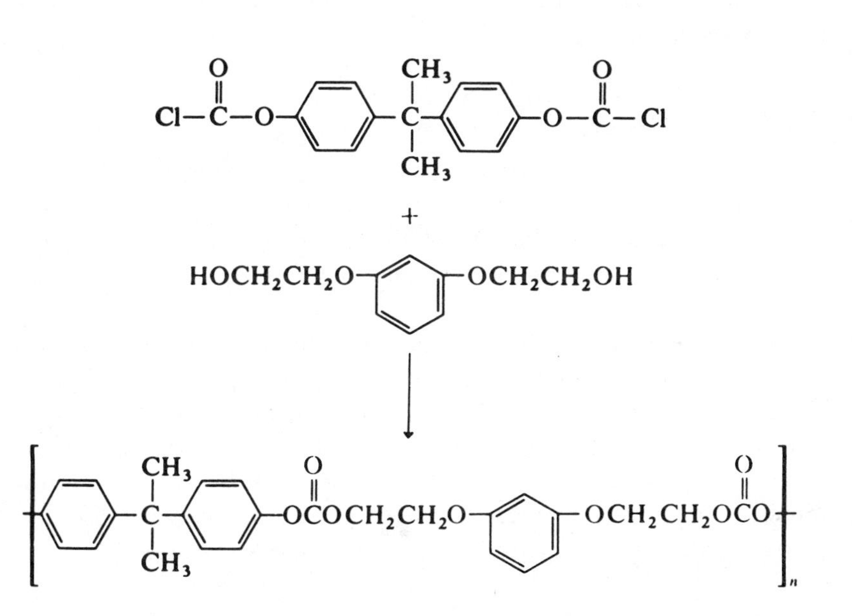

Submitted by T. Sulzberg (1)
Checked by Z. G. Gardlund (2)

1. Procedure

A. Pyridine Catalysis in 1,2-Dichloroethane

Caution! Halogenated materials may be toxic and should be handled with care.

Bisphenol A dichloroformate (35.32 g., 0.10 mole) (Note 1), 2,2'-(*m*-phenylenedioxy)diethanol (19.82 g., 0.10 mole) (Note 2), and 400 ml. of anhydrous 1,2-dichloroethane (Note 3) are placed in a 1-1., four-necked flask equipped with a mechanical stirrer, a condenser, a gas inlet tube, a drying tube, and a serum stopper. The mixture is purged with nitrogen for 10 min. and then heated to 65-70°. Pyridine (31.6 g., 0.40 mole, 32 ml.) is added all at once. The mixture begins to reflux almost immediately and within 1 min. a viscosity increase is noted. After 30 min. under reflux, the viscous solution is cooled, coagulated by pouring it into 2.5 1. of methanol in an explosion-proof high-speed blender, and the resulting solid is collected by filtration. The polymer is washed with two 1.5-1. portions of water in the blender, collected, and dried in an oven for 24 hr. at 75° under reduced pressure (approx. 30 torr) The yield of polycarbonate is 45.2 g. (94%) with an inherent viscosity of 0.64 dl./g. (Note 4).

B. 3-Picoline Catalysis in Toluene

A mixture of 35.32 g. (0.10 mole) of bisphenol A dichloroformate, 19.82 g. (0.10 mole) of 2,2′-(*m*-phenylenedioxy)diethanol, 0.93 g. (0.01 mole) of 3-picoline (Note 5), and 400 ml. of anhydrous toluene is placed in a 1-1., three-necked flask equipped with a mechanical stirrer, a condenser, a gas inlet tube (directly into the liquid), and a drying tube. After the flask is purged with nitrogen for 10 min., the reaction mixture is heated to reflux. Rapid evolution of hydrogen chloride (*Hood!*) takes place during the initial stages of heating but is gradually diminishes to nothing after 20 hr. The hot, pale yellow, viscous polymer solution is coagulated by pouring it into 2 1. of methanol in an explosion-proof blender, and the resulting solid is collected by filtration. By the same procedure described above 44.6 g. (93%) of the colorless polycarbonate, with an inherent viscosity of 0.59 (Note 4) is obtained.

2. Notes

1. Bisphenol A dichloroformate, m.p. 95-96°, is prepared from bisphenol A and phosgene as reported previously (3) and recrystallized from *n*-heptane.
2. The 2,2′-(*m*-phenylenedioxy)diethanol, obtained from Matheson, Coleman, and Bell, is recrystallized from a benzene-ethanol mixture (100 g. of compound/ 500 ml. of benzene plus 175 ml. of ethanol) and melts at 88°.
3. The 1,2-dichloroethane was obtained from Matheson, Coleman and Bell and dried with molecular sieves (Linde, type 5A).
4. The inherent viscosity is determined in chloroform at 25° with 0.2 g./100 ml.

5. The 3-picoline was obtained from Matheson, Coleman, and Bell and dried with molecular sieves (Linde, type 5A).

3. Merits of the Preparations

These procedures are also generally applicable, with little or no modification, to the preparation of aryl and alkyl polycarbonates, polyesters, polyamides, and polyurethanes. The important advantages of these methods are their utility for preparations in which one of the monomers is sensitive to high temperature, long reaction time, or an excess of base catalyst.

Method A combines an excellent polymer solvent, low temperature, and very fast reaction time in one procedure. This eliminates unwanted side reactions which cause gel formation, discoloration, termination, and low molecular weight. The rate of reaction approaches that of interfacial polymerizations.

Method B is important in preparations in which one of the monomers might react or complex with an excess of base. Pyridine could not be used because of the volatility of pyridine hydrochloride.

4. Methods of Preparation

Polycarbonates of this nature can also be prepared by ester exchange form the diol and diphenyl carbonate.

5. References

1. Union Carbide Corporation, Chemicals and Plastics, Bound Brook, New Jersey 08805; Current Address: Sun Chemical Corporation, Corporate Research Laboratories, Carlstadt, New Jersey 07072.
2. Polymers Department, Research Laboratories, General Motors Technical Center, Warren, Michigan 48090.
3. H. Schnell and L. Bottenbruch, *Makromol. Chem.*, 57, 1 (1962).

High Molecular Weight Atactic and Isotactic Polyepichlorohydrin

$$\underset{\diagdown O \diagup}{CH_2-CH}-CH_2Cl \xrightarrow[Et_3Al-0.6\,H_2O]{} (-CH_2-\underset{\underset{CH_2Cl}{|}}{CH}-O-)_n$$

Submitted by E. J. Vandenberg (1)
Checked by M. L. Senyek (2)

1. Procedure

A. Triethylaluminum-0.6 Water Catalyst

A 1-1., three-necked, round-bottomed flask is fitted with a stirrer (half-moon Teflon® blade, Trubore bearing, about 600 r.p.m.), a thermometer, a nitrogen inlet (over the surface of the liquid), a nitrogen outlet through a reflux condenser to a bubbler containing *n*-heptane, a rubber serum stopper for injecting ingredients with hypodermic equipment, and a heating mantle (Note 1). The air in the flask is replaced with nitrogen by passing a fast stream of nitrogen through it for 30 min. Then 385 g. of C.P. anhydrous diethyl ether (Note 2) is added, and the nitrogen sweeping is continued for 15 min. While stirring is continued and a slow stream of nitrogen is used, 261 ml. (186 g., 0.4 mole) of 1.53 *M* triethylaluminum in *n*-heptane (*Caution!*) (Note 3) is added dropwise for 30 to 60 min. with hypodermic equipment (Note 4), the temperature being allowed to rise until reflux. Then 4.3 ml. of water is added dropwise for 2 hr. (gas evolution occurs immediately after each water addition and then stops) at reflux (36-38°). Refluxing is continued by external heating for 2 hr. After the contents of the flask are cooled to 20°, they are transferred to a nitrogen-filled, 28-oz. pressure bottle fitted with a crown cap having two

1/8-in. holes and a Buna N self-sealing liner (Note 5). The transfer is accomplished by connecting the flask and the 28-oz. bottle with a long length of hypodermic tubing (18G) and then applying vacuum to the bottle (Note 6). When the transfer is completed, the bottle is pressurized with nitrogen to 15 p.s.i.g. (Note 7). This clear catalyst solution is stable to room-temperature storage. It is 0.58 *M* with respect to aluminum.

B. Polymerization

A 1-1., five-necked resin kettle is equipped in the same manner as the catalyst preparation flask, except that a stainless steel anchor stirrer (about 200 r.p.m.) and a Dry Ice condenser are used. The reactor is placed in a water bath at 25-30°. After the air is displaced with nitrogen as described above (except that the nitrogen passes through a Dry Ice trap before entering the reactor), 125 g. of commercial epichlorohydrin (*Caution!*) (Note 8) and 550 ml. of C.P. absolute diethyl ether (Note 2) are added and nitrogen sweeping is continued for 15 min. Then, while stirring and the introduction of a slow stream of nitrogen are continued, 22.0-ml. portions of the catalyst solution are added at 0, 0.5, 1.0, and 1.5 hr. While the reaction mixture is stirred for 13 hr. (Note 9), the water bath is adjusted to keep the reaction temperature at 30° ± 1°. The polymerization is stopped by the addition of 54 ml. of anhydrous ethanol. Most of the polymer is in the form of a slurry, although a small amount is on the walls of the reactor. After being stirred an additional 30 min. the reaction mixture, plus 200 ml. of ether, is filtered; the filter cake is washed twice by slurrying with 700 ml. of ether, followed by filtration (*Caution!*) (Note 10). The ether-insoluble material is slurried in an explosion-proof blender with 700 ml. of methanol, and the slurry is transferred to a beaker containing 80 ml. of 10% methanolic hydrochloric acid (2 1. of methanol plus 612 ml. of concentrated hydrochloric acid). After standing for 1.5 hr. the mixture is filtered and the solid is washed until neutral with methanol (four washes). The polymer is then slurried with 700 ml. of a 0.2% solution of Santonox (Note 11) in methanol, and the resulting slurry is filtered. The product is dried for 16 hr. at 80° under water-pump pressure to give 97 g. (78% conversion) of a white powder which is a mixture of high molecular weight atactic and isotactic polyepichlorohydrins (Note 12).

C. Separation

Atactic Polyepichlorohydrin. The crude product, plus 4 1. of commercial acetone, is agitated overnight at room temperature (in a stirred flask or in a bottle on a tumbler). The acetone-insoluble polymer is collected by filtration, agitated for 2 hr. with an equal amount of acetone and recollected by filtration. The combined acetone washes are concentrated under water-pump pressure in a rotating evaporator until viscous (about 440 ml.), and an equal volume of methanol containing 0.2% of Santonox is added. The insoluble material is collected, washed twice with the precipitant, and dried for 16 hr. at 80° under water-pump pressure to give 29 g. (30% of crude) of a tough rubber, η_{sp}/c of 1.9 dl./g. (0.1% in α-chloronaphthalene, 100°), $\overline{M}_w$ = 600,000 (Note 13), essentially completely amphorous by x-ray (Note 14). It is soluble in acetone, benzene, methylene chloride, and chloroform; insoluble in aliphatics and alcohols.

Isotactic Polyepichlorohydrin. The acetone-insoluble fraction obtained as described above and 7 l. of acetone are placed in a three-necked, 12-l. flask fitted as in the catalyst preparation. Then, under a nitrogen atmosphere and with stirring, the temperature is raised to reflux with a hot water bath (70°). After the polymer is dissolved (3 hr. of reflux), it is recrystallized by allowing the solution to cool spontaneously to room temperature during 16 hr. of stirring. The recrystallized polymer is collected by filtration on a Büchner funnel with a medium-weight cotton filter cloth and washed twice with acetone and once with acetone containing 0.05% Santonox. It is dried to constant weight (30 hr. at 80°) under water-pump pressure to give 26 g. (27% of crude) of crystalline isotactic polyepichlorohydrin, η_{sp}/c = 5.8 dl./g. (0.1% in α-chloronaphthalene, 100°), $\overline{M}_w$ = 2,000,000 (Note 13), m.p. 117° (Note 15). It is not soluble in any common solvent at room temperature.

2. Notes

1. Apiezon N lubricant is used for all joints, stopcocks, and stirrer. Rubber serum stoppers are available from Arthur H. Thomas Company, Philadelphia, Pennsylvania, Catalog No. 8826.
2. Mallinckrodt Analytical Reagent anhydrous ether may be used without further treatment if the container has been newly opened.
3. Extreme care must be exercised in handling triethylaluminum. This heptane solution (approximately 25%, from Texas-Alkyls, Inc.) is very reactive, although not pyrophoric. Contact with the skin must be avoided.
4. It is convenient to store aluminum alkyl solutions and reactive catalysts under nitrogen in 8-oz. or 28-oz. pressure bottles (Note 5) fitted with a crown cap having two 1/8-in. holes and a Buna N self-sealing liner (Note 6). Aliquots

are then withdrawn with an appropriate nitrogen-filled hypodermic syringe (10-100 ml.), fitted with a metal stopcock (No. FL/Z, Propper Manufacturing Company, Long Island City, New York) and hypodermic needle (20G), and are injected into the reaction flask through the rubber stopper. Larger quantities are more conveniently measured in a pressure bottle and then forced by nitrogen pressure into the flask through a stopcock (or needle valve such as No. 323, stainless steel, 1/8-in. N.P.T., Teflon packing, Hoke Inc., Englewood, New Jersey), fitted with a hypodermic needle on each end via an appropriate adapter.

5. Standard, crown-capped beverage bottles of the returnable type are satisfactory, if they are not exposed to more than 50° thermal shock. Scratched bottles should not be used.

6. The general procedure for using puncture-sealing gaskets has been described (3), particularly as a convenient method for polymerizing in the absence of air. In general, an 8-oz. pressure bottle (Note 5) is charged with relatively nonvolatile, air-stable solvents and/or reactants. The free space is then swept with nitrogen, and the bottle is capped with a metal cap having two 1/8-in. holes and fitted with a special Buna N rubber liner. The bottle is placed in a protective metal shield (Note 7). Air is further excluded by alternately applying vacuum and nitrogen pressure through a hypodermic needle (20G), which is inserted through the cap liner and attached by pressure tubing through a three-way stopcock to an oil pump and a source of nitrogen at 15 p.s.i.g. pressure. Vacuum is applied for 1 min. and then followed by nitrogen until pressure reaches 15 p.s.i.g. This procedure is repeated once. The evacuation time for 28-oz. bottles should be increased to 4 min. Molded Buna N liners from Firestone Rubber and Latex Products Company, 1620 S. 49 Street, Philadelphia, Pennsylvania, are satisfactory. They are extracted for 3 days in a Soxhlet extractor with benzene and dried to their original size.

7. For pressures at/or above 15 p.s.i.g. it is recommended that the bottles be provided with protective metal cans, which are conveniently made from an appropriate size of brass tubing drilled with enough ¼-in. holes near the top and bottom to permit water circulation when placed in the polymerization bath.

8. Epichlorohydrin is toxic and should be handled with rubber gloves, preferably in a good hood (4). Eastman Kodak White Label, Shell Chemical Corporation, or Union Carbide Corporation epichlorohydrin is satisfactory. Excessive water in the epichlorohydrin caused by improper storage will lead to low conversions.

9. A 7-hr. reaction time will give approximately 50% conversion to crude

product. Somewhat better conversion to a similar crude product should be obtained in 7 hr. by operating at reflux.

10. The ether washes contain about 3% conversion of low molecular weight polymer and some unreacted epichlorohydrin (use rubber gloves).

11. This is Monsanto Chemical Company's 4,4′-thiobis(6-*t*-butyl-*m*-cresol) antioxidant.

12. An alternative procedure, which gives essentially the same results, is to charge the ingredients (0.8 of the quantities given) into a 28-oz. capped beverage bottle with a self-sealing rubber liner, place the bottle in a protective metal shield (Note 7), replace the air with nitrogen (Note 6), equilibrate it in a 30° water bath on a rotating rack, inject the catalyst in the same manner, and tumble the vessel in the 30° bath for 19 hr.

13. These values are based on the following relationships between η_{sp}/c and $\overline{M}_w$(5) for either isotactic or atactic polymer:

$$\log \eta_{sp}/c = \log [\eta] + 0.15 [\eta] c$$

$$c = \text{concentration in g./100 ml.}$$

$$[\eta] = 8.93 \times 10^{-5} \overline{M}_w^{0.731}$$

The checkers obtained only 19% of the atactic fraction.

14. This fraction can be further purified by dissolving the polymer in acetone, as described in the isotactic polyepichlorohydrin section, followed by crystallization at -30° (16 hr.) and filtration. The acetone-soluble portion is then recovered by methanol precipitation as described for atactic polyepichlorohydrin.

15. The melting point was obtained by measuring the temperature at which birefringence disappeared on a sample that had been melted and then recrystallized by cooling to eliminate strain birefringence. The checkers, using differential thermal analysis, observed a m.p. of 120°.

3. Methods of Preparation

High molecular weight atactic polyepichlorohydrin has been reported only with the water-modified organometallic-type catalyst described here (6). Very low molecular weight (450 to 2000), ether- and methanol-soluble, liquid atactic polyepichlorohydrin has been sold commercially (7). Crystalline polyepichlorohydrin, apparently of substantially lower molecular weight than that described here but with the same x-ray pattern, has been prepared with a ferric chloride catalyst (8,9) and with an aluminum isopropoxide–zinc chloride catalyst (10). This crystalline polymer is presumably isotactic, because the catalysts used give isotactic polymer with propylene oxide. The structure has been investigated by x-ray diffraction studies (10).

4. References

1. Hercules Research Center, Wilmington, Delaware 19899; Hercules Research Center Contribution No. 1507.
2. Research Division, The Goodyear Tire & Rubber Company, Akron, Ohio 44316.
3. S. A. Harrison and E. R. Meincke, *Anal. Chem.*, **20**, 47 (1948).
4. Shell Chemical Company, 500 Fifth Avenue, New York, New York 10036, "Epichlorohydrin Booklet," p. 35.
5. R. Chiang, private communication based on light scattering studies at Hercules Research Center, Wilmington, Delaware 19899.
6. E. J. Vanderberg, *J. Polym. Sci.*, **47**, 486 (1960).
7. Dow Chemical Company and Shell Chemical Company, no longer available.
8. J. M. Baggett and M. E. Pruitt (Dow Chemical Company), U.S. Patent 2,871,219 (January 27, 1959) [*C.A.*, **53**, 8708c (1959)].
9. S. Ishida, *Bull. Chem. Soc. Japan*, **33**, 727 (1960).
10. J. R. Richards, Doctoral Dissertation, University of Pennsylvania, 1961.

Isotactic Poly(phenyl glycidyl ether)

$$\underset{\text{O}}{\overset{}{CH_2-CH}}-CH_2OC_6H_5 \xrightarrow{Et_3Al-0.6\ H_2O} (-CH_2-\underset{\substack{|\\ CH_2\\ |\\ OC_6H_5}}{CH}-O-)_n$$

Submitted by E. J. Vandenberg (1)
Checked by M. L. Senyek (2)

1. Procedure

A. Polymerization

This polymerization is conducted in a 1-1., five-necked resin kettle by the general procedure, triethylaluminum–0.6 water catalyst, and other reagents previously described (3). Diethyl ether (630 ml.) and phenyl glycidyl ether (200 g.) (Note 1) are added to the reactor. While this mixture is stirred under a slow stream of nitrogen, 34.6-ml. portions of the 0.58 *M* triethylaluminum–0.6 water catalyst are added at 0, 0.5, 1.0, and 1.5 hr. The reaction mixture is stirred for 5.5 hr., with the water bath adjusted to keep the reaction temperature at 31 ± 1°. The polymerization is stopped by the addition of 100 ml. of anhydrous ethanol. The polymer slurry, plus 300 ml. of ether, is filtered through a sintered-glass funnel. The filter cake is washed twice by reslurrying it with 1 1. of ether and refiltering the slurry. The product is further purified by slurrying it with 1 1. of methanol plus 25 ml. of concentrated hydrochloric acid. After 15 min. of standing the mixture is filtered and the solid is washed until neutral with methanol and then washed once with methanol containing 0.4% Santonox. It is then dried for 16 hr. at 80° under water-pump pressure to give 108 g. [54% conversion, 67% yield (Note 2)] of crude isotactic poly(phenyl glycidyl ether), η_{sp}/c = 2.2 dl./g. (0.1% in α-chloronaphthalene, 135°) (Notes 3-5).

B. Purification

Benzene Extraction. Crude isotactic poly(phenyl glycidyl ether) (42.0 g.) and C. P. benzene (1680 ml.) are mixed in a bottle and placed on a shaker overnight at room temperature. The benzene-insoluble material is collected and washed once with the same amount of benzene and once with benzene containing 0.05% Santonox antioxidant. It is dried for 16 hr. at 80° under water-pump pressure to give 40.5 g. (96.5% of crude) of purified polymer, η_{sp}/c = 2.2 dl./g. crystalline melting point 203°, % Santonox, 0.3 (Notes 4, 5, and 6).

Xylene Recrystallization. Crude isotactic poly(phenyl glycidyl ether) (62.1 g.) plus 3 1. of xylene (Note 7) are placed in a 5-1., three-necked flask fitted with a thermometer, a stirrer, a nitrogen inlet, and a nitrogen outlet through a reflux condenser. Under a nitrogen atmosphere the polymer slurry is heated to reflux (136°). After the polymer is dissolved (about 1 hr. of heating), it is recrystallized by allowing the solution to cool spontaneously to room temperature while being stirred for 16 hr. The crystallized polymer is collected on a Buchner funnel with an Orlon filter cloth. It is then washed once with 2 1. of xylene and once with 2 1. of benzene containing 0.03% Santonox. It is dried for 16 hr. at room temperature under water-pump pressure and then to constant weight at 80° under oil-pump pressure (approximately 28 hr.). This procedure gives 42 g. of a hard, tough, somewhat opaque solid (68% of crude), η_{sp}/c = 2.0, dl./g. crystalline melting point 195°C. and % Santonox, 0.6 (Notes 4, 5, and 6).

2. Notes

1. Phenyl glycidyl ether from Shell Chemical Company is satisfactory.
2. The yield is based on the sum of isolated crude isotactic polymer plus the atactic polymer in the combined ether filtrates, as determined by the total solids in an aliquot.
3. This preparation is also conveniently carried out in a capped pressure bottle fitted with a self-sealing liner by the general procedure previously described (4). In an 8-oz. bottle, into which 40 g. of phenyl glycidyl ether, 80 ml. of diethyl ether, and 28 ml. of catalyst are added all at once, 84%

conversion and 92% yield of crude isotactic poly(phenyl glycidyl ether), having essentially the same properties as described, are obtained in 19 hr. at 30°.

4. The checkers obtained reduced specific viscosities of 2.0, 1.9, and 1.6 dl./g. on the crude polymer, the benzene-extracted polymer, and the xylene-recrystallized polymer, respectively.

5. The melting points were determined by measuring the temperature at which birefringence disappeared on a sample which had been melted and then recrystallized to eliminate strain birefringence. The checkers, using differential thermal analysis, observed T_m 190° and T_g 9° for the benzene-extracted polymer and T_m 194° and T_g 9° for the xylene-recrystallized polymer.

6. The Santonox is determined by ultraviolet analysis.

7. Industrial xylene from Allied Chemical Corporation is satisfactory.

3. Methods of Preparation

Crystalline poly(phenyl glycidyl ether) was first prepared by Noshay and Price (5) who used an aluminum isopropoxide–zinc chloride catalyst or a triethylaluminum catalyst. The present method (6) produces much higher yields and higher molecular weights. This crystalline polymer is presumably isotactic, because the catalysts used give isotactic polymer with propylene oxide. The stereochemistry has not been verified.

4. References

1. Hercules Research Center, Wilmington, Delaware 19899; Hercules Research Center Contribution No. 1508.
2. Research Division, The Goodyear Tire & Rubber Company, Akron, Ohio 44316.
3. E. J. Vandenberg, *Macromol. Syn.*, this volume, p. 433.
4. E. J. Vandenberg, *Macromol. Syn.*, this volume, Note 6, p. 436.
5. A. Noshay and C. C. Price, *J. Polym. Sci.*, 34, 165 (1959).
6. E. J. Vandenberg, *J. Polym. Sci.*, 47, 486 (1960).

Poly(bisphenol A-4,4′-biphenyldisulfonate)

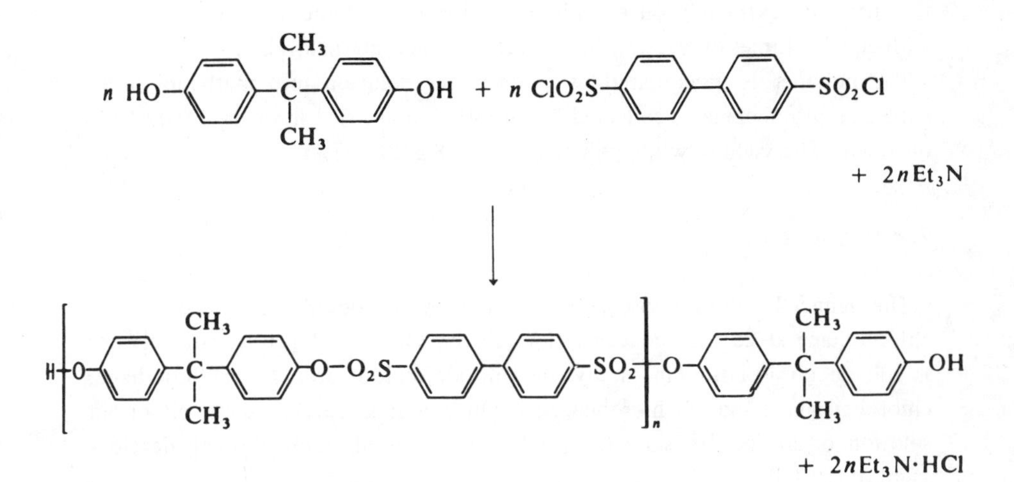

Submitted by R. J. Schlott, F. Scardiglia, E. P. Goldberg, and D. F. Hoeg (1)
Checked by G. B. Butler, (2a) C. Lawson Rogers, and A. J. Sharpe, Jr. (2a, b)

1. Procedure

Into a carefully dried, 100-ml. resin reaction flask fitted with a paddle stirrer, a thermometer, a dropping funnel, and a reflux condenser are placed 7.022 g. (0.0200 mole) of 4,4′-biphenyldisulfonyl chloride (BPDSC) (Note 1), 4.563 g. (0.0200 mole) of bisphenol A (Note 2), and 50 ml. of dry methylene chloride (*Caution! Many halogenated materials are toxic and should be used in an efficient hood.*)

The reactor is fitted with an inert-gas inlet tube at the top of the condenser to exclude moisture, and the stirrer is started to effect suspension of the insoluble solid monomers. Through the dropping funnel, 4.848 g. (0.0480 mole) of freshly dried triethylamine (Note 3) is added slowly over a 45-min. period. The exothermic reaction provides sufficient heat to maintain a gentle reflux of the methylene chloride at this rate of addition. As the amine addition proceeds, the bisphenol monomer dissolves and is followed by the BPDSC as the reaction progresses. Finally, near the end of the amine addition triethylamine hydrochloride separates as fine white needles.

After completion of the amine addition external heating is applied with a heating mantle to continue the refluxing for 1 hr. The viscous polymer solution is then cooled and poured into an equal volume of dilute (1%) aqueous hydrochloric acid. After the mixture is shaken, the organic layer is separated, extracted once with an additional volume of acid, and then with equal volumes of water until the water extracts are neutral.

The polymer is precipitated by pouring the solution into methanol in an explosion-proof blender, collected by filtration, and dried in vacuo at 100-120° overnight. The yield of white polymer is 9.3-9.8 g. (92-97%).

2. Characterization

The reduced solution viscosity of the polymer determined in sym-tetrachloroethane at 25° at a concentration of 0.1 g./dl. is 0.6-1.1.dl./g. The polymer is soluble on isolation in methylene chloride, chloroform, tetrachloroethane, chlorobenzene, and dichlorobenzene. On standing several days either in solution or in the dry state the product becomes insoluble through development of crystalline character.

Colorless transparent films may be pressed between aluminum foils at 260-300°. These films have a tensile yield strength of 9600 p.s.i., an ultimate tensile strength of 9820 p.s.i., and an elongation of 12%. The heat distortion temperature is 175° at 10-mil deflection under 264 p.s.i. load. Molded bars are quite resistant toward hydrolysis, losing <0.3 wt. % on treatment in refluxing 10% aqueous caustic for 20 hr. Cast films from chlorinated hydrocarbons are nearly opaque and quite brittle. These exhibit crystalline domains and show the typical "Maltese cross" effect under the microscope between crossed polarizing elements.

3. Notes

1. The 4,4′-biphenyldisulfonyl chloride used was prepared in our laboratory by sulfonation of biphenyl with concentrated sulfuric acid, isolation of the

dipotassium salt, and subsequent chlorination with phosphorus oxychloride. The product, when crystallized from toluene, melts at 206-207°, to give a clear, colorless melt.

2. Bisphenol A (4,4′-isopropylidenediphenol) is readily available in high purity from commercial sources. It should be thoroughly dry and have a m.p. of 155.5-157°.

3. Triethylamine is most satisfactory if freshly distilled and stored over silica gel. Substantial amounts of color bodies and/or water lowers the molecular weight of the final product.

4. Methods of Preparation

Poly(bisphenol A-4,4′-biphenyldisulfonate) may be prepared by stirred interfacial condensation (3).

5. References

1. R. C. Ingersoll Research Center, Borg-Warner Corporation, Des Plaines, Illinois 60018.
2a. Department of Chemistry, Unviersity of Florida, Gainesville, Florida 32601.
2b. Current Address: Calgon Corp., Calgon Center, Pittsburgh, Pennsylvania 15320.
3. P. W. Morgan, *Condensation Polymers by Interfacial and Solution Methods,* Wiley-Interscience, New York, 1965, p. 500.

Bisphenol A-Polythiocarbonate

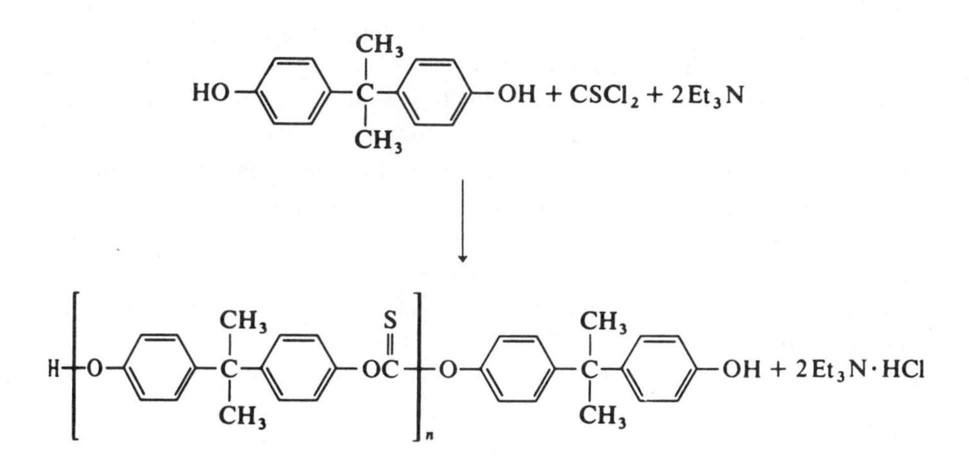

Submitted by R. J. Schlott, F. Scardiglia, E. P. Goldberg, and D. F. Hoeg (1)
Checked by G. B. Butler (2a) C. L. Rogers, and A. J. Sharpe (2a, b)

1. Procedure

A carefully dried, 250-ml. resin reaction flask, fitted with a paddle stirrer, a thermometer, a dropping funnel and a reflux condenser, is charged with 11.41 g. (0.050 mole) of bisphenol A and 100 ml. of dry methylene chloride. (*Caution! Many halogenated compounds are toxic and should be used in an efficient hood.*) To the stirred mixture, under a nitrogen blanket, are added 15.15 g. (0.15 mole) of dry triethylamine. The resulting solution is cooled to $-20°$ (Note 1) and held with stirring while 5.75 g. (0.050 mole) of thiophosgene (*Caution!*) (Note 2) in 50 ml. of dry methylene chloride is added through the addition funnel over a 1-hr. period.

The viscous light-yellow mixture is then warmed to room temperature and quenched by pouring into an equal volume of 1% aqueous hydrochloric acid.

The organic phase is separated and extracted with equal volumes of water until the aqueous phase is neutral to pH paper.

The polymer is coagulated by pouring the solution into rapidly stirred methanol in an explosion-proof blender and dried in vacuo at 100-120° overnight. The yield is 13.0-13.3 g. (96-99%).

2. Characterization

The reduced viscosity (η_{sp}/c) of the polymer determined in *sym*-tetrachloroethane at 25° at a concentration of 0.1 g./dl. is 0.3-0.8. dl./g. The polymer is soluble in chlorinated hydrocarbons, which include methylene chloride, chloroform, chlorobenzene, tetrachloroethane, and dichlorobenzene. Clear transparent films can be cast from these solvents, or the polymer may be pressed at 250-270° into tough films and bars. These specimens have ultimate tensile strengths of 9-10,000 p.s.i. and tensile elongations of 15-20%. The heat distortion temperature is 163° at 10-mil deflection under a 264 p.s.i. load. Though sensitive to amine hydrolysis in solution, molded thiocarbonate bars are quite resistant toward caustics, losing $<$0.2 wt. % on treatment in refluxing 10% aqueous sodium hdyroxide for 20 hr.

3. Notes

1. The temperature is important. At higher temperatures ($>$0°), low molecular weight dark products are obtained, and even at 0° the product is considerably darkened by the presence of reaction side products.

2. Thiophosgene was freshly distilled at 72-73° just before use. A good hood should be employed, because this monomer is reported to be toxic and has an offensive odor.

4. Methods of Preparation

Bisphenol A-polythiocarbonate has been prepared by stirred interfacial polycondensation (3).

5. References

1. R. C. Ingersoll Research Center, Borg-Warner Corporation, Des Plaines, Illinois 60018.

2a. Department of Chemistry, University of Florida, Gainesville, Florida 32601.
2b. Current Address: Calgon Corp., Calgon Center, Pittsburgh, Pennsylvania 15320.
3. P. W. Morgan, *J. Polym. Sci.,* **A2**, 432 (1964).

Polydimethylsiloxane Gum

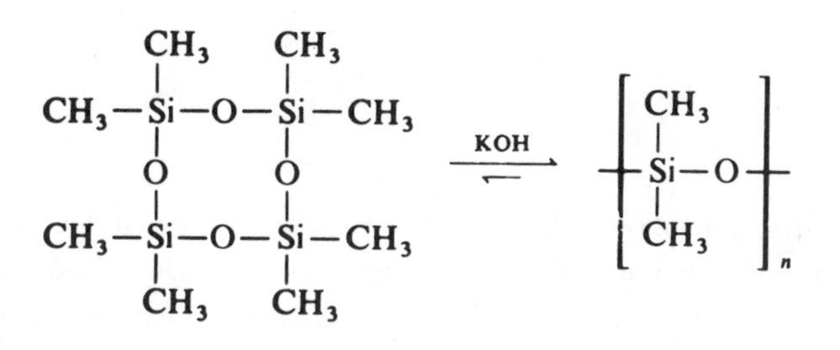

Submitted by R. A. Mansfield (1)
Checked by T. W. Brooks (2)

1. Procedure

In a 1-l. resin kettle, fitted with a mechanical stirrer (Note 1), a nitrogen inlet tube, and a distilling trap surmounted by a condenser protected with a drying tube, is placed 800 g. of octamethylcyclotetrasiloxane (Note 2). The vessel is placed in an oil bath maintained at 160° and the contents are rapidly stirred. Trace amounts of water that might be present in the starting material are removed from the reaction vessel by directing a moderate stream (about 30 ml./min.) of dry nitrogen onto the surface of the liquid for 1 hr. (Note 3). A pellet of U.S.P. potassium hydroxide is crushed and 0.08 g. is added to the kettle. To control the viscosity of the final product 0.10 ml. (0.085 g.) of decamethyltetrasiloxane, a chain stopper, is added (Note 4).

Within 3 hr. an increase in the viscosity of the liquid will become evident (Note 5). The maximum viscosity is reached about 2 hr. after this increase is first apparent (Note 6). At this stage the material is completely transparent, free of entrapped nitrogen bubbles, and is of a soft gumlike consistency. To ensure complete equilibration the reaction is allowed to proceed for an additional 2 hr.

The kettle is then removed from the oil bath and the mixture is allowed to cool to room temperature. An essentially quantitative yield of polymer with a voscosity between 3 and 7 million cP. is obtained (Note 7). This product contains the equilibrium amount of low molecular weight, volatile siloxanes (Note 8). If desired, they may be removed after the equilibration catalyst is deactivated. This is most conveniently accomplished by placing a 1-in. layer of the polymer in a shallow container and exposing it overnight to an atmosphere of carbon dioxide. When the resulting polymer is heated in a vacuum oven for 2 hr. at 135° (5 torr) (Note 9), 12% of the linear polymer-cyclic mixture is volatilized and the viscosity of the residue is increased by a factor of 1.7 (Note 10).

2. Notes

1. A motor capable of developing very high torque at low speeds is required. The checker reported that a stirring motor with a rated torque of 4.7 in.-lb. at 500 r.p.m. was stopped when the reaction mixture became viscous. An anchor-shaped, stainless steel stirrer is recommended.

2. The preparation (3) and properties (4) of octamethylcyclotetrasiloxane have been described. Minimum purity should be at least 99.8%. In view of the usual method of synthesis of the precursor (dimethyldichlorosilane) of this intermediate, it is possible that the impurities in this difunctional siloxane oligomer contain monofunctional or trifunctional siloxane moieties, which, even in trace amounts, will affect the molecular weight and the molecular weight distribution of the polymeric product. Their presence in larger quantities may adversely affect the physical properties of silicone compositions incorporating such polymers. The checker used redistilled octamethylcyclotetrasiloxane, b.p. 174-175.5°, n_D^{22} 1.3958. Gas chromatographic analysis of the monomer on a packed silicone SE-30 column showed only a single peak.

3. The dry nitrogen stream is continued throughout the polymerization to remove any water introduced with the catalyst and to prevent the entrance of water vapor or oxygen into the system.

4. Other linear siloxanes, such as hexamethyldisiloxane, and octamethyltrisiloxane, may be used to prepare the same type of polymer. Their preparation, and that of decamethytetrasiloxane, is described by Patnode and Wilcock (3).

5. This time interval is chiefly dependent on the efficiency of water removed and the degree of subdivision of the catalyst. A method of preparing a more finely divided potassium hydroxide catalyst is described by Osthoff and Grubb (5).

6. As the viscosity increases, the agitation rate should be reduced to prevent cavitation. The checker, using extremely pure monomer (Note 2), noted a marked increase in viscosity after 2 hr., which was followed within 15 min. by the attainment of a sufficiently high viscosity to stop the stirrer motor.

7. The final viscosity is dependent on the purity of the starting material (Note 2) and the degree of exclusion of water. Varying the amount of chain stopper changes the viscosity in a regular manner (6).

8. About 13% of the product boils at less than 250° (1 atm.). Approximately half of this is starting material, the remainder being chiefly the cyclic pentamer and hexamer (6).

9. The vacuum should be applied cautiously to prevent the viscous gum from "bumping".

10. The checker obtained a yield of 640 g. of clear, colorless gum with an intrinsic viscosity of 1.46 dl./g. in benzene at 25°.

3. Methods of Preparation

This method is essentially that of Hurd, Osthoff, and Corrin (7). Alternative decatalization procedures have been described (8,9). Catalysts that can be destroyed by heat, such as quaternary ammonium, and phosphonium hydroxides, have also been used (10). Reference to other basic catalysts, acidic catalysts, and other types of chain stoppers can be found in the articles already cited.

4. References

1. Silicone Products Department, General Electric Co., Waterford, New York 12188.
2. Peninsular ChemResearch Inc., Gainesville, Florida 32601.
3. W. Patnode and D. F. Wilcock, *J. Amer. Chem. Soc.*, **68**, 358 (1946).
4. R. C. Osthoff and W. T. Grubb, *J. Amer. Chem. Soc.*, **76**, 399 (1954).
5. R. C. Osthoff and W. T. Grubb, *J. Amer. Chem. Soc.*, **77**, 1405 (1955).
6. N. Kirk, *Ind. Eng. Chem.*, **51**, 515 (1959).
7. D. T. Hurd, R. C. Osthoff, and M. L. Corrin, *J. Amer. Chem. Soc.*, **76**, 249 (1954).
8. R. C. Osthoff, A. M. Bueche, and W. T. Grubb, *J. Amer. Chem. Soc.*, **76**, 4659 (1954).
9. R. G. Linville (General Electric Company), U.S. Patent 2,739,952 (Mar. 27, 1956) [*C.A.*, **50**, 9784g (1956)].
10. A. R. Gilbert and S. W. Kantor, *J. Polym. Sci.*, **40**, 35 (1959).

Preparation and Polymerization of a Cyclic Disulfide

$$ClC_2H_4OC_2H_4Cl + Na_2S_x \longrightarrow -\!\!\left(C_2H_4OC_2H_4S_x\right)\!\!-_n + 2NaCl$$

$$-\!\!\left(C_2H_4OC_2H_4S_2\right)\!\!-_n \rightleftharpoons O\begin{matrix} \diagup CH_2-CH_2-S \\ \diagdown CH_2-CH_2-S \end{matrix} \quad (I)$$

Submitted by F. O. Davis (1)
Checked by C. G. Overberger (2a) and Y. Shimokawa (2a, b)

1. Procedure

A. Preparation of a Cyclic Disulfide Monomer

In a 3-1., round-bottomed flask equipped with a stirrer, a reflux condenser, a dropping funnel, and a thermometer 972 ml. of a 3.09 *M* aqueous solution of $Na_2S_{4.27}$ (3.0 moles) is mixed at room temperature with 1 g. of sodium alkylnaphthalenesulfonate (Note 1), 8 g. of sodium hydroxide and 25 g. of magnesium chloride hexahydrate, all used as approximately 25 wt. % aqueous solutions. The mixture is heated to 85°, and 386 g. (2.7 moles) of bis(2-chloroethyl) ether (Note 2) is added dropwise to maintain the reaction temperature at 85~90° (Note 3). After complete addition the reaction mixture is heated at about 93° for 30 min. and then steam-distilled to collect 1 1. of distillate (Note 4).

The residue is washed once with water by decantation and dispersed in about 1 1. of water. Then 180 g. (4.5 moles) of sodium hydroxide is added and the resulting mixture heated for 1 hr. at 80~85° (Note 5). The latex is washed free of polysulfide solution with water by decantation (Note 6) and again steam-distilled. The distillate is a cloudy emulsion with droplets of a pale yellow, oily compound, the cyclic disulfide (I). Decantation gives approximately 5 g. of the cyclic disulfide per 500 ml. of distillate. This amount could be somewhat increased by extraction of the distillate with ethyl ether (Note 7). The cyclic disulfide is stable for years when completely dry.

B. Polymerization of a Cyclic Disulfide Monomer

1. To 100 ml. of a 0.5 *M* aqueous solution of sodium tetrasulfide (0.05 mole), 1 ml. of a 5 wt.-% aqueous solution of sodium alkylnaphthalenesulfonate, and 1 ml. of an aqueous dispersion of magnesium hydroxide, prepared by the interaction of 0.5 ml. of a 50 wt.-% aqueous solution of magnesium chloride hexahydrate and 0.5 ml. of a 20 wt.-% aqueous solution of sodium hydroxide, are added with stirring. The mixture is heated at 80~85°, and 10 g. (0.0735 mole) of the cyclic disulfide is added slowly with stirring. After complete addition the mixture is heated for 30 min. at 80~85° and then washed free of polysulfide solution with water by decantation. The resulting latex is coagulated by acidification to give the rubbery polymer.

2. The cyclic disulfide (10 g., 0.0735 mole) is mixed at room temperature with 0.2 ml. of a 25 wt.-% methanol solution of sodium methylate (Note 8) in a 20-ml. beaker. Polymerization begins immediately without heating and in several hours a rubbery polymer is obtained. The polymer thus obtained has η_{sp}/c = 0.2~0.4 dl./g. ($CHCl_3$, c = 0.5 g./dl.) (Note 9).

2. Notes

1. The sodium alkylnaphthalenesulfonate is available as NEKAL BX from GAF Corporation, 140 West 51st St., New York, New York.

2. Other cyclic disulfides may be made in variable quantities, as stated above, from linear polymers that have the following repeating units prepared from the corresponding dichlorides:

$$-\!\!(SC_2H_4SC_2H_4S)\!\!-$$
$$-\!\!(SC_2H_4OCH_2OC_2H_4S)\!\!-$$
$$-\!\!(SC_2H_4OC_2H_4OC_2H_4S)\!\!-$$
$$-\!\!(SCH_2CH_2CH_2S)\!\!-$$

3. The feed period is approximately 90 min. The polymer is produced as a dispersion in the reaction mixture.

4. The congeneric 1,4-thioxane formed in the reaction is removed by this steam distillation.

5. It is desirable to use a high sulfur content of sodium polysulfide, e.g., $Na_2S_{4.3}$, to obtain the maximum quantity of polymer. Lower sulfur content gives a greater yield of 1,4-thioxane. This treatment with sodium hydroxide removes much of the excess sulfur atoms above $-S_2-$.

6. The polymer may be examined at this point, if desired, by coagulation of the latex with dilute acid when free of polysulfide.

7. Although the rate of formation of the cyclic disulfide, as evidenced by the rate of removal in the steam distillation, is slow, the rate continues almost unchanged for a considerable period.

8. A variety of catalysts, including alkali alcoholates and alkaline sulfides (e.g., the sulfides, hydrosulfides and polysulfides of sodium, potassium, ammonium, calcium, and barium) may be used. Even water alone acts as a catalyst. The polymerization reaction is exothermic.

9. A small part of the polymer may not be soluble.

3. References

1. Thiokol Chemical Corporation, Bristol, Pennsylvania 19007.
2a. The University of Michigan, Ann Arbor, Michigan 48103.
2b. Current Address: Toyo Rayon Co., Pioneering R. & D. Laboratory, Ohtsu, Japan.
3. For further information see: (a) F. O. Davis (Reconstruction Finance Corp.), U.S. Pat. 2,657,198 (Oct. 27, 1953) [*C.A.*, 48, 4247g (1954)], and (b) F. O. Davis and E. M. Fettes, *J. Amer. Chem. Soc.*, 70, 2611 (1948).

Alternating Copolymer of Dimethylketene with Acetone

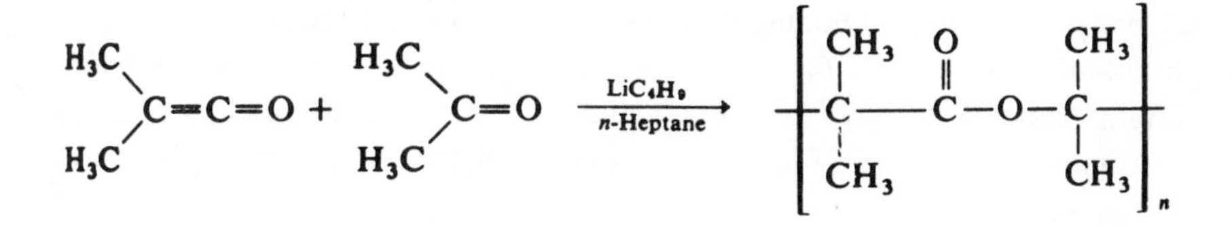

Submitted by G. Natta, G. F. Pregaglia, and M. Binaghi (1)

Checked by L. J. Fetters (2)

1. Procedure

Caution! In the presence of air, dimethylketene yields highly explosive peroxides. Vapors are harmful and irritating, and handling must be carried out in an efficient hood with suitable shielding. All apparatus that has come in contact with dimethylketene must be thoroughly washed with methanol soon after use.

A 100-ml., three-necked flask is fitted with a mechanical stirrer and a tube connected to a vacuum-nitrogen line. A nitrogen atmosphere is attained in the apparatus by repeated evacuation with vacuum release (Note 1). The apparatus is kept under a slight nitrogen overpressure. The flask is charged with 30 ml. of anhydrous *n*-heptane (Note 2) and cooled externally with a Dry Ice-saturated methanol bath. Freshly distilled dimethylketene (5 ml.) (Note 3) and anhydrous acetone (5 ml.) (Note 4) are added from a pipet connected to a syringe. The stirring of the solution is started soon after and is followed by the addition of 0.6 ml. of *n*-butyllithium (Note 5). The polymerization starts almost immediately with the evolution of heat and precipitation of the copolymer. After 2 hr. the reaction is terminated by adding 10 ml. of methanol

(Note 6), and the contents of the flask are transferred into a beaker containing 200 ml. of methanol. The polymer is crushed with a glass rod and allowed to stand for a few hours, then collected by filtration, washed on the filter with methanol, and dried under vacuum. The dry polymer is treated for 1 hr. with 250 ml. of refluxing *n*-heptane, collected by suction, and dried again *in vacuo*. The residual white powdery polymer weighs 4-5 g. (55-68% yield) (Note 7).

2. Characterization

Dimethylketene-acetone alternating copolymer, m.p. (capillary) 162-165°, is soluble in chloroform, boiling benzene and dioxane. Its intrinsic viscosity in phenol-tetrachloroethane (1:1) is 0.1 to 0.2 dl./g. The x-ray powder diffraction pattern shows a highly crystalline structure, and the infrared absorption spectrum obtained from a film cast on a sodium chloride disk from a chloroform solution shows a characteristic band at 1724 cm.$^{-1}$ (3).

3. Notes

1. Prepurified or lamp-grade nitrogen containing no more than 10 p.p.m. of oxygen or water should be used.
2. Reagent-grade *n*-heptane heated under reflux over sodium-potassium alloy and distilled under nitrogen, is used.
3. For synthesis, handling, storage, and purification of dimethylketene see (4).
4. Pure reagent-grade acetone, dried over anhydrous calcium sulfate and distilled under nitrogen, is used.
5. A commercial 15% solution (1.6 *M*) of *n*-butyllithium in *n*-hexane is used.
6. If the reaction mixture is still strongly yellow, the addition of methanol causes a vigorous reaction because of the presence of large amounts of unreacted dimethylketene. In this case methanol must be added very slowly form a dropping funnel. The presence of dimethylketene 1 hr. after initiation of the reaction indicates that the reactants, in particular dimethylketene, are not sufficiently pure. Dimethylketene can be further purified by distilling over trialkylaluminum.
7. The checkers also used a high vacuum apparatus and obtained a 75-80% yield of product, m.p. 167-171° (capillary), with an intrinsic viscosity of 0.9 to 1.2 dl./g.

4. Methods of Preparation

By methods very similar to that reported here, crystalline alternating copolymers can be obtained from dimethyl ketene with aliphatic (5-7) and aromatic (6) aldehydes and with alkyl formates (5).

5. References

1. Istituto di Chimica Industriale del Politecnico, Piazza Leonardo Da Vinci, 32, Milan, Italy.
2. Institute of Polymer Science, The University of Akron, Akron, Ohio 44304.
3. G. Natta, G. Mazzanti, G. Pregaglia, and M. Binaghi, *J. Amer. Chem. Soc.*, **82**, 5511 (1960).
4. G. F. Pregaglia and M. Binaghi, *Macromol. Syn.*, **3**, 152 (1969); Coll. Vol I, p. 367.
5. R. G. J. Miller, E. Nield, and A. Turner-Jones *Chem. Ind.* (London), 181 (1963).
6. G. Natta, G. Mazzanti, G. F. Pregaglia, and A. Pozzi, *J. Polym. Sci.*, **58**, 1201 (1962).
7. K. Hashimoto and H. Sumitomo, *Polym. J.* (Japan) **1**, 190 (1970).

Polystyrene with Predictable Molecular Weights and Uniform Molecular Weight Distributions

$$\underset{\displaystyle C_6H_5}{CH_2{=}CH} \xrightarrow[C_6H_6-C_4H_8O]{n\text{-}C_4H_9Li} \left[-CH_2-\overset{\displaystyle H}{\underset{\displaystyle C_6H_5}{C}}- \right]_n$$

Submitted by L. J. Fetters and M. Morton (1)
Checked by D. J. Worsfold (2)

1. Procedure

An apparatus is required with which it is possible to remove all impurities that can react with the organolithium initiator or the active chain-end of the polymer. Some impurities, i.e., water or peroxides, can be removed by conventional means. For the reaction presented here, however, even trace impurities must be eliminated. This is necessary because the total initiator concentration is usually in the range of 10^{-3} to 10^{-5} *M*. Hence the use of a high-vacuum line is necessary. A description of this apparatus is available elsewhere (3,4) along with a presentation of the basic high-vacuum techniques employed in these homogeneous, termination-free polymerizations.

A. Solvent Purification

Tetrahydrofuran. Tetrahydrofuran (Eastman Organic Chemicals) is placed

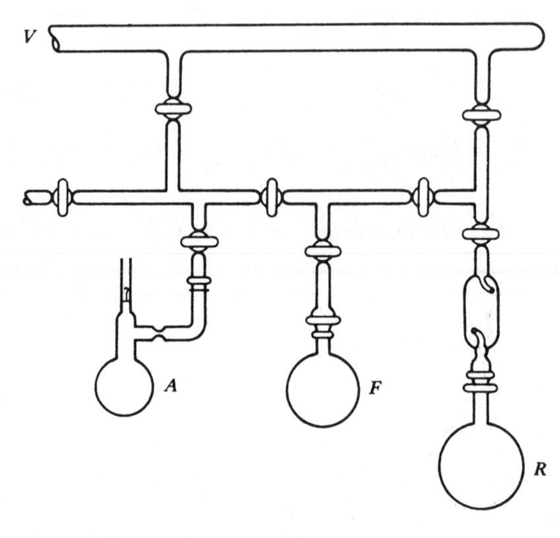

Fig. 1. Solvent purification apparatus.

in contact with pulverized calcium hydride (Flask *R*) and stirred for several days on the vacuum line (Fig. 1). After the tetrahydrofuran (THF) is degassed, it is distilled with the aid of a Dry Ice-alcohol bath into a 1-l. flask (*F*). This flask contains approximately ¼ g. of naphthalene (Eastman recrystallized) and several grams of sodium dispersion. The confluence of the ether with the sodium and naphthalene causes the formation of the dark green sodium-naphthalene complex. The appearance of this complex serves as a visual indication of the purity of the ether.

Measured quantities of tetrahydrofuran can be distilled, as needed, into evacuated and weighed ampoules *A*. Shielding tetrahydrofuran from sunlight or any ultraviolet source is practiced as a precaution against peroxide formation and other deleterious side-reactions.

Benzene. The apparatus used for the purification of benzene is similar to that illustrated in Fig. 1, and the following description refers to it. Benzene (Merck Analyzed Reagent Grade) is stored over sulfuric acid, with stirring, for 1 week. The solvent is then thoroughly degassed and flash-distilled to *R*, which contains sodium dispersion (ca. 2 g.). Stirring is then continued for several hours in flask *R*. Then, as needed, portions are distilled to a sodium-coated flask *F*. From flask *F*, the solvent is distilled to an identical flask containing *n*-butyllithium, and polystyryllithium. If desired, the benzene can be distilled into a measuring cylinder before collection in an ampoule *A* or distilled directly into the polymerization reactor.

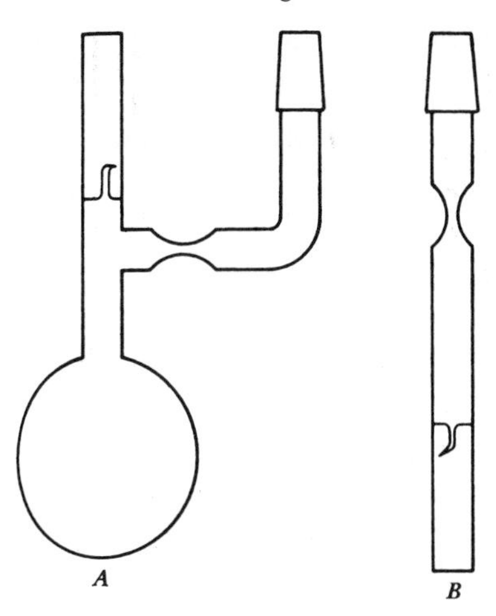

Fig. 2. Monomer-solvent ampoules.

Figure 2 depicts typical monomer-solvent ampoules. The capacity of these ampoules may be varied according to need. The breakseals (Eck and Krebs) are usually made from 12-mm. diameter tubing. A convenient size of ground glass joint is 14/35. Care must be taken, if "homemade" seals are used, that they are well annealed, thick enough to withstand thermal and slight mechanical shock, but thin enough to be broken at the proper time.

B. Monomer Purification

Styrene is purified in a vacuum apparatus: A flask *R*, Fig. 1, containing a slurry of finely ground calcium hydride in styrene is prepared. The mixture is then thoroughly evacuated and degassed. The contents (~100 to 150 ml.) are then agitated with a magnetic stirrer for several days with periodic degassing. During this time the viscosity of the mixture will usually increase slightly, an indication of the formation of a small amount of low molecular weight polymer. At the end of this treatment the styrene is distilled into a flask in which the monomer is brought into contact with several millimoles of hydrocarbon-soluble dialkylmagnesium, e.g., a mixture of di(*n*-butyl)magnesium and di(*sec*-butyl)magnesium (Lithium Corporation). This organometallic compound will purge the styrene of any residual impurities but, unlike organoalkali species, will not initiate the polymerization of styrene. After several hours exposure at room temperature the monomer can be collected into a weighed ampoule or a measured amount distilled directly into the polymerization reactor.

C. Preparation of Butyllithium Initiator Ampoules

The two most commonly used organolithium initiators are *n*-butyllithium or *sec*-butyllithium. The synthesis of *n*-butyllithium from di-*n*-butylmercury is presented elsewhere (5). In some respects *sec*-butyllithium is preferred, because this compound can be purified by a short-path high-vacuum distillation (6). Although *sec*-butyllithium decomposes slowly at room temperature [about 1.2% loss per week (7) at 25°], it can be kept indefinitely at ca. −10 to −20° Thus, if desired, commercial initiator can be used and freed from hydroxide and alkoxide by distillation. The commercial butyllithium can be transferred from the bottle to an ampoule in a dry-box. The ampoule is then placed on the vacuum system, frozen, and degassed.

The dilution of butyllithium can be effected as follows: the flask *B* containing the solution, is sealed onto a manifold containing approximately ten ampoules, as shown in Fig. 3. The manifold is then connected to a high-vacuum apparatus, evacuated, leak tested, and flamed in the usual manner. After the manifold is removed from the high-vacuum line by sealing at the constriction *A*, the breakseal of the solution is ruptured. Flask *B* is removed by sealing it at the constriction, which has been rinsed with refluxing solvent. The manifold is then inverted several times to achieve complete mixing, and is allowed to stand for approximately 24 hr. so that each constriction has sufficient time to drain free of solution. The refluxing of solvent onto the constrictions is not recommended, because distillation of solvent among the ampoules will cause fluctuations in the organolithium concentration.

Additional dilutions of the butyllithium solution can be performed as needed. Dilution of the initiator is preferably done in an apparatus, the walls of which are washed in the same way as described for the reaction vessel. An ampoule of solution is sealed to a flask equipped with a breakseal. An appropriate amount of *n*-hexane is distilled into the flask, and the organolithium solution is added. This solution is then subdivided in a manifold. The procedure is repeated until the desired concentration level is attained. Analysis of butyllithium content is then accomplished according to the method of Gilman (8,9) or by the thermometric titration method of Everson (10).

2. Polymerization Procedure

It is recognized (5,11) that butyllithium does not react rapidly with styrene in pure hydrocarbon media. The relatively slow rate of initiation is apparently the result of the tendency of the initiating organolithium and the

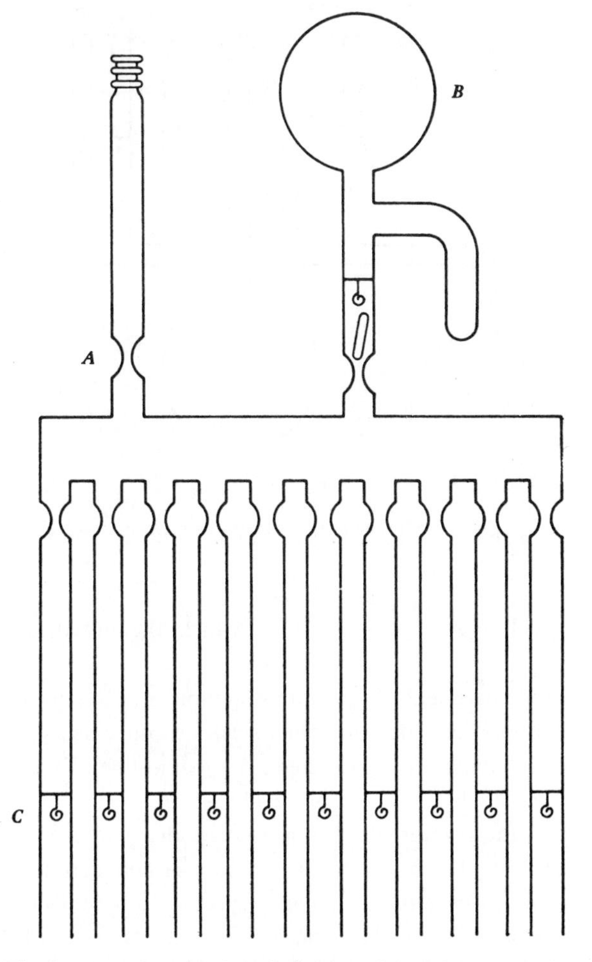

Fig. 3. Apparatus for the sub-division of the initiator solution.

polystyryllithium to form various inactive associated species. Kinetic evidence (11) indicates that the initiation reaction exhibits a low fractional order with repsect to the butyllithium in benzene, whereas the propagation reaction is one-half order in polystyryllithium. Hence the initiation and propagation processes are competitive. From a synthesis standpoint the simultaneous occurrence of the two reaction steps can lead to relatively broad molecular weight distributions, i.e., $\overline{M}_w/\overline{M}_n \simeq 1.3$ to 1.4 (12), when the molecular weight of the polymer is low. As described later, the addition of a small amount of

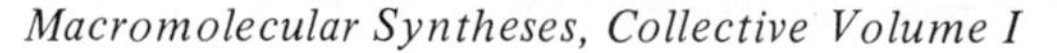

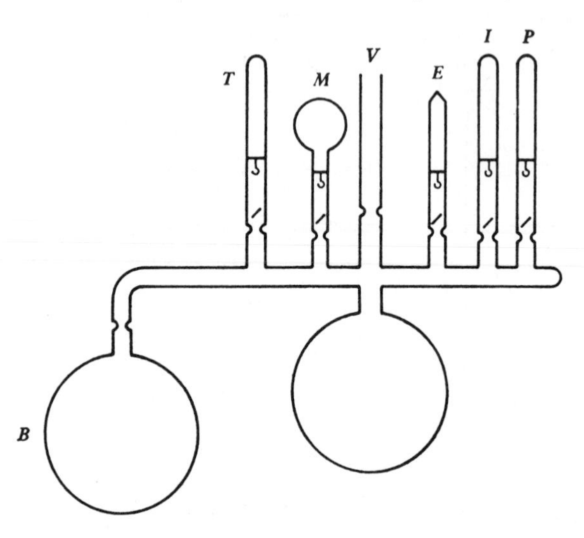

Fig. 4. Styrene polymerization apparatus.

tetrahydrofuran will serve as a method by which the initiation rate can be accelerated (13-15).

To perform an experiment, the various ampoules are sealed to a reactor (Fig. 4) so that the separate components can be conveniently added to the mixture while maintaining a closed system. A vacuum line (*V*), consisting of a standard-taper ground glass joint at one end and a constriction at the other, is also attached to the reactor vessel in a position that will evenly support the weight of the reactor. The vacuum line is the sole connection and the only support holding the reactor to the vacuum system. It is imperative therefore that the weight of the reaction flask and its components be well balanced. The repsective ampoules pictured contain monomer *M*, initiator, ether *E*, terminator *T*, and purge *P*. The terminator may consist of several drops of well-degassed water or methanol. The reactor is usually a 500- to 1000-ml. flask.

Once the reactor is assembled, as shown in Fig. 4, it is connected to the high-vacuum line and evacuated. After several minutes the reactor is tested with a Tesla coil for pinholes or cracks, especially in the areas on which the components are sealed. If no pinholes are found or after those that are found are sealed, the reactor is evacuated until a pressure of 10^{-6} torr is obtained. All the glass surfaces except the ampoules containing the reaction ingredients are flamed strongly with a yellow-blue flame of a hand torch. The reactor is continuously under evacuation during this procedure. After 1 hr. or so the vacuum is again checked and, if the pressure of 10^{-6} torr is attained, the reactor is ready for use.

The styrene and benzene may then be added to the reactor by distillation from graduated cylinders or from ampoules attached directly to the reactor. If the latter technique is used, the reactor is sealed from the vacuum line before the addition of the reactants.

The ether (THF) used to accelerate the initiation step is added with the styrene and benzene, but its concentration is 20 to 30 times greater than the initiator concentration. This ether concentration will cause the initiation reaction to become virtually instantaneous while not changing the propagation rate by any significant degree (13,14). Depending on temperature (0-30°) and the concentrations of tetrahydrofuran and butyllithium, complete conversion will generally be achieved in 30 min. to 8 hr. Usually the organolithium concentration will be 10^{-2} to 10^{-4} *M*.

The reactor, monomer, and solvent may then be simultaneously purged by the addition of several millimoles of the dialkylmagnesium solution, which is allowed to remain at room temperature for 1 hr. or so before it is transferred to the flask *B*. The remainder of the reactor is rinsed by back-distilling benzene and then pouring it into the purge flask *B*. This cycle is repeated four or five times, whereupon the solution is redistilled into the reactor and the purge flask is removed while the styrene-benzene mixture is frozen by a liquid nitrogen bath. By use of this procedure, with tetrahydrofuran as the solvent, a polystyrene sample with an $\overline{M}_w$ of 4.4 x 10^7 has been prepared (16) with only one addition of dilute initiator. This demonstrates that the dialkylmagnesium compound can eliminate virtually all terminating impurities.

The reaction mixture is then brought to the desired temperature (usually between 0° to 30°) and the initiator is introduced. When the initiator concentration is greater than 10^{-3} *M*, a reaction temperature of about 0° is recommended. This will prohibit the onset of an exotherm during polymerization. The initiator is then rinsed with refluxing solvent. The exact volume of initiator solution used can be determined at the end of the reaction by sealing the ruptured breakseal and titrating to the previous level.

The monomer charge is usually composed of 5-15% by volume of the mixture, with 5-20 g. being a typical amount. This, depending on the projected moelcular weight, can vary. The stoichiometric molecular weight is governed by the ratio of monomer to initiator, i.e.,

$$\overline{M}_s = \text{g. of monomer}/[\text{BuLi}]$$

at 100% conversion. Thus virtually any molecular weight starting from the dimer can be made by manipulating this ratio of monomer and initiator.

It is obvious that the destruction of a fraction of the butyllithium will result in a higher molecular weight than predicted, without, however, interfering with the attainment of a monodisperse molecular weight. The molecular weight distribution will, however, be markedly affected by any termination of the growing polymer chains. The only guarantee that a high degree of monodispersity has been attained can be obtained by a determination of the number and weight average molecular weights along with a gel permeation chromatogram. Intrinsic viscosity measurements, however, afford a simple means of determining whether the desired molecular weights have been obtained. The following equation (15) has proved to be valid over a molecular weight range of 0.05 to 150 x 10^4:

$$[\eta] = 8.4 \times 10^{-4} M_v^{0.50}$$

where the solvent is cyclohexane and the measurement temperature 34.8°.

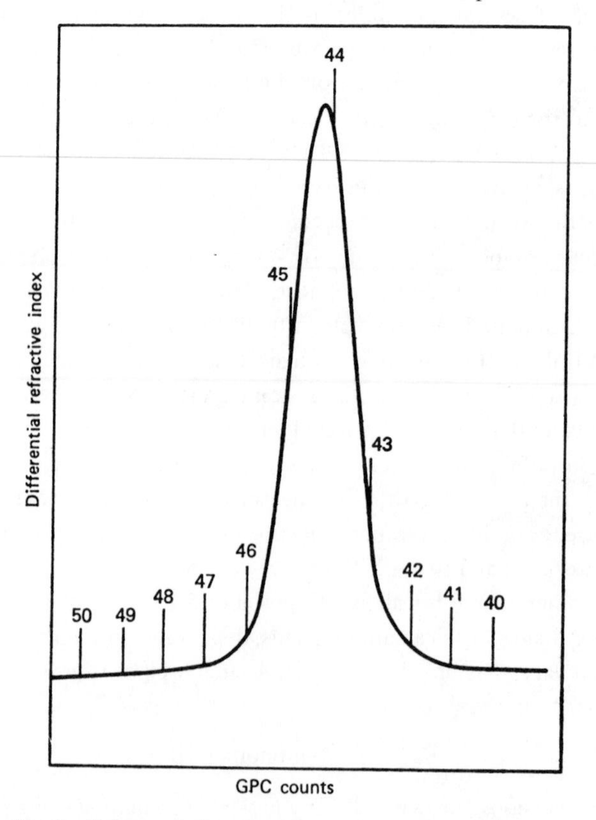

Fig. 5. Gel permeation chromatogram of polystyrene.

TABLE 1. EXPERIMENTAL DATA FOR STYRENE POLYMERIZATION AT 30°.

Monomer:	10.5 g.	$\overline{M}_s = 7.9 \times 10^4$ g./mole
Solvent:	250 cm³.	$\overline{M}_n = 8.0 \times 10^4$ g./mole
Initiator:	5.1×10^{-4} *M*	$\overline{M}_w = 8.1 \times 10^4$ g./mole
Tetrahydrofuran:	1.5×10^{-2} *M*	

In Table I a listing of the experimental parameters pertaining to a typical polymerization is presented. Figure 5 shows the gel permeation chromatogram for this polystyrene sample prepared by these procedures. The closeness of the molecular weight values in Table I and the sharpness of the GPC trace demonstrate the efficacy of these techniques. [Ed. Note: A comparable synthesis of a similarly prepared poly(isoprene) has been described (17).]

3. References

1. Department of Polymer Science, Akron University, Akron, Ohio 44325.
2. Division of Applied Chemistry, National Research Council of Canada, Ottawa, Canada.
3. M. Morton, R. Milkovich, D. McIntyre, and L. J. Bradley, *J. Polym. Sci.*, **A1**, 443 (1963).
4. L. J. Fetters, *J. Res. Nat. Bur. Stand.*, **70A**, 421 (1966).
5. M. Morton, A. Rembaum, and J. L. Hall, *J. Polym. Sci.*, **A1**, 426 (1963).
6. S. Bywater and D. J. Worsfold, *J. Organometal. Chem.*, **10**, 1 (1967).
7. R. O. Bach, C. W. Kamienski, and R. B. Ellestad, *Encyclopedia of Chemical Technology*, Vol. 12, Wiley-Interscience, New York, 1967, p. 529.
8. H. Gilman and A. H. Haubein, *J. Amer. Chem. Soc.*, **66**, 1515 (1944).
9. H. Gilman and F. K. Cartledge, *J. Organometal. Chem.*, **2**, 447 (1964).
10. W. L. Everson, *Anal. Chem.*, **36**, 854 (1964).
11. D. J. Worsfold and S. Bywater, *Can. J. Chem.*, **38**, 1891 (1960).
12. L. Gold, *J. Chem. Phys.*, **28**, 91 (1958).
13. S. Bywater and D. J. Worsfold, *Can. J. Chem.*, **40**, 1564 (1962).
14. J. F. Meier, Ph.D. Dissertation, University of Akron, 1963.
15. T. Altares, D. P. Wyman, and V. R. Allen, *J. Polym. Sci.*, **A2**, 4533 (1964).
16. D. McIntyre, L. J. Fetters, and E. Slagowski, *Science*, **176**, 1041 (1972).
17. N. Hadjichristidis and J. E. Rovers, *J. Polym. Sci., Polym. Phys. Ed.*, **12**, 2521 (1974).

Vinyl Isocyanate Polymers

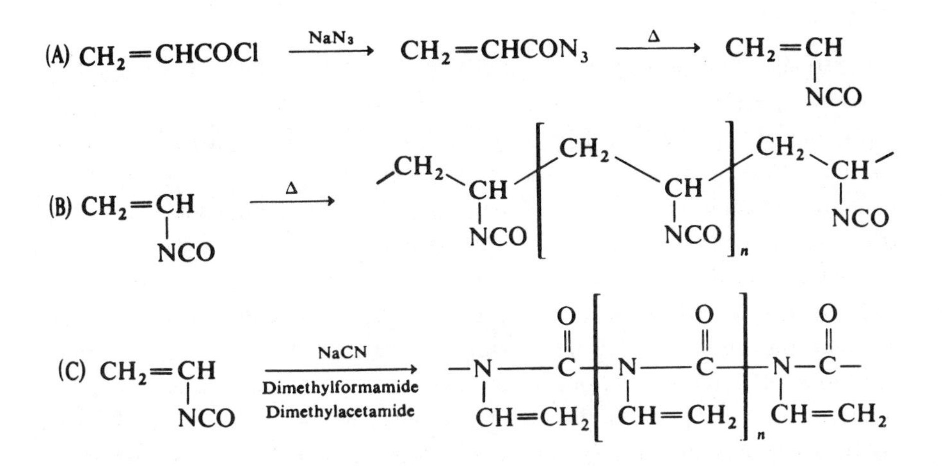

Submitted by C. G. Overberger (1a) and C. J. Podsiadly (1a, b)
Checked by C. Schuerch and W. Lindenberger (2)

1. Procedure

A. Vinyl Isocyanate

Caution! This material and its precursors are potent lachrymators.

A 1-1. flask is equipped with a mechanical stirrer, a thermometer, and a dropping funnel (Note 1). To 91 g. (1.4 moles) of sodium azide in 250 ml. of water, cooled to 0°, is slowly added, with rapid stirring, 106.5 g. (1.2 moles) of acrylyl chloride in 300 ml. of toluene over a period of 1 hr. After the addition the temperature is kept at 0-10° for 20 min. The mixture is separated and the toluene layer (Note 2) is thoroughly washed with sodium bicarbonate-ice water until neutral (Note 3). The toluene layer is then dried over sodium sulfate for 1 day in a cool place.

A 1-1., four-necked flask is equipped with a thermometer, a dropping funnel, and a 30-cm. Vigreux column. The toluene solution of the azide is added to a mixture of 200 ml. of dry toluene and 6.5 g. of dinitrobenzene heated at 85° at such a rate that the temperature remains at 90-95° (Note 4). The distillate formed was collected at the top of a Vigreux column to give 55 g. (66%) of vinyl isocyanate, b.p. 38.0-41.0° (Note 5). Redistillation produced 51 g. of vinyl isocyanate, b.p. 38.7-39.5° (Note 6).

B. N-Vinyl-1-nylon

A 250-ml., three-necked flask is equipped with a two-holed sidearm adapter, a low-temperature thermometer, a stirrer, a nitrogen-inlet tube, and a serum cap (Note 7). The flask is then flamed while being swept with pure nitrogen. After the flask is allowed to cool, 21 ml. of dimethylformamide and 9 ml. of dimethylacetamide (Note 8) are added, and the mixture is cooled to -61°. Nitrogen is passed through the system for 20 min.

Vinyl isocyanate (6.3 g.) is added while the temperature is maintained at -61°. A drop of catalyst solution (Note 9) is added through a serum cap with a hypodermic needle. [The temperature of the solution rises suddenly from -61 to -54° and then drops very slowly to -61°. The viscosity of the solution rises and stirring by motor becomes impossible. Stirring is then continued by hand (*Caution!*) (Note 10).] After the mixture is stirred for 3 min. methanol (150 ml.) is added gradually at -61° to precipitate the polymer The temperature is allowed to rise to room temperature and the polymer is collected by filtration, washed with 600 ml. of methanol, and dried at 30° under reduced pressure for 24 hr.; 6.2 g. (almost 100%) of polymer is obtained.

C. Poly(vinyl isocyanate)

Vinyl isocyanate, 10 g. is kept at room temperature, 25° for 24 hr. without inhibitor or solvent (Note 11). A white precipitate that increases in quantity with time then appears. After 3 days the remaining monomer is removed under reduced pressure to leave 5.5 g. of polymer.

2. Characterization

N-vinyl-1-nylon was insoluble in water, dioxane, benzene, ethanol, chloroform, and carbon disulfide, but soluble gradually in hot dimethylformamide and dimethylacetamide; $[\eta]^{25°}$=0.04 in dimethylformamide (Note 12). The polymer becomes brown at 300°, black at 340°, and does not decompose at 395° even though the weight of the polymer seems to decrease.

Poly(vinyl isocyanate) does not dissolve in any known solvent and is probably crosslinked.

3. Notes

1. A more detailed procedure for preparing the monomer is reported in the literature (3).

2. During this operation the temperature is maintained at about 10°. Loss of product occurs if the extractant is not kept cold.

3. Neutralization is best achieved when the toluene layer is washed with four portions of a 10% sodium bicarbonate solution followed by a single washing with ice water.

4. This decomposition should be effected slowly with small quantities of the toluene solution to keep the reaction under control.

5. The Vigreux column is connected to two gas traps in a mixture of Dry Ice-acetone. If only one gas trap is used, the yield drops to 38%.

6. The compound was stored at Dry Ice temperature. It can also be stabilized if kept cold in the presence of dinitrobenzene (0.1%).

7. The procedure is similar to that described by Shashoua (4) for the preparation of *N*-alkyl-1-nylons.

8. Dimethylformamide (DMF), the purest grade from the Fisher Scientific Company, was purified by distillation over phosphorous pentoxide through a 30-cm. Vigreux column to give pure material, b.p. 153°. A forecut of 5% of the amount charged and a residue of 10% was discarded, even though both gave a constant boiling point.

Dimethylacetamide (DMAA), the purest grade from Fisher Scientific Company, was distilled over phosphorous pentoxide through a 30-cm. Vigreux column to give pure material, b.p. 65.3° (20 torr). A forecut, b.p. 65.2° (20 torr) (10 g.), was discarded.

9. A sodium cyanide (*Caution! Sodium cyanide is toxic and may be very rapidly absorbed through the skin from solutions in solvents such as DMF.*) solution was prepared by dissolving the dry compound in dry dimethylformamide to give a saturated solution containing 0.68% sodium cyanide. This solution was stored under nitrogen and used as required.

10. The viscosity increased markedly in the first 5 min. and remained constant for the next 20 min. The solution remained homogeneous in contrast

to the polymerization of an alkyl isocyanate, in which case the polymer precipitated.

11. The polymerization should be conducted in closed but not tightly sealed containers. Minor explosions have been experienced by both the submitters and checkers if the tubes used were completely sealed. The explosions may possibly be caused by the high vapor pressure of the monomer or by the generation of decomposition products.

12. This value was obtained 1 day after mixing the polymer with solvent. Depolymerization has probably occurred.

4. Methods of Preparation

Shashoua reported the first polymerization of an isocyanate to polymer by using sodium cyanide as a catalyst (4). The polymerization of vinyl isocyanate to *N*-vinyl-1-nylon, described here, has been published (5). Similar work has also been reported by R. C. Schulz (6). The insolubility of poly(vinyl isocyanate) has been observed by other workers (3,7). A new method for the stabilization of vinyl isocyanates has been disclosed (8).

5. References

1a. Department of Chemistry, The University of Michigan, Ann Arbor, Michigan 48104.
1b. Current Address: 3M Co. St. Paul, Minnesota 55101.
2. Department of Forest Chemistry, State University College of Forestry at Syracuse University, Syracuse, New York 13210.
3. R. Hart, *Bull. Soc. Chem. Belg.*, **65**, 291 (1956).
4. V. E. Shashoua, W. Sweeny, and R. F. Tietz, *J. Amer. Chem. Soc.*, **82**, 866 (1960).
5. C. G. Overberger, S. Ozaki, and H. Mukamal, *J. Polym. Sci.*, **B**, 627 (1964).
6. R. C. Schulz and R. Stenner, *Makromol. Chem.*, **72**, 202 (1964).
7. D. D. Coffman (E. I. du Pont de Nemours & Company), U.S. Patent 2,334,476 (Nov. 16, 1943) [*C.A.*, **38**, 2772 (1944)].
8. D. H. Heinert (Dow Chemical Co.) U.S. Pat. 3,728,369 (April 17, 1973).

Solid Phase Synthesis of N-Benzyloxycarbonyl-S-Benzyl-L-cysteinyl-L-prolyl-L-leucylglycinamide

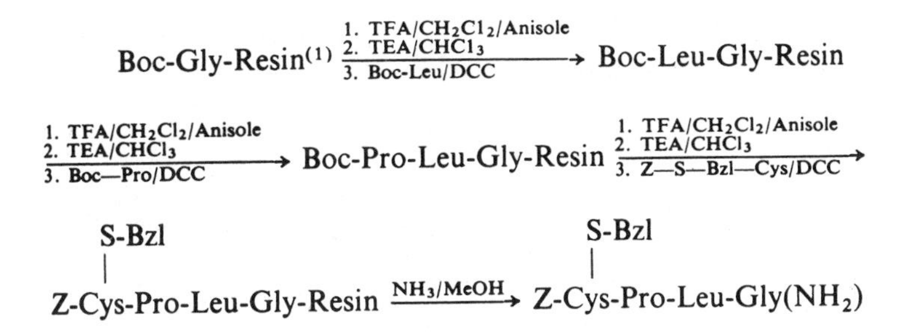

Submitted by V. J. Hruby (2a) and L. E. Barstow (2a, b)
Checked by L. C. Dorman and J. Kadlabitsky (3)

1. Procedure

N-t-Butyloxycarbonylglycine resin ester, which contained 0.32 mmoles/g. of *N-t*-butyloxycarbonylglycine substitution, was purchased from Mann Research (Note 1). The resin ester (2.00 g.) is placed in a 35-ml. Merrifield-type (4) reaction vessel (Note 2) and the cycles of deprotection, neutralization, and coupling described in Table I are followed for the introduction of each new amino acid residue (5).

For the washing steps 17 ml. of the appropriate solvent is used. In the deprotection step 17 ml. of trifluoroacetic acid/methylene chloride (*Caution! Many halogenated compounds are toxic and must be used with care in an efficient hood.*) (2:3 by volume) containing 1% anisole is used, and for the neutralization step 17 ml. of triethylamine/chloroform (1:9 by volume) is used. In the coupling step 1.6 mmoles (a 2.5-fold excess) of the appropriate *N-t*-butyloxycarbonyl or

N-benzyloxycarbonyl amino acid (Note 3) in 7 ml. of methylene chloride and 1.6 mmoles (0.330 g.) of *N,N'*-dicyclohexylcarbodiimide (*Caution! Irritant. Skin rashes may be induced in susceptible individuals.*) in 7 ml. of methylene chloride are used. Reaction times of 5 hr. are used for the coupling steps.

After incorporation of the *N*-benzyloxycarbonyl-*S*-benzyl-L-cysteine residue, the protected polypeptide resin ester is further washed with three 17-ml. portions of methylene chloride. The product is dried *in vacuo* over potassium hydroxide pellets to yield 2.25 g. of resin.

The peptide resin (2.25 g.) is suspended in 160 ml. of anhydrous methanol (freshly distilled from magnesium methoxide). Anhydrous ammonia (*Hood!*) (freshly distilled from sodium) is bubbled through the stirred suspension for 3 hr. at 0°. The stirring is continued at 0-5° for 36 hr. and subsequently at room temperature for 3 hr. (Note 4). The ammonia and methanol are removed under aspirator vacuum with the exclusion of water, and the residue is then dried *in vacuo* over potassium hydroxide pellets.

TABLE 1.

Step	Solvent or Reagent	Volume (ml.)	Shake Time (min.)	Number of Times
1	Methylene chloride	17	1.5	3
2	Trifluoroacetic acid/ methylene chloride (2:3) with 1% anisole	17	30	1
3	Methylene chloride	17	1.5	3
4	Ethanol	17	1.5	3
5	Chloroform	17	1.5	3
6	Triethylamine/chloroform (1:9)	17	10	1
7	Chloroform	17	1.5	3
8	Methylene chloride	17	1.5	4
9	Protected amino acid in methylene chloride	7	–	1
10	*N,N'*-Dicyclohexylcarbo-diimide in methylene chloride	7	300	1
11	Methylene chloride	17	1.5	3
12	Ethanol	17	1.5	3

The cleaved polypeptide material is extracted from the resin with two 25-ml. portions of methanol and three 25-ml. portions of methylene chloride. The spent resin is removed by filtration, the solvents are removed *in vacuo,* and the product is dried overnight *in vacuo.* The peptide material is twice reprecipitated from 8 ml. of methanol/water (5:3) (Notes 5 and 6) to give the pure tetrapeptide derivative, m.p. 167-168° corr. [lit. (6) m.p. 170-171.5°].

2. Notes

1. The *N-t*-butyloxycarbonylglycine resin ester may be purchased from the Mann Research Laboratories, 136 Liberty Street, New York, New York 10006. The resin may also be prepared by published procedures (7) from *N-t*-butyloxycarbonylglycine, triethylamine, and chloromethylated polystyrene which is crosslinked 2% with divinylbenzene.

2. The synthesis may be carried out manually with a simple reaction vessel and shaker (5,8) or it may be facilitated by the use of a semiautomated apparatus (5,9) or by a completely automated machine (4,10).

3. The *N-t*-butyloxycarbonyl-L-amino acids and the *N*-benzyloxycarbonyl-*S*-benzyl-L-cysteine can be readily purchased in highly purified form from a number of commercial sources, including Schwarz BioResearch, Inc., Orangeburg, New York 10962, Mann Research Laboratories, 136 Liberty Street, New York, New York 10006, and Fox Chemical Company, 1650 East 18th Street, Tucson, Arizona 85719. *N-t*-Butyloxycarbonyl-L-amino acids can be conveniently prepared by the Schnabel (11) method from the corresponding L-amino acids. *N*-Benzyloxycarbonyl-*S*-benzyl-L-cysteine can be conveniently prepared from *S*-benzyl-L-cysteine by the Harrington and Mead method (12). It is important that these materials be checked for purity by m.p., thin layer chromatography and, if necessary, optical rotation before use (5).

4. During the ammonolysis the reaction mixture is protected from moisture at all times with a calcium sulfate drying tube.

5. The polypeptide powder was dissolved in the hot methanol/water mixture and cooled in the refrigerator overnight. The product was collected by filtration and washed with a few ml. of cold methanol/water (5:3). The product was dried *in vacuo* and the entire procedure was repeated.

6. In four separate syntheses of the peptide the submitters obtained an average yield of 90%. Beginning with 1.98 g. of *N-t*-butyloxycarbonylglycine resin ester, which contained 0.35 mmole of *N-t*-butyloxycarbonylglycine

substitution, the checkers obtained a yield of 95% of material, m.p. 170-171° (uncorrected).

3. Methods of Preparation

N-Benzyloxycarbonyl-*S*-benzyl-L-cysteinyl-L-prolyl-L-leucylglycinamide has been prepared by a variety of methods (6,13) with solution techniques of polypeptide synthesis. In general, intermediate peptides were isolated and purified in these preparations.

4. Merits of Preparation

The highly purified *N*-benzyloxycarbonyl-*S*-benzyl-L-cysteinyl-L-prolyl-L-leucylglycinamide prepared by the solid phase method is obtained in high yield and with the expenditure of a minimum of laboratory time.

5. References

1. Abbreviations are those in standard use: *Biochemistry,* **5**, 2485 (1966); *J. Biol. Chem.,* **241**, 2491 (1966).

2a. Department of Chemistry, University of Arizona, Tucson, Arizona 85721.

2b. Current Address: Vega-Fox Biochemicals, Tucson, Arizona 85719.

3. Chemical Biology Research, The Dow Chemical Company, Midland, Michigan 48640.
4. R. B. Merrifield, J. M. Stewart, and N. Jernberg, *Anal. Chem.,* **38**, 1905 (1966).
5. For a thorough discussion of the methodology of solid phase polypeptide synthesis see J. M. Stewart and J. D. Young, *Solid Phase Peptide Synthesis,* Freeman, San Francisco, California, 1969.
6. M. Bodanszky and V. duVigneaud, *J. Amer. Chem. Soc.,* **81**, 2504 (1959).
7. R. B. Merrifield and M. A. Corigliano in W. E. M. Lands, Ed., *Biochemical Preparations,* Vol. 12, Wiley, New York, 1968, p. 98.
8. R. B. Merrifield, *J. Amer. Chem. Soc.,* **85**, 2149 (1963). A commercial shaker can be purchased from Mann Research Laboratories, 136 Liberty Street, New York, New York 10006.
9. A semiautomated apparatus similar to those described by R. B. Samuels and L. D. Holybee on pp. 68-70 in (5) can be readily constructed. A commercial semiautomated instrument can be purchased from Chromatronix, Inc., 2743 Eighth Street, Berkeley, California 94710.
10. V. J. Hruby, L. E. Barstow, and T. Linhart, *Anal. Chem.,* **44**, 343 (1972); A. B. Robinson, Ph.D. Dissertation, University of California, San Diego, 1967; A. Loffet and J. Close and K. Brunfeldt, J. Halstrom, and P. Poepstorff, and G. W. H. A. Mansveld, H. Hindriks, and H. C. Beyerman in E. Bricas, Ed., *Peptides* 1968, North-Holland, Amsterdam, 1968, pp. 189-200. Commercial instruments are also available from

Schwarz Bio-Research, Inc., Orangeburg, N.Y. 10962; Beckmann Instruments Inc., Palo Alto, California 94304; and Vega Engineering, Tucson, Arizona 85717.

11. E. Schnabel, *Ann.*, **702**, 188 (1967).
12. C. R. Harington and T. H. Mead, *Biochem. J.*, **30**, 1598 (1936).
13. See, for example, M. Zaoral and J. Rudinger, *Collection Czech. Chem. Commun.*, **20**, 1183 (1955); H. C. Beyerman, J. S. Bontekoe, and A. C. Koch, *Rec. trav. chim.*, **78**, 935 (1959).

Cellulose p-Toluenesulfonyl Carbamate

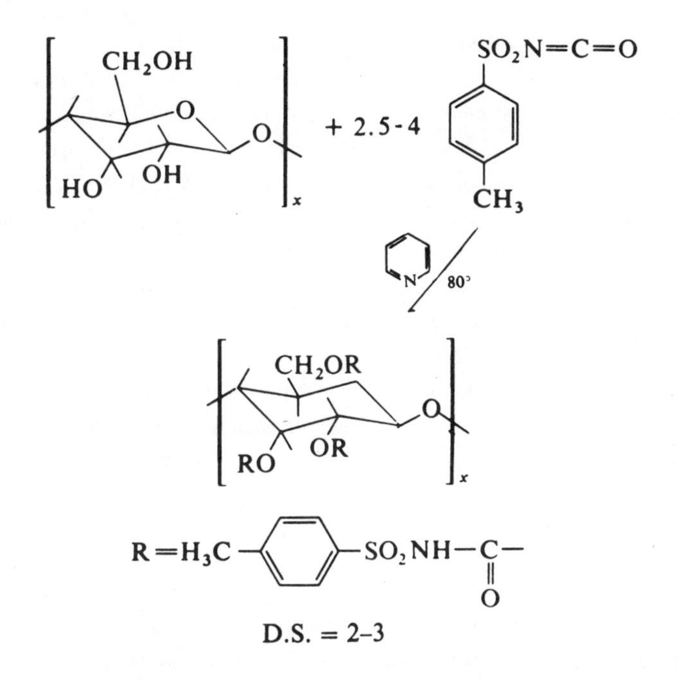

Submitted by R. W. Rosseau (1), J. G. Dillon and W. H. Daly (2)
Checked by V. T. Stannett and R. Boyette (1)

1. Procedure

A dried 8-oz. beverage bottle (Note 1) is charged with 2.0 g. (0.01235 mol. eq.) of cotton linters (Note 2) and 150 ml. of pyridine (Note 3). The bottle is flushed with nitrogen while 6.5 ml. (0.0495 mole) of *p*-toluenesulfonyl isocyanate (*Caution! This compound is extremely reactive and is a potent irritant.*) (Note 4) is injected and then sealed with a pressure cap. The bottle is immersed in a constant temperature bath (Note 5) at 80° and agitated intermittently. Swelling of the cotton linters is apparent within 30 min. and dissolution usually occurs within 2 hr. to produce an extremely viscous solution. The degree of substitution (D.S.) is controlled by the reaction time; a D.S. of $\simeq$2 is reached within 3 hr., but the solution must be heated for 18 hr. to achieve a D.S. of 3. The cellulose *p*-toluenesulfonyl carbamate is isolated by the slow addition of the pyridine solution to 1 1. of ethanol stirred vigorously in an explosion-proof blender (Note 6). The bottle is washed with several portions of fresh pyridine and the washings are added to the ethanol precipitant to ensure quantitative recovery of the product. The flocculent precipitate is allowed to settle, the supernatant liquid is decanted, and the polymer is recovered by filtration. After the polymer is washed with 1 1. of fresh ethanol in the blender to remove low molecular weight by-products (Note 7), the polymer is isolated and dried at 40° *in vacuo* over phosphorus pentoxide. The yield of white powdery material is 6.89 g. (D.S. = 2).

The stable sodium salt of the carbamate may be prepared by stirring, under nitrogen, 6.47 g. of the previously prepared carbamate in 150 ml. of a 2% (wt./vol.) sodium hydroxide solution at room temperature. The medium becomes increasingly viscous with time. The reaction may be terminated after 24 hr. and the remaining suspended particles removed by centrifugation and subsequent decantation of the clear supernatant into 1 1. of ethanol. The salt is collected by filtration and dried *in vacuo* at 50°. The yield of white powdery salt is 6.25 g.

2. Characterization

The solubility of the cellulose derivative is dependent on the degree of substitution. Derivatives with a D.S. range of 1.1 to 2.8 are soluble in pyridine and 2% aqueous sodium hydroxide and are swollen by dimethylformamide, dimethyl sulfoxide, and 10% sodium hydroxide (Note 8). Derivatives in this range prepared from wood cellulose are soluble in acetone. The infrared spectrum (Note 9) displays the following characteristic absorptions: 3300 $cm.^{-1}$ (N–H); 1750-1650 $cm.^{-1}$ (carbamate); 1340, 1170 $cm.^{-1}$ ($-SO_2-$); 1080 $cm.^{-1}$ (–OH).

The derivatives do not exhibit a glass transition temperature or melting point (Note 10). The DTA indicates an endothermic process at 110° (loss of water) and an exotherm beginning at 250° that contains an endotherm at 276°. The products of this thermal decomposition are carbon dioxide and *p*-toluenesulfonamide (3). The DTA of the sodium salt of cellulose *p*-toluenesulfonyl carbamate is simply an endothermic absorption at 325° that indicates the salt is thermally stable. Thus it is advisable to store the derivatives in the salt form.

The degree of substitution was determined by weight gain by the following relationship:

$$\text{D.S.} = \frac{(\text{weight gain})/(\text{mol. wt. of substituent})}{(\text{initial wt.})/(\text{mol. wt. of cellulose unit})}$$

These results were confirmed by nitrogen analysis; the refractory nature of the samples necessitated the use of neutron activation techniques to obtain accurate nitrogen assays.

3. Notes

1. A 200-ml., three-necked, round-bottomed flask equipped with a mechanical stirrer, a condenser, a septum, and a nitrogen inlet may be used.

2. The ground cotton linters are dried at 90-110° *in vacuo* over phosphorus pentoxide for 4 to 5 hr. Phosphorus pentoxide is required because the more common desiccants (calcium sulfate, calcium chloride, etc.) cannot prevent moisture regain in cellulose. No further purification of cotton linters is required. Purified wood cellulose may be substituted for the linters with no change in the reaction conditions.

3. Reagent-grade pyridine is dried by heating under reflux over calcium hydride followed by distillation under a nitrogen atmosphere. A generous forerun (~10% of the initial volume) was discarded and the distillate was stored over molecular sieves (4A) under nitrogen.

4. *p*-Toluenesulfonyl isocyanate (Aldrich Chemical Company) is used without further purification or prepared from *p*-toluenesulfonamide by the method of Ulrich and Sayigh (4). The major impurity in this reagent is *p*-toluenesulfonyl chloride, which appears to be inert under the experimental conditions.

5. A heating mantle may produce hot spots that lead to charring of the product.

6. Alternatively, the reaction mixture may be precipitated in 1 1. of ethanol containing 25 ml. of concentrated hydrochloric acid to aid in the removal of pyridine from the polymer. Filtration must be carried out as rapidly as possible to prevent alcoholysis of the cellulose substrate.

7. *p*-Toluenesulfonamide is soluble in ethanol.

8. Solubility tests were carried out by placing approximately 0.1 g. of the cellulose derivative in a test tube with 10-15 ml. of the solvent. The mixture is shaken intermittently for about 24 hr. Solution was noted by a significant

increase in the viscosity of the mixture as well as the disappearance of a major portion of the fibers. If solution does not occur, the mixture is heated at 60° for several hours. The derivative was then classified as soluble, swollen (distinct gel-like particles form but the over-all viscosity of the solvent does not change), and insoluble.

9. Infrared spectra were measured on potassium bromide pellets or on films cast from pyridine.

10. Differential thermal analysis was carried out on a Dupont Model 900 DTA instrument.

4. Merits of the Preparation

This procedure represents a new method of introducing acidic ionogenic groups on a cellulose substrate because the sulfonyl carbamate function exhibits a pK_a of 3.7. Crosslinked cellulose sulfonylcarbamates have been evaluated as ion-exchange resins (5).

5. References

1. Department of Chemical Engineering, North Carolina State University, State College Station, Raleigh, North Carolina 27607.
2. Department of Chemistry, Louisiana State University, Baton Rouge, Louisiana 70803.
3. L. C. Roach and W. H. Daly, *Chem. Comm. (London)*, 606 (1970).
4. H. Ulrich and A. A. R. Sayigh, *Angew. Chem. Int. Ed.*, 5, 704 (1966).
5. R. W. Rousseau, C. D. Callihan and W. H. Daly, *Macromolecules*, 2, 502 (1969).

Alternating Conjugated Diene-Maleic Anhydride Copolymers

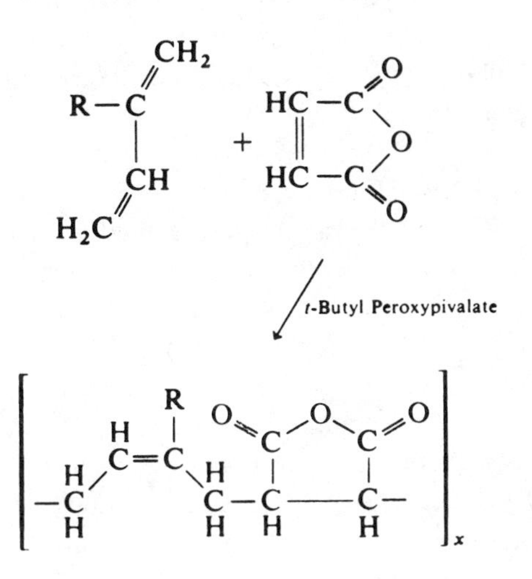

Submitted by N. G. Gaylord (1a) and A. Takahashi (1a, b)
Checked by H. Pledger, Jr, J. W. Schwietert, and G. B. Butler (2)

A. Poly(isoprene-alt-Maleic anhydride)

1. Procedure

A 500-ml., round-bottomed, three-necked flask equipped with a mechanical stirrer, a thermometer, and a reflux condenser is charged with 98 g. (1.0 mole) of maleic anhydride (*Caution! Maleic anhydride is toxic and care should be taken to avoid skin contact with, or inhalation of dust from, this material.*) (Note 1) and 80 ml. of dioxane (Note 2). Solution is achieved by heating the mixture to 80° with continuous stirring on a hot water bath. A solution containing 102 ml. (1.02 moles) of distilled isoprene (Note 3) and 1 g. of *t*-butyl peroxypivalate (*Caution! Peroxidic materials are strong oxidants and should be handled with care.*) (TBPP) (Note 4) is added carefully through the open end of the condenser over a total period of 20 min. (Note 5) with continuous stirring. The reaction is highly exothermic and external cooling is required to maintain the temperature at 80 ± 2° (Note 6). The onset of polymerization is instantaneous and copolymer formation is visible at the surface of the reaction mixture. The solution becomes extremely viscous and the stirrer speed is gradually increased to maintain efficient stirring. After approximately one-half of the isoprene-TBPP solution has been added, the viscosity increase is so pronounced that periodic addition of dioxane is necessary to keep the mixture fluid. A total of 60 ml. of dioxane is added for this purpose. After addition of all the isoprene-TBPP solution stirring is continued for an additional 45 min. to complete the reaction. The viscous solution is diluted with 60 ml. of acetone (Note 7), and this mixture is slowly poured with stirring into 2000 ml. of dry benzene. The copolymer is precipitated in fibrous form, collected by filtration, washed with petroleum ether, and dried at 50° *in vacuo* overnight. The dried, tan-colored, fluffy copolymer is obtained in a yield of 84 g. (50%). The copolymer can be purified by solution in acetone (Note 8) and then by the addition of the acetone solution in a thin stream into excess benzene (four volumes of benzene per volume of acetone solution) with vigorous stirring. The copolymer is washed with petroleum ether and dried in a vacuum oven at 50°. The recovery of purified copolymer is 75-80%.

2. Characterization

The purified copolymer is soluble in polar solvents, such as *N,N*-dimethylformamide, *N,N*-dimethylacetamide, tetrahydrofuran, dioxane, acetonitrile, nitrobenzene, acetone, methyl ethyl ketone, and cyclohexanone (Note 9) and is insoluble in aliphatic, aromatic, and chlorinated hydrocarbons. The intrinsic viscosity of a 0.5% solution in cyclohexanone at 30° is 0.80 dl./g. (Note 10). The copolymer has a stick temperature of 145-150°.

Anal. Calcd. for 1:1 copolymer: C, 65.1%, H, 6.1%. Found: C, 64.7%, H, 6.1%.

Determination of unsaturation by chemical analyses gives unusual results. Attempts to hydrogenate the copolymer with platinum oxide or palladium/

charcoal as catalysts are unsuccessful. Ozonolysis indicates 85% unsaturation; titration with iodine monochloride indicates 70-85% unsaturation.

Nuclear magnetic resonance analysis of the polymer in acetone-d_6 or deuterated acetic acid with tetramethylsilane as the internal standard indicates the presence of

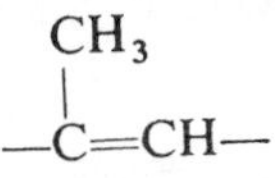

groups and 85% of 1,4-structure. The infrared spectra of polymeric films cast from acetone solution indicate less than 10% unsaturation. The presence of the cyclic anhydride units in the copolymers is readily detected from infrared absorption bands at 1220, 1775, and 1855 $cm.^{-1}$

B. Poly(butadiene-alt-maleic anhydride)

1. Procedure

A 500-ml., round-bottomed flask equipped with a dropping funnel, a mechanical stirrer, a thermometer, a Dry Ice, condenser and a nitrogen inlet and outlet is charged with 49 g. (0.5 mole) of maleic anhydride (Note 1) and 76 g. of distilled cyclohexanone. The mixture is heated to 60-70° under nitrogen to effect solution. A solution containing 35 g. of butadiene (Note 11), 76 g. of cyclohexanone, and 1.5 g. of *t*-butyl peroxypivalate (TBPP) (Note 4) is added to the reaction flask through the dropping funnel over a period of 40 min. (Note 5) with continuous stirring while the temperature is maintained at 60-70° (Note 6). The solution becomes extremely viscous within 10 min. after the addition of the monomer-catalyst solution is started. The solution is stirred for 1 hr. after the monomer addition is completed and then cooled to room temperature with external cooling. The viscous solution is poured with stirring into 1000 ml. of dry benzene to precipitate the copolymer. The latter is collected by filtration, washed with 200 ml. of petroleum ether, and dried to constant weight *in vacuo* at 50°. The yield of copolymer is 40 g. (53% based on the maleic anhydride).

2. Characterization

The copolymer is soluble in the same polar solvents as poly(isoprene-altmaleic anhydride) (Note 9) and is similarly insoluble in hydrocarbons or chlorinated hydrocarbons. The intrinsic viscosity of the copolymer in cyclohexanone at 25° is 0.7 dl./g. (Note 10). The copolymer has a softening point of 150-155° on a melting point bar.

Anal. Calcd. for 1:1 copolymer: C, 62.6%, H, 5.5%. Found: C, 63.0%, H, 5.3%.

The infrared spectra of polymeric films cast from acetone solution indicate 85% *cis*-1,4, 10% *trans*-1,4, and 5% 1,2-vinyl structures (3,4). The anhydride groups are detectable at 1220, 1775, and 1855 cm.$^{-1}$. Nuclear magnetic resonance analysis of the polymer in acetone-d_6 with tetramethylsilane as the internal standard indicates 85-90% of 1,4-structure.

3. Notes

1. Maleic anhydride, m.p. 56°, is sublimed before use to remove any maleic acid formed during storage. Maleic acid, with reluctance, undergoes copolymerization or the Diels-Alder reaction with conjugated dienes and the copolymers formed in the presence of radical catalysts have a different structure from that of the diene-maleic anhydride copolymers.
2. Dioxane, b.p. 101°, which undergoes peroxide formation on exposure to air, should preferably be distilled before use. Distilled solvent stored for 2 or 3 days in a capped container may be used; however, any peroxide formed on storage provides additional free radicals during polymerization.
3. Isoprene, b.p. 34°, is purified and distilled and kept in the refrigerator (5-10°). A small excess (2%) of isoprene is added to compensate for monomer loss during the addition.
4. When a 75% solution of TBPP in mineral spirits is used, 1 g. of the solution is equivalent to 0.75 g. of pure TBPP. Benzoyl peroxide, *t*-butyl perbenzoate and other radical catalysts may also be used (Note 6).
5. The catalyst solution, i.e., TBPP in monomer, is added as quickly as possible while controlling the temperature of the reaction mixture (Note 6). If the catalyst solution is added slowly over an extended period, the yield of copolymer is decreased and the formation of the Diels-Alder adduct is enhanced.
6. The reaction temperature is chosen so that the half-life of the catalyst is of the order of 1 hr. or less, preferably 30 min. Because the reaction is highly exothermic, the flask should be cooled in an ice-water bath periodically to control the temperature. The temperature of the polymerization reaction is important, because the molecular weight of the copolymer is influenced

more by the reaction temperature than by the catalyst concentration. The preferred catalyst is, therefore, selected on the basis of the desired molecular weight of the copolymer.

7. Acetone used for diluting the viscous copolymer solution should preferably be distilled and dry, because any moisture present may hydrolyze the anhydride groups of the copolymer. The exact volume of acetone is not critical. If too little acetone is added, however, difficulty in pouring the solution and stirring the precipitated polymer may be encountered. If too much acetone is added, there may be a decrease in polymer yield.

8. The volume of acetone required depends on the molecular weight of the polymer. The usual range is 5-8 ml. of acetone per 1 g. of dried polymer.

9. The solubility of the isoprene-maleic anhydride copolymer in a particular solvent depends on its molecular weight. Thus a copolymer with an intrinsic viscosity of 0.8-1.0 dl./g. at 30° in cyclohexanone is soluble in acetone and methyl ethyl ketone but insoluble in methyl isobutyl ketone, whereas a sample with an intrinsic viscosity of 0.1-0.3 is soluble also in methyl isobutyl ketone.

10. The solvents used for viscosity measurements should be free of moisture. *N,N*-Dimethylformamide is a good solvent for the copolymer, but its high miscibility with water makes it unsuitable. In the presence of water the polymer behaves as a typical polyelectrolyte and the η_{sp}/c vs. c (c = concentration) plot yields a parabolic curve. Cyclohexanone has little miscibility with water, and is recommended as a solvent for viscosity measurements.

11. Butadiene is passed successively through columns packed with solid potassium hydroxide pellets and Linde 4A molecular sieves and then condensed in an Erlenmeyer flask. An excess of butadiene (8 g.) is used in the procedure.

4. Methods of Preparation

Alternating equimolar copolymers of conjugated dienes and maleic anhydride have been prepared with free radical initiators (5,6) and gamma radiation (7).

5. References

1a. Gaylord Research Institute Inc., Newark, New Jersey 07105.
1b. Current Address: Hooker Chemical Co., Niagara Falls, New York.
2. Center for Macromolecular Science, University of Florida, Gainesville, Florida 32601.
3. R. S. Silas, J. Yates, and V. Thornton, *Anal. Chem.*, **31**, 529 (1959).

4. E. Oikawa and A. Takahashi, *Kogyo Kagaku Zasshi,* 72, 1940 (1969) [C.A., 72, 32983a (1970)].
5. N. G. Gaylord (Borg-Warner Corporation), U.S. Pat. 3,491,068 (Jan. 20, 1970) [*C.A.,* 72, 56272u (1970)].
6. Y. Yamashita, S. Iwatsuki, and T. Kokubo, *J. Polym. Sci.,* C23, 753 (1968).
7. Y. Tsuda, T. Sakai, and Y. Shinohara, *IUPAC International Symposium on Macromolecular Chemistry, Preprints, Vol.* III, Tokyo-Kyoto, 1966, p. 44.

Poly(ethylene N-phenylcarbamate)

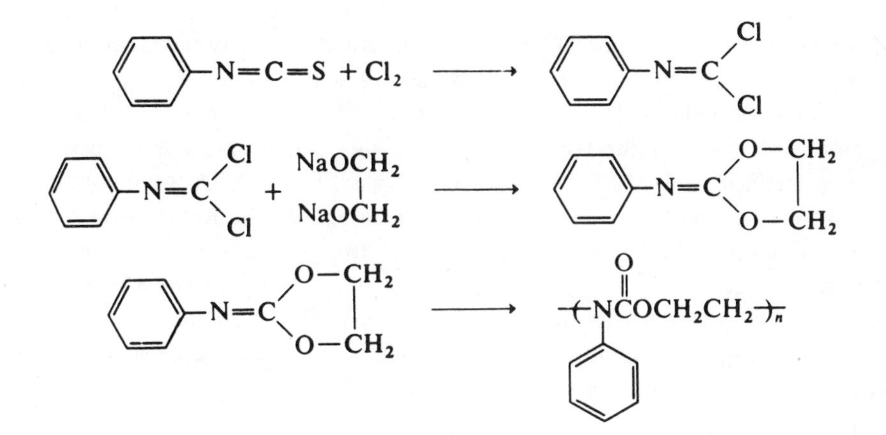

Submitted by T. Mukaiyama (1) and T. Fujisawa (2)
Checked by M. Mitoh and Y. Minoura (3)

1. Procedure

A. Phenyliminophosgene (4) (Caution!) (Note 1)

A 500-ml. flask equipped with a condenser protected from atmospheric moisture with a calcium chloride tube and a gas inlet tube is charged with 135 g. (1.0 mole) of phenyl isothiocyanate (Note 2) and 125 ml. of carbon tetrachloride. (*Caution! Many halogenated solvents are toxic and should be used with care.*) Chlorine (*Caution! Chlorine is a toxic gas and should be used in an efficient hood.*) is passed into the reaction mixture at a slow rate until about 150 g. of chlorine has been absorbed (Note 3). During this introduction evolution of heat occurs, and the reaction vessel is cooled in an ice bath. The carbon tetrachloride and sulfur dichloride are removed by distillation at atmospheric pressure. The fraction, boiling at 100-110° (30 torr), is collected and refractionated at 104° (27 torr). The yield of pale yellow phenyliminophosgene is 140-156 g. (80-90%).

B. Disodium Glycolate (5)

In a 500-ml. flask a solution of sodium ethylate is prepared by the addition of 23.0 g. (1.0 g.-atm.) of sodium to 300 ml. of absolute ethanol. To this solution 31.0 g. (0.5 mole) of ethylene glycol (Note 4) is added. A suction tube with a stopcock is connected to the flask and the ethanol is removed by evaporation under reduced pressure and collected in a trap cooled with a Dry Ice-acetone mixture (Note 5). During this evaporation the temperature is maintained below 100°. After about 250 ml. of ethanol has been collected, the reaction vessel is cooled to room temperature and flushed with nitrogen. The resulting solid mass is rapidly crushed to a powder. The flask is then evacuated at 1 torr and gradually heated to 160° to remove the ethanol completely (*Caution!*) (Note 6). The white powder of disodium glycolate is cooled at room temperature and used as such in the following preparation. The yield of product is 50-53 g. (95-100%).

C. Ethylene N-Phenyliminocarbonate (6)

A 1-l. flask is equipped with a magnetic stirrer, a condenser, and a dropping funnel and is protected from moisture. The flask is charged with 53 g. (0.5 mole) of disodium glycolate suspended in 300 ml. of benzene. From the dropping funnel is added a solution of 80 g. (0.5 mole) of phenyliminophosgene in 200 ml. of benzene over a 10- to 15-min. period. The heat evolved is sufficient to reflux the benzene. After the addition is completed, the reaction mixture is heated under reflux for 2 hr. (Note 7). The reaction mixture is cooled and the sodium chloride is extracted by washing with two 500-ml. portions of water. The benzene solution is dried over calcium chloride. The benzene is removed under reduced pressure to leave a light yellow solid. Recrystallization from ether gives 50-60 g. (62-75%) of colorless needles, m.p. 74-76° (Note 8).

D. Polymerization of Ethylene N-Phenyliminocarbonate (7)

Bulk Polymerization. Ethylene *N*-phenyliminocarbonate (10 g.) is dried at 50° *in vacuo* and is placed in a previously dried 30-ml. flask protected from moisture with a drying tube. The monomer is melted by heating the flask in a 100° oil bath under a nitrogen atmosphere. To the melt is added a drop of sulfuric acid from a capillary (about 0.05% by weight of monomer) (Note 9). Polymerization proceeds instantly to give a white solid mass. The resulting polymer is mixed with 150 ml. of hot chloroform. The resulting solution is filtered to remove insoluble material and the filtrate is poured into 1 l. of acetone to precipitate the polymer. The polymer is collected by filtration and then dried in vacuum at 100° for 5 hr. The yield of white polymer is 6-7 g. (60-70%). Its intrinsic viscosity is 1.26 dl./g. in chloroform at 30°.

Solution Polymerization. A 100-ml. flask is charged with 10 g. of ethylene *N*-phenyliminocarbonate and 50 ml. of dry benzene (Note 10). While the mixture is shaken, about 5 mg. of titanium tetrachloride is added to the solution with a hypodermic syringe (Note 11). The flask is immersed in an oil bath and heated to reflux for 12 hr. The polymer separates gradually from the solvent. After the heating is complete, the polymer is isolated by removal of the benzene by evaporation and reprecipitated from a chloroform solution with acetone. The yield of polymer is about 9 g. (90%) (Note 12). Its intrinsic viscosity is 0.96 dl./g. in chloroform at 30°.

2. Characterization

The polymer is soluble in chloroform and ethyl acetate, and insoluble in ether, benzene, acetone, alcohol, and petroleum ether. It has a polymer melt temperature of 180°. Clear films can be melt-pressed at 180°. Fibers can be melt-spun at 200°. The infrared spectrum of the polymer measured in a potassium bromide pellet shows the characteristic peak of the urethane linkage at 1720 cm.$^{-1}$, of the monosubstituted benzene ring at 1600, 760, and 695 cm.$^{-1}$, of the methylene linkage at 1410 cm.$^{-1}$, and of the ether linkage at 1150 cm.$^{-1}$.

3. Notes

1. The vapor of phenyliminophosgene is toxic and lachrymatory. The reaction vessel used should be washed with base after phenyliminophosgene is reacted.
2. Phenyl isothiocyanate was available from Tokyo Kasei Kogyo Company, Ltd. and was used without purification.
3. The solution soon turns a deep cherry red and ultimately becomes yellow.

4. Ethylene glycol was available from Tokyo Kasei Kogyo Company, Ltd. and was purified by distillation to produce material, b.p. 92-3° (10 torr).

5. A white solid is soon precipitated. To carry out the evaporation smoothly a rotary evaporator is used or the vessel is shaken by hand.

6. If the reaction mixture is still hot and is opened to the air to crush the solid mass, a black spot that burns gradually appears in the mass.

7. Prolonged reflux leads to polymeric side products.

8. A further recrystallization gives material melting at 76-77°.

9. About 3 μl. of a practical grade of 96% sulfuric acid is added with a capillary.

10. The benzene is dried over sodium wire and then distilled.

11. A practical grade of titanium tetrachloride is used without purification.

12. When boron trifluoride etherate is used instead of titanium tetrachloride, the white polymer separates from the solvent after 1 hr. at room temperature. Its intrinsic viscosity in chloroform at 30° is 0.33. dl./g.

4. Methods of Preparation

Poly(ethylene *N*-phenylcarbamate) of high molecular weight can also be made with such cationic initiators as Lewis and protonic acids, phosphorus pentoxide (7), and phosphorus pentafluoride (8). The ring-opening polymerization has also been carried out without catalyst (7). Thermal stabilization of the polymer by end-group acetylation has been reported (8). *N*-Substituted ethylene iminocarbonates have also been reported to give *N*-substituted polyurethanes by a similar type of polymerization (9). *N*-Unsubstituted poly(ethylene carbamate) has been obtained by the ring-opening polymerization of ethylene *N*-chloroiminocarbonate followed by reduction (10).

5. References

1. Laboratory of Organic Chemistry, Tokyo Institute of Technology, Ohkayama, Tokyo, Japan.
2. Sagami Chemical Research Center, Ohnuma, Sagamihara-shi, Kanagawa, Japan.
3. Department of Chemistry, Osaka City University, Osaka, Japan.
4. E. Kühle, B. Anders, and G. Zumach, *Angew. Chem. Int. Ed.*, **6**, 649 (1967).
5. D. Vorländer, *Ann.*, **280**, 167 (1894).
6. T. Mukaiyama, T. Fujisawa, and T. Hyugaji, *Bull. Chem. Soc. Japan*, **35**, 687 (1962).
7. T. Mukaiyama, T. Fujisawa, H. Nohira, and T. Hyugaji, *J. Org. Chem.*, **27**, 3337 (1962).
8. W. R. Sorenson and T. W. Campbell, *Preparative Methods of Polymer Chemistry*, 2nd ed., Wiley-Interscience, New York, 1968, p. 365.
9. T. Fujisawa, Y. Tamura, and T. Mukaiyama, *Bull. Chem. Soc. Japan*, **37**, 793 (1964).
10. T. Fujisawa, H. Koda, and T. Mukaiyama, *Bull. Chem. Soc. Japan*, **40**, 190 (1967).

Poly(hexamethylene-1,3-benzenesulfonamide)

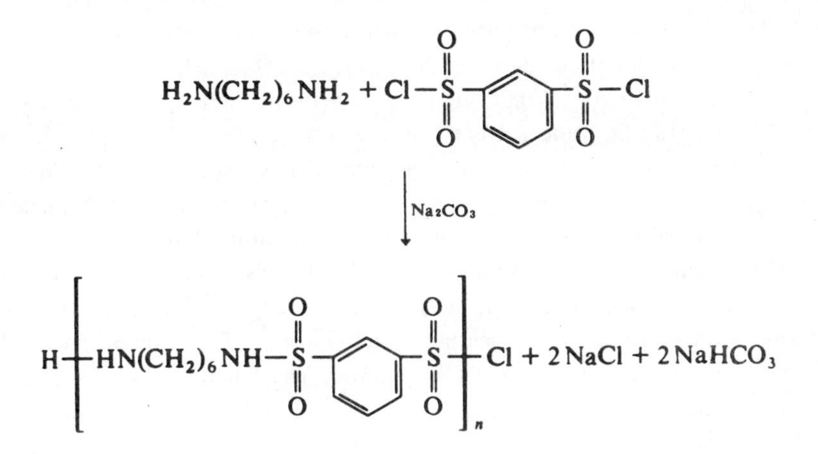

Submitted by S. L. Kwolek and P. W. Morgan (1)
Checked by J. E. McGrath (2a, b)

1. Procedure A

In a 1-quart explosion-proof blender (Note 1) is placed a solution consisting of 170 ml. of distilled water, 20 ml. of 10% aqueous Duponol® ME surface-active agent (Note 2), 3.02 g. (0.026 mole) of hexamethylenediamine (Note 3), and 5.30 g. (0.05 mole) of sodium carbonate. The blender is turned to high speed with a rheostat and a solution of 6.88 g. (0.025 mole) of *m*-benzenedisulfonyl chloride (Note 3) in 200 ml. of methylene chloride (*Caution! Many halogenated compounds are toxic and should be used with care.*) is added rapidly. The mixture is stirred for 15 min. at moderately high speed, Precipitation of polymer is effected by the addition of 100 ml. of alcohol. The polymer is collected on a fritted-glass filter (Note 4) and washed several times with water and alcohol. After it is dried in a vacuum oven at 80°, the product has an inherent viscosity of 2.71 dl./g. The yield is 78% (3).

2. Procedure B

In a round-bottomed flask equipped with a stirrer, a condenser, and a dropping funnel, are placed 5.81 g. (0.05 mole) of hexamethylenediamine (Note 3) and 33 ml. of a 90:10 (by volume) mixture of tetramethylene sulfone and alcohol-free chloroform (Note 5). When complete solution has occurred, 9.26 g. (0.125 mole) of powdered calcium hydroxide is added.

A solution of 13.76 g. (0.05 mole) of *m*-benzenedisulfonyl chloride (Note 3) in 18 ml. of the same solvent mixture is added from a dropping funnel over a period of 2 min. with stirring and without cooling to control the appreciable heat of reaction (Note 6). The mixture is immediately heated with a water bath at about 100° and stirring is continued for 15 min. (Note 7). The polymer is isolated by slowly pouring the viscous solution into water stirred rapidly in a blender. The polymer is washed in the blender with two portions of 5% aqueous acetic acid and three portions of hot water and finally rinsed several times on the funnel with water. The inherent viscosity is 1.29 dl./g. and the yield is 89% (4).

3. Characterization

Solubility. The polymer is soluble in solvents such as hot *m*-cresol and sulfuric acid. It is also soluble in electron-donor solvents, such as dimethylformamide, dimethylacetamide, and *N*-methylpyrrolidone. Solubility in 10% sodium hydroxide is attributable to the acidic character of the sulfonamide group. Tetramethylene sulfone, dimethyl sulfoxide, and diethyl sulfoxide are also solvents for the polymer.

Dilute Solution Viscosity. The inherent viscosity is determined in sulfuric acid at 30° with 0.5 g. of polymer per 100 ml. of solution.

Melt Temperature. The polymer melt temperature on a hot metal surface is approximately 200°.

4. Notes

1. The home blender ensures rapid and efficient stirring. The stirrer bearing,

unless it is sealed and lubricated with an inert substance, such as "Celvacene" medium vacuum grease (Distillation Products Industries), may allow leakage of liquids on long usage. The top of the blender is covered with aluminum foil, over which is placed a plastic cap. A wide powder funnel is inserted through a ¾-in. hole in the center of the cap and foil (5).

2. "Duponol" ME is a Du Pont trademark for sodium lauryl sulfate.

3. Reactants of high purity are necessary to obtain polymer of high molecular weight.

Hexamethylenediamine is nylon-production grade of proven quality. It may also be purchased from chemical supply houses as a 70% aqueous solution or a 97% solid. It may be further purified by distillation under reduced pressure.

m-Benzenedisulfonyl chloride is synthesized from the acid by the action of phosphorus pentachloride in phosphorus oxychlordie solvent. *m*-Benzenedisulfonic acid (425 g.) is added over a period of 30 min. to a mixture of phosphorus pentachloride (1687 g.) and phosphorus oxychloride (680 g.). This mixture is heated under reflux for 3.5 hr. The excess phosphorus oxychloride is removed under reduced pressure. The residue is poured into ice water with stirring while a temperature of 0-15° is maintained. After this cold mixture is stirred for 20 min., it is filtered. The precipitate is dissolved in benzene and the solution is washed with three portions of a 5% aqueous sodium bicarbonate solution and finally with water. After the benzene solution is dried over Drierite® desiccant, it is diluted with *n*-hexane and the mixture is cooled. The crystalline product is collected by filtration, washed with *n*-hexane, and dried in a desiccator under reduced pressure. The yield of product, m.p. 62.0-62.5°, is 76% (3).

Methylene chloride is reagent grade.

4. A fritted-glass filter is used to avoid contamination with cellulose fiber.

5. Tetramethylene sulfone is commercially available. It is purified by distillation at reduced pressure to give material, b.p. 79.5-81.5° (0.5 mm.).

Chloroform is A.C.S. reagent-grade which is freed of alcohol stabilizer by washing with three equal volumes of water. It is first dried with calcium chloride and then stored over a mixture of equal parts of potassium carbonate and calcium chloride in a dark bottle. Washed chloroform, which is well dried in the preliminary step, is stable for several weeks. Decomposing chloroform yields hydrogen chloride and phosgene which may be detected by moist litmus paper in the vapors or by the immediate development of turbidity on the addition of a drop of ethylenediamine (5).

6. When this preparation was carried out with cooling to 30° during the acid chloride addition, the inherent viscosity of the product was 0.95 (4).

7. Samples of polymer from this preparation were isolated at different reaction

times. After 6.5 min. of heating the inherent viscosity of the polymer was 1.18 dl./g.; at 15 min., 1.29 dl./g.; thereafter there was no change (4)

5. Methods of Preparation

Procedure A is taken from the work of Sundet, Murphey, and Speck (3). In 1954 Jones and McFarlane published a process for preparing aliphatic polysulfonamides that is much like the interfacial polymerization method described herein but no molecular weights are given (6). Work on polysulfonamides published through 1964 is described in the book, *Condensation Polymers: By Interfacial and Solution Methods* (7). Since then publications have appeared on the preparation of aromatic polysulfonamides (8,11), water-soluble polysulfonamides (9), and piperazine polysulfonamides (10).

6. References

1. Pioneering Research Division, Textile Fibers Department, E. I. du Pont de Nemours & Company, Wilmington, Delaware 19899.

2a. Union Carbide Corporation, Bound Brook, New Jersey 08805.

2b. Current Address: Virginia Polytechnic Institute, Blacksburg, Virginia 24060.

3. S. A. Sundet, W. A. Murphey, and S. B. Speck, *J. Polym. Sci.,* **40**, 389 (1959).
4. S. L. Kwolek and P. W. Morgan, *J. Polym. Sci.,* **A2**, 2693 (1964).
5. P. W. Morgan and S. L. Kwolek, *J. Polym. Sci.,* **A2**, 181 (1964).
6. W. D. Jones and S. B. McFarlane (Celanese Corporation of America), U.S. Pat. 2,667,468 (Jan. 26, 1954) [*C.A.,* **48**, 5553g (1954)].
7. P. W. Morgan, *Condensation Polymers: By Interfacial and Solution Methods,* Wiley-Interscience, New York, 1965, Chapter VII.
8. F. E. Arnold, S. Cantor, and C. S. Marvel, *J. Polym. Sci.,* **A-1, 5**, 553 (1967).
9. H. A. Smith (Dow Chemical Company), U.S. Pat. 3,371,073 (Feb. 27, 1968) [*C.A.,* **68**, 96369e (1968)].
10. R. C. Evers and G. F. L. Ehlers, *J. Polym. Sci.,* **A-1, 5**, 1797 (1967).
11. S. L. Kwolek (E. I. du Pont de Nemours & Co.), U.S. Pat. 3,591,559 (July 6, 1971) [*C.A.* 75, 130452h (1971)].

Alternating Poly(trans-2,5-dimethylpiperazine-4,4'-methylenediphenylurea) by Low Temperature Solution Polymerization

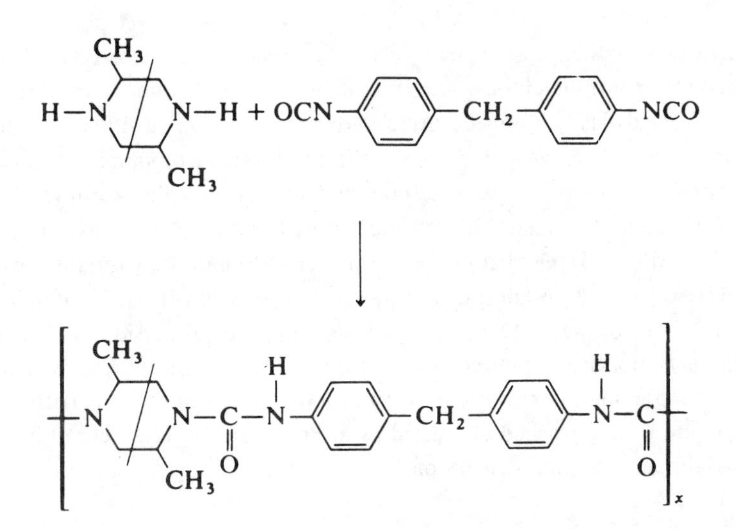

Submitted by S. L. Kwolek and P. W. Morgan (1)
Checked by J. E. McGrath (2a, b)

1. Procedure A

Methylenebis(4-phenyl isocyanate) (3.253 g.) (Note 1) is dissolved in 79 ml. of tetramethylene sulfone/chloroform (70/30 by volume) (Note 2) and placed in a 1-qt. explosion-proof blender (Note 3). The blender is turned to high speed with a rheostat, and *trans*-2,5-dimethylpiperazine (1.484 g.) (Note 1) in 79 ml. of tetramethylene sulfone/chloroform (70/30 by volume) is added rapidly. The resulting viscous solution is stirred at high speed (Note 4) for 7 min. The polymerization reaction is then quenched by stirring for 2 min. with 4 ml. of *n*-butylamine in 96 ml. of water (Note 5). More water is added to precipitate the polymer, which is collected by filtration on a medium-pore fritted-glass funnel (Note 6). The polymer is washed with two portions of acetone/water (1/1 by volume) and several portions of water by stirring in a blender. It is dried in a vacuum oven at 80°. The yield of polymer is 100% with an inherent viscosity of 2.95 dl./g. (3).

2. Procedure B

Methylenebis(4-phenyl isocyanate) (2.453 g.) (Note 1) is dissolved in 60 ml. of tetramethylene sulfone/chloroform (80/20 by volume) (Note 2) in a 500-ml., round-bottomed flask equipped with a stirrer and a dropping funnel. While the solution described above is vigorously stirred, 1.119 g. of *trans*-2,5-dimethylpiperazine (Note 1) in 50 ml. of tetramethylene sulfone/chloroform (80/20 by volume) is added dropwise. This is followed by a 9-ml. solvent rinse. The total time of addition of the diamine and rinse is 85 minutes. After the reaction solution is stirred for an additional 5 min. it is quenched with 4 ml. of *n*-butylamine in 96 ml. of water (Note 5). The resulting oily dispersion is stirred for 10 min. The mixture is poured into a 1-quart blender and stirred with more water. The polymer is collected by filtration on a medium-pore fritted-glass funnel (Note 6) and washed and dried as in Procedure A. The yield of polymer is 100% with an inherent viscosity of 2.34 dl./g. (3).

3. Characterization

Solubility. The polymer is soluble in dimethylformamide from which tough and flexible films can be cast.

Dilute Solution Viscosity. The inherent viscosity (η_{inh}) is determined in sulfuric acid at 30° with 0.5 g. of polymer per 100 ml. of solution. The intrinsic viscostiy can be approximated from the equation $\eta_{sp}/c = [\eta] + 0.34 [\eta]^2 c$.

Melt Temperature. The polymer melt temperature on a hot metal bar is 314°.

Number Average Molecular Weight. $\overline{M}_n$ determined for a polymer with $[\eta] = 1.34$ by the osmotic method is 23,200. The number average molecular weight can be determined approximately from the equation $\overline{M}_n = 15{,}480\ [\eta]^{1.40}$.

4. Notes

1. Both methylenebis(4-phenyl isocyanate) and *trans*-2,5-dimethylpiperazine are commercially available. Methylenebis(4-phenyl isocyanate) is subjected to two vacuum distillations. The colorless fraction which distills at 163-166° (0.3 torr) is used, *trans*-2,5-Dimethylpiperazine, m.p. 118° (3), is obtained by recrystallization from acetone.

2. Tetramethylene sulfone is distilled under reduced pressure through a 15-in. Vigreux column. The desired fraction distills at 101-102° (0.3 torr) and $n_D^{31.8°}$ is 1.4813.

The chloroform is A.C.S. reagent-grade that is freed of alcohol stabilizer by washing with three equal volumes of water. It is dried over anhydrous calcium chloride for several hours, filtered, and then stored over a mixture of approximately equal parts of anhydrous calcium chloride and potassium carbonate. Washed chloroform requires at least an overnight drying period before use. Hydrogen chloride and phosgene, which are present in decomposing chloroform, may be detected by the development of a precipitate on the addition of a drop or so of an aliphatic diamine or by the insertion of moist litmus paper into the vapors (4).

3. Before use the blender is thoroughly dried and the stirrer bearing is lubricated with "Celvacene" (Note 7) medium vacuum grease. The top is covered with aluminum foil and over this is placed a plastic cap which has a ¾-in. hole in the center. A powder funnel is inserted in this hole and a hole made directly below it in the foil.

4. High stirring speed is necessary when the reaction is of short duration or when the polymer precipitates; i.e., when chloroform alone is used as the solvent.

5. The polymerization reaction can also be quenched with ammonium hydroxide. When equivalents of reactants are employed in the preparation of the polyurea, similar inherent viscosities are obtained, regardless of the final work-up of the polymers. When excess isocyanate is present, however, the polymer increases in inherent viscosity and even becomes insoluble in sulfuric acid unless it is quenched. Both chain extension and crosslinking can take place during the isolation step when water is added to precipitate the polymer (5).

6. Paper filters contaminate the polymer with cellulose fibers which interfere in characterization tests made on the polymer.

7. Celvacene is a trademark of Distillation Products Industries.

5. Methods of Preparation

Polyureas have been prepared from diamines and diisocyanates in *m*-cresol, dimethylformamide, acetone, and alcohols. Diamines with phosgene have also produced polyureas. A third method used to prepare polyureas involves a reaction between biscarbamyl chlorides and equivalents of the same or different diamines. Literature references to these methods of preparation for the period through 1964 are summarized in *Condensation Polymers: By Interfacial and Solution Methods* (6). Since then much work has been published on polyureas, copolyureas, and modified polyureas. One method, for example, describes the preparation of polyureas from a diamine and biscarbamyl chloride with powdered magnesium as a catalyst (7).

6. References

1. Pioneering Research Division, Textile Fibers Department, E. I. du Pont de Nemours & Company, Wilmington, Delaware 19899.
2a. Union Carbide Corporation, Bound Brook, New Jersey 08805.
2b. Current Address: Virginia Polytechnic Institute, Blacksburg, Virginia 24060.
3. S. L. Kwolek, *J. Polym. Sci.*, **A2**, 5149 (1964).
4. P. W. Morgan and S. L. Kwolek, *J. Polym. Sci.*, **A2**, 181 (1964).
5. R. G. Arnold, J. A. Nelson, and J. J. Verbanc, *Chem. Rev.*, **57**, 51 (1957).
6. P. W. Morgan, *Condensation Polymers: By Interfacial and Solution Methods*, Wiley-Interscience, New York, 1965, Chapters 4 and 5.
7. M. Matzner, R. P. Kurkjy, R. J. Cotter and R. Barclay, Jr., *J. Polym. Sci.*, **B3**, 389 (1965).

Poly(phthalaldehyde)

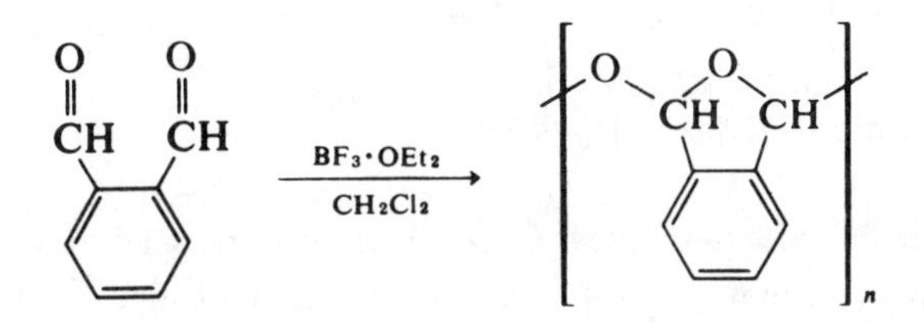

Submitted by C. Aso and S. Tagami (1)
Checked by F. L. Hedberg (2)

1. Procedure

A 100-ml., three-necked, round-bottomed flask is equipped with a thermometer and a nitrogen inlet tube. The third opening is stoppered with a rubber serum-bottle cap and used for the introduction of the reactants. In the flask, which is flushed with dry nitrogen in advance, are placed 5 g. of phthalaldehyde (Note 1) and 50 ml. of dry methylene chloride (*Caution! Many halogenated compounds are toxic and should be used with care.*) (Note 2). After the flask is swept again with dry nitrogen, the flask is capped and shaken gently by hand under a positive nitrogen pressure (Note 3) until the phthalaldehyde has dissolved. The color of the solution is light yellow-green. The flask is cooled externally to -78° in a Dry Ice-methanol bath and 0.047 ml. (1 mole % on monomer) of distilled boron trifluoride-diethyl ether complex is added dropwise with a hypodermic syringe inserted through the serum cap as the flask is shaken by hand (Note 4). During the addition the temperature of the reaction mixture should not rise above -70°. After the addition is completed, the polymerization is continued at $-75 \sim -78^{\circ}$ (Note 5). As polymerization proceeds, the color of the solution fades gradually.

At the end of 3 hr. the polymerization is terminated by addition of 1 ml.

of distilled pyridine with a hypodermic syringe while the flask is shaken by hand. After the flask is removed from the cooling bath, the reaction mixture is poured into 2 l. of distilled methanol with stirring. The resulting white powder is separated by filtration on a sintered-glass funnel, washed with methanol several times, transferred to a suitable container, and dried under vacuum (< 1 mm.) at room temperature for 24 hr. The product is 45.2 g. (90%) of poly(phthalaldehyde) which has a molecular weight of 11,000 and a softening point of 138-140°.

2. Characterization

Poly(phthalaldehyde) is soluble in benzene, chloroform, and dimethyl sulfoxide and insoluble in carbon tetrachloride and softens with decomposition at 110-140° (Note 6). Brittle films may be cast from solutions in methylene chloride or benzene. Because the polymer decomposes slowly to monomer in the presence of moisture, it should be stored in a desiccator.

The number average molecular weight $\overline{M}_n$ of the polymer is measured in benzene at 37° with a vapor pressure osmometer (Mechrolab. Company, Model 301 A). Intrinsic viscosity $[\eta]$ is obtained in benzene at 30°. The relationship between $[\eta]$ and molecular weight, $[\eta] = 7.8 \times 10^{-4} \overline{M}_n^{0.59}$, has been described in the published literature (3).

An infrared spectrum of the polymer shows the absorption bands of the acetal linkage in the region of 850-1180 $cm.^{-1}$ and the absence of carbonyl bands (1700 $cm.^{-1}$) which are observed in the monomer spectrum (3). An nmr spectrum of the polymer in dimethyl sulfoxide has three peaks: peak *A* (3.41 τ) and peak *B* (3.10 τ) may be assigned to the methine protons and peak *C* at 3.57 τ to the phenyl protons. The area ratio $C/(A + B)$ is in good agreement with the theoretical value of 4/2 (Note 7). These results strongly indicate that the polymer is composed entirely of the 1,3-dioxyphthalan repeat units.

3. Notes

1. Phthalaldehyde is prepared from *o*-xylene by the method of Bill and Tarbell (4). It is recrystallized from ligroin to give yellow-green crystals, m.p. 54.5-55.0°.

2. Methylene chloride is washed with an aqueous solution containing 5% potassium carbonate, dried overnight over phosphorus pentoxide and distilled over calcium hydride.

3. To provide an approximately constant positive pressure of nitrogen (approximately 10 mm.), the nitrogen inlet line is attached to the flask with a T-tube, the third end of which is dipped in mercury.

4. Because the polymerization is exothermic, the rapid addition of catalyst should be avoided. A solution of boron trifluoride-diethyl ether complex in methylene chloride may also be used.

5. The cationic polymerization of phthalaldehyde is an equilibrium polymerization, and the conversion and the molecular weight of the resulting polymer decrease with increasing polymerization temperature. The ceiling temperature of polymerization is -43° (3).

6. The decomposition of polymer to monomer at 180° in an evacuated ampoule has been described in the published literature (3).

7. The configuration of the dioxyphthalan ring in polymers is determined by comparing nmr spectra of polymers with those of *cis*- and *trans*-dialkoxyphthalans, as shown in the literature (5). Thus peaks A and B in the nmr spectrum of the polymer are presumed to correspond to the methine proton of the *cis*- and *trans*-cyclic units, respectively. The *cis*-content of the polymer repeat unit can be given by $A/(A + B)$.

4. Methods of Preparation

Poly(phthalaldehyde) can be obtained with other cationic catalysts, anionic catalysts, and coordination catalysts and by γ-ray irradiation of a methylene chloride solution of the monomer (3,5). The configuration of the ring in polymers changes with the catalysts used. *cis*-Rich polymers are obtained with cationic catalysts and *trans*-rich polymers are obtained with triethylaluminum-titanium tetrachloride catalyst (5).

5. References

1. Department of Organic Synthesis, Faculty of Engineering, Kyushu University, Fukuoka, 812, Japan.
2. Wright-Patterson Air Force Base, Ohio 45433.
3. C. Aso, S. Tagami, and T. Kunitake, *J. Polym. Sci.*, **A-1**, 7, 497 (1969).
4. J. C. Bill and D. S. Tarbell, *Org. Syn.*, Coll. Vol. 4, Wiley, New York, 1963, p. 803.
5. C. Aso and S. Tagami, *Macromolecules*, **2**, 414 (1969).

Enaminopolyamides

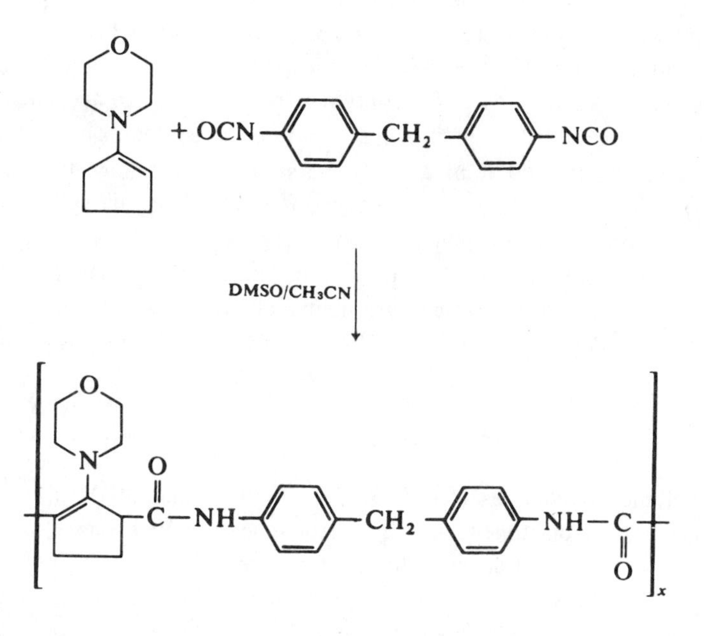

Submitted by W. H. Daly (1)
Checked by H. Panzik and J. E. Mulvaney (2)

1. Procedure

A 100-ml., three-necked flask or resin kettle is equipped with a precision ground stirrer, a nitrogen inlet, and a rubber serum stopper. The flask is flamed and purged with nitrogen. To provide a positive pressure of nitrogen the inlet line is attached to the flask with a T-tube that is also connected to a vented trap containing an inert dry hydrocarbon, such as mineral oil. Alternatively, the inlet may be attached to a small balloon filled with nitrogen. Dimethyl sulfoxide (DMSO) (Note 1) (10 ml.) and 10 ml. of a 1.5 *M* solution of 1-*N*-morpholino-1-cyclopentene (Note 2) in acetonitrile are injected through the serum cap. The reaction flask is immersed in a 20° constant temperature bath and vigorous stirring is initiated. A solution of 3.75 g. (0.015 mole) of methylenebis(4-phenyl isocyanate) (Note 3) in 10 ml. of DMSO is injected over a 5-min. interval. Although the reaction is initially exothermic, the exotherm subsides within 15 min. so that an external source of heat is required.

The polymerization flask is immersed in a 60° oil bath for 20 min. At this point the addition of 10 ml. of DMSO is required to reduce the viscosity and prevent gelation (Note 4). The polymerization is completed by heating for 120 min. As the viscosity continues to increase, 10-ml. aliquots of DMSO should be added to maintain a homogeneous system. The first aliquot can be injected after 60 min. at 60°; the second may follow after 90 min. (Note 5). After 2 hr. at 60° the oil bath is removed. The pale yellow, viscous solution of polyamide is poured immediately into 700 ml. of methanol agitated in an explosion-proof blender to precipitate the polymer. The polymer is collected by filtration, washed with methanol and dried *in vacuo* at 50° for 24 hr. (Note 6). A yield of 5.3-6.0 g. (87-99%) of white polyamide is obtained.

2. Characterization

The polymer is soluble in DMSO, dimethylformamide (DMF), dimethylacetamide, γ-butyrolactone, *N*-methylpyrrolidone, pyridine, cresol, and conc. sulfuric acid. The intrinsic viscosity ranges from 0.5 to 0.7 in DMF at 30°. A 20% by weight solution of the polymer in pyridine can be cast into pale yellow flexible films which exhibit the following absorption maxima in UV and IR: 318, 258 nm.; 1670, 1520 $cm.^{-1}$ (amide) and 1600 $cm.^{-1}$ (enamine $C = C$). The nmr spectrum of a 10% solution in hexafluoroacetone sesquihydrate consisted of a series of poorly resolved multiplets at 7.47 τ cyclopentane H; 6.72 τ and 6.08 τ morpholino H; 6.48 τ, ϕ-$CH_2\phi$; 2.67 τ phenyl H; 1.43 τ, NH. The polymer begins to decompose rapidly at 200° and does not melt. Because films of the enamino-polyamides darken and become brittle on exposure to light, the polymer should be stored in dark bottles to prevent photochemical degradation. The enamine structural unit can be hydrolyzed to the corresponding ketone derivative by dissolving 2 g. of polymer in 25 ml. of DMSO and adding 2 ml. of 98% formic acid. After 40 hr. at room temperature the polyketone can be isolated by pouring the reaction mixture into 400 ml. of methanol. The influence exerted by the enamine function on solubility is evidenced by the failure of the polyketone to dissolve in organic solvents other then DMSO. The polyenamine can also be converted directly to the corresponding 2,4-dinitrophenylhydrazone by treating 0.5 g. of polymer in 10 ml. of DMSO with 0.5 g. of 2,4-dinitrophenylhydrazine in 2 ml. of DMSO and 5 drops of conc. hydrochloric acid. The solution is allowed to stand at room temperature overnight and is then poured into 500 ml. of ethanol containing 50 ml. of 10% hydrochloric acid. The bright yellow polymer isolated begins to decompose at 202° and melts between 225-230°.

3. Notes

1. Both DMSO and acetonitrile are dried and deoxygenated by refluxing over calcium hydride with a nitrogen ebullator and distilled in a nitrogen atmosphere. DMSO, b.p. 88-89° (20 torr), $n_D^{20°}$ = 1.4783, is obtained by distillation under reduced pressure to minimize decomposition.

2. Commercially available 1-*N*-morpholino-1-cyclopentene is distilled over nitrogen to obtain material, b.p. 114° (20 torr), $n_D^{20°}$ = 1.5120. An approximately 1.5-*M* stock solution (23 ml. of enamine brought to 100 ml. of solution with acetonitrile) is prepared in a volumetric flask sealed with a serum cap and stored under nitrogen at −20°. The precise concentration of enamine is determined by injecting 3-ml. aliquots of the solution (after it is allowed to warm to room temperature) into 30 ml. of a 1:1 methanol-water mixture and titrating to the bromophenol blue end point with 0.1 *N* hydrochloric acid. Because the major impurity is cyclopentanone, the titration is a more accurate measure of the enamine concentration than gravimetric techniques.

3. Commercially available methylenebis(4-phenyl isocyanate) is vacuum distilled under nitrogen to give material, b.p. 148-150° (0.2 mm.). The distillate is recrystallized from petroleum ether (b.p. 50-80°). A slight excess (1-2 mole %) of diisocyanate may be used without severely affecting the molecular weight of the polyamide formed. Aromatic diisocyanates dimerize rather rapidly at room temperature; storage under nitrogen at −20° is recommended.

4. Maximum molecular weight is achieved by maintaining a high concentration of reactants during the initial stages of the polymerization. However, gelation occurs if the polymerization is heated at 60° for more than 1 hr. when the weight-percent of solids is greater than 8%. If gelation does occur, the gel can be redissolved by adding 25 ml. of DMF and heating at 95° for 2 hr. The time required for dissolution of the gel at 95° must be held to a minimum because thermal degradation of the polymer occurs rapidly at temperatures above 60°.

5. The quantity of DMSO required to prevent gelation depends on the power of the stirring apparatus. The checkers found that even three aliquots of DMSO did not prevent gelation. Nevertheless, treatment with DMF (Note 4) resulted in the yield and molecular weight reported here.

6. The polymer undergoes thermal degradation at temperatures above 60°. Therefore long drying times at low temperature are recommended. Precautions should also be taken against excessive exposure of the sample to light to prevent photodegradation.

4. Methods of Preparation

The polymerization of enamines is described in detail in the literature (3) and in a series of patents (4). The enamine structure as well as the diisocyanate can be varied to produce a variety of enamino-polyamides.

5. References

1. Louisiana State University, Baton Rouge, Louisiana 70803.
2. Department of Chemistry, The University of Arizona, Tucson, Arizona 85721.
3. W. H. Daly and W. Kern, *Makromol. Chem.*, **108**, 1 (1967).
4. G. A. Berchtold (E. I. duPont de Nemours & Company), U.S. Patent, 3,314,921 (Apr. 18, 1967) [*C.A.*, **67**, 22421y (1967)]; U.S. Patent 3,314,922 (Apr. 18, 1967) [*C.A.*, **67**, 12367w (1967)].

Poly-N-(β-propionamideoacrylamide

(A) $CH_2{=}CH(CO{-}NH_2) \xrightarrow{n\text{-BuLi}} CH_2{=}CH(CO{-}NH{-}CH_2{-}CH_2{-}CO{-}NH_2)$

(B) $CH_2{=}CH(CO{-}NH{-}CH_2{-}CH_2{-}CO{-}NH_2) \xrightarrow{AIBN} {-}[{-}CH_2{-}CH(CO{-}NH{-}CH_2{-}CH_2{-}CO{-}NH_2){-}]{-}_n$

Submitted by A. Leoni and S. Franco (1)
Checked by Y. Sasaki and J. E. Mulvaney (2)

1. Procedure

A. Synthesis of N-(β-propionamido)acrylamide

In a 3-1. flask 200 g. (2.82 moles) of acrylamide (freshly crystallized from chloroform and thoroughly dried) (Note 1) and 0.5 g. of *N*-phenyl-2-naphthylamine (Note 2) are dissolved in 1600 ml. of anhydrous dioxane (Note 3). A solution of *n*-butyllithium in hexane (1.6 *M*, 25 ml.) is added (Note 4) to the vigorously stirred solution in an atmosphere of pure dry nitrogen. The flask is then stoppered and kept for 24 hr. at room temperature with intermittent stirring. The solid, which separates, is collected by filtration, washed with small amounts of dioxane and ethyl ether, and dried to yield 140-160 g. of crude product.

The crude product is placed in a 2-1. flask, and 0.8 1. of anhydrous dioxane (dried with sodium and distilled) containing 0.4 g. of *N*-phenyl-2-naphthylamine is added. The mixture is heated to boiling for 10 min. The product is kept in contact with the boiling dioxane for 4-5 min. and the mixture is filtered while hot. The filtrate is then cooled to 15° and allowed to stand for 10 hr. at this temperature. After the *N*-(β-propionamido)acrylamide has crystallized, it is collected on a filter. The solid residue from the first extraction is treated twice with 0.8 l. of anhydrous dioxane as already described. The crystalline product obtained by cooling the three dioxane solutions is combined and recrystallized: it is placed in 0.8 l. of anhydrous dioxane (dried with sodium and distilled) containing 0.4 g. of *N*-phenyl-2-naphthylamine and the mixture is heated at 90-95° until almost complete dissolution is obtained. The mixture is filtered hot and the filtrate is kept at 15° for 24 hr. The crystalline solid is collected by filtration, washed on the filter with three 50-ml. portions of cold dioxane and three 50-ml. portions of ethyl ether, and dried under reduced pressure to yield 23-35 g. (12-18%) of *N*-(β-propionamido)acrylamide, m.p. 149-150° (Note 5).

B. *Polymerization of N-(β-propionamido)acrylamide*

Into a 1-l. flask, equipped with a stirrer, a nitrogen-inlet tube, and a reflux condenser are introduced 40 g. of *N*-(β-propionamido)acrylamide and 400 ml. of dry methanol. After 0.4 g. of α,α′-azobis(isobutyronitrile) has been added, the solution is heated under reflux in a stream of pure nitrogen for 4 hr. After the methanol has been decanted, the polymer, which has separated as a soft, sticky mass, is dissolved in 200 ml. of water. The solution is saturated with acetone, and the polymer is recovered by precipitation in acetone. The product is collected by filtration and dried to yield 35-39 g. (87-97%) of poly-*N*-(β-propionamido)acrylamide, a white powder readily soluble in water, with an intrinsic viscosity of 0.42 dl./g. in aqueous 1*M* sodium nitrate solution at 30°.

2. Notes

1. Acrylamide is a product commercially available. In some cases it is sublimed at reduced pressure (0.1 torr) according to the usual technique. No differences are observed when the sublimation is carried out.

2. *N*-Phenyl-2-naphthylamine is added as an inhibitor for vinyl polymerization. This is particularly useful for the extractions at elevated temperatures.

3. The dioxane is dried with sodium and distilled; it is then treated with potassium-benzophenone ketyl and distilled again. Potassium-benzophenone

ketyl has been prepared directly in the dioxane by allowing potassium and benzophenone to react at room temperature.

4. *n*-Butyllithium in hexane supplied by the Foote Mineral Company, Route 100, Exton, Pennsylvania, was used. The *n*-butyllithium is transferred from the bottle through the rubber plug with a glass syringe and quickly transferred into the reaction flask; the addition of the catalyst requires 1 to 2 min. When the catalyst contacts the monomer solution, an exothermic reaction takes place but efficient stirring dissipates the heat and cooling is not required. When the addition of catalyst has been completed, the electromagnetic stirrer is stopped and the flask is stoppered tightly. The titer of the *n*-butyllithium was determined shortly before use by the Gilman and Aubein (3) method.

5. The submitters have obtained a 39% yield of pure monomer.

3. Methods of Preparation

The method given here is described by S. Franco, A. Leoni, and M. Marini (4). The *N*-(β-propionamido)acrylamide has been also obtained by D. S. Breslow, G. E. Hulse, and A. S. Matlack (5) and later by H. Nakayama, T. Higashimura, and S. Okamura (6).

4. References

1. Research Division, Ferrania 3M, 17016 Savona, Italy.
2. Department of Chemistry, The University of Arizona, Tucson, Arizona 85721.
3. H. Gilman and A. H. Haubein, *J. Amer. Chem. Soc.*, **66**, 1515 (1944).
4. S. Franco, A. Leoni, and M. Marini (Ferrania Societa per Azioni), Belgian Patent 685,690 (1966); Italian Patent 731,800 (1965); French Patent 1,489,646 (July 21, 1967) [*C.A.*, **68**, 60066q (1968)].
5. D. S. Breslow, G. E. Hulse, and A. S. Matlack, *J. Amer. Chem. Soc.*, **79**, 3760 (1957).
6. H. Nakayama, T. Higashimura, and S. Okamura, *Polymer Previews,* **3**, 27 (1967).

Alternating Styrene-Methyl Methacrylate Copolymer

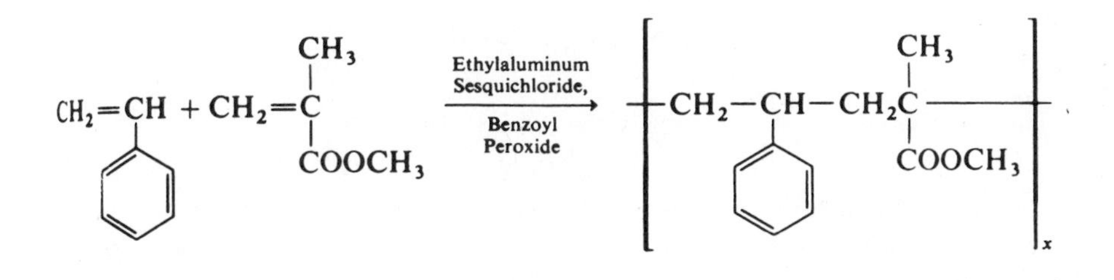

Submitted by B. K. Patnaik and N. G. Gaylord (1)
Checked by J. C. Falk and M. A. Benedetto, Jr. (2)

1. Procedure

A 250-ml., three-necked flask equipped with a nitrogen inlet and outlet, a thermometer, and a mechanical stirrer is charged with a mixture of 10.4 g. (0.1 mole) of distilled styrene, 10.0 g. (0.1 mole) of distilled methyl methacrylate, 0.24 g. (1 mmole) of benzoyl peroxide (*Caution! Benzoyl peroxide is a strong oxidant.*) (Note 1), and 50 ml. of toluene (Note 2). The reaction vessel is flushed with dry nitrogen for 20-25 min. and then placed in a constant temperature water bath at 40°. Ethylaluminum sesquichloride (50 mmoles, 25% solution in toluene) (*Caution!*) (Notes 3,4) is added dropwise with a hypodermic syringe. A mild exothermic reaction occurs, and the rate of addition of the ethylaluminum sesquichloride is controlled to maintain the temperature at 40° (Note 5). After the reaction is carried out for 5 hr. at 40°, the contents of the flask are poured into 500 ml. of methanol containing 25 ml. of 6 *N* hydrochloric acid. The product is collected by filtration, washed with two 50-ml. portions of methanol, and dried *in vacuo* at 40°. The crude yield of copolymer is 19.0 g. (93%). The crude copolymer is dissolved in toluene, the solution is filtered, and the filtrate is added to a large excess of methanol. The purified copolymer is obtained in a yield of 16.7 g. (82%).

2. Characterization

The copolymer has an intrinsic viscosity $[\eta] = 0.4\text{-}0.6$ dl./g. in toluene at 30° which corresponds to a molecular weight of 76,000-133,000 as calculated from the equation (3):

$$[\eta] = 1.09 \times 10^{-4} M_w^{0.73}$$

The copolymer has a softening point of 135-140°.

The alternating copolymer is characterized distinctively by the nmr spectrum (Fig. 1) obtained with a 10% deuterochloroform solution with tetramethylsilane as the internal standard at 100 MHz and 70°. Some of the characteristic features that distinquish the equimolar alternating copolymer from the free radical-catalyzed equimolar copolymer are the following:

1. The diamagnetic shift of the ortho protons is absent.
2. The methoxy-proton resonance appears in three sharp peaks centered at 6.66, 7.06, and 7.65 τ. These peaks represent the coisotactic, coheterotactic, and cosyndiotactic configuration, respectively (Note 6).
3. The methylene- and methine-proton resonances appear in the range 7.9-9.0 τ but the peaks are well defined.
4. The α-methyl-proton resonance appears in three distinct peaks at 9.05, 9.35, and 9.50 τ.

3. Notes

1. The reaction can also be carried out spontaneously in the absence of a radical catalyst, in which case the copolymer is obtained in a lower yield but has a higher molecular weight. When the procedure described herein is carried out in the absence of benzoyl peroxide, the crude yield is 39%; the purified copolymer obtained in a yield of 34% has $[\eta] = 2.35$ dl./g. in toluene at 30°.

2. Toluene must be dry and preferably freshly distilled. The reaction can also be carried out in bulk, but care must be taken to control the exothermic reaction. The amount of toluene can be varied. In general, the use of less

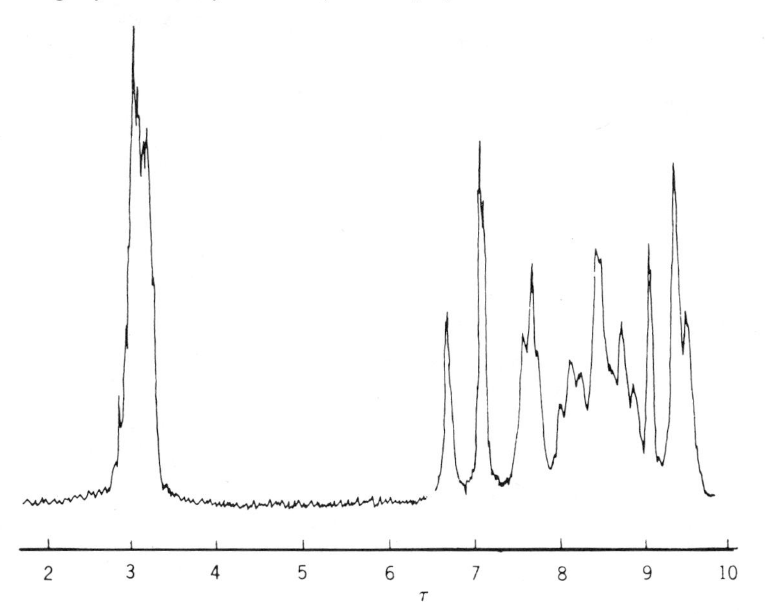

Fig. 1. N.m.r. spectrum of alternating styrene-methyl methacrylate copolymer in deuterochloroform at 100 MHz and 70°.

toluene results in a faster polymerization and a higher molecular weight copolymer.

3. The 25% solution of ethylaluminum sesquichloride in toluene, supplied by Texas Alkyls Company, is used with the usual precautions in handling the material. If a solution in another solvent, e.g., heptane or benzene, is used, the results will differ because of the difference in the solubility of the growing polymer chain.

4. The amount of ethylaluminum sesquichloride can be varied. In general, decreasing the amount of the sesquichloride results in a slower rate of polymerization and the formation of a copolymer of lower molecular weight.

5. The polymerization temperature can be varied over a wide range. The reactions carried out at lower temperatures result in a slow reaction, but the copolymer has a higher molecular weight. To obtain the copolymer in a reasonable yield and a medium molecular weight, a reaction temperature of 25-50° is recommended.

6. The cotactic configurations can be determined by measuring the areas under the corresponding peaks. The cotacticity appears to be dependent on the structure of the charge transfer complex (4,5) and the nature of the metal halide used. The tacticities, however, are not so dependent on the temperature of polymerization.

4. Methods of Preparation

Alternating styrene-methyl methacrylate copolymers can be prepared by use of other metal halides, such as stannic chloride, aluminum chloride, boron trifluoride (4,5), and zinc chloride (6). The polymerization can be carried out also by initially preparing the methyl methacrylate-metal halide complex, which may be isolated if desired, and by reacting this complex with styrene (4,5).

5. References

1. Current Address: 28 Newcomb Drive, New Providence, New Jersey 07974.
2. Borg-Warner Corporation, Research Center, Des Plaines, Illinois 60018.
3. T. Kotaka, T. Tanaka, H. Ohnuma, Y. Murakami, and H. Inagaki, *Polymer J.,* **1**, 245 (1970).
4. T. Ikegami and H. Hirai, *J. Polym. Sci.,* **A-1**, 8, 195 (1970).
5. T. Ikegami and H. Hirai, *J. Polym. Sci.,* **A-1**, 8, 463 (1970).
6. S. Yabumoto, K. Ishii, and K. Arita, *J. Polym. Sci.,* **A-1**, 8, 295 (1970).

Poly(9-vinyladenine)

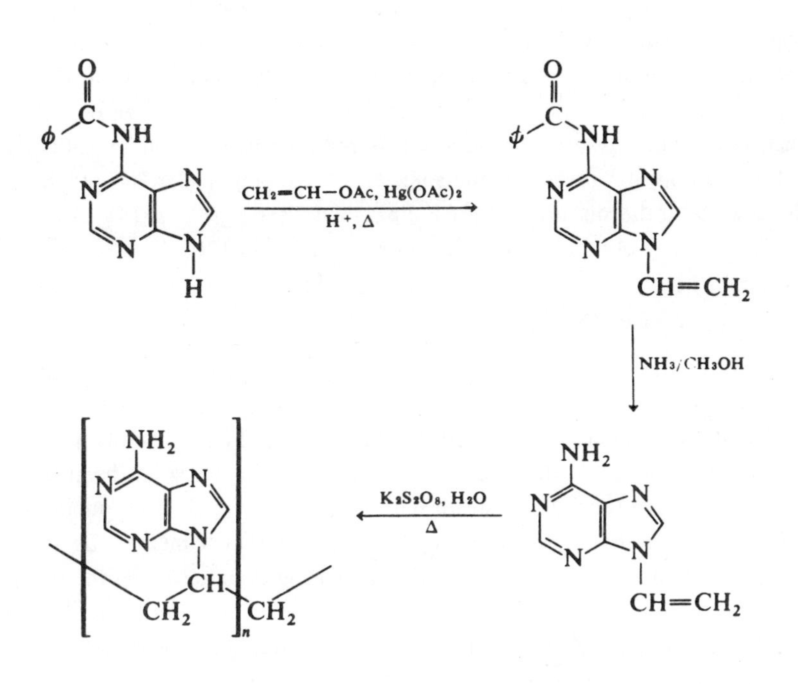

Submitted by H. Kaye (1a, b)
Checked by W. L. Hahn (2)

1. Procedure

A. 6-Benzoyladenine (3)

A mixture of adenine (14.8 g., 0.110 mole) and benzoic anhydride (74.5 g., 0.330 mole) is melted and then heated at 180° for 15 min. On cooling, the melt solidifies. The product is dissolved in boiling ethanol (900 ml.) and the solution is treated with decolorizing carbon and filtered. The filtrate is cooled to yield 19.4 g. (62%) of white needles of 6-benzoyladenine, m.p. 237-239°.

B. 6-Benzoyl-9-vinyladenine

6-Benzoyladenine (19.4 g., 0.0812 mole) is heated under reflux with vinyl acetate (200 ml.) (Note 1), mercuric acetate (2.5 g.), and 50% aqueous sulfuric acid (0.2 ml.) in a 1-1. flask equipped with an efficient condenser. After 24 hr. most of the 6-benzoyladenine dissolves. After 3 days a light brown solution remains and thin layer chromatography (Note 2) indicates nearly complete conversion to product. The reaction mixture is filtered while hot and the insoluble residue is extracted with two 50-ml. portions of boiling vinyl acetate. The combined solutions are placed in a freezer overnight to yield crude tan crystals of product. This material is dissolved in 1500 ml. of hot methanol-water (1:5) and the solution is treated with Darco G-60 absorbing charcoal. After the mixture is filtered, the cooled filtrate gives 10-14 g. (46.5-65%) of white needles of 6-benzoyl-9-vinyladenine, m.p. 168-170°.

C. 9-Vinyladenine (Note 12)

6-Benzoyl-9-vinyladenine (9 g., 0.034 mole) is magnetically stirred at room temperature with a saturated methanolic solution of ammonia (200 ml.) in a 500-ml., stoppered, round-bottomed flask. After 24 hr. a white precipitate forms but immediately dissolves after some of the ammonia has been removed by heating the mixture on a steam bath (Note 3). Thin layer chromatography indicates quantitative conversion to product (Note 4). The methanolic ammonia is removed under reduced pressure with a rotating evaporator and the residue is crystallized from 1 1. of hot benzene (Note 5) to yield 5-6 g. of product, m.p. 170-190°. This material is recrystallized from 600 ml. of hot benzene by allowing the solution to stand at room temperature for 24 hr. to yield 3.82 g. (70%) of 9-vinyladenine, as white needles, m.p. 198-200°; λ_{max} (0.1 *N* sodium hydroxide) 261 nm. (ϵ_{max} 11,861).

D. Poly(9-vinyladenine) (Note 12)

9-Vinyladenine (0.9 g., 0.0056 mole) and 90 ml. (Note 6) of distilled water are placed in a polymerization tube containing a Teflon® magnetic stirring bar, a side arm, and a rubber serum cap. A 12-in. stainless steel syringe needle is inserted through the serum cap so that the tip is well below the surface of the liquid. The mixture is purged with prepurified nitrogen for 1 hr., after which time 1.12 ml. of a 2.5 x 10^{-2} *M* potassium persulfate solution (Note 7) is added with a syringe through the serum cap. The side arm is closed and the nitrogen purge needle withdrawn. After the polymerization tube is sealed below the side arm, it is placed in a constant-temperature oil bath at 95° for 4 days. At the end of this time the tube is cooled and opened to give a faintly turbid solution that has increased in viscosity. The solution is filtered through a cellulose ester Millipore filter (RAWP 047 00), but the faint turbidity persists. The polymer is precipitated from 1 l. of dust-free methanol, and most of the methanol is removed by decantation (Note 8). After consecutively washing the polymer with methanol and ether, it is collected by centrifugation and dried at 100° (0.1 torr) for 24 hr. to yield 0.80 g. (89%) of poly(9-vinyladenine) as a white powder (Note 9): $\overline{M}_n$ = 106,000, $[\eta]$ = 0.25 (Note 10); λ_{max} (0.01 *M* sodium chloride and 0.01 *M* trishydroxymethylaminomethane solution, pH 7.4), 254 nm (ϵ_{max} 9,038). The polymer is hygroscopic and should be stored in a desiccator over phosphorus pentoxide.

Differential thermal analysis indicates a glass transition temperature at 300°, and thermogravimetric analysis indicates thermal stability to 375° (Note 11). The polymer is amorphous by x-ray diffraction and can be cast into brittle transparent films from aqueous solutions. It is insoluble in common organic solvents. The unperturbed dimensions of poly(9-vinyladenine) (7) and its e-mission spectrum (8) have been determined.

2. Notes

1. Commercial vinyl acetate containing 0.2% diphenylamine as a free radical inhibitor was used.

2. The reaction was followed by thin layer chromatography on Merck 254G silicic acid-covered microscope slides, and the chromatograms are developed with 4.75% methanol in chloroform (by volume). Spots were detected with short wavelength ultraviolet radiation.

3. If a less concentrated ammonia solution or a greater volume of methanolic ammonia is used, no precipitation will occur.

4. The thin layer chromatogram is developed on silicic acid with 10% methanol in chloroform.

5. Filtration of this solution must be carried out in a heated, jacketed funnel to avoid crystallization on the filter paper.

6. If the polymerization is carried out at higher concentrations of monomer, low yields of polymer are obtained.

7. The potassium persulfate solution was freshly prepared and purged with nitrogen. The checker used distilled water which was boiled and transferred under nitrogen.

8. The checker found it more convenient to collect the polymer on a fine sintered-glass funnel.

9. A faintly discolored granular product has been obtained on a few occasions. The checker found that this occurs when the moist polymer is allowed to dry before it is washed with methanol and ether.

10. The molecular weight was determined on a Mechrolab osmometer at 34.6° in 0.5 *M* sodium chloride and 0.01 *M* sodium cacodylate, pH 7.3, with a B-20 membrane. The intrinsic viscosity was determined in the same solvent system at 34.6° with a Cannon semimicro Ubbelhode viscometer.

11. These measurements were carried out under nitrogen with Stone DTA and TGA equipment.

12. This material is commercially available from the Vertizon Chemical Co., P.O. Box 11669, Houston, Texas 77039.

3. Methods of Preparation

An earlier account of this method of synthesis has been reported (4). Since then two alternate synthetic approaches have been published (5,6).

4. Applications

Poly(9-vinyladenine) complexes with polyuridylic acid (9); it inhibits cell-free protein synthesis (10) and murine leukemia virus replication (10).

5. References

1a. Department of Chemistry, Texas A amd M University, College Station, Texas 77840.
1b. Current Address: Howard Kaye & Associates, P.O. Box 39086, Houston, Texas 77039.
2. Textile Fibers Department, E. I. du Pont de Nemours & Company, Waynesboro, Virginia 22980.
3. J. Prokop and D. H. Murray, *J. Pharm. Sci.,* **54**, 359 (1965).
4. H. Kaye, *J. Polym. Sci.,* **B7**, 1 (1969).
5. K. Kondo, H. Iwasaki, N. Nakatani, N. Ueda, K. Takemoto, and M. Imoto, *Makromol. Chem.,* **125**, 42 (1969).
6. P. M. Pitha and J. Pitha, *Biopolymers,* 9, 965 (1970).
7. H. Kaye and H. J. Chou, *J. Poly. Sci.,* Polym. Phys. Ed., **13**, 477 (1975).
8. Y. Inaki, T. Renge, K. Kondo, and K. Takemoto, *Makromol. Chem.* **176**, 2683 (1975).
9. H. Kaye, *J. Am. Chem. Soc.* **92**, 5777 (1970).
10. F. Reynolds, D. Grunberger, J. Pitha, and P. M. Pitha, *Biochem.* **11**, 3261 (1972).
11. P. M. Pitha, N. M. Teich, D. R. Lowry, and J. Pitha, *Proc. Natl. Acad. Sci.,* U.S.A., **70** 1204 (1973).

Vinyl Chloride-Vinylidene Chloride Copolymer

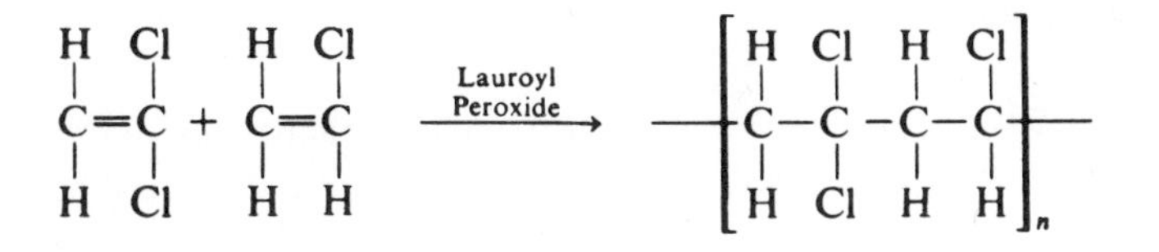

Submitted by M. R. Meeks (1)
Checked by R. F. Reinhart (2)

1. Procedure

A pressure polymerization vessel (Note 1) of approximately 1-gallon capacity is charged with 1400 g. of distilled or deionized water, 3.5 g. of lauroyl peroxide (*Caution! Peroxides are strong oxidants and should be used with care.*) (Note 2), and 100 g. of 2% methyl hydroxypropyl cellulose in water solution (Note 3). The vessel is evacuated to 28 in. of mercury with slow stirring (*Caution!*) (Note 4). Vinylidene chloride, 845 g., is sucked into the reactor after the polymerization inhibitor has been removed (*Caution!*) (Note 5), and 155 g. of vinyl chloride (Note 6) is forced into the reactor under nitrogen pressure (3). The suspension is stirred at 300 r.p.m. for 10-15 min. at room temperature to complete solution of the lauroyl peroxide. The slurry is then brought to 70° by passing hot water or steam through the coil or jacket. Stirring is continued at 300 r.p.m. throughout the polymerization. At operating temperature the polymerization becomes exothermic and heat must be removed (Note 7) to maintain 70°. The initial pressure, as the system reaches 70°, will be about 70 p.s.i. As the polymerization proceeds, the unpolymerized monomers become richer in vinyl chloride content and the pressure will steadily rise to a maximum of 118 to 120 p.s.i. When the pressure has dropped 2 p.s.i. below its peak, the excess monomers are vented to an adequate disposal system and the reactor is cooled with a flow of cold water. Polymerization time to a 2-p.s.i. pressure drop will be about 60 hr. The granular polymer is removed from the reactor, collected by filtration with suction, and washed thoroughly with cool water (Note 8). It is dried in an air-circulating oven at 50°. Total weight of the copolymer formed is about 90% of the total monomers loaded, and the monomers vented are high in vinyl chloride content (Note 9).

2. Characterization

Transitions. This copolymer shows two melting transitions, one at 155°, the other at 170°. The glass transition temperature is near 0°. The above values may be determined by differential thermal analysis at 20° temperature increase per minute.

Composition. Vinyl chloride content is 9-11% by weight and depends upon the degree of conversion. Pyrolysis-gas chromatography (4) can be used to determine total vinyl chloride content as well as its distribution in triads.

Flammability. This copolymer is recognized for its resistance to burning. Its high chlorine content (about 70%) makes it useful in applications where low flammability is required.

Raman Scattering. Laser-excited Raman spectra of this and other similar copolymers have been obtained by Meeks and Koenig (5). Intensities of certain scattering peaks were found to correlate with comonomer sequence concentrations.

Solubility. This copolymer is recognized for its impermeability and insolubility. Slight solubility is reported in hot tetrahydrofuran, and thionyl chloride has been used as a solvent at 80-90° (6). It will also dissolve in o-dichlorobenzene at 120°.

Dilute Solution Viscosity. A 2.0% solution of this copolymer in o-dichlorobenzene has an absolute viscosity of 0.86 to 0.90 cp. at 120° (7) (Note 10).

Molecular Weight. The number average molecular weight, as determined by gel permeation chromatography, is 35,000 to 45,000.

Absorption Spectrum. The infrared absorption spectrum of this copolymer has been published by Krimm and Liang (8).

3. Notes

1. This pressure vessel may be glass or stainless steel. It must be equipped with a stirrer, a pressure gauge, a thermocouple, and either a cooling jacket or coil to remove heat of polymerization. *The reactor must withstand pressures up to 120 p.s.i. at 70° and should contain a pressure release device for safe operation.* A suitable vessel has been described by Sutherland and McKenzie (9).

2. Lauroyl peroxide of commercial grade may be used without purification.

3. The 400-cP. methyl hydroxypropyl cellulose powder is added to a small quantity of hot water and stirred to form a suspension. Enough ice is added to make the concentration 2%, and stirring is continued until solution is complete at low temperature.

4. *The entire polymerization vessel should be in a hood and the stirring motor should be air-driven or explosion-proof.*

5. Inhibitor in vinylidene chloride is removed by washing with caustic. The monomer is shaken for several minutes with about one-half its volume of a 10% sodium hydroxide solution in a large separatory funnel. After the monomer layer has settled, it is separated and washed with two portions of distilled or deionized water and added immediately to the reactor. (*Caution! The inhibitors are phenolic and should be handled with care.*) Unhibited vinylidene chloride must be stored at −10° in the absence of air and light. (*Caution! Polymer and dangerous peroxides can form if the monomer is in contact with air at room temperature. Neither of the monomers should come into contact with copper or brass, because explosive copper acetylides may form. Vinylidene chloride is toxic and flammable and should be handled in a hood with proper eye protection.*)

6. *Vinyl chloride is a suspected cancer-causing agent! Persons handling this monomer should become thoroughly familiar with the government (OSHA) regulations concerning its use (10).*

Vinyl chloride may be obtained without polymerization inhibitor. It is best added to the polymerizer by volume from a calibrated shot tank containing a pressure sight glass (11). A stainless steel pipe serves well as such a vessel. (*Caution! Vinyl chloride is flammable and low boiling, b.p. −14.6°, and must be handled in a pressure vessel.*)

7. Temperature control is achieved with the thermocouple and a controller-recorder such as a Brown instrument with pneumatic output. The air output operates a pneumatic valve that controls the flow of cooling water through the jacket or coil.

8. A bowl-shaped vessel about 6 x 6 in. with a three-blade agitator of 1¾ in.

radius turned at 300 rpm will give granular polymer of 20 to 100 mesh size.

9. The checker carried out the reaction in a 15 gal. reactor and obtained a 78% conversion.

10. The checker found an inherent viscosity of 0.38 dl./g. in a 0.2% cyclohexanone solution at 30°. This corresponds to a number average molecular weight of 34,000.

4. Methods of Preparation

The procedure described here employs a suspension of monomer droplets, containing initiator, in water. Coalescence of the droplets is prevented by use of a water soluble cellulose derivative. Both poly(vinyl alcohol) and gelatin have been suggested as protective colloids (12). A method for the preparation of a vinyl chloridevinylidene chloride copolymer emulsion with Aersol MA emulsifier and water soluble initiators has appeared in the patent literature (13).

5. References

1. Dow Chemical Company, Midland, Michigan 48640.
2. B. F. Goodrich Chemical Company, Development Center, Avon Lake, Ohio 44012.
3. L. C. Friedrich, J. W. Peters, and M. R. Rector (Dow Chemical Company), U.S. Patent 2,968,651 (1961); Ger. Patent 1,069,384 (Nov. 19, 1959) [C.A., 55, 9961g (1961)].
4. S. Tsuge, T. Okumoto, T. Takeuchi, *Makromole ku lare Chemie.* **123**, 123-9, (1969).
5. M. R. Meeks, J. L. Koenig, *J. Polym. Sci.,* Part A-2, 9, 717-729 (1971).
6. Y. Yamashita, K. Ito, H. Ishie, S. Hoshino, and M. Kai, *Macromolecules,* **1** (6), 529 (1968).
7. ASTM-D729-57, *Book of ASTM Standards,* Part 26, American Society for Testing and Materials, 1916 Race St., Philadelphia, Pa., 1968, p. 43.
8. S. Krimm and C. Y. Liang, *J. Polym. Sci.,* **22**, 95 (1956).
9. J. D. Sutherland and J. P. McKenzie, *Ind. Eng. Chem.,* **48** (1), 17 (1956).
10. Federal Register 39, (194), 35890-35898, (1974) (OSHA Standard for Vinyl Chloride).
11. Jergusen Gage and Valve Company, Burlington, Massachusetts 01803.
12. R. A. Wessling and F. G. Edwards in H. F. Mark, N. G. Gaylord, and N. M. Bikales, Eds., *Encyclopedia of Polymer Science and Technology, Vol. 14,* Wiley-Interscience, New York, 1971.
13. P. K. Isacs and A. Trafimow (W. R. Grace and Company), U.S. Patent 3,033,812 (May 8, 1962) [C.A., 57:2430a (1962)].

Poly(1,2-dimethyl-5-vinylpyridinium salts)

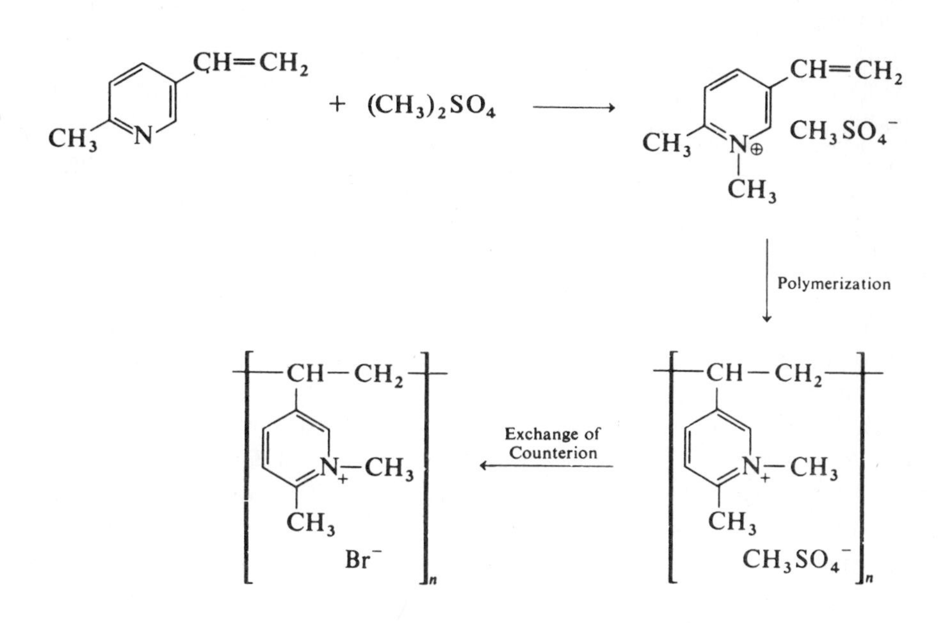

Submitted by W. P. Shyluk (1)
Checked by H. Schuttenberg (2a, b) and R. C. Schulz (2a, c)

1. Procedure

A. Monomer Synthesis

A solution of 59.5 g (0.50 mole) of 2-methyl-5-vinylpyridine (*Caution!*) in 200 ml. of acetone (anhydrous) is stirred magnetically in a screw-capped jar placed in an ice-water bath. An equimolar amount, 63.0 g., of dimethyl sulfate (*Caution!*) (Note 1) in 100 ml. of acetone is added, and the crystalline

1,2-dimethyl-5-vinylpyridinium methyl sulfate begins to separate in about 5 min. After 1 hr. the ice bath is removed and the stirring is continued for another hour. A sintered-glass filter stick is used to remove the supernatant solution. The pyridinium salt is washed three times by adding 150 ml. of acetone, stirring for 5 min. and removing the acetone through the filter stick. A mixture of 600 ml. of anhydrous ethanol and 6 g. of activated carbon is brought to boiling and the salt is added. After the mixture has been stirred for 5 min. near the boiling point, it is filtered through filter paper in a heated funnel and the filtrate is collected in a beaker placed in an ice-water bath. Crystallization is allowed to occur overnight at 5°. The crystals are collected on a filter funnel, resuspended in 100 ml. of cold ethanol, and recollected by filtration. A second recrystallization in 500 ml. of ethanol by a similar procedure gives 74-83 g. of 1,2-dimethyl-5-vinylpyridinium methyl sulfate (Note 2).

B. Homopolymerization

A solution of 7.00 g. of 1,2-dimethyl-5-vinylpyridinium methyl sulfate in 13.3 ml. of a 0.027% potassium persulfate solution in water is sealed in a 250-ml. pressure bottle with a two-holed crimped cap and a self-sealing Buna N liner. By using a two-way stopcock and a hypodermic needle the air is displaced by 15 cycles of evacuation, followed by repressurization to 10 p.s.i.g. with nitrogen. The bottle is kept in a 35° bath until the reaction mixture becomes viscous (about 50 min.). After 50 g. of water is added, the mixture is shaken until it is uniform and the mixture is dialyzed (Note 3). The conversion to polymer is determined by ultraviolet analysis (Section C) and it should be 10-20%. Other anions may be exchanged for the methyl sulfate by a dialysis technique; e.g., the poly-1,2-dimethyl-5-vinylpyridinium methyl suflate may be converted to poly(1,2-dimethyl-5-vinylpyridinium bromide) by the procedure in Note 4. It is recommended that the polymer be stored in solution in a refrigerator. The dry polymer may be obtained by freeze-drying the dialyzed solution. Isolation of the polymer by precipitation with a non-solvent such as acetone is usually complicated by the tendency of the polymer to separate in a colloidal form.

2. Properties

A. Variation of Molecular Weight

The procedure just described gives poly(1,2-dimethyl-5-vinylpyridinium bromide) with an intrinsic viscosity of 1.0-2.0 in 0.50 *M* potassium bromide at 25° with a No. 1 Ubbelhode viscometer. A plot of log η_{sp}/c at five polymer concentrations is used to determine the intrinsic viscosity according to Martin's equation:

$$\log \eta_{sp}/c = \log [\eta_{sp}/c]_{c=0} + kc\ [\eta_{sp}/c]_{c=0}$$

A relationship between the intrinsic viscosity and the molecular weight is not available. The intrinsic viscosity depends on the simple electrolyte that is used to repress the polyelectrolyte effect and on its concentration (4).

To decrease the intrinsic viscosity it is recommended that the initiation rate be increased. Poly(1,2-dimethyl-5-vinylpyridinium bromide), with an intrinsic viscosity of 0.80 in 0.50 *M* potassium bromide, is obtainable by carrying out a polymerization as described above for 5 min. with 7.00 g. of 1,2-dimethyl-5-vinylpyridinium methyl sulfate in 6.7 ml. of a 0.27% potassium persulfate solution and 6.7 ml. of a 0.10% sodium bisulfite solution. Unless the conversions are very low, increasing the initiation rate by increasing the temperature is not recommended because an increased effect of branching-crosslinking on the molecular weight as a result of chain transfer to the polymer is indicated (5). A decrease in the monomer concentration in the polymerization of the pyridinium salt at 35° without the addition of an initiator can also be used to decrease the intrinsic viscosity (3,6). Because high monomer concentrations are used in the latter "spontaneous" polymerizations, crosslinked poly(1,2-dimethyl-5-vinylpyridinium methyl sulfate) may be formed at a high conversion of monomer to polymer. Higher intrinsic viscosities are obtainable at higher monomer concentrations but the conversions must be kept very low to avoid branching-crosslinking.

B. Copolymerization

Low-conversion copolymerizations (below 10%) are recommended because the generally higher reactivity of 1,2-dimethyl-5-vinylpyridinium methyl sulfate (3) would be expected to give a changing copolymer composition with conversion. The reactivity of the monomer salt is also affected by the dielectric constant of the solvent (7).

C. Ultraviolet Analysis

The concentrations of monomer and polymer are determined by an ultraviolet analysis on a 0.005% solution in water. The following equations give the concentrations expressed as grams of poly(1,2-dimethyl-5-vinylpyridinium methyl sulfate) per liter:

$$\text{monomer} = 0.0308\Delta_{310}^{274} - 0.0344\Delta_{246}^{274}$$
$$\text{polymer} = 0.0537\Delta_{310}^{274} - 0.00463\Delta_{246}^{274}$$

where Δ_{310}^{274} is the difference in absorbance between 274 and 310 nm. and Δ_{246}^{274} is the difference in the absorbance between 274 and 246 nm. When the counterion is not the methyl sulfate, the apparent concentration is determined as described above and then corrected by multiplying by the ratio of the molecular weight of the monomer with the different counterion to the molecular weight of 1,2-dimethyl-5-vinylpyridinium methyl sulfate.

3. Notes

1. The 2-methyl-5-vinylpyridine and the dimethyl sulfate can be purchased from Phillips Petroleum, Inc., and Matheson, Coleman and Bell, respectively. These reactants are toxic and must be handled with extreme care.

2. The chemical composition can be checked by elemental analysis (3). This monomer is less hygroscopic than most of the vinylpyridinium monomers, but it should be kept dry to prevent spontaneous polymerization.

3. The diluted polymerization mixture is transferred to a 2-ft. length of 1 $^{1}/_{8}$-in. dialysis tubing (No. 4465-A2 from A. H. Thomas Company, presoaked in water for about an hour) that is sealed with a knot about 4 in. from the end. A knot is tied at the other end and the tube is looped into a 1-1. jar filled with distilled-deionized water (demineralizer was supplied by Atlas Electric Devices Company, Chicago, Illinois). While a rubber stopper holds the ends above the knots in place, the bottle is rotated end-over-end on a tumbler at 30 r.p.m. After 4 hr. of tumbling (or other type of agitation) the water surrounding the dialysis tube is replaced and the tumbling is continued for another 4 hr. This procedure is repeated until there are four water changes. There is a twofold increase in the volume of the solution in the dialysis sac.

4. The diluted polymerization mixture is dialyzed by the procedure in Note 3 three times against 1 1. of 0.4 *M* potassium bromide and five times against 1 1. of distilled-deionized water. On the basis of the ultraviolet analysis for polymer in Procedure C, the bromide concentration can be checked. [The author analyzed for bromide by the x-ray emission technique (8).]

4. References

1. Research Center, Hercules, Inc., Wilmington, Delaware 19899, Hercules Research Center Contribution No. 1519.
2. Institut für Makmolekulare Chemie der Technischen Hochschule, Darmstadt, West Germany.

2b. Current Address: Gonnet, La Plata, Argentina.

2c. Current Address: Institute fur Organische Chemie, Universitat Mainz, D65 Mainz, West Germany.

3. W. P. Shyluk, *J. Polym. Sci.,* **A2**, 2191 (1964).
4. W. P. Shyluk, *J. Appl. Polym. Sci.,* 8, 1063 (1964).
5. T. G. Fox and S. Gratch, *Ann. N. Y. Acad. Sci.,* 57, 367 (1953).
6. V. A. Kabanov, T. I. Patrikeeva, and V. A. Kargin, *Dokl. Phys. Chem. (USSR),* **168**, 405 (1966).
7. D. J. Monagle, *Amer. Chem. Soc., Div. Polym. Chem., Preprints,* **10** (2), 705 (1969).
8. H. A. Liebhafsky, H. G. Pfeiffer, E. H. Winslow, and P. D. Zemany, *X-Ray Absorption and Emission in Analytical Chemistry,* Wiley, New York, 1960.

Trifluoronitrosomethane-Tetrafluoroethylene Copolymer

$$n\,CF_3NO + n\,CF_2{=}CF_2 \longrightarrow {-}\!\!\left[N(CF_3)OCF_2CF_2 \right]\!\!{-}_n$$

Submitted by C. D. Padgett and E. C. Stump (1)
Checked by G. H. Crawford and D. E. Rice (2)

Caution! Tetrafluoroethylene, particularly in the uninhibited state, has been reported to detonate unpredictably. All operations involving this compound should be carried out behind a suitable barricade. For detailed information concerning the use of tetrafluoroethylene see Kirk-Othmer, *Encyclopedia of Chemical Technology (3). Although toxicity data on trifluoronitrosomethane has not been reported, it should be considered as extremely toxic and handled accordingly.*

1. Procedure

A. Suspension Polymerization

A 2-1. stirred autoclave with an internal cooling coil is charged with water (1000 g.), lithium bromide (530 g.), and magnesium carbonate (35 g.) (Note 1); it is then cooled to −183° evacuated, and charged with pure trifluoronitrosomethane (94 g., 0.095 mole) (Note 2) and pure tetrafluoroethylene (95 g., 0.95 mole) (Note 3). The autoclave is sealed and allowed to warm to −23°, at which temperature it is maintained while the contents are stirred for 36 hr. It is then vented in an efficient hood, allowed to warm to ambient temperature, and opened (Note 4). The product mixture is carefully acidified by the addition of

1:1 hydrochloric acid; there is no coagulation of polymer until the mixture becomes acidic. The polymer is removed from the reaction mixture and again washed with 1:1 hydrochloric acid to remove occluded magnesium carbonate. It is then dried in a vacuum desiccator to give 104 g. (56% conversion) of an opaque white elastomeric gum. Attempts to effect solution in Freon 113 (1,1,2-trichloro-1,2,2-trifluoroethane) and FC-43 (perfluorotributylamine) were not successful, because some insoluble material remained even after stirring for several days (Note 5 and 6).

B. Bulk Polymerization

A 900-ml., soft-iron cylinder is evacuated, cooled to -183°, and charged with pure trifluoronitrosomethane (53.5 g., 0.54 mole) (Note 2) and pure tetrafluoroethylene (54.0 g., 0.54 mole) (Note 3) and then sealed and placed in a -35° bath for 24 hr. The cylinder is then vented while it is allowed to warm to ambient temperature, openend, and found to contain 82 g. (76% conversion) of a clear colorless gum (Note 4). Intrinsic viscosity in FC-43, was found to be 1.15 dl./g. and, represents a molecular weight of 1.5×10^6 (Notes 6 and 7).

2. Notes

1. The lithium bromide serves as an antifreeze and the magnesium carbonate as a suspending agent. No initiator is required, because the diradical trifluoronitrosomethane (4) itself serves as an initiator.

2. Unless the purity of the trifluoronitrosomethane is above 99%, low molecular weight polymers result. One of the major impurities that must be removed is trifluoronitromethane. Commercial trifluoronitrosomethane was purified by passing the gas through a train of traps containing first a 5% aqueous sodium hydroxide solution, then calcium chloride, and finally No. 4A Linde Molecular Sieve maintained at -78°. The purified trifluoronitrosomethane is collected in a trap cooled in liquid air.

3. Commercial tetrafluoroethylene contains a terpene inhibitor that must be removed before polymerization. Pure tetrafluoroethylene was obtained by passing the commercial product through concentrated sulfuric acid and condensing the gas in a trap cooled in liquid air. A laboratory preparation of tetrafluoroethylene is given in Kirk-Othmer, *Encyclopedia of Chemical Technology* (3).

4. The unreacted monomers should be vented as the reaction mixture warms because they may react explosively at room temperature.

5. The insoluble material is polymeric, but it probably contains particles of

magnesium carbonate that are extremely difficult to remove. A discussion of the problem connected with effecting solution of the copolymer is given by Ball *et al.* (5).

6. Nmr analysis showed that the product was a 1:1 copolymer consisting primarily of alternating units of trifluoronitrosomethane and tetrafluoroethylene. A study of structural irregularities frequently found in a copolymer by nmr analysis has been reported (6).

7. A determination of molecular weight in relation to intrinsic viscosity in Freon 113 or FC-43 has been reported by the 3M Company (7).

3. Methods of Preparation

As first reported by Barr and Haszeldine (8), trifluoronitrosomethane and tetrafluoroethylene react exothermically at room temperature to give mostly perfluoro-2-methyl-1,2-oxazetidine with some low molecular weight copolymer of alternating trifluoronitrosomethane and tetrafluoroethylene units. At lower temperatures the formation of higher molecular weight copolymer is favored.

The reaction was thoroughly investigated by the 3M company (4, 7, 9, 10) and later by Thiokol Chemical Corporation (11), where, after carrying out polymerization in solution, bulk, and suspension systems, more than 200 lb. of copolymer was made in successive suspension polymerizations. As much as 39 lb. of copolymer was produced in a single run.

At Peninsular ChemResearch, Inc. (12), it was found that for the production of small quantities (100 g.) of the copolymer the bulk system was perferable because of the simplicity of the system and the easy isolation and purification of the product. The copolymers produced in these small-scale bulk systems were comparable to those produced in the suspension systems. The suspension system, however, is easier to scale up to larger amounts. Bulk polymerizations on a multipound scale have been successfully carried out at PCR, Inc. using a "thin film" polymerization technique (13). In this process the monomers are charged into metal cylinders that are then rotated about their longitudinal axis in a cold bath. As the polymer is formed it is deposited as a film on the cylinder wall. In this way no large amounts of insulation are formed between unreacted monomer and coolant and the heat of polymerization is efficiently dissipated.

4. References

1. PCR, Inc., Gainsville, Florida 32601.
2. Central Research Laboratories, 3M Company, St. Paul, Minnesota 55101.

3. S. Sherratt in Kirk-Othmer, *Encyclopedia of Chemical Technology,* 2nd ed., Vol. 9, Wiley-Interscience, New York, 1966, pp. 807-812.
4. G. H. Crawford, D. E. Rice, and B. F. Landrum, *J. Polym. Sci.*, **A1**, 565 (1963).
5. G. L. Ball, III, I. O. Salyer, J. V. Pustinger, and H. S. Wilson, "Physical and Rheological Properties of Fluoronitroso Rubbers," Technical Report 67-63-CM, U.S. Army Natick Laboratories, Natick, Massachusetts (13).
6. D. D. Lawson and J. D. Ingham, *J. Polym. Sci.*, **B6**, 181 (1968).
7. G. H. Crawford, D. E. Rice, and W. J. Frazer, "Arctic Rubber," final report under contract DA-19-129-QM-1684 (AD 418638) (13); G. A. Morneau, P. I. Roth, and A. R. Shultz, *J. Polym. Sci.*, **55**, 609 (1961).
8. D. A. Barr and R. N. Haszeldine, *J. Chem. Soc.*, 1881 (1955).
9. H. A. Brown, N. Knoll, and D. E. Rice, "Arctic Rubber," final report under contract DA-19-129-QM-1684 (AD 418638) (13).
10. D. E. Rice (3M Company), U.S. Patent 3,213,050 (Oct. 19, 1965) [*C. A.*, 59, 14126f (1963)]; G. H. Crawford, (3M Company), U.S. Patent, 3,399,180 (Aug. 27, 1968) [*C. A.*, **60**, 7009e (1964)].
11. J. Paustian et al., "Nitroso Rubber Research, Development and Production," final report under contract DA-19-129-AMC-69(X) (13).
12. E. C. Stump and C. D. Padgett, "Synthesis of New Fluorine-Containing Nitroso Compounds, Copolymers and Terpolymers," final report under contract DA-19-129-AMC-152(N) (AD 666801) (13).
13. C. D. Padgett and J. R. Patton, "Manufacturing Methods for Carboxy Nitroso Rubber," AFML-TR-72-185, final report under contract F-33615-71-C-1166, 1972.
14. Copies of these Government reports may be obtained by writing to Defense Documentation Center, Cameron Station, Alexandria, Virginia 22314.

Preparation of a Stereoregular Polysaccharide, [1→6]-α-D-Glucopyranan

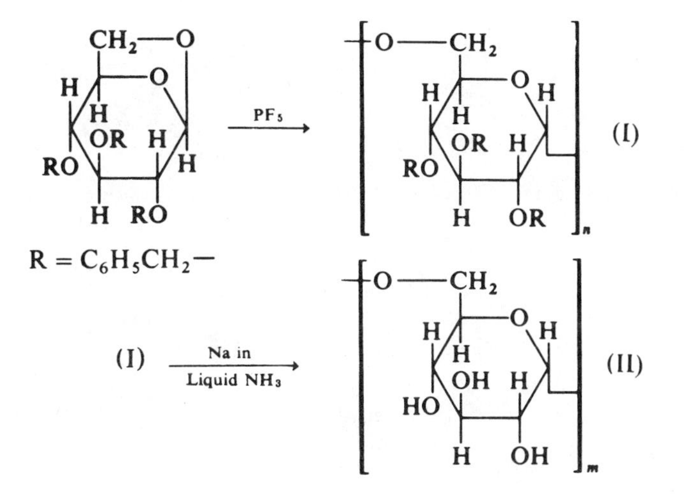

Submitted by C. Schuerch and T. Uryu (1)
Checked by John W.-P. Lin (2)

1. Procedure

A. Polymerization

The reaction vessel shown in Fig. 1 is constructed in three parts: a reaction tube (*A*) of approximately 50 ml. capacity with ground joint attached at 4, a solvent ampoule (*B*) of approximately 20-ml. total volume with a similar ground joint attached at 1 and a breakseal at 2, and a catalyst ampoule (*C*) of about 20-ml. volume with a Vigreux column at *E*, a breakseal at 6, and a ground joint

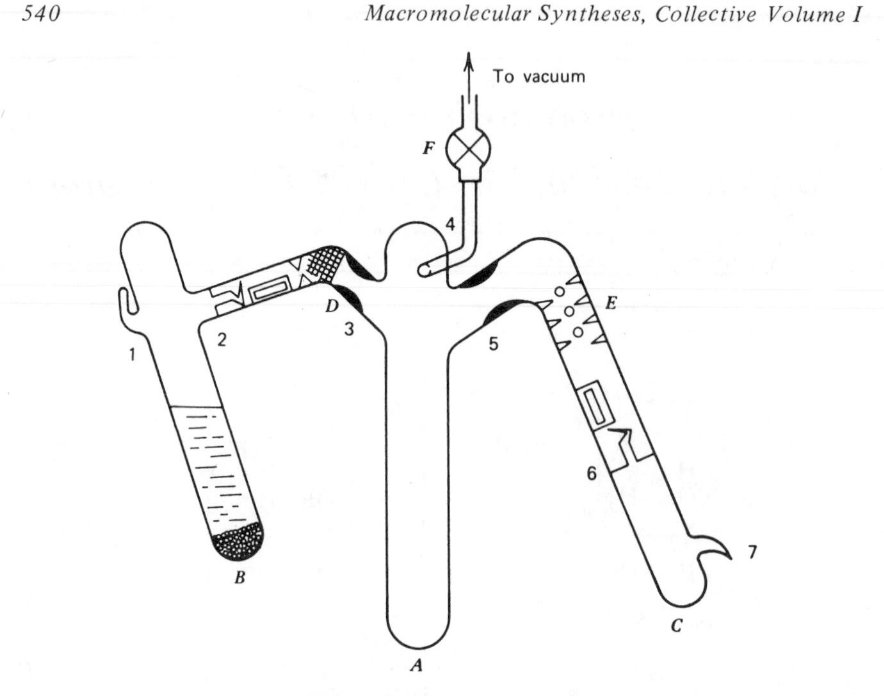

Fig. 1. Polymerization vessel.

at 7. (It is also convenient to have a medium-porosity glass filter situated near *D* to avoid having drying agent carried over into reaction tube during solvent transfer.) The two vessels *B* and *C* are separately attached to stopcocks on a high vacuum line (~10^{-5} torr), evacuated for a few hours with intermittent heating with a Bunsen flame and then filled with dry nitrogen (Note 1). The top of the solvent ampoule is removed, and 10 ml. of methylene chloride (*Caution! Many halogenated compounds are toxic and should be handled with care.*) (Note 2) is introduced with a few calcium hydride particles under a stream of nitrogen. The tube is chilled and the top is resealed after a short period of drying while excluding air with moderate nitrogen pressure. Methylene chloride is degassed three times by freezing with liquid air, evacuating the ampoule, closing the stopcock, and thawing. The solvent ampoule is then sealed at 1 with the solvent frozen and under high vacuum. The solvent is kept at least overnight before use. *p*-Chlorobenzenediazonium hexafluorophosphate (Notes 1 and 3) (17.0~34.0 mg., 6~12 $\times 10^{-5}$ mole) is added similarly to catalyst ampoule *C* and evacuated overnight before sealing at 7. Ampoules *B* and *C* are connected to the reaction vessel *A*, as shown (Fig. 1), with a glass wool plug *D* placed between *A* and *B* and constrictions at 3 and 5. The polymerization vessel is now evacuated, heated with a flame, and cooled, as were the other ampoules, and 5 g. (1.15×10^{-2} mole) of 1,6-anhydro-2,3,4-tri-*O*-benzyl-β-D-glucopyranose (Note 4) is added, as were previous reagents, and dried overnight on the vacuum rack.

Methylene chloride is chilled before breaking the breakseal (2) and distilled into the reaction vessel with stopcock *F* closed. When the monomer has completely dissolved, the solution is frozen in liquid air and the solvent ampoule is sealed at 3. The catalyst breakseal is broken next, and the catalyst precursor is decomposed by heating cautiously with a yellow flame. After the catalyst has condensed and frozen in the polymerization ampoule, the catalyst tube is sealed at 5 and the polymerization tube is sealed at 4. The reaction vessel is now allowed to thaw with shaking at a temperature not above $-60°$ and then allowed to stand with occasional shaking at $-60°$, for 3 hr. in a Dewar flask containing isopropyl alcohol. Gelation occurs within 1 hr.

The reaction vessel is opened and excess methanol is added at $-60°$ to terminate the polymerization. The precipitated polymer is dissolved in about 100 ml. of chloroform, and this solution is neutralized with at least 10% excess of sodium bicarbonate solution and washed with two portions of water. After the solution is dried over anhydrous sodium sulfate, and filtered, the filtrate is concentrated to about 15 ml. The polymer is precipitated by pouring the concentrate into 250 ml. of ligroin. Reprecipitations and performed twice from chloroform into ligroin. The polymer is freeze-dried from about 15 ml. of benzene and dried under high vacuum. The yield is about 4.8 g. (6%).

B. *Debenzylation*

Dry ammonia (200 ml.) is condensed at $-78°$ in a 250-ml., three-necked flask equipped with a magnetic stirrer, a coldfinger with a gas outlet, a gas inlet, and a dropping funnel. Sodium metal, 3.7 g. (0.161-g. atom), is put into liquid ammonia with stirring under a stream of nitrogen. 2,3,4-Tri-*O*-benzyl-[1→6]-α-D-glucopyranan (2.5 g., 5.78×10^{-3} mole) is dissolved in 40 ml. of 1,2-dimethoxyethane, which had been dried by passage through an alumina column, or in 1,2-dimethoxyethane-toluene mixtures (1:1).

The polymer solution is added dropwise to the liquid ammonia containing sodium at $-78°$ with stirring for 30 min. under a gentle stream of nitrogen. After a further 3.5 hr. of reaction ammonium chloride is added until the blue color of reaction solution completely disappears.

After 30 ml. of distilled water is added to the flask cautiously, the ammonia is allowed to evaporate under a stream of nitrogen overnight. About 220 ml. of distilled water is added, and the solution is shaken with methylene chloride and separated three times to remove any unreacted polymer. The polymer solution is dialyzed in cellulose tubing with running distilled water until ion-free (Note 5). The dialyzed solution is concentrated in a rotatory evaporator to about 25 ml. and freeze-dried, and the polymeric residue is dried under high vacuum. The yield of [1→6]-α-D-glucopyranan is 0.79 g. (84%) (Note 6).

2. Characterization

The specific rotation, $[\alpha]_D^{25}$, of 2,3,4-tri-*O*-benzyl (1) was +113.5° in chloroform solution (c, 1) (3). That of [1→6]-α-D-glucopyranan (II) was taken in dimethyl sulfoxide solution (*c*, 1) (Note 7).

The intrinsic viscosities of I and II were determined in chloroform and in dimethyl sulfoxide (protected from atmospheric moisture), respectively, at 25°. They were 0.95 and 0.29 dl./g.

Number averaged molecular weights were determined with a Hewlett-Packard 503 high-speed membrane osmometer. For polymer I toluene was used as solvent for measurements at 37° to give $\overline{M}_n = 3.33 \times 10^5$ ($\overline{DP} = 769$). The intrinsic viscosity and number average molecular weight are related by the equation (4)

$$[\eta] = 2.12 \times 10^{-4}\ \overline{M}_n^{\,0.66}$$

For polymer II the molecular weight was determined in dimethyl sulfoxide solutions at 37° to give $\overline{M}_n = 2.42 \times 10^4$ ($\overline{DP} = 149$).

Elemental analysis of the benzylated polymer invariably is within close limits of the calculated value. Initially, several samples of the glucan gave analyses corresponding to a hemihydrate.

Anal. Calcd. for $[C_6H_{10}O_5)_2 \cdot H_2O]_n$: C, 42.10; H, 6.48. Calcd. for $(C_6H_{10}O_5)_n$: C, 44.44; H. 6.22. Found: C, 40.76, H, 6.99; Ash, none.

This analysis presumably corresponds to a higher degree of hydration, and some variation in optical rotation of the glucan may reflect such differences.

3. Notes

1. This procedure for polymerization requires precautions to ensure the ab-

sence of impurities, especially water or hydroxylic solvents. When impurities are present, larger quantities of catalyst are required and slightly lower optical rotations and viscosities may be obtained (3, 4). Optimum polymer physical properties can be obtained at catalyst to monomer ratios of up to 1.3 mole %. Temperature control is also important, and no portion of the reaction vessel should be allowed to become warmer than -60° during polymerization.

2. Methylene chloride is purified by successive washings with concentrated sulfuric acid (several times), dilute sodium hydroxide solution, and water. It is dried over calcium chloride, heated to reflux over calcium hydride, and distilled.

3. *p*-Chlorobenzenediazonium hexafluorophosphate (Ozark Mahoning Company, 1870 South Boulder, Tulsa, Oklahoma; trademark Phosfluorogen A) is recrystallized from water, dried, and kept in a dark bottle.

4. 1,6-Anhydro-2,3,4-tri-*O*-benzyl-β-D-glucopyranose is prepared from β-D-glucose pentaacetate as a starting material (5). The crude monomer is recrystallized three times from absolute ethanol and finally from petroleum ether or methylene chloride-petroleum ether mixtures to give pure monomer, $[\alpha]_D^{25}$ -30.8° (*c*,1, in chloroform). Recrystallization from a hydrocarbon is essential to remove any traces of a hydroxylic solvent. The monomer should be stored in a dark bottle, preferably under nitrogen, to prevent peroxidation and benzaldehyde formation (odor!).

5. Seamless cellulose dialysis tubing (Fisher Scientific Company) is used. The dialysis can be done conveniently by suspending the tubing in a graduated cylinder and allowing a slow trickle of distilled water to pass through for 3 or more days. Loss of inogranics may be monitored with a conductivity cell or by a test for chloride ion in the polymer solution.

6. Similar conditions can be used to prepare a stereoregular mannan or galactan (6, 7) with comparable results. The galactan will require higher concentrations of monomer and considerably longer polymerization time.

7. The water content of dimethyl sulfoxide was undetectably small by nmr analysis i.e., $<0.1\%$ by volume. The optical rotation of the free polysaccharide is somewhat less reproducible than the benzyl derivative because of variations in water content. Usually the polysaccharide contains 5.25% water, one water molecule per two glucose units, and the specific rotation $[\alpha]_D^{25}$ (corrected by this amount) is 204°.

4. References

1. Chemistry Department, State University College of Forestry at Syracuse University, Syracuse, New York, 13210.

2. Chemistry Department, State University College of Forestry at Syracuse University, Syracuse, New York 13210, independent evaluation.
3. J. Zachoval and C. Schuerch, *J. Amer. Chem. Soc.*, **91**, 1165 (1969).
4. T. Uryu and C. Schuerch, *Macromolecules,* **4**, 342 (1971).
5. E. M. Montgomery, N. K. Richtmyer, and C. S. Hudson, *J. Amer. Chem. Soc.*, **64**, 690 (1942).
6. J. Frechet and C. Schuerch, *J. Amer. Chem. Soc.*, **91**, 1161 (1969).
7. T. Uryu, H. Libert, J. Zachoval, and C. Schuerch, *Macromolecules,* **3**, 345 (1970).

Polythiosemicarbazide from Methylenebis (4-phenyl isothiocyanate) and N,N'-Diaminopiperazine

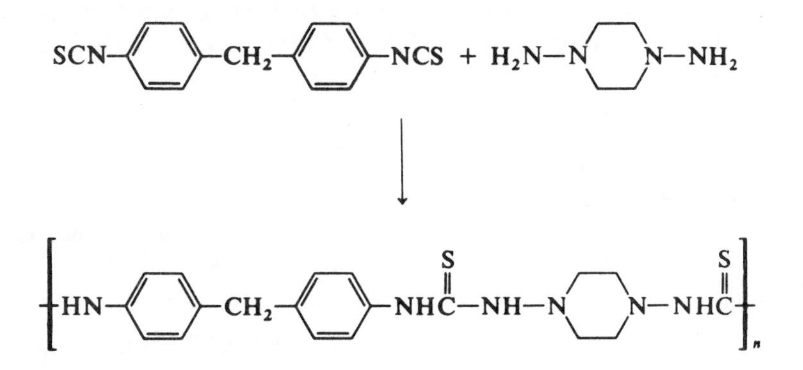

Submitted by V. S. Foldi (1) and T. W. Campbell (1)
Checked by W. J. Bailey and A. DeMilo (2)

1. Procedure

A. N,N'-Dinitrosopiperazine

To a solution of 194 g. (1 mole) of piperazine hexahydrate in 500 ml. of water and 250 ml. of concentrated hydrochloric acid at 15-25° is added dropwise and with stirring a solution of 150 g. (2.1 moles) of 97% sodium nitrite (*Caution! Note 5.*) in 300 ml. of water. After the mixture is stirred for 2 hr. the solid product is collected on a filter and washed with water to yield 128.5 g. (89%) of dinitrosopiperazine, m.p. 162-164° [lit. (3) 158°].

B. N,N′-Diaminopiperazine Dihydrochloride

To a stirred mixture of 144 g. (1 mole) of dinitrosopiperazine, 272 g. (4.16 g-. atoms) of zinc dust, and 1000 ml. of water contained in a 3-1., three-necked flask equipped with a stirrer, a condenser, and dropping funnel, is added dropwise, over a period of 2.5 hr., 600 ml. (10.5 moles) of glacial acetic acid at a temperature maintained at 20-30°. The mixture is stirred at room temperature overnight and at 80-85° for 1 hr. When the mixture is cooled, zinc acetate precipitates. This salt is removed by filtration and washed with 500 ml. of cold ethyl alcohol. The washings and another 500-ml. portion of alcohol are added to the filtrate, and 400 g. (11 moles) of hydrogen chloride gas is passed into the alcohol solution. When the mixture has cooled to 0°, the resulting solid product is collected by filtration to yield 117.6 g. (62%) of *N,N′*-diaminopiperazine dihydrochloride.

C. N,N,′-Diaminopiperazine (3)

To 150 ml. of ethanol contained in a 1-1., three-necked flask equipped with a stirrer, a condenser, and a powder funnel is added 45.3 g. (0.81 mole) of potassium hydroxide. The mixture is stirred until solution is effected. To this is added, in portions, 28.35 g. (0.15 mole) of diaminopiperazine dihydrochloride. The mixture is heated under reflux for 2 hr. and the insoluble inorganic material is removed by filtration and discarded. On concentration of the filtrate by evaporation to a small volume and cooling of the residue on ice, a light brown, pasty solid precipitates. This is collected and dried under nitrogen on a suction filter and recrystallized from 500 ml. of a 1:1 alcohol-ether mixture to give 14 g. (85%) of tan crystals. Sublimation of the tan product gives 10.4 g. (63%) of white *N,N′*-diaminopiperazine, m.p. 117-119°. The product could be further purified by recrystallization from chlorobenzene.

*D. Methylenebis(4-phenyl isothiocyanate) (4) (*Hood!*)*

In a 5-1., three-necked flask cooled with an ice bath is placed 150 g. of thiophosgene (*Caution! Thiophosgene is a highly reactive, toxic substance and should by handled with care.)* and 1 1. of ice water. A solution of 87 g. of 4,4′-diaminodiphenylmethane in 1 1. of chloroform is added with stirring over about 1 hr. The mixture is stirred at 0-10° for an additional 2 hr., then at room temperature overnight. The chloroform layer is separated and evaporated to dryness under a stream of nitrogen. The solid residue is dissolved in a mixture consisting of 400 ml. of benzene and 800 ml. of cyclohexane at the boiling point. The solution is decolorized, filtered, and allowed to cool. The fine needlelike precipitate is collected by filtration and washed with cold cyclohexane, as described above, the yield 84 g. (66%) of pure methylenebis(4-phenyl isothiocyanate), m.p. 141-142° (Note 1).

Anal. Calcd. for $C_{15}H_{10}N_2S_2$: C, 63.83; H, 3.52. Found: C, 63.76: H, 3.60.

E. *Polymerization of N,N'-Diaminopiperazine and Methylenebis(4-phenyl isothiocyanate)*

To a solution of 23.2 g. (0.20 mole) of *N,N'*-diaminopiperazine in 600 ml. of dimethyl sulfoxide (Note 2) in a 1-l., round-bottomed, three-necked flask equipped with a stirrer and a reflux condenser is added at about 50° 56.4 g. (0.20 mole) of powdered methylenebis(4-phenyl isothiocyanate). The mixture rapidly becomes viscous, and heating and stirring are discontinued after 2 hr.

The polymer is isolated by pouring the viscous solution into water. The tough polymer is subsequently washed with three-to-four 300-ml. portions of water in a blender and dried in a vacuum oven at 90° (Note 3). The yield is quantitative; the polymer has an inherent viscosity of 1.1-1.8 dl./g. at room temperature in dimethyl sulfoxide (0.5 g./100 ml. of solvent). The polymer melt temperature is about 230°.

The polymer is readily soluble in dimethyl sulfoxide; clear and tough films can be cast from such solutions, or fibers can be obtained by wetspinning a 12% solution into 50% aqueous dimethyl sulfoxide (Note 4).

2. Notes

1. The reaction products are ordinarily contaminated by traces of unidentified by-products with extremely repulsive odors. Rubber gloves were worn at all times until the product was obtained in a reasonable state of purity, at which point it was essentially odorless (4).

2. Dimethyl sulfoxide, b.p. 66° (5 torr), is purified by distillation. It is necessary to use highly purified dimethyl sulfoxide, because the molecular weight and color of the final polymer is strongly affected by the purity of the solvent.

3. Prolonged heating at 90° sometimes produces a slight discoloration of the polymer.

4. This polymer chelates various metals, particularly cupric ions. Trace amounts (parts per billion) of this element can be removed from water by filtration through a fibrous mat of the polymer (5).

5. 1,4-dinitrosopiperazine and sodium nitrite are listed in *The Toxic Substance List*, 1974 ed., U.S. Dept. of Health, Education and Welfare, p. 627 and 536, respectively (TL63000 and RA12250).

3. References

1. Pioneering Research Division, Textile Fibers Department, E. I. du Pont de Nemours & Company, Wilmington, Delaware 198999.
2. Department of Chemistry, University of Maryland, College Park, Maryland 20742.
3. A Schmidt and G. Wichmann, *Ber.*, **24**, 3245 (1891).
4. T. W. Campbell and E. A. Tomic, *J. Polym. Sci.*, **62**, 379 (1962).
5. E. A. Tomic, T. W. Campbell, and V. S. Foldi, *J. Polym. Sci.*, **62**, 387 (1962).

Poly(ferrocenylmethyl acrylate)

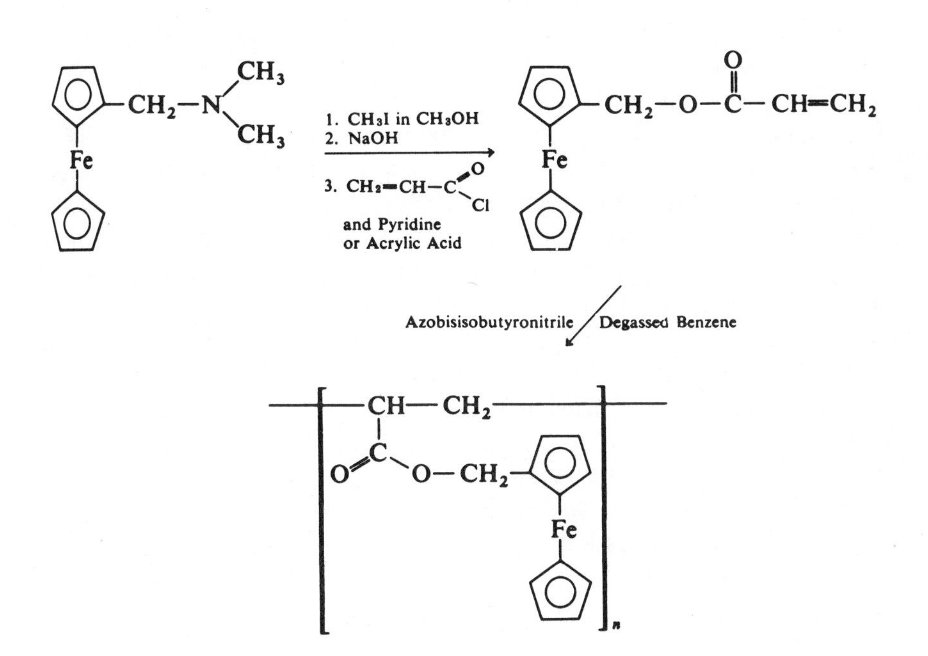

Submitted by C. U. Pittman, Jr., and J. C. Lai (1)
Checked by R. T. Conley (2)

1. Procedure

A. Ferrocenylmethyl Acrylate (3) (Note 14)

Because this key monomer is easily solvolyzed in methanol-water solvents, it should be used only in very dry solvents. A 1-1., three-necked, round-bottomed flask, equipped with a pressure-equalizing funnel, a condenser, and a mechanical stirrer, is placed in an external ice bath. The flask is charged with 48.6 g. (0.2 mole) of *N,N*-dimethylaminomethylferrocene (Note 1) in 40 ml. of absolute methanol, and a solution of 45.2 g. (0.3 mole) of iodomethane in 40 ml. of absolute methanol is added dropwise through the pressure-equalizing funnel. The clear solution is heated under reflux for 5 min., and 500 ml. of diethyl ether is then added. The quaternary salt immediately precipitates and is collected by filtration on a Buchner funnel, washed with ether until colorless, and dried to yield 73 g. (95%) of the methiodide as yellow crystals, which decompose slowly on heating to 220°.

A 2-1., one-necked flask, equipped with a condenser, is charged with 200 g. (0.52 mole) of the methiodide salt and 700 ml. of 3 *N* sodium hydroxide solution. The mixture is heated under relux for 6 hr. with evolution of trimethylamine from the top of the condenser (Note 2). After the mixture has been allowed to cool, the resulting oily material is extracted into 500 ml. of diethylether. The ether solution is washed with distilled water until the washings are neutral to litmus and dried over anhydrous sodium sulfate; the mixture is then filtered. The solvent is removed from the filtrate by rotary evaporation and the residual powder is recrystallized from hexane to give 101 g. (90%) of hydroxymethylferrocene as yellow plates, m.p. 81-82° (4).

Hydroxymethylferrocene is esterified by two methods: (a) with acrylyl chloride and pyridine, (b) direct esterification with acrylic acid (Note 3).

(a) Into an oven-dried, 500-ml., three-necked flask, equipped with a condenser, a pressure-equalizing funnel, and a mechanical stirrer and placed in an external ice bath, 10.8 g. (0.05 mole) of hydroxymethylferrocene and 7.0 ml. of pyridine are mixed into 300 ml. of anhydrous diethyl ether (Note 4). A solution of 6.4 g. (0.06 mole) of acrylyl chloride (Note 5) in 50 ml. of anhydrous ether is added dropwise from the pressure-equalizing funnel over 30. min. The precipitation of pyridine hydrochloride is immediate. After 2 hr. the mixture is diluted with 200 ml. of ether, washed with two 200-ml. portions of cold aqueous 0.2 *N* sodium bicarbonate solution and three portions of an excess sodium choride solution. After the ether layer is dried over anhydrous sodium sulfate, the mixture is filtered, the solvent is removed from the filtrate by rotary evaporation, and the residual powder is recrystallized from hexane to give 7.45 g. (55.5%) of ferrocenylmethyl acrylate, m.p. 42-43° (Note 6). Key infrared bands are observed at 3110, 2980-2860, 1720, 1625, 1635, 1460, 1400, 1385, 1280, 1190, 1115, 1050, 994, 955, 937, 820, and 740 $cm.^{-1}$ in a potassium bromide pellet. Nmr bands are unsubstituted cyclopentadienyl ring hydrogens, 5.96τ (s); substituted ring hydrogens, 6,00τ (tr) and 583τ (tr), J = 1.5 Hz; CH_2 5.15τ (s); and the vinyl hydrogens, exhibit an ABC pattern, 4.49-3.60τ below internal tetramethylsilane.

(b) A 500-ml., three-necked flask, equipped with a condenser, a nitrogen inlet, and a mechanical stirrer, is charged with 10.8 g. (0.05 mole) of hydroxymethylferrocene, 18.0 g. (0.25 mole) of unpurified commercial acrylic acid, 0.1 g. of hydroquinone, 0.05 g. of *p*-toluenesulfonic acid, and 300 ml. of methylene chloride. The mixture is heated under reflux for 5.0 hr. and then cooled. The solution is filtered and the filtrate is washed with 250 ml. of a 20% aqueous sodium carbonate solution at 0°, then several times with distilled water, and dried over anhydrous sodium sulfate. The mixture is then filtered and the filtrate is evaporated to dryness. The residual powder is recrystallized from hexane to give 1..4 g. (85%) of ferrocenylmethyl acrylate.

B. Polymerization of Ferrocenylmethyl Acrylate

Ferrocenylmethyl acrylate (15 g. 0.088 mole), prepared shortly before use and stored at −15° in the dark, is weighed into a Fischer-Porter aerosol compatability tube (Note 7). Reagent grade benzene, 20 ml., distilled over phosphorus pentoxide shortly before use, and 0.1 g. (6.1×10^{-4} mole) of commercial azobisisobutyronitrile (Note 8) are added. The tube is then equipped with its valve, and the solution is degassed at 10^{-3} torr by three alternate freeze-thaw cycles (Note 9). After the degassing is complete, the tube is placed in a constant temperature water bath, controlled to ± 0.01°, at 60° for 120 hr. The tube is then removed, cooled, and the polymer precipitated on the addition of the benzene solution to 800 ml. of 30-60° petroleum ether with vigorous stirring. The polymer is collected by filtration, redissolved in benzene, and reprecipitated twice more. The solvent is removed under vacuum to give 7.0 to 9.3 g. (47 to 62%) of poly(ferrocenylmethyl acrylate). The molecular weight was $\overline{M}_n$ = 17,300, $\overline{M}_w$ = 36,800. This method can also be used to produce polymers of ferrocenylmethyl methacrylate.

The copolymerizations of 1-ferrocenylmethyl acrylate and methacrylate have been described for the following comonomers: methyl acrylate (10), styrene, (10) methyl methacrylate (10), maleic anhydride (11) and N-vinyl-2-pyrrolidone (11). In addition, the solution homopolymerization (12) and copolymerization (13) of 2-ferrocenylethyl acrylate and methacrylate have been described.

2. Characterization

The polymer is an orange-yellow solid, which is easily ground into a powder and may be cast into films from benzene solutions. The polymer may be produced as described above at polymerization temperatures from 60 to 120°

for 60 to 80 hr. in yields above 40%. The yields vary, depending on the amount of benzene used, and can be raised to 90% by removing some of the benzene during the polymerization. The formation of a benzene insoluble, crosslinked fraction was noted at higher temperatures and low benzene concentrations.

The molecular weights are determined by vapor pressure osmometry and gel permeation chromatography (Note 10). The polymers had values of $\overline{M}_n$ between 5600 and 24,000 and $\overline{M}_w$ between 9000 and 82,000 when prepared by this method. The GPC curves indicated homogeneous homopolymers had been produced. The value of $[\eta]$ in benzene at 25° was .085 dl./g., when $\overline{M}_n$ = 17,300 and $\overline{M}_w$ = 36,800. The Mark-Houwink equations of $[\eta] = 6.84 \times 10^{-3} [M_n]^{0.75}$ and $[\eta] = 1.16 \times 10^{-1} [M_w]^{0.42}$ fitted the experimental data (Note 11). The ultraviolet spectrum of poly(ferrocenylmethyl acrylate) exhibits λ max at 430 and 237 nm. with extinction coefficients of 105 and 4950 in methylene chloride solution. These values are close to those of ferrocene and demonstrate that the ferrocene nucleus is not oxidized (6).

The Mössbauer spectrum (Note 12) exhibits a doublet with an isomer shift of 0.78 mm./sec. and a quadrupole splitting of 2.42 mm./sec. These values are very close to those of ferrocene itself (5). The infrared spectrum was devoid of the carbon-carbon double bond stretch at 1625-1635 $cm.^{-1}$ found in the monomer. Bands were found at 3095, 2290-2860, 1457, 1360, 1218, 1103, 1038, 1020, 997, 805, and 670 $cm.^{-1}$ Differential scanning calorimetry gave broad curves for these polymers from which a "glass transition" temperature of 190-210° may be suggested.

The kinetics of these homopolymerizations, measured by dilatometry (5, 8, 9) were first order in monomer concentration and one-half order in azobisisobutyronitrile concentration (Note 13). The Arrhenius activation energy for the polymerization was 18.7 kcal./mole between 50 and 70°. The kinetics indicate a classic first-order polymerization.

3. Notes

1. *N,N*-Dimethylaminomethylferrocene was obtained from Arapahoe Chemical Company, Boulder, Colorado 80302. It may be used without further purification.

2. Trimethylamine is readily detected both by its odor and by its reaction with moist red litmus paper.

3. Higher yields were always obtained by the direct esterification route in the preparation of ferrocenylmethyl acrylate. However, the yields of the related monomer, ferrocenylmethyl methacrylate, were about 85% by either route.

4. Pyridine and diethyl ether are distilled from powdered calcium hydride before use.

5. Acrylyl chloride was obtained from Borden Monomer-Polymer Laboratories, 5000 Langdon Street, Philadelphia, Pennsylvania. It was vacuum-distilled before use.

6. Ferrocenylmethyl acrylate appears pure by thin layer chromatography (silica gel/benzene-chloroform and alumina/chloroform) after a second recrystallization from hexane which does not change the melting point appreciably. After six careful crystallizations from hexane a rapid increase in the melting point to 69.5-70° occurs. This material exhibits the same TLC, infrared spectrum, n.m.r. spectrum, and polymerization behavior as the lower melting material. The elemental analysis was also unchanged within experimental error.

Anal. Calcd. for $C_{14}H_{14}FeO_2$: C, 62.26; H, 5.22; Fe, 20.67. Found: C, 62.57; H, 5.36; Fe, 20.31.

7. Obtained from Fischer-Porter Company, Lab-Crest Scientific Division, Warminster, Pennsylvania 18974.

8. The azobisisobutyronitrile was recrystallized three times from absolute methanol to produce material, m.p. 102-103° with decomposition, and stored at − 15° in the dark.

9. Liquid nitrogen was used as a freezing medium.

10. A Waters Associates Model 200 gel permeation chromatograph, packed with a bank of four styragel columns and standardized with commercial monodisperse polystrene fractions, was used to obtain the GPC curves. A Mechrolab vapor pressure osmometer was used to obtain the absolute measure of $\overline{M}_n$ and this value was used to calibrate the GPC curve at the count number corresponding to the number average chain length. A Q factor of 90.8 was determined for poly(ferrocenylmethyl acrylate) (3,5).

11. These equations were determined on unfractionated polymers and each sample had a different value of $\overline{M}_w/\overline{M}_n$. Thus the values of K and α must be considered preliminary.

12. The Mössbauer spectra were collected in a Nuclear Data, series 2200, multichannel analyzer operated in the multiscaling mode. The preamplifier, linear amplifier, linear gate, and Mössbauer drive unit (Model 54) were supplied by Austin Science Associates. The spectra were taken at 77°K. on sample sizes in the range of 25-38 mg./cm^2. The isomer shifts are reported relative to nitroprusside (7).

13. The specific volume of this monomer was 0.7943 ml./g. and that of the polymer was 0.7302 ml./g. to give a ΔV/mole of 17.3 ml./mole in this homopolymerization.

14. This monomer is now sold by Polyscience Corp., Paul Valley Industrial Park, Warrington, Pennsylvania 18976. The related monomer 1- and 2-ferrocenylethyl acrylate, 1- and 2-ferrocenylethyl methylacrylate are also available from the same source.

4. References

1. Department of Chemistry, University of Alabama, University, Alabama 35486.
2. Dean of Science and Engineering, Wright State University, Dayton, Ohio 45431.
3. C. U. Pittman, Jr., J. C. Lai, D. P. Vanderpool, M. Good, and R. Prados, *Macromolecules,* **3**, 746 (1970); J. C. Lai, T. D. Rounsefell, and C. U. Pittman, Jr., *Macromolecules,* **4**, 155 (1971); C. U. Pittman, Jr., J. C. Lai, and D. P. Vanderpool, *Am. Chem. Soc., Div. Polymer Chem., Preprints,* **11** (2), 1149 (1970).
4. J. K. Lindsay and C. R. Hauser, *J. Org. Chem.,* **22**, 355 (1957).
5. J. C. Lai, Ph.D. Dissertation, University of Alabama, 1970.
6. M. Rosenblum, *Chemistry of the Iron Group Metallocenes,* Part I, Wiley, New York, 1965, p. 40.
7. Samples were analyzed by Dr. M. Good at Louisiana State University in New Orleans.
8. C. U. Pittman, Jr., J. C. Lai, and D. P. Vanderpool, *Macromolecules,* **3**, 105 (1970).
9. C. U. Pittman, Jr., J. C. Lai, D. P. Vanderpool, M. Good, and R. Prados, *Macromolecules,* **3**, 746 (1970).
10. J. C. Lai, T. D. Rounsefell, and C. U. Pittman, Jr., *Macromolecules,* **4**, 115 (1971).
11. O. E. Ayers, S. P. McManus, and C. U. Pittman, Jr., *J. Polym. Sci., Chem. Ed.,* **11**, 1201 (1973).
12. C. U. Pittman, Jr., R. L. Voges, and W. R. Jones, *Macromolecules,* **4**, 291 (1971).
13. C. U. Pittman, Jr., R. L. Voges, and W. B. Jones, *Macromolecules,* **4**, 298 (1971).

Polyorthocarbonate

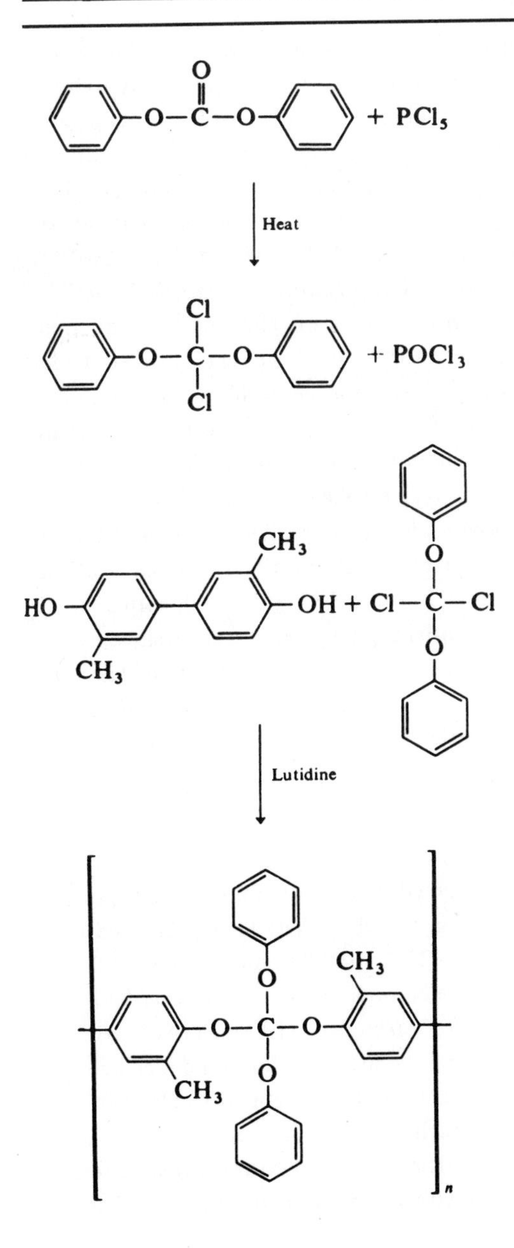

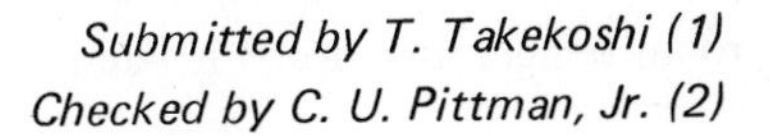

Submitted by T. Takekoshi (1)
Checked by C. U. Pittman, Jr. (2)

1. Procedure

A. Dichlorodiphenoxymethane (3)

This key intermediate is extremely sensitive to moisture. Accordingly, the entire procedure must be conducted in an absolutely dry atmosphere. A 100-ml., three-necked flask is fitted with a thermometer and a distillation apparatus equipped with a 15-cm. packed column (Note 1). The flask is flushed with dry nitrogen and charged with diphenyl carbonate (42.8 g., 0.2 mole) and phosphorus pentachloride (*Caution! Phosphorus halides evolve hydrogen chloride on contact wtih moisture.*) (41.7 g., 0.2 mole). The flask is heated at 175-180° for 5 hr., during which time phosphorus oxychloride (*Caution! Phosphorus halides evolve hydrogen chloride on contact with moisture.*) is removed by distillation. The reaction mixture is then heated at 200° for 1 hr. and distilled under vacuum. After the removal of unreacted phosphorus pentachloride (Note 2) two fractions boiling at 170-174° (10 torr) (18 g.) and 175-180° (10 torr) (24 g.) are collected. The first fraction consists mainly of the unreacted diphenyl carbonate and the last is the crude dichlorodiphenoxymethane, m.p. 42-45° (sealed tube), in a 44% yield. The purity of the product is 94.1%, as determined by titration (Note 3). The crude dichlorodiphenoxymethane is redistilled through a 30-cm. packed column (Note 1) to give material, b.p. 178-179° (8.5 torr), with a purity of 96.35% (Note 4). Alternately, substantially purer material is obtained by use of a Teflon spinning-band column, if the apparatus is kept leak-free. The pure dichlorodiphenoxymethane, m.p. 44-45° (sealed tube), is obtained by recrystallization of the redistilled material (Note 5) at low temperature in a drybox.

B. Polymerization

A 250-ml., three-necked flask, equipped with a thermometer, a powder funnel, and a magnetic stirrer, is placed in a drybox. The flask is charged with 4,4′-dihydroxy-3,3′-dimethylbiphenyl (6.427 g., 0.0300 mole) (Note 6), 2,6-lutidine (10 ml.) (Note 7), and methylene chloride (*Caution! Many chlorinated materials are toxic and should be handled with care.*) (100 ml.) (Note 8). The mixture is well dispersed by stirring at room temperature for 10 min. and cooled below 0°. Dichlorodiphenoxymethane (8.235 g., 0.0306 mole) is added in portions over a period of 0.5 hr., during which time an exothermic reaction takes place with the formation of a red color. The reaction mixture is then stirred at room temperature for 3 hr. and diluted with 100 ml. of methylene chloride; the polymer is recovered by precipitation by the addition of the solution to 2 l. of methanol. The polymer is redissolved in 200 ml. of methylene chloride and is precipitated by the addition of this solution to 3 l. of methanol. The yield of the polymer is 11.1 g. (90%). The intrinsic viscosity of the benzene solution is 0.45 to 0.70 dl./g. at 25°.

2. Characterization

The polymer is soluble in aromatic or chlorinated hydrocarbons, such as benzene, chlorobenzene, methylene chloride, and chloroform. Flexible, clear films can be cast from the solutions of the polyorthocarbonate. The characteristic infrared absorption band is 1080 cm.$^{-1}$ (ether). The n.m.r. spectrum in chloroform-*d* shows a sharp singlet at 8.04τ (CH_3, 6H) and broad peak at 2.70τ (aromatic, 16H). The glass transition temperature of the polyorthocarbonate is 125°, as determined by DSC (Note 9). The polymer decomposes above 366° (TGA) (Note 10). Additional information on other polyorthocarbonates is described elsewhere (4).

3. Notes

1. A 14-mm. O.D. column packed with 1/8-in. glass helices and equipped with an electric heating jacket is used.

2. Phosphorus pentachloride sublimes and tends to clog the distillation head. A hot-air blower may be used to remove this solid.

3. The titration procedure is as follows: dichlorodiphenoxymethane (ca. 0.4 g.) is weighed precisely in a 250-ml. pressure-resistant Erlenmeyer flask with a ground joint stopper. The flask is briefly cooled in a Dry Ice-acetone bath. A magnetic stirring bar and 20 ml. of tetrahydrofuran containing 5% water is added at once. The flask is immediately closed tightly and the mixture is stirred magnetically at room temperature for 10 min. The solution is neutralized by addition of sodium carbonate (164 mg.) dissolved in 10 ml. of water. The resulting solution is titrated with a 0.2 *N* silver nitrate solution with sodium chromate as indicator. The amount of sodium carbonate should be adjusted as precisely as possible to the molar equivalent of net dichlorodiphenoxymethane. A preliminary titration may be necessary to determine the required amount. When the solution is too basic, the end point may be obscured by precipitation of silver hydroxide. However, when the solution is too acidic, chromate is converted to bichromate that does not exhibit any recognizable color.

4. If the distillation is properly performed, diphenyl carbonate is the sole impurity. Therefore material with a purity higher than 95% can be used for the polymerization without further purification if its purity is precisely determined before use (Note 3).

5. The impure material (10 g.) is dissolved in about 20 to 30 ml. of *n*-heptane (distilled from calcium hydride) in a 50-ml. Erlenmeyer flask. The flask is clamped and held to a wide-mouth Dewar flask partially filled with liquid nitrogen. The top parts of the flasks are covered with glass wool and the temperature of the solution is maintained at about −10 to −25° by adjusting the height of the Erlenmeyer flask. Colorless crystals are formed in a few hours. The mother liquor is removed by decantation and the recrystallization may be repeated once. The pure crystals are dried under vacuum at room temperature. The entire procedure should be carried out in a drybox.

6. The commercially available material (Eastman Organic Chemicals) is recrystallized from benzene solution and treated with decolorizing carbon to obtain 4,4′-dihydroxy-3,3′-dimethylbiphenyl, m.p. 162.5-164°.

7. 2,6-Lutidine (Eastman Organic Chemicals) is treated with calcium hydride and distilled to obtain pure material. b.p. 144.5°.

8. Methylene chloride is washed with a 5% sodium carbonate solution and with water, dried several times over fresh calcium chloride, and distilled.

9. A heating rate of 40° per minute is used.

10. A heating rate of 2.5° per minute in nitrogen is used.

4. References

1. General Electric Research and Development Center, Schenectady, New York 12305.
2. Chemistry Department, University of Alabama, University, Alabama 35486.
3. H. Gross, A. Rieche, and E. Hoft, *Chem. Ber.*, 94, 554 (1961).
4. T. Takekoshi, *J. Polym. Sci.*, **10**, 3509 (1972).

Poly-p-xylylene Tetrasulfide and Related Polysulfide Polymers of Uniform Repeating Structure

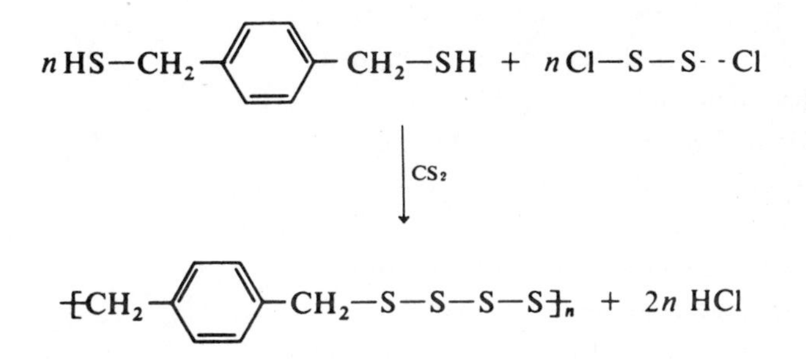

Submitted by R. M. Fitch (1) and D. C. Helgeson (2)
Checked by W. J. Bailey and H. J. Booth (3)

1. Procedure

A solution of *p*-xylylenedithiol (*Caution!* p-*Xylylene dithiol may cause severe dermatitis to susceptible individuals.*) in carbon disulfide (*Caution! Carbon disulfide is extremely flammable and toxic. It should be used only in an efficient hood.*) containing approximately 1.5-2.0 meq. of thiol per ml. is prepared (Note 1). [The exact normality can be checked by iodimetric titration of an aliquot (Note 2).] This solution is placed in the buret of the apparatus shown in Fig. 1. Pure sulfur monochloride (0.700-1.000 g.) (Note 3) in a glass ampoule (Note 4) is placed in the 8-in. test tube which constitutes the lower part of the apparatus, and 10 ml. of carbon disulfide (dried over "Drierite") is added. The exact amount of sulfur monochloride in the ampoule will determine the volume of thiol solution to be added (ca. 10-20 ml.).

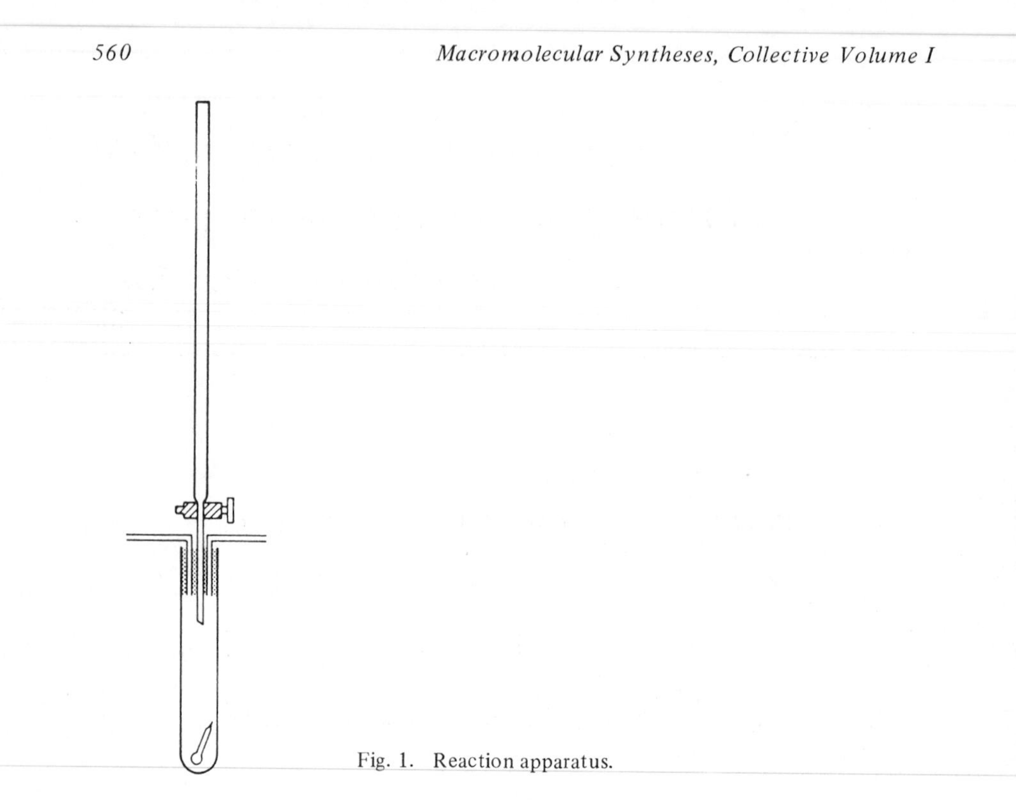

Fig. 1. Reaction apparatus.

The test tube is then attached to the stopper containing the buret and the inlet and outlet tubes, and flushed with thoroughly dried nitrogen (Note 5). The nitrogen stream in momentarily stopped while the tube is shaken vigorously to break the ampoule. The calculated amount of dithiol is added from the buret as rapidly as possible while avoiding excessive foaming of the solution caused by the evolved hydrogen chloride. During this period the test tube is held in an ice bath and its contents are gently swirled to absorb the heat of reaction. When the addition of thiol solution is complete, the system is flushed with nitrogen and agitated for another 5 min.

The test tube is now placed under vacuum (water aspirator) of approximately 15 torr for 0.5 hr. to remove dissolved hydrogen chloride and force the reaction to a conversion as complete as possible. The reaction mixture is poured into 250 ml. of methanol under vigorous stirring. The polymer which separates is washed with two portions of methanol and dried *in vacuo* at room temperature.

The product, which is obtained in almost quantitative yield, is colorless and has a glass transition temperature of 3° and a crystalline melting temperature of 96°. Its number average molecular weight, as determined by vapor pressure osmometry, is around 4300, which represents a degree of polymerization of 20.

The n.m.r. spectrum shows two principal peaks: (a) the aromatic protons at 2.92τ and (b) the methylenic protons at 5.94τ. The latter peak exhibits some multiplicity which indicates the presence of small amounts of di, tri and penta-sulfides.

Other polysulfide polymers of uniform repeating structure may be prepared in this manner: the substitution of sulfur dichloride for sulfur monochloride yields poly-*p*-xylylene trisulfide, whereas substitution of decamethylenedithiol for *p*-xylylenedithiol yields corresponding polymers containing a flexible organic moiety (6).

2. Notes

1. The reactants must be extremely pure and the dithiol is resublimed before use. This should be sufficient to achieve a purity of 100 ± 0.06%. A sample of the dithiol is carefully weighed into a volumetric flask to which is added carbon disulfide to the correct volume.

2. A modification of the Kimball, Kramer, and Reid method (4) was followed. Flushing the titration flask with nitrogen during addition of both the thiol and the standard iodine solutions was found to be a necessary precaution to avoid air oxidation.

A 1-ml. aliquot of the thiol solution (~2 meq. of thiol) in carbon disulfide is added to a mixture of 25 ml. of benzene and 25 ml. of methanol. An appropriate excess of a 0.1000 *N* standard silver nitrate solution is added with stirring and 5 ml. of a 6 *N* nitric acid solution, which has recently been boiled to remove nitrogen oxides, and 10 ml. of a ferric alum indicator solution are added. (The ferric alum indicator solution is prepared by the addition of 40 g. of ferric alum to 100 ml. of 1 *N* nitric acid; the solution is boiled to remove oxides of nitrogen, cooled, and diluted with three volumes of distilled water.) While the solution is vigorously stirred, it is titrated to a reddish-brown end point with 0.1000 *N* standard ammonium thiocyanate solution. (The apparent end point fades after approximately 30 sec.; one additional drop gives a permanent end point.) (The checkers found that this titration gave somewhat erratic results; however they also found that the use of a standard thiol solution prepared from a weighed sample of the pure dithiol gave the indicated results on polymerization.)

3. Sulfur monochloride, which is commercially available, may be purified readily by fractional distillation within a 1° range (135-136°). Precautions, which must be observed, include drying the glassware in an oven at 200° for at least 4 hr., distilling under nitrogen, and maintaining a thoroughly dry atmosphere by use of phosphorus pentoxide/vermiculite ("Aquasorb") drying tubes.

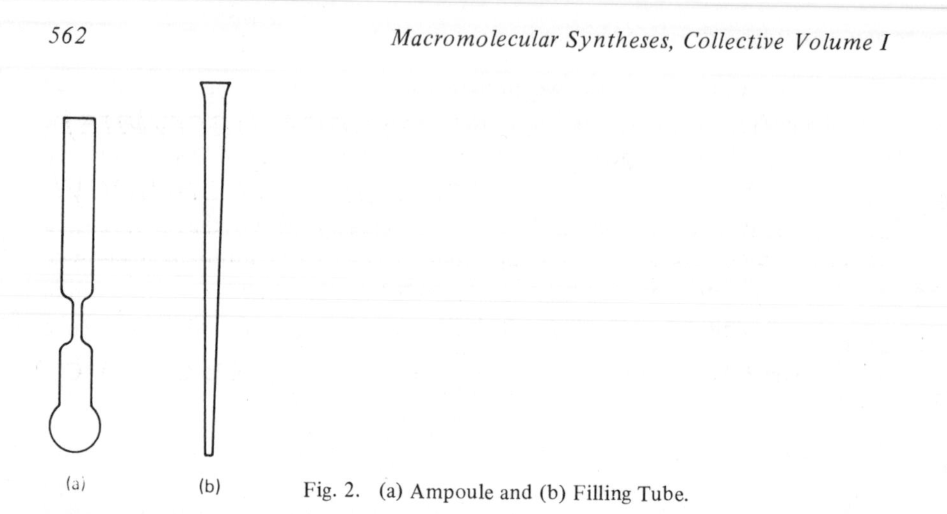

Fig. 2. (a) Ampoule and (b) Filling Tube.

Analysis for purity can best be performed by the method of Feher and co-workers (4).

4. Ampoules should be made from 6-in. sections of 6-mm. diameter *soft* glass tubing to have the shape shown in Fig. 2. They are oven-dried at 200° for 4 hr., cooled in a desiccator, and filled, with a glass capillary pipet similar to that shown in Fig. 2. The ampoules must be sealed without cooling of the contents to avoid condensation of water vapor. A scratch is made on the ampoule before it is placed in the reaction tube to facilitate breaking.

5. Passage of the nitrogen through a drying tube containing phosphorus pentoxide/vermiculite ("Aquasorb") is recommended.

3. References

1. Department of Chemistry, University of Connecticut, Storrs, Connecticut 06268.
2. Plastics Department, E. I. du Pont de Nemours & Company Inc., Parkersburg, West Virginia 26101.
3. Department of Chemistry, University of Maryland, College Park, Maryland 20742.
4. J. W. Kimball, R. L. Kramer, and E. E. Reid, *J. Amer. Chem. Soc.*, **43**, 1199 (1921).
5. F. Feher, K. Naused, and H. Weber, *Z. Anorg. Allg. Chem.*, **290**, 303 (1957).
6. R. M. Fitch and D. C. Helgeson, *J. Polym. Sci.*, **C22**, 1101 (1969).

Polymerization of π-(Benzyl acrylate)-chromium Tricarbonyl

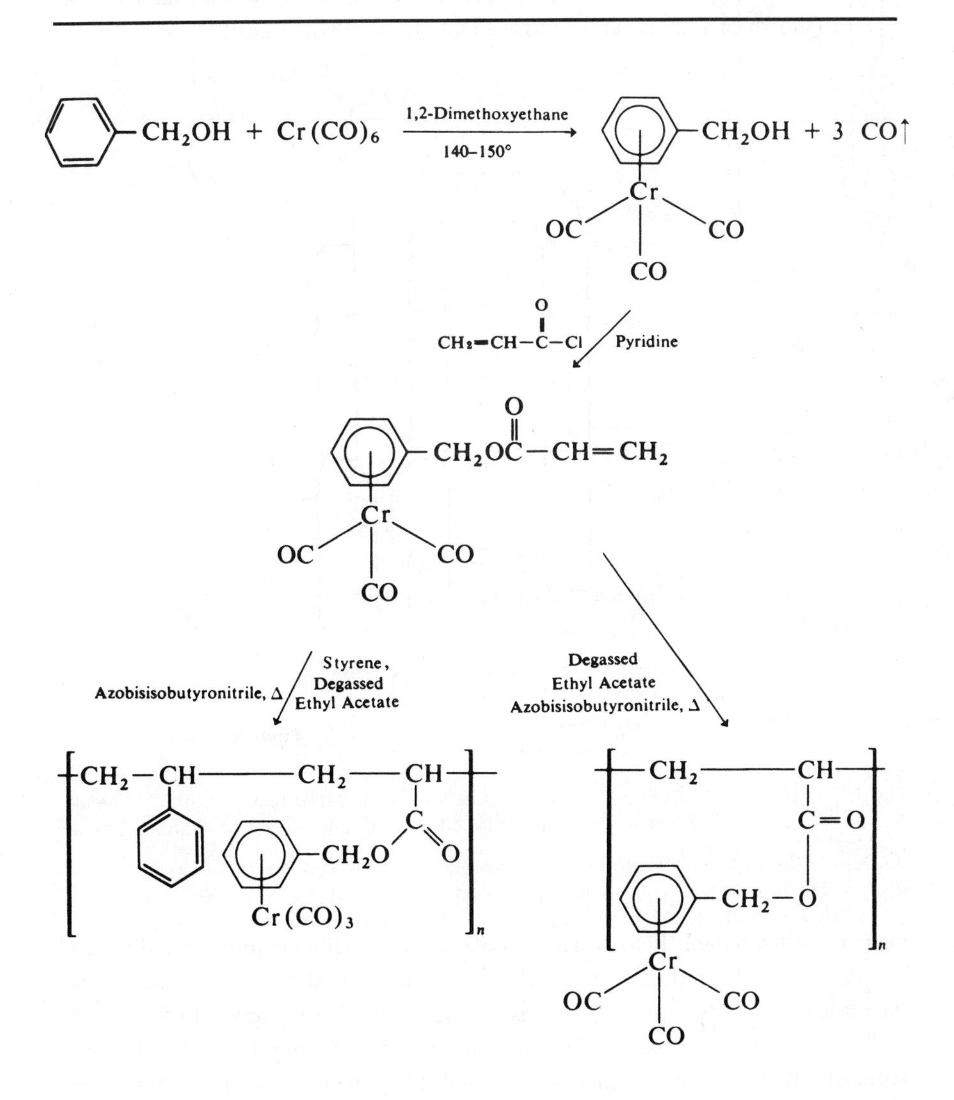

Submitted by C. U. Pittman, Jr., and R. L. Voges (1)
Checked by W. J. Bailey and M. Isogawa (2)

1. Procedure

A. π-(Benzyl acrylate)chromium Tricarbonyl

Chromium hexacarbonyl, 50 g. (0.227 mole), is placed in a Strohmeier (3) reactor (shown in Fig. 1) under nitrogen (Note 1). Benzyl alcohol, 366 g. (3.4

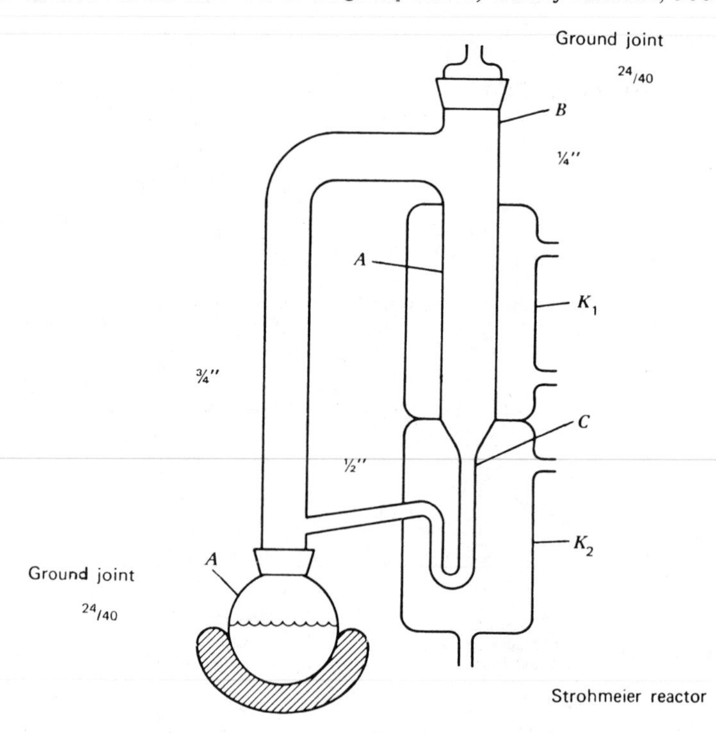

Fig. 1. *A*, 1000 ml reaction pot; *B*, 55/60 male joint with 24/40 female joint; *C*, U bend; K_1, Condenser No. 1. Input water temperature 60°; K_2, Condenser No. 2. Water temperatures 85°; K_3, Cold water.

moles), is then added. Benzyl alcohol is dried over calcium chloride and distilled before use. 1,2-Dimethoxyethane, 180 g. (2.0 moles), distilled from sodium, is then added. (1,2-Dimethoxyethane is an inert solvent that serves to lower the boiling point of the chromium hexacarbonyl-benzyl alcohol mixture.) A temperature of 140-150° is maintained in a flask until no white deposits of chromium hexacarbonyl are evident in the condensing portion (Fig. 1, *A*) of the Strohmeier apparatus (144 hr.). By this time almost all of the chromium hexacarbonyl has reacted with the benzyl alcohol and has turned the color of the solution from colorless to light yellow and finally to dark yellow-green. To isolate the π-(benzyl alcohol) chromium tricarbonyl the reaction products are evaporated to dryness with a rotary evaporator [100° (0.1 torr)]. After the crude product has been dissolved in 500 ml. of ether, this mixture is filtered and the filtrate is washed with three portions of an aqueous sodium chloride solution and then with distilled water. After the etheral solution has been dried over sodium sulfate and the mixture has been filtered, the solution is evaporated to dryness with a rotary evaporator. Recrystallization from heptane afforded 48-52 g. (87-94%) of π-(benzyl alcohol)chromium tricarbonyl as light yellow crystals, m.p. 95-96° [lit. (3) 95.5-96.5°]. Key I.R. bands observed are 3600-3140, 3100, 2960-2880, 1975, 1885, 1460, 1410, 1163, 1152, 1080, 1062, 1008, 998, 949, 868, 851, 810, 660, 630, 530, 470 $cm.^{-1}$ N. m. r. bands observed are aromatic protons 4.46τ (s, broad); methylene protons 7.63τ (s), OH 8.0τ (s) downfield from tetramethylsilane.

B. *Operating Procedure for the Strohmeier Reactor*

At the temperatures necessary for the reactor to be efficient (140-150°), chromium hexacarbonyl sublimes readily with the refluxing solvent vapor and separates below the location at which the solvent condenses. In the Strohmeier reactor the solvent condenses first in the upper portion of the condenser K_1, and in the lower portion the hexacarbonyl then crystallizes. In this way the condensed solvent in the condenser K_1 rinses the hexacarbonyl over the "U" tube back into the reaction vessel *A*. Because the solubility of the hexacarbonyl at room temperature in the condensed solvent is small, most of it would remain in condenser K_1. Therefore the water temperature in K_1 is kept at a minimum of 60° so that the hot saturated solution flows through the "U" bend back into the reaction vessel. To avoid the possibility that chromium hexacarbonyl might precipitate from the warm saturated solution while going through the "U" bend and plug it the water in K_2 is 20-25° warmer than the water in K_1. Increasing the diameter of the tubing in the "U" bend is not practical because too much solvent would be out of the reaction vessel.

It is advantageous during the heating of the reactor pot first to turn on cold water in condenser K_1. By this method the hexacarbonyl will be deposited as a tight crust in condenser K_1. After the solvent has started to condense and return to the pot, water of the desired temperature is started in condenser K_1. Over the joint *B* a condenser K_3 is added so that no solvent can escape. Cold water flows through this condenser.

The reactor allows easy recovery of the unreacted hexacarbonyl because at the end of the reaction time cold water is passed through K_1 and any unreacted hexacarbonyl will condense on the walls of condenser K_1.

C. *Reaction of π-(Benzyl alcohol)chromium Tricarbonyl with Acrylyl Chloride* (Note 2)

Pyridine, 7 ml. (0.086 mole) is added to a cold (0°) solution of 10 g. (0.041 mole) of π-(benzyl alcohol)chromium tricarbonyl in 400 ml. of anhydrous ether in a 1-1., three-necked flask fitted with a condenser (with nitrogen inlet), a mechanical stirrer, and a pressure-equalizing funnel (Note 3). Then 7 ml. (0.082 mole) of acrylyl chloride in 4 ml. of anhydrous ether is added through the funnel, over 30 min. (Note 4). Pyridine hydrochloride precipitates immediately as each drop of acrylyl chloride enters the solution. After a reaction time of 3 hr. during which time the temperature of the flask is gradually allowed to reach room temperature, the contents of the flask are diluted with 400 ml. of anhydrous ether and the contents are filtered to remove the pyridine hydrochloride. The pyridine hydrochloride is washed with ether until no yellow color remains in the ether. The combined ethereal solutions are washed with five 500-ml. portions each of an aqueous sodium bicarbonate solution, an aqueous sodium chloride solution, and distilled water. The ethereal solution is then dried over sodium sulfate and the mixture is filtered. Rotary evaporation of the solvent from the filtrate left a light tan oil. The oil is dissolved in carbon disulfide and the mixture is filtered. Recrystallization from carbon disulfide yields 2.8-6.2 g. (23-51%) of π-(benzyl acrylate)chromium tricarbonyl, m.p. 51-52°.

Anal. Calcd. for $C_{13}H_{11}CrO_5$; C, 52.35; H, 3.38; Cr, 17.44. Found (after one recrystallization): C, 51.27; H, 3.88; Cr, 18.65; (after three recrystallizations): C, 52.01; H, 3.45; Cr. 17.62.

Key infrared bands observed are 3110, 2860-2940, 1975, 1885, 1730, 1665, 1625, 1460, 1410, 1300, 1265, 1178, 1070, 1043, 1008, 980, 801, 650, 620, 525, 427 $cm.^{-1}$ N. m. r. bands observed are aromatic protons 4.79τ (s, broad); CH_2 5.22τ (s); vinyl hydrogens, ABC pattern 4.28-3.51τ downfield from tetramethylsilane.

D. *Polymerization Reactions*

The procedures for both homo- and copolymerization reactions are essentially the same. Polymerization is carried out in freshly distilled ethyl acetate (Note 5).

Azobis(isobutyronitrile) is used as the free radical initiator. Commercial azobis-(isobutyronitrile) is recrystallized three times from methanol before use to give material, m.p. 102-103°, with decomposition. Monomer(s) are stored at −15° in the dark before use and are freshly prepared before use.

Fischer-Porter aerosol compatibility tubes (Note 6), equipped with valves, are used for polymerization. Weighed amounts of monomer, solvent, and initiator are added to the tubes, and the tubes are sealed and degassed at 10^{-3} torr by three alternate freeze-thaw cycles. Liquid nitrogen is used as the freezing medium. After the mixture has been degassed, the tubes are placed in a Haake constant temperature bath (Model 1280-3, ± 0.01°) for a specified time. After removal from the bath the polymer is precipitated in a rapidly stirred large excess of 30-60° pet. ether. The polymer solution is diluted with more solvent when necessary so that fine particles of polymer will be precipitated rather than large chunks. This process is repeated three times to ensure that all monomers and other contaminants have been washed from the polymer. After the last precipitation the polymer is collected by filtration and dried in a vacuum drying oven at 60° for 24 hr. and weighed. When 2 g. of π-(benzyl acrylate)chromium tricarbonyl and 0.01 g. of azobis(isobutyronitrile) are charged to the tubes and the polymerization is carried out at 70° for 24 hr. in 5 ml. of ethyl acetate, the homopolymer exhibited $\overline{M}_n$ = 17.500 (Note 7).

2. Characterization

Homopolymers of π-(benzyl acrylate)chromium tricarbonyl are light yellow solids and are obtained in 60-90% yields (Note 8). Molecular weight determinations are carried out by vapor pressure osmometry and gel permeation chromatography (Note 9). Homopolymers with $\overline{M}_n$ between 7,700 and 60,000 (4) are obtained and their intrinsic viscosities are measured in dimethylformamide at 30° (Note 10). When $\overline{M}_n$ = 13,800 and $\overline{M}_w$ = 20,800, the value of $[\eta]$ = .09 dl./g. The Mark-Houwink equation, $[\eta] = 3.95 \times 10^{-3}\ \overline{M}_n^{0.82}$, fits the π-(benzyl acrylate)chromium tricarbonyl homopolymer experimental data. The homopolymers exhibit infrared bands at 3100, 2970-2910, 1975, 1885, 1740, 1460, 1430, 1260, 1165, 1070, 1045, 1018, 998, 815, 750, 660, 628, 531, and 470 $cm.^{-1}$ (Note 11). The carbonyl stretching bands at 1975 and 1885 $cm.^{-1}$ are particularly intense, and the ester carbonyl stretch at 1740 $cm.^{-1}$ is strong. No trace of C=C stretch, present in the monomer at 1625 $cm.^{-1}$, has been detected. Differential scanning calorimetry indicated the homopolymers underwent an endothermic process, maximizing at 245°.

Smooth solution copolymerizations were carried out with styrene and methyl acrylate. The monomer ratios in the copolymers were calculated from elemental analysis data. By construction of composition-conversion curves at two different initial monomer ratios and by application of the integrated form of the copolymer equation (5, 6), the relative reactivity ratios were obtained (Note 12). These ratios were $r_1 = 0.10$, $r_2 = 0.34$ and $r_1 = 0.56$, $r_2 = 0.63$ for the styrene and methyl acrylate copolymerizations, respectively (π-(benzyl acrylate)chromium tricarbonyl = M_1). Infrared bands in the chromium monomer-styrene copolymers appeared at 3100, 2960-2860, 1974, 1885, 1735, 1601, 1489, 1440, 1260, 1155, 1070, 1030, 1013, 995, 812, 760, 698, 655, 625, 530, and 470 $cm.^{-1}$ Bands for the chromium monomer methyl acrylate polymers appeared at 3100, 2960-2860, 1975, 1885, 1735, 1485, 1440, 1260, 1155, 1070, 1030, 1013, 995, 810, 652, 622, 563, and 525 $cm.^{-1}$.

3. Notes

1. Chromium hexacarbonyl was obtained from Pressure Chemical Company, Pittsburgh, Pennsylvania, and used without further purification.

2. Attempts to esterify π-(benzyl alcohol)chromium tricarbonyl with acrylic acid in methylene chloride, catalyzed by small amounts of *p*-toluenesulfonic acid, resulted in 70-80% yields of the bis-π-benzyl chromium tricarbonyl ether, m.p. 122-123°, after recrystallization from pet. ether.

3. Pyridine and diethyl ether are distilled from powdered calcium hydride before use.

4. Acrylyl chloride was obtained from Borden Monomer-Polymer Laboratories, 5000 Langdon Street, Philadelphia, Pennsylvania. It was vacuum distilled before use.

5. The polymer was not sufficiently soluble in benzene to use benzene as the solvent. Polymerizations were also carried out in tetrahydrofuran.

6. Obtained from Fischer-Porter Company, Lab-Crest Scientific Division, Warminster, Pennsylvania.

7. The molecular weights were lower when tetrahydrofuran was used as the solvent; for example, when 3.2 g. of π-(benzyl acrylate)chromium tricarbonyl and 0.03 g. of azobis(isobutyronitrile) were heated to 70° for 96 hr. in 6 ml. of degassed tetrahydrofuran, a polymer with $\overline{M}_n = 7{,}750$ was obtained.

8. These yields were obtained with ethyl acetate as the solvent. The yields in tetrahydrofuran for equivalent reaction conditions were much lower.

9. A Waters Associates Model 200 gel permeation chromatograph, packed with a bank of four styragel columns and standardized with commercial monodisperse polystyrene fractions, was used to obtain the GPC curves. A Mechrolab

vapor pressure osmometer was used to obtain the absolute mearure of $\overline{M}_n$ and this valve was used at the GPC curve position of the number average chain length. A GPC Q factor of 94 was thus determined for π-(benzyl acrylate)-chromium tricarbonyl homopolymers. GPC curves were obtained in tetrahydrofuran while osmometry measurements were performed in dimethylformamide.

10. Viscosity measurements were made with a Cannon-Ubbelohde semimicro dilution viscometer (50L631).

11. The infrared spectra were obtained on potassium bromide pellets on a Perkin-Elmer Model 237 spectrometer.

12. The computer programs developed by Montgomery and Fry (6) were utilized in these calculations.

13. A closely related monomer, η^6-(2-phenylethyl acrylate) tricarbonyl chromium has been synthesized, homopolymerized and copolymerized with styrene, methyl acrylate, acrylonitrile, and 2-phenyl ethyl acrylate. The exposure of polymers of this monomer (or the title monomer) to sunlight or ultraviolet light in air caused deposition of Cr_2O_3 within polymer films (7).

4. References

1. Department of Chemistry, University of Alabama, University, Alabama 35486.
2. Department of Chemistry, University of Maryland, College Park, Maryland 20742.
3. W. Strohmeier, *Chem. Ber.*, **94**, 2490 (1961).
4. C. U. Pittman, Jr., R. L. Voges, and J. Elder, *Macromolecules*, **4**, 302 (1971).
5. F. R. Mayo and F. M. Lewis, *J. Amer. Chem. Soc.*, **66**, 1594 (1944).
6. D. R. Montgomery and C. E. Fry, *J. Polym. Sci.*, **C25**, 59 (1968).
7. C. U. Pittman, Jr. and G. V. Marlin, *J. Polym. Sci. Chem. Ed.*, **11**, 2753 (1973).

Poly(N-vinylcarbazole) by a Cation-Radical Initiator, Tris-p-bromophenylaminium Hexachloroantimonate

Submitted by A. Ledwith (1a) and D. C. Sherrington (1a, b)
Checked by A. M. Toothaker (2)

1. Preparation of Initiator Salt

The most convenient aminium salt to prepare and use as a catalyst is tris-(*p*-bromophenyl)aminium hexachloroantimonate (3). Tris-(*p*-bromophenyl)amine is prepared first from triphenylamine by the Walter method (4). The product, m.p. 144-145° [lit. (4) 144.5-145.5°], is purified by Soxhlet extraction with *n*-heptane. *Tris*-(p-bromophenylamine (2.4 g.) is dissolved in distilled, dry benzene (20 ml.) (Note 1). A solution of freshly distilled antimony pentachloride (1 ml.) in benzene (10 ml.) is added slowly. The reaction is instantaneous as shown by the formation of the characteristic deep-blue color. On cooling in an ice-bath, a high yield of fine, needle-like crystals is produced. The blue crystals are collected by filtration, and washed thoroughly with benzene before being dissolved in purified methylene chloride (Note 1). The solution is filtered through a fine fritted glass filter and the solvent is removed. Finally, the salt is recrystallized from a cooled benzene/methylene chloride mixture, collected by suction filtration and dried under vacuum, m.p. 141-143°, log ϵ (CH_2Cl_2) 4.56 (725 nm.). Yield, 80%.

2. Polymerization Procedure

The initiator salt (0.0035 g.) is dissolved with slight warming in methylene chloride (5 ml.) (Note 1). At the same time *N*-vinylcarbazole monomer (1 g.) (Notes 2 and 3) is dissolved in the same solvent (45 ml.). The monomer solution is retained in a 250-ml. Erlenmeyer flask at room temperature open to the air and stirred magnetically as the catalyst solution is rapidly introduced. Almost immediately the deep blue color of the aminium salt is discharged and polymerization to 100% conversion is complete within seconds (Note 4). The polymer is precipitated by the addition of methanol (100 ml.) to the stirred polymerization mixture. After the polymer has been collected by filtration and washed with methanol, the yield is >96%. After the polymer has been dried, it is found to be a pure white powdery solid with a pronounced tendency to retain static charge. It is most soluble in methylene chloride but also readily dissolves in benzene, toluene, and tetrahydrofuran. Molecular weights are most conveniently determined by viscometry with benzene solutions. Typically, an intrinsic viscosity of ~0.16, dl./g. obtained for a polymer prepared at room temperature, corresponds to a molecular weight of ~10,000 (5).

3. Reaction Mechanism

Initiation by a stable aminium salt must involve electron transfer to produce the monomer cation radical (I), which rapidly forms a propagating dicationic species (II). Proof of the intermediacy of (I) and the dication (II) is readily obtained by carrying out the reactions in methanol-containing solvents: the aminium salt (0.96 g.) is added to a solution of *N*-vinylcarbazole (0.30 g.) in a 50:50 methylene chloride/methanol mixture (20 ml.). This heterogeneous mixture is agitated, and after a few minutes the salt dissolves and its color is discharged. The solution is shaken with an aqueous alkali solution (10 ml., 10%) and ether (10 ml.). The organic layer is separated, washed with water, and dried with sodium sulfate. Removal of the solvents on a rotary evaporator yields a brown tacky solid. The neutral tris-(*p*-bromophenyl)amine is removed by extraction with petroleum ether and the remaining solid (~0.3 g. yield, ~90%) is recrystalized from the methanol/acetone mixture. This produces a pure white solid, m.p. 208°, with an infrared spectrum consistent with that of a pure sample of 1,4-dicarbazol-9-yl-1,4-dimethoxybutane (III) prepared independently (6).

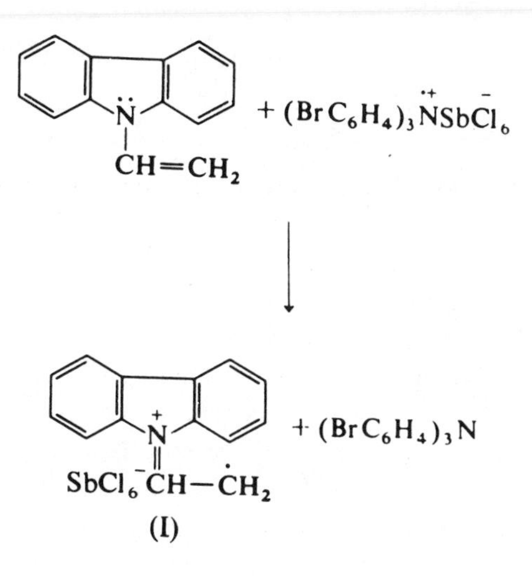

N
CH=CH$_2$
+ (BrC$_6$H$_4$)$_3$NSbCl$_6$
N
SbCl$_6$ CH—CH$_2$
+ (BrC$_6$H$_4$)$_3$N
(I)

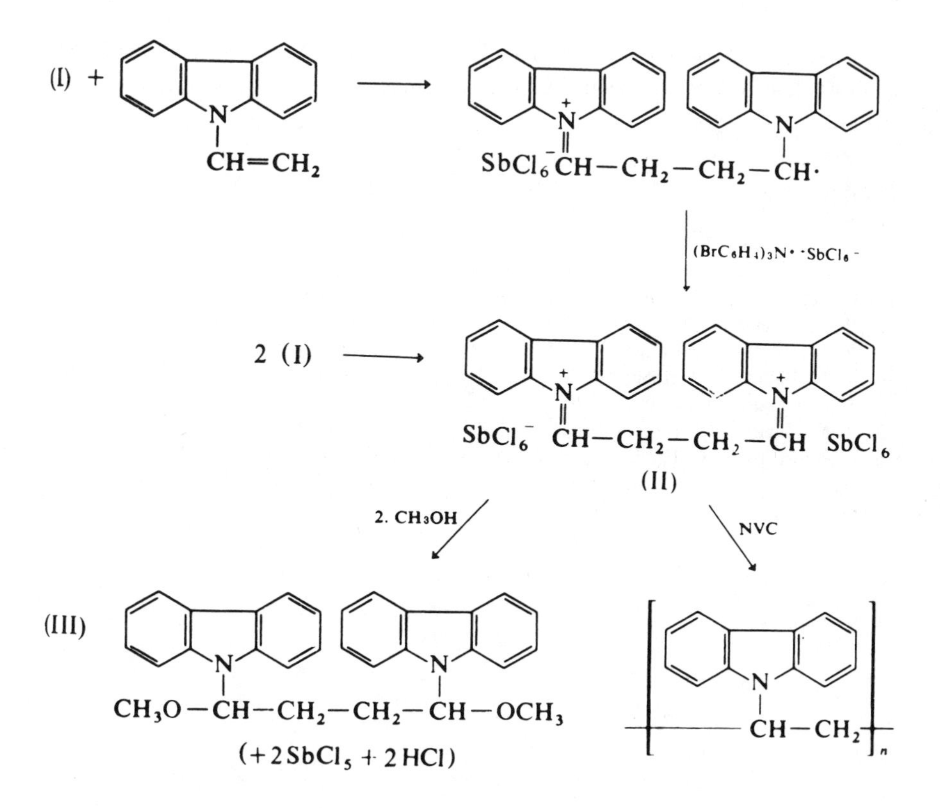

(I) +
N
CH=CH$_2$
SbCl$_6$ CH—CH$_2$—CH$_2$—CH·
(BrC$_6$H$_4$)$_3$N· ·SbCl$_6$$^-$
2 (I)
SbCl$_6$ CH—CH$_2$—CH$_2$—CH SbCl$_6$
(II)
2. CH$_3$OH
NVC
(III)
CH$_3$O—CH—CH$_2$—CH$_2$—CH—OCH$_3$
(+2SbCl$_5$ + 2HCl)
CH—CH$_2$
n

4. Notes

1. In the preparation of the catalyst salt and the polymerization reaction the solvents must be pure and dry. Careful fractionation from calcium hydride is adequate in this instance and avoids the need for high-vacuum technique.

2. *N*-vinylcarbazole monomer is readily purified by recrystalization from methanol. Care must be taken to ensure that all the solvent is subsequently removed, because this is an effective terminating agent of active cationic centers.

3. The over-all initial catalyst concentration used is $\sim 10^{-4}$ *M* for an initial monomer concentration of $\sim 10^{-1}$ *M*. These values have been found to be optimum. A much lower initiator concentration results in less than 100% conversion to polymer) presumably because some termination reaction becomes relevant. Increasing the initiator concentration, on the other hand, produces a polymer crosslinked in 3,3′ positions of spacially adjacent carbazole units, probably by methylene bridges, similar to the crosslinking of this polymer produced by excess boron trifluoride or ferric chloride in methylene chloride (7).

4. Any residual monomer can readily be detected in the polymer by infrared analysis. The olefin has a strong C=C absorbance at 1655 $cm.^{-1}$ that is absent in the polymer.

5. References

1. Donnan Laboratories, University of Liverpool, Liverpool, England L693BX.
1b. Current Address: University of Strathclyde, Glasgow, Scotland.
2. Research and Development Center, Schenectady, New York, 12301.
3. F. A. Bell, A. Ledwith, and D. C. Sherrington, *J. Chem. Soc., C,* **1969**, 2719.
4. T. N. Baker, W. P. Doherty, Jr., W. S. Kelley, W. Newmeyer, J. E. Rogers, Jr., R. E. Spalding, and R. I. Walter, *J. Org. Chem.,* **30**, 3714 (1965).
5. A. M. North and J. Hughes, *Trans. Faraday Soc.,* **62**, 1866 (1966).
6. P. Beresford, M. C. Lambert, and A. Ledwith, *J. Chem. Soc., C,* **1970**, 2508.
7. A. Ledwith, A. M. North, and K. E. Whitelock, *European Polym. J.,* **4**, 133 (1968).

Poly(2,5-Dimethyl-p-Xylyleneadipamide)

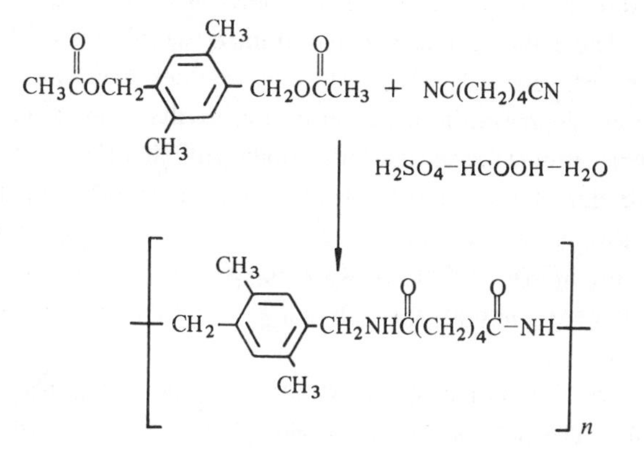

Submitted by L. T. C. Lee (1a) and E. M. Pearce (1b, c)
Checked by G. D. Cooper (2)

1. Procedure

In a 1-qt. stainless steel blender (Note 1) is placed a solution of 4.40 g. (17.6 mmoles) of 2,5-dimethyl-1,4-bis(acetoxymethyl)benzene (Note 2) and 1.89 g. (17.5 mmoles) of adiponitrile in 70 ml. of chloroform (*Caution! Many halogenated solvents are toxic and should be used with care.*) (Note 3), and high-speed stirring is initiated. A solution of 30 ml. of concentrated sulfuric acid (80%) and 8 ml. of formic acid (98%) is added over a period of 10 min. (Note 4). The solution is kept below 30° during the addition. The mixture is stirred for 2 hr. and is then poured into 150 ml. of an ice-water mixture and stirred for an additional 60 min. (Note 5). The chloroform is separated from the solution in a separatory funnel (Note 6), and the aqueous layer is neutralized with dilute sodium carbonate solution. The precipitated polymer is colledted on a fritted glass filter and washed with distilled water until free of salt (Note 7). The polymer is dried in a vacuum oven at a temperature below 100° for 10 hr. The yield is about 4.72 g. (98%).

2. Characterization

The polymer is insensitive to most common low-boiling organic solvents but is soluble in formic acid, concentrated sulfuric acid, trifluoroacetic acid, phenol, and *m*-cresol. The reduced viscosity determined at 25° on a 0.5% solution in *m*-cresol, is 0.47 dl./g.

The infrared spectrum of the polymer shows the absorption bands of the secondary linear amide group at 1535, 1640, and 3300 cm^{-1}, of the aliphatic group at 2950 $cm.^{-1}$, and of the 1,2,4,5 substituted phenyl group at 1282 $cm.^{-1}$.

The melting temperature (T_m) of the polymer is 283°, and the glass transition temperature (T_g) is 78°. Both were measured with a Differential Thermal Analyzer (DuPont Model 900) with a heating rate of 10°/min.

The thermal stability of the polymer is measured by thermogravimetric analysis on an Ainsworth balance in N_2 at a heating rate of 10°/min. and with 10 mg. samples. The polymer shows weight losses of 0.5% and 3.5% at 300° and 350°, respectively.

3. Notes

1. Vigorous stirring with high shearing action is needed to disperse the reactants. Inefficient stirring will reduce the molecular weight and yield of the polymer.
2. 2,5-Dimethyl-1,4-bis(acetoxymethyl)benzene is prepared according to the method described by Rhoad and Flory (5), using 2,5-dimethyl-1,4-bis(chloromethyl)benzene (6) and silver acetate in glacial acetic acid.
3. Other organic solvents can be used, such as 1,2-dichloroethane, dichloromethane, and nitrobenzene.
4. Addition may be facilitated by adding the solution through a dropping funnel inserted in a hole made in the cover of the blender.
5. The chloroform can be distilled from the solution before pouring into the ice-water mixture. However, it is more convenient to remove it at a later stage.
6. Complete removal of the chloroform can also be accomplished by distillation under vacuum.
7. The washing may be hastened by the use of an organic solvent such as

acetone or alcohol or a mixture of these solvents with water. This will help to remove the water-immiscible solvent. Stirring will also speed the washing.

4. Methods of Preparation

The preparation of poly(2,5-dimethyl-*p*-xylyleneadipamide) by interfacial polycondensation is described in ref. 3; the method is based on the Ritter reaction (4).

5. References

1a. Corporate Research Center, Allied Chemical Corporation, Morristown, New Jersey 07960.
1b. Camille Dreyfus Laboratory, Research Triangle Institute, P. O. Box 12194, Research Triangle Park, North Carolina 27709.
1c. Polytechnic Institute of New York, Brooklyn, New York, 11201.
2. Plastics Department, General Electric Company, Selkirk, New York 12158.
3. L. T. C. Lee and E. M. Pearce, *J. Polym. Sci.*, **A-1**, 9, 557 (1971).
4. J. J. Ritter and P. P. Minieri, *J. Amer. Chem. Soc.*, **70**, 7048 (1948).
5. M. J. Rhoad and P. J. Flory, *J. Amer. Chem. Soc.*, **72**, 2218 (1950).
6. J. V. Braun and J. Nelles, *Chem. Ber.*, **67B**, 1094 (1934).

Sodium Tetraethylaluminate-Isopropyl Mercaptan Catalyst for Polyacrylonitrile

$$NaAl(C_2H_5)_4 + i\text{-}C_3H_7SH \longrightarrow NaAl(C_2H_5)_3S(i\text{-}C_3H_7) + C_2H_6$$

$$nCH_2{=}\underset{\displaystyle CN}{\underset{|}{CH}} \xrightarrow[NaAl(C_2H_5)_3S(i-C_3H_7)]{} \left[CH_2{-}\underset{\displaystyle CN}{\underset{|}{CH}} \right]_n$$

Submitted by R. Chiang, J. H. Rhodes, and R. A. Evans (1)
Checked by K. C. Kauffman (2)

1. Procedure

A. Catalyst

Sodium tetraethylaluminate, $NaAlEt_4$, the basic material for this catalyst, is prepared by reacting metallic sodium and $AlEt_3$ (3) (*Caution!*) (Note 1). A dry, clean flask (Note 2) equipped with a rubber serum cap is charged with 0.668 g. (29.0×10^{-3} g.-atom) of sodium (finely cut pieces or dispersion) and 2 ml. of purified, degassed toluene (Note 3). To the stirred suspension, 17 ml. of a 25% solution of $AlEt_3$ in toluene (Note 4) containing 2.93 g. (25.6×10^{-3} mole) of $AlEt_3$ is added gradually over a period of about 10 min. The mixture is heated to 100° for 1 hr. with adequate agitation. The hot solution containing $NaAlEt_4$ is removed, with a syringe, from the unreacted sodium and aluminum and is filtered under nitrogen at 90° into a pressure tube modified to include a fine-filter side arm (4). Upon cooling, $NaAlEt_4$ crystallizes from the toluene in

30-35% yield. It is purified by recrystallization from toluene. After recrystallization and drying, the melting point of the $NaAlEt_4$ needles, as determined in a sealed tube placed in an oil bath, is 113-114°. After the melt is cooled, larger crystals are formed which have a melting point of 125.4 ± 0.2°, in agreement with the value reported by Baker and Sisler (5).

The active catalyst, $NaAl(C_2H_5)_3S(i{-}C_3H_7)$, is prepared *in situ* by reacting 2 $\times$ 10^{-3} mole of $NaAlEt_4$ in 10 ml. of toluene with a stoichiometric equivalent of of isopropyl mercaptan. Although the compound formed from these two reagents nominally contains a 1:1 ratio of aluminum to sulfur, a slight excess of $NaAlEt_4$ is preferred in actual practice. Aluminum is determined as the 8-hydroxyquinolate, and sulfur by radioactive assay. The reaction product is dissolved in about 10 ml. of dimethylformamide (Note 5) and used as the polymerization catalyst.

B. Polymerization

A dry reaction flask is filled with nitrogen and charged with 100 ml. of DMF and 10 ml. (0.15 mole) of acrylonitrile (Note 6). After the contents have been cooled to -78° in a Dry Ice-methanol bath, an aliquot of the catalyst solution prepared as above and containing 0.7×10^{-3} mole of the catalyst is injected through a syringe needle into the flask. Polymerization takes place immediately, as shown by the increase in viscosity of the mixture, and is allowed to proceed until a clear, thick gel is obtained (Note 7). This normally requires a period of about 4 hr. The reaction is then terminated while still under nitrogen by adding 5 ml. of a DMF solution containing a few drops of methanol and HCl. The polymer is isolated by pouring the reaction mixture into a large excess of methanol (Note 8). The polymer is filtered, washed thoroughly, and dried in a vacuum oven at 60° overnight. The yield ranges from 90 to 100%. The intrinsic viscosity measured in DMF at 25° varies from 1.5 to 2.5 dl./g. (Note 9).

2. Characterization

The polymer thus prepared is characterized by its high dissolution temperature, low solubility in organic solvents, and high crystallizability. For example, the dissolution temperature measured in propylene carbonate is 165°, as compared with 125° for a free-radical sample. The polymer crystallizes from a dilute propylene carbonate solution at a temperature as high as 125°, whereas the free-radical polymer does not crystallize at temperatures above 100° (4). The linear growth rate of the polymer, measured in propylene carbonate at 100° by the electron microscopic method, is about 50 times higher than that of free-radical polyacrylonitrile. The polymer does not fluoresce in dilute solutions, nor does it show absorption peaks at 265 and 275 nm in the ultraviolet region, attributed to the –C=N– and keto groups, respectively, beta to the nitrile group in the free-radical polymer (6). The polymer appears to be free from structural impurities such as branching points and cyanoethyl groups.

3. Notes

1. *Caution! Triethylaluminum ignites instantly upon exposure to air. Even in solution, it should be handled with care.* Although $NaAlEt_4$ reacts violently with pure oxygen, the reaction is smooth when oxygen is diluted with an inert gas.

2. The flasks are flamed and degassed under high vacuum, flushed with nitrogen, and kept under positive pressure of nitrogen. All transfers should be conducted under an inert atmosphere.

3. Mallinckrodt reagent grade toluene is purified by washing serveral times with concentrated sulfuric acid until no further discoloration occurs. The toluene is washed with dilute $NaHCO_3$ solution and with water, then dried with $MgSO_4$, and distilled over $MgSO_4$.

4. $AlEt_3$ (as a 25% toluene solution) was obtained from Texas Alkyls, Inc., Weston, Michigan, and used without further purification.

5. Dimethylformamide is rigorously purified according to a procedure reported by Thomas and Rochow (7). The distilled DMF is further dried by passing through a column packed with molecular sieves directly into the flask before use. The molecular sieves are activated by heating overnight in a furnace at about 350°.

6. Monsanto polymerization grade acrylonitrile is redistilled azeotropically with benzene to remove moisture. Distillation should be carried out under nitrogen, using hydroquinone to prevent thermal polymerization. Only middle fractions of the distilled acrylonitrile are collected; these are stored at 0° under nitrogen in the dark.

7. If all the reagents are carefully purified, the gel is water white; no trace of yellow color should be observed.

8. Often the solution is so viscous that it is necessary to dilute it with DMF before it is poured into the methanol. This can be carried out conveniently in an explosion-proof blender.

9. The molecular weight of the polymer is increased markedly if the monomer is dried over molecular sieves.

4. Methods of Preparation

More details on this procedure can be found in ref. 4.

5. References

1. Chemstrand Research Center. Inc., P. O. Box 731, Durham, North Carolina 27702.
2. Research Division, The Goodyear Tire and Rubber Company, Akron, Ohio 44316.
3. A. V. Grosse and J. M. Mavity, *J. Org. Chem.*, **5**, 106 (1940).
4. R. Chiang, J. H. Rhodes, and R. A. Evans, *J. Polym. Sci.*, A-1, **4**, 3089 (1966).
5. E. B. Baker and H. H. Sisler, *J. Amer. Chem. Soc.*, **75**, 5193 (1953).
6. J. Brandrup, J. R. Kirby, and L. H. Peebles, *Macromolecules*, **1**, 59 (1968).
7. A. B. Thomas and E. G. Rochow, *J. Amer. Chem. Soc.*, **79**, 1843 (1957).

Amorphous, High Molecular Weight Poly(propylene oxide)

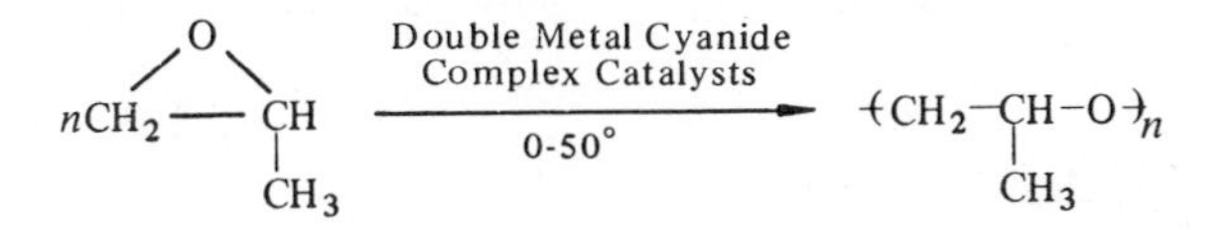

Submitted by R. J. Herold (1)
Checked by K. E. Steller (2)

1. Procedure

A. Catalyst Preparations (3)

Zinc Hexacyanochromate · $xZnCl_2$ · *yGlyme* · zH_2O *(I)* (Note 1). A solution of 4.50 g. (0.0125 mole) of potassium hexacyanochromate (BDH Laboratory Chemicals) dissolved in 17.5 ml. of water is passed through 25 g. of cation exchange resin (Rohm and Haas A-15). The effluent and that from a 25 ml. water wash of the resin are passed immediately (*Caution!*) (Note 2) into a solution of 2.55 g. (0.0188 mole) of zinc chloride (Note 3) in 5 ml. of water with stirring. A brick red precipitate forms, and 25 ml. of glyme (dimethoxyethane) (Note 4) is added to the slurry. After stirring for 3 hr., the slurry is centrifuged, whereupon it is alternately suspended in, and centrifuged from, 70 ml. of 30% glyme in water (three times) and 70 ml. of 90% glyme in water (twice). It is then vacuum-dried at less than 1 torr at room temperature overnight. The resulting solid is easily crushed into a fine powder with a spatula. Analysis: Zn, 24.7; Cr, 10.8; N, 15.5 wt.%.

To make highly active iron and cobalt catalysts, the following variations of the above procedure are recommended.

Zinc Hexacyanocobaltate (Valence 3) · x$ZnCl_2$ · yGlyme · zH_2O *(II)*. (a) Using 0.0125 mole of potassium hexacyanocobaltate (Shepherd Chemical Company), increase the amount of water to 100 ml.; (b) increase the amount of zinc chloide to 3.82 g. (a 50% excess).

Zinc Hexacyanoferrate (Valences 2 and 3) · x$ZnCl_2$ · yGlyme · zH_2O *(III)*. (a) Use 0.0125 mole of potassium hexacyanoferrate dissolved in 42 ml. of water and precipitate directly without passing it through an ion exchange resin; (b) increase the amount of zinc chloride to 3.82 g. (50% excess) and the amount of water to 60 ml.; (c) increase the initial amount of glyme to 35 ml.; (d) stir only 20 min; (e) change washing to two times with 90% glyme and once with pure glyme.

B. *Polymerization*

The requisite amount of one of the catalyst powders (e.g., ~0.02 g. of III, 0.005 g. of II, or 0.05 g. of I) (*Caution!*) (Note 5) is weighed in air (Note 6) and transferred to a 12-oz. Pyrex® (Note 7) crown-cap bottle. The bottle is capped and evacuated at 0.1 torr for 1 hr. (Note 8). Thereafter 50 g. of dry propylene oxide (Note 9) is added from a storage cylinder through a hypodermic needle. Sufficient nitrogen (Note 10) is then introduced to bring the pressure slightly above atmospheric, and the rubber cap is replaced quickly by one covered with an unbroken Teflon® liner Polymerization is then carried out by rotating the bottles in a 30° constant temperature bath. With the concentrations of catalyst suggested above, yields of > 70, > 61, and > 59% were obtained for the cobalt, chromium, and iron catalyst, respectively.

Poly(propylene oxide) is very prone to oxidation, especially in light. Therefore it is imperative to have anitoxidants present before placing the polymers in contact with the atmosphere. This is accomplished by dissolving or "soaking" the polymer in solvent containing antioxidant. Even more conveniently, some of the antioxidant can be added along with the catalyst before polymerization (Note 11). Usually 0.10% of amine or phenolic antioxidant is sufficient for room temperature storage. Polymers prepared in soft glass bottles are removed by cracking the bottles with a hot wire, whereas use of the Pyrex® bottles generally requires solution of the polymer. Vacuum drying to remove the solvent and excess monomer is then carried out in trays in a vacuum oven.

2. Characterization

Typical intrinsic viscosities of polymers prepared as noted with the iron, cobalt, and chromium catalysts are 10-5.3, 6.5-4.1, and 4.1-2.1 dl./g. respectively (Note 12), For reference, the number average molecular weight (Note 13) of a polymer made with the iron catalyst, having an intrinsic viscosity of 7.62 dl./g., is 131,000. The molecular weight distributions (Note 14) of polymers made with the iron and cobalt catalysts are both quite broad. Whereas the distribution of polymer prepared with the iron catalyst is symmetrical without excessive tailing, that of polymer prepared with the cobalt catalyst is bimodal with a plateau between the peaks.

Poly(propylene oxide) prepared in this manner has very little retractive force when extended. Whereas the polymers made with the iron and chromium catalysts are slightly yellow, the polymer prepared with the cobalt catalyst is colorless; all were clear (Note 15). The amorphous nature of these polymers is also demonstrated by the lack of x-ray diffraction peaks corresponding to the crystalline poly(propylene oxide) and by their solubility in acetone at $-20°$. (Note 16).

3. Notes

1. The valence of the chromium was not determined. The valences noted for iron and cobalt are those found in preparations starting with the trivalent cyanide complexes of these metals and in the presence of ethers and other organic compounds as noted below.

2. *Caution!* Immediate precipitation with zinc chloride is specified because hexacyanochromic acid (valence 3) formed in this exchange is not thermally stable and evolves hydrogen cyanide.

3. Zinc chloride is present in the final catalyst compositions as shown above. Catalysts made with other zinc salts such as the nitrate, acetate, and even the bromide and iodide are not as active.

4. Glyme (the dimethyl ether of ethylene glycol) is one of a large group of organic complexing agents that have been found to enhance greatly the activity of these catalysts. In general, water-soluble, oxygen-containing organic compounds, including ethers, ketones, esters, and amides, are useful. Of those tried, glyme and diglyme (the dimethyl ether of diethylene glycol) are preferred; acetone and dioxane have also been used with considerable success.

5. *Caution!* These are extremely active catalysts. The amounts given are for the most active forms found to date. Use of excessive amounts has led to explosively rapid reaction.

6. Such catalysts have been stored without the exclusion of air for periods of years without adversely affecting their catalytic properties.

7. Polymerizations with these catalysts set in suddenly after initial induction periods. The heat shock so generated has been responsible for breaking a number of soft glass bottles. When the activities of particular catalyst preparations are known, it is possible to develop recipes which are safe for use in soft glass bottles.

8. Evacuation is used to bring the catalysts to a reproducible stage of hydration.

9. Passage of liquid propylene oxide through a bed of Linde 3A molecular sieves has been found to be a satisfactory method of drying.

10. The reaction can be carried out in air, but an inert gas blanket is used to avoid variations in laboratory atmosphere.

11. There is some evidence that phenyl-β-naphthylamine increases the intrinsic viscosity of the polymer obtained when added at the same weight per cent as the catalyst.

12. The intrinsic viscosities reported were determined in benzene at 25°.

13. Measurement was made in a Mechrolab dynamic osmometer, Model 503.

14. The molecular weight distributions were determined by gel permeation chromatography. The conditions used were as follows: (1) polymer from an iron catalyst – 1.0% solution in DMF, 60-sec. injection, columns 10^7 Å + 10^6 Å + 3×10^4 Å + 8×10^3 Å + 10^3 Å; (2) polymer from a cobalt catalyst – 0.5% solution in DMF, 60 sec. injection, columns 10^6 Å + 1.5×10^5 Å + 10^5 Å + 1.5×10^4 Å + 8×10^3 Å.

15. These characteristics set these high molecular weight propylene oxide homopolymers apart from those made with other catalysts that have been used by the writer of this preparation. When metal alkyl-water reaction products (4, 5), the ferric chloride-propylene oxide complex (6), or a combination of aluminum isopropoxide and zinc chloride (7) is used, the polymers are cloudy even when free of catalyst residue, and the high molecular weight portions have considerable "nerve" or "gum strength."

16. It has been shown that crystalline isotactic poly(propylene oxide) fractions are insoluble under these conditions (7).

4. References

1. General Tire and Rubber Company Research Center, Akron, Ohio 44309.
2. Hercules Research Center, Wilmington, Delaware 19899.
3. J. Milgrom (General Tire and Rubber Company), U.S. Pats. 3,278,457 and 3,427,256; R. A. Belner (General Tire and Rubber Company, U.S. Pats. 3,278,458 and 3,427,334; R. J. Herold (General Tire and Rubber Company, U.S. Pats, 3,278,459 and 4,327,335.

4. J. Furukawa. T. Tsuruta, R. Sakata, T. Saegura and A. Kawaski, *Makromol. Chem.*, **32**, 90 (1959).
5. E. J. Vandenberg, U.S. Pat. 3,135,706 (1963).
6. M. E. Pruit and J. M. Baggett, U.S. Pat. 2,706,189 (1955).
7. M. Osgan and C. C. Price, *J. Polym. Sci.*, **34**, 153 (1959).

Poly (phenyl-as-triazines)

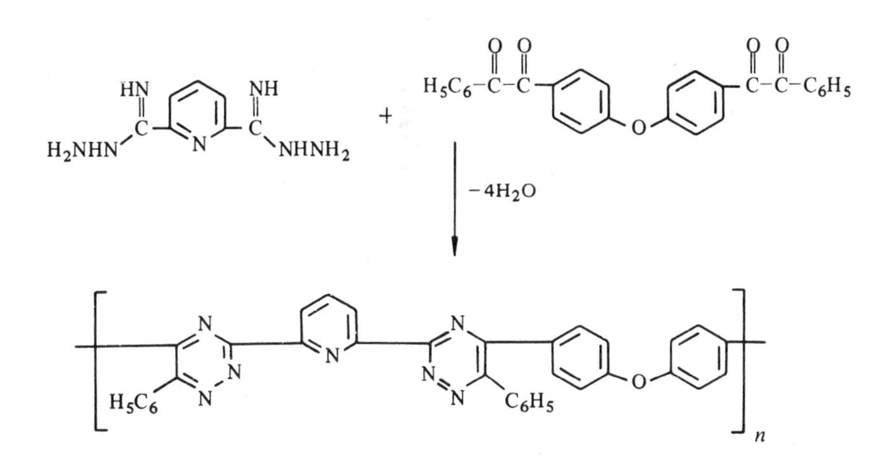

Submitted by P. M. Hergenrother (1a, b)
Checked by W. J. Wrasidlo (2)

1. Procedure

To a stirred solution of 9.66 g. (0.05 mole) of 2,6-pyridinediyl diamidrazone (Note 1) in 200 ml. of a 1:1 mixture of *m*-cresol and xylene (Note 2), 21.72 g. (0.05 mole) of 4,4′-oxydibenzil (Note 3) as a fine powder is added over a period of about 3 min. Residual dibenzil is washed in with 53 ml. of the solvent mixture, and the reaction temperature is maintained at $< 35°$ by cooling in a cold water bath (Notes 4 and 5). After the initial surge in temperature, the reaction mixture is stirred at ambient temperature for 18 hr. This procedure provides a 10% solution by weight which has a viscosity of 4000 to 25,000 cP., depending on the purity of the starting reactants (Note 6).

2. Characterization

Poly(phenyl-*as*-triazines) prepared by this procedure have an inherent viscosity (η_{inh}) of 0.5-2.8 dl./g. at 0.5% solution in *m*-cresol at 25°. The polymer solutions exhibit excellent shelf life and can be stored at room temperature in a capped bottle.

Films can be prepared by spreading the poly(phenyl-*as*-triazine) solution onto glass plates with a doctor knife and drying in a forced air oven at 80° for 2 hr. followed by drying *in vacuo* at 130° for 4 hr. The transparent yellow film, depending on the thickness, may contain a small amount of solvent. This can be removed by further drying *in vacuo* at 200° for 4 hr. The film can be removed from the glass plate by running water over the film.

3. Notes

1. 2,6-Pyridinediyl diamidrazone can be prepared by the addition of hydrazine to a solution of 2,6-dicyanopyridine in ethanol (3). Oxalamidrazone may also be used.

2. Technical grade *m*-cresol and xylene can be used without purification. other solvents such as chloroform and *sym*-tetrachloroethane may also be used.

3. 4,4′-Oxydibenzil can be prepared through a known procedure (3) or obtained from Research Organic/Inorganic Chemical Corporation, Sun Valley, California. Alcohol should be used as the recrystallization solvent to obtain high purity 4,4′-oxydibenzil. Other dibenzils may also be used (4-6).

4. Upon the addition of the dibenzil, an orange color is observed; this changes to yellow after stirring for about 2 hr.

5. Best results are obtained when the reaction temperature is maintained between 20° and 35°. Above 35°, the viscosity of the solution may increase appreciably, making stirring difficult.

6. More concentrated solutions (e.g., 15-20% solids) can be prepared by adding the last 3% of the dibenzil over a period of about 1 hr. or by adding a 1% excess of dibenzil. However, in some cases, the solution viscosity increases appreciably (e.g., $>$ 50,000 cP.).

4. References

1a. Materials Section, Aerospace Group, The Boeing Company, Seattle, Washington.
1b. Current Address: NASA Langley Research Center, Hampton, Virginia 23665.

2. Gulf General Atomic Company, San Diego, California 92112.
3. P. M. Hergenrother, *J. Polym. Sci.,* A-1, 7, 945 (1969).
4. P. M. Hergenrother, *J. Macromol. Sci.-Chem.,* A5(2), 365 (1971).
5. P. M. Hergenrother, *J. Macromol. Sci.-Revs. Macromol. Chem.,* C6(1), 1 (1971).
6. P. M. Hergenrother, *Macromolecules,* 7, 575 (1974).

Poly(phenylquinoxalines)

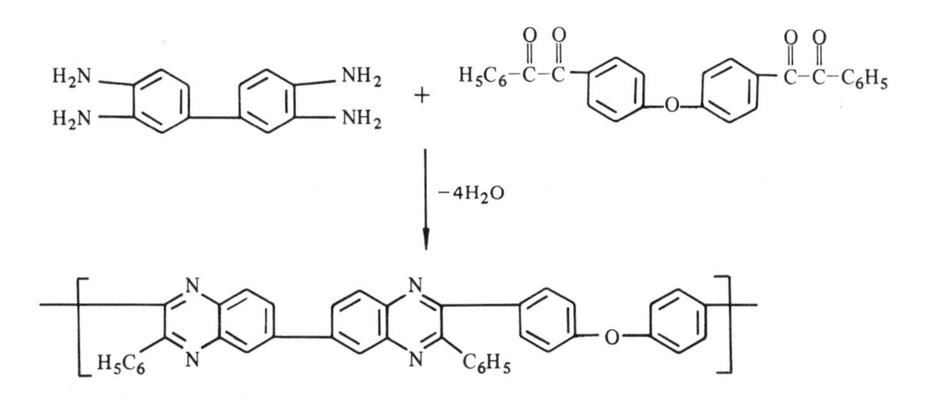

Submitted by P. M. Hergenrother (1a, b)
Checked by J. K. Stille, J. Wittman, and L. Mathias (2)

1. Procedure

Finely powdered 4,4′-oxydibenzil (21.72 g., 0.05 mole) (Note 1) is added over a period of ~3 min. to a stirred slurry of finely powdered 3,3′-diaminobenzidine (10.71 g., 0.05 mole) (Note 2) in 200 ml. of a 1:1 mixture of *m*-cresol and xylene (Note 3). Residual dibenzil is washed in with 74 ml. of a 1:1 mixture of *m*-cresol and xylene; during this addition, the reaction temperature is maintained at <35° by cooling in a cold water bath (Note 4). After the initial surge in temperature, the reaction mixture is stirred at ambient temperature for 18 hr. (Note 5).

2. Characterization

This procedure yields a 10% solution by weight that has a viscosity of 5000 to 30,000 cP., depending on the purity of the reactants (Note 6). Poly(phenylquinoxalines) prepared in this manner exhibit η_{inh} of 1.0-3.5 dl./g. at 0.5% solution in *m*-cresol at 25°. The polymer solutions have excellent shelf life and can be stored at room temperature in a capped bottle. Films can be readily prepared by spreading the poly(phenylquinoxaline) solution onto glass plates with a doctor knife and drying in a forced air oven at ~80° for 2 hr. Further drying can be accomplished *in vacuo* at ~130° for 4 hr. Depending on the film thickness, the transparent yellow film may retain a small amount of solvent (e.g., 3.0-mil-thick film has ~4% residual solvent), which can be removed by further drying *in vacuo* to temperatures of 200° for 4 hr. The film may be removed from the glass plate by running water over the film.

3. Notes

1. 4,4′-Oxydibenzil can be prepared by following a known procedure (5) or obtained from Research Organic/Inorganic Chemical Corporation, Sun Valley, California. Alcohol should be used as the recrystallization solvent to obtain high purity 4,4′-oxydibenzil. Other dibenzils may also be used (3).

2. 3,3′-Diaminobenzidine can be recrystallized from deoxygenated water 2.5 g./100 ml.) under an inert atmosphere, using a pinch of sodium hydrosulfite to produce near-white material. Other aromatic bis(*o*-diamines) may also be used (3).

3. Technical grade *m*-cresol and xylene can be employed without purification. Other solvents such as chloroform and *sym*-tetrachloroethane can also be used (4).

4. Best results are obtained when the reaction temperature is maintained at 20-35°. Above 35°, the viscosity of the solution may increase appreciably, making stirring difficult.

5. A reddish orange color is observed upon the addition of the dibenzil. This color changes first to a yellowish orange and then to a yellowish brown.

6. More concentrated solutions (e.g., 15-20% solids) can be prepared by adding the last 3% of dibenzil over a period of 1 hr. or by adding a 1% excess of dibenzil. However, the solution viscosity increases appreciably in some cases.

4. Methods of Preparation

The literature (6-9) describes the preparation of poly(phenylquinoxalines) by melt polycondensation of aromatic bis(*o*-diamines) with aromatic and aliphatic dibenzils.

5. References

1. Materials Section, Aerospace Group. The Boeing Company, Seattle, Washington 98101.
2. Department of Chemistry, University of Iowa, Iowa City, Iowa 52240.
3. P. M. Hergenrother, *J. Macromol. Sci.-Revs. Macromol. Chem.*, C6(1), 1 (1971).
4. W. Wrasidlo and J. M. Augl, *J. Polym. Sci.*, A-1, 7, 3393 (1969).
5. P. M. Hergenrother, *J. Polym. Sci.*, A-1, 7, 945 (1969).
6. P. M. Hergenrother and H. H. Levine, *J. Polym. Sci.*, A-1, **5**, 1453 (1967).
7. P. M. Hergenrother, *J. Polym. Sci.*, A-1, **6**, 3170 (1968).
8. P. M. Hergenrother, *Macromolecules*, 7, 575 (1974).
9. V. V. Korshak, E. S. Kronganz, A. M. Berlin, O. Neilands, and J. Skuja, *Vysokomol. Soedin.*, Ser A. **16**, 1770 (1974), (CA **82**: 4685t).

Copolymerization of Terephthalonitrile Oxide and 1,4-Diethynylbenzene: 1,3-Dipolar Cycloaddition

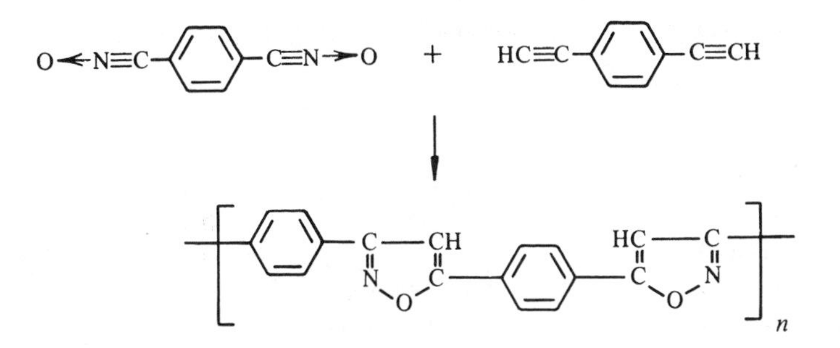

Submitted by C. G. Overberger (1a) and R. A. Nelson (1a, b)
Checked by K. Lawson and J. Preston (2)

1. Procedure

A. Terephthalaldehyde Dioxime (3)

To a warm solution of terephthalaldehyde, 22.1 g. (0.174 mole) in 63 ml. of ethanol, is added a warm solution of hydroxylamine hydrochloride, 27.6 g. (0.419 mole) in 32 ml. of water. Then a solution of sodium hydroxide, 19.8 g. (0.495 mole) is 35 ml. of water, is added (Note 1), and the mixture is refluxed for 2.5 hr. After cooling to room temperature, 83 g. of crushed ice is added, and the solution is saturated with carbon dioxide by adding Dry Ice. After standing at 0° overnight, the precipitate is filtered, washed with water, dried, and recrystallized from ethanol to give 20.5 g. (72% yield) of product, m.p., 217-218°.

B. Terephthalodihydroxamyl Chloride (4)

To a solution of terephthalaldehyde dioxime, 21.6 g. (0.132 mole) in 400 ml. of carbon tetrachloride, is introduced chlorine gas (*Hood!*) for 30 min. keeping the reaction temperature between −10° and 0°. The color of the solution at this point becomes mint green. An ultraviolet lamp is placed over the flask, and the reaction mixture is subjected to light for 30 min. The lamp is then removed, and the mixture is stirred while covered for 48 hr. and then stirred uncovered for an additional 2 hr. to allow excess chlorine gas to escape. The color of the solution is now a milky white. The product is filtered, washed with carbon tetrachloride, dried, and recrystallized from benzene to give 22 g. (71% yield) of product, m.p. 186-187°.

C. Terephthalonitrile Oxide (5)

To a solution of terephthalodihydroxamyl chloride, 1.16 g. (6 mmoles) in 100 ml. of methanol, is slowly added, without agitation at room temperature, a solution of triethylamine, 1.11 g. (11 mmoles) in 100 ml. of methanol. The colorless crystals that precipitate are filtered, washed with methanol, and dried to give 0.70 g. (86% yield) of product with a decomposition point of 155°.

D. α,α′,β,β′-Tetrabromo-1,4-Diethylbenzene (6)

To a solution of 1,4-divinylbenzene, 140 g. (1.08 moles) in 225 ml. of chloroform, is added slowly 243 g. (1.52 moles) of bromine, while keeping the reaction temperature between 5° and 15°. The white precipitate is filtered, washed with cold chloroform, and recrystallized from chloroform to give 73 g. (15% yield) (Note 2) of product, m.p. 159-160°.

E. 1,4-Diethynylbenzene (6)

t-Butanol is dried over molecular sieve 5A and distilled into a round-bottomed flask containing potassium tertiary butoxide (Note 3), 41.0 g. (0.368 mole), until a total volume of 750 ml. has been distilled. Into this solution is introduced 39.2 g. (0.087 mole) of $\alpha,\alpha',\beta,\beta'$-tetrabromo-1,4-diethylbenzene, and the solution is heated under mild reflux for 2 hr. The reaction mixture is diluted to 3.2 l. with ice water, and a pale yellow precipitate is isolated, 10.0 g. (92% yield), m.p. 90-92°. Sublimation under reduced pressure at 50-55° at 0.1 torr gives a colorless, crystalline product, 9.5 g. (87% yield), m.p. 96°.

F. Polymerization

Terephthalonitrile oxide, 160 mg. (1 mmole) (Note 4), is suspended in 10 ml. of benzene (Note 5) in a 50-ml. round-bottomed flask (Note 5) equipped with a reflux condenser and magnetic stirring bar. To this suspension is rapidly added a solution of 1,4-diethynylbenzene, 126 mg. (1 mmole) (Note 6) in 10 ml. of benzene. The mixture is stirred at 25° for 20 hr., followed by heating at the refluxing temperature of the solvent for 24 hr. A yellow-brown precipitate is gradually formed; after dilution with methanol, this is filtered, washed with methanol, and dried under vacuum to give 263 mg. (92% yield) (Note 7). The intrinsic viscosity of the polymer in concentrated sulfuric acid at 25° is 1.94 dl./g. (Note 8).

2. Notes

1. A substantial amount of heat is evolved upon addition. It is recommended that the addition be made through the barrel of the condenser.
2. The yield of tetrabromide from commercial divinylbenzene is low because of isomer contamination. Commercial divinylbenzene contains 40% of a mixture of *m*- and *p*-divinylbenzene.
3. Potassium *t*-butoxide is available from Mine Safety Appliances Corp.
4. Terephthalonitrile oxide should be freshly prepared. It can be stored for a short period of time (2 weeks) in a refrigerator.
5. The benzene and the flask are dried before use.
6. 1,4-Diethynylbenzene should be sublimed before use; 50° at 1.0 torr produces maximum purity.
7. Average yield based on three consecutive polymerizations.
8. Approximately 10-12 days under constant shaking are required to prepare a clear 0.1 g./100 ml. solution.

3. Methods of Preparation

This polymerization has been published (7). The checkers obtained comparable results by adding solid diethynylbenzene to a suspension of terephthalonitrile oxide in 20 ml. of benzene.

4. References

1a. University of Michigan, Ann Arbor, Michigan 48104.
1b. Current Address: Xerox Corp., Webster, New York 14580.
2. Monsanto Triangle Park Development Center, Research Triangle Park, North Carolina 27709.
3. H. Gilman, ed., *Organic Syntheses,* Coll. Vol. II. Wiley, New York, p. 622.
4. A. Ricca, *Gazz. Chim. Ital.,* **91**, 83 (1961).
5. Y. Iwakura, M. Akiyama, and K. Nagakubo, *Bull. Chem. Soc. Japan,* **37**, 767 (1964); J. M. Craven and R. Wehr, E. I. du Pont de Nemours and Company, private communication.
6. A. S. Hay, *J. Org. Chem.,* **25**, 637 (1960).
7. C. G. Overberger and S. Fujimoto, *J. Polym. Sci.,* B, **3**, 735 (1965); C. G. Overberger and S. Fujimoto, *J. Polym. Sci.,* C, **16**, 4161 (1968).

Preparation and Polymerization of Thiocarbonyl Fluoride

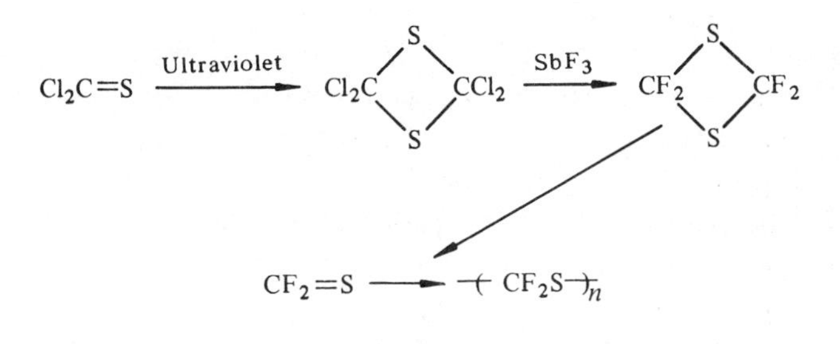

Submitted by W. H. Sharkey and H. W. Jacobson (1)
Checked by P. D. Schuman (2)

1. Procedure

A. 2,2,4,4-Tetrachloro-1,3-dithietane

2,2,4,4-Tetrachloro-1,3-dithietane is formed by ultraviolet dimerization of thiophosgene (3). A Pyrex® glass test tube 4 in. I.D. and 18 in. long (4) is equipped with a water jacket and mounted vertically on a tripod. The top of the tube is closed with a rubber stopper fitted with a stirrer. The stirrer should be a sturdy glass rod to which a paddle of Teflon® fluorocarbon resin is attached at the lower end. A vent tube is also inserted through the stopper and protected with a Drierite drying tube. Six ultraviolet lamps (GE-H85-C3), each fitted with a parabolic reflector and connected to 550-volt transformers, are placed vertically around the reaction tube about ¼ in. from its water jacket. Then 4000 g. of thiophosgene (*Caution!*) (Note 1) is placed in the tube and stirred slowly (Note 2) with the stirrer paddle about 3 in. below the top of the liquid. At the start, the lamps (*Caution!*) (Note 3) are arranged so that their tops are level with the surface of the liquid thiophosgene. Water at 20° of slightly below is circulated through the jacket to keep the contents of the tube cool. The lamps are turned on and, as irriadation proceeds, crystals of 2,2,4,4-tetrachloro-1,3-dithietane form. The lamps are adjusted to maintain their tops even with the liquid level as it drops. The stirrer paddle is also lowered.

After approximatley 215 hr. dimerization is about 75% complete. At this point, the stirrer is stopped to allow the crystals to settle. Irradiation is continued for an additional 2 hr. to form a crust over the solid. The temperature of the water circulating in the jacket is then reduced to 5°, the lamps are shut off, and the stirrer is removed. The liquid fraction is poured into a brown bottle for recovered thiophosgene. The tube is placed at an angle of 30° for about 30 min. to allow more liquid to drain. The solid 2,2,4,4-tetrachloro-1,3-dithietane is then removed from the tube and stored in a brown bottle until it is used in the next step. This solid fraction should weigh 2930 g. (75% of the theoretical amount).

B. *2,2,4,4-Tetrafluoro-1,3-dithietane (5)*

A three-necked, 12-1. flask is equipped with a glass stirrer with a Teflon® paddle, a thermometer, a nitrogen, inlet, and a condenser connected to three cold traps in series. The first two traps are cooled with ice, and the third trap is cooled with solid carbon dioxide in acetone. The flask is charged with 7500 g. of tetramethylene sulfone. Air is displaced by passage of dry nitrogen through the system. Nitrogen flow is maintained during the addition of all ingredients and until the apparatus is closed. After adding 4500 g. of antimony trifluoride, the 2,2,4,4-tetrachloro-1,3-dithietane from the first reaction (3000 g.) is added through the condenser, and up to 900 g. of additional tetramethylene sulfone is used to wash all of it into the reaction flask. The reaction flask is heated to 67° with a Glas-Col mantle. The heat source is removed when bubble formation indicates that the reaction has started, and stirring is continued until bubbling decreases. Heating is then resumed at a rate such that the inside temperature reaches 180° over a 1-hr. period. Crude tetrafluorodithietane fractions that collect in traps are combined, dried with magnesium sulfate (1 g./100 g. of product), filtered, and distilled. Distillation is easily done with a column 24 in. long and 1.25 in. in diameter, packed with ⅛-in. glass helices, with an attached 3-1. pot. The fraction that boils at 45-50° is collected for further purification.

Impurities are oxidized and removed by the addition of hydrogen peroxide. A 5-l., three-necked flask is equipped with a condenser, a thermometer, and a glass stirrer with a Teflon® paddle. Tetrafluorodithietane from the above distillation is added to the flask. As the diethietane is vigorously stirred, a solution of 20 cc. of 30% hydrogen peroxide in 100 ml. of 10% sodium hydroxide is added through the condenser over a period of about 3 min. The flask temperature rises with the addition of the caustic-peroxide mixture, but the temperature is maintained at 35-40° with an ice bath. Additional caustic-peroxide mixture is added frequently to maintain the temperature of the reaction mixture between 35° and 40° with continual ice-bath cooling. When the color in the heavy liquid layer is only a faint yellow, an additional 120 ml. of the caustic-peroxide mixture is added and stirring is continued for 3 hr. There are two layers in the reaction mixture; the heavy lower layer is separated and dried with magnesium sulfate, and the upper layer is discarded. The dried, lower layer is distilled, and the fraction boiling at 47.3-47.7° is collected as pure 2,2,4,4-tetrafluoro-1,3-dithietane. The yield is 1300 g. (Note 4), which is 65% of the theoretical amount.

C. *Thiocarbonyl Fluoride*

Caution! Although toxicity studies have not been made, thiocarbonyl fluoride is undoubtedly dangerous because it quickly hydrolyzes to hydrogen fluoride and carbon oxysulfide.

An unpacked platinum tube 0.5 in. in diameter (Note 5) and 20 in. long is mounted at a 30° angle and fitted at the upper end with a T-tube that is connected to a dropping funnel and a nitrogen source. The lower end is fitted to a coiled tube trap of stainless steel that is cooled with an ice bath. The outlet from this trap is connected to a glass trap cooled with solid carbon dioxide in an acetone bath. The outlet from the second trap is protected with a drying tube. The platinum tube is heated to 500° over a 12-in. section, and nitrogen is passed through the tube at a rate of about 100 ml./min. to purge all moisture. Then 100 g. of 2,2,4,4-tetrafluoro-1,3-dithiethane is added dropwise to the tube from the funnel over a 5-hr. period. Thiocarbonyl fluoride, which condenses in the trap cooled with solid carbon dioxide and acetone, is purified by fractionation through a 2 ft. × 9 mm. vacuum-jacketed column, which is packed with glass helices and equipped with a still head kept at a low temperature with solid carbon dioxide in acetone. Pure monomer boils at −54°, and the amount obtained is at least 90 g. This compound must be protected from moisture and should be kept very cold until used. Because impurities are slowly formed by this very reactive monomer, storage for more than a week is not advisable, even at low temperatures.

D. Anionic Polymerization (6)

A 300-ml., round-bottomed flask having two necks, one with a 29/26 spherical ground glass joint and the other with a ₸ 10/30 joint, is connected through the larger neck to a glass tee made from 27-mm. tubing. The tee is constructed so that one end of the crossbar is bent 90° parallel to the stem and is terminated with a 29/26 ground glass male joint. The other end of the crossbar is connected by a three-way stopcock to a vacuum pump and a source of dry nitrogen. A glass-coated stirrer bar is added to the flask, and the entire assembly dried by flaming as nitrogen is passed through the system. Nitrogen flow is continued as the equipment cools, and then 100 ml. of diethyl ether dried over sodium is added through the 10/30 neck with a large hypodermic syringe. (The syringe is previously dried by baking in a vacuum oven at 125° and cooled in a stream of dry nitrogen.) A serum stopper is immediately placed in the 10/30 neck, the nitrogen stream to the tee and flask is closed, and a 100-ml. flask with 100 g. of thiocarbonyl fluoride is attached to the 29/26 male ground glass joint of the tee. The contents of the polymerization flask and the monomer flask are frozen with liquid nitrogen, and the system is evacuated. Then the monomer-containing flask is allowed to warm, and the thiocarbonyl fluoride distills into the polymerization flask. After all the monomer has been transferred, the liquid nitrogen bath around the polymerization flask is replaced with a solid carbon dioxide-acetone bath. The contents of the flask melt as they warm; when they are completely melted, dry nitrogen is admitted until the pressure inside the system is at atmospheric level. The stirrer is then started, and 5 drops of dimethylformamide is added from a No. 22 hypodermic needle. Polymer formation, which begins almost immediately, is completed after 2 hr. The white, spongy polymer is removed from the flask and boiled in about 200 ml. of water which contains 5 ml. of 50% nitric acid. After drying, the weight of the polythiocarbonyl fluoride is 95-98 g. The molecular weight of the polymer is very high as indicated by an inherent viscosity of 4-6 dl./g. for a 0.5% solution in chloroform (Note 6).

E. Free-Radical Polymerization

Free radicals can be used to homopolymerize thiocarbonyl fluoride and to

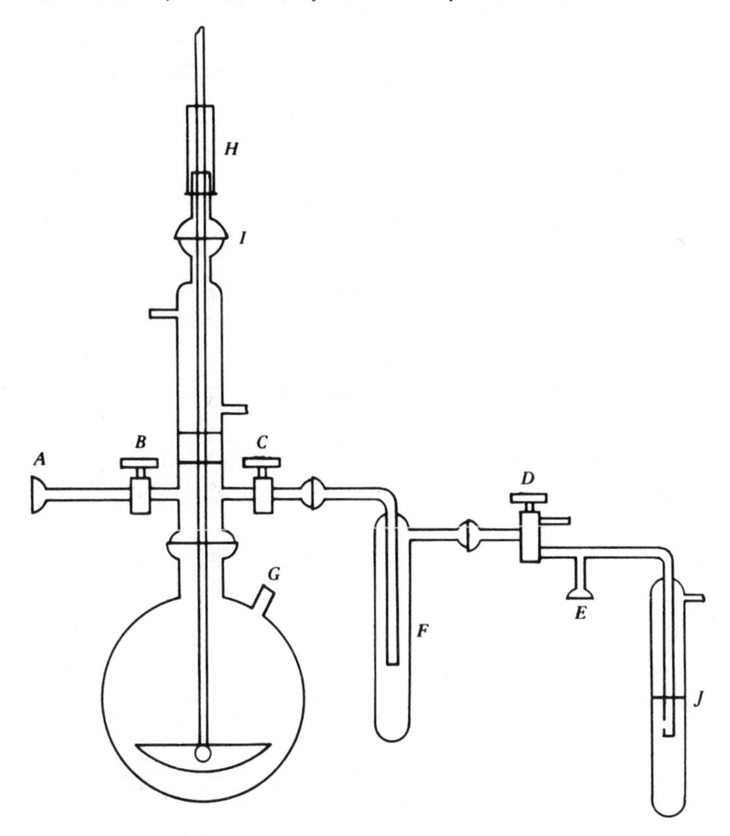

Fig. 1. Apparatus for free-radical polymerization of thiocarbonyl fluoride.

copolymerize thiocarbonyl fluoride with a variety of unsaturated compounds (7). Copolymerization with propylene is illustrative of this process.

The apparatus for this copolymerization is sketched in Fig. 1. The main elements are a 500-ml. polymerization flask having a small neck at *G* for serum stopper and a True-Bore stirrer with a Teflon® paddle modified to hold a vacuum of about 0.3 mm. This modification involves sealing a 28/12 outer spherical joint to the top of the True-Bore condenser at *I* and slipping the corresponding inner spherical joint over the stirrer shaft and sealing it to the shaft with a rubber sleeve at *H* (Note 7). This joint is clamped together when the system is under vacuum and must be allowed to run free when the stirrer is operating. At *A* there is an 18/9 outer spherical joint; *B* and *C* are stopcocks, *D* is a two-way stopcock connected to a helium cylinder at *E* and to a vacuum pump, and *J* is a helium bubbler filled with mineral oil to about the level indicated. Before polymerization, all the glassware to the left of stopcock *D* is dried overnight in a vacuum oven at 125°. The apparatus is assembled hot under a stream of dry helium entering at *E*. The helium stream is maintained throughout preparation for polymerization and allowed to exit through the bubbler when not needed to fill the polymerization flask.

A cylinder of dichlorodifluoromethane (Freon®-12) is connected at *A* with a stainless steel line having an 18/9 stainless steel spherical joint. Stopcock *D* is turned to the vacuum pump, and the system is evacuated to about 0.4 torr. Next, the system is filled with helium by reversing stopcock *D*. This process is then repeated (Note 8). After being reevacuated, the polymerization flask is surrounded with a solid carbon dioxide-acetone bath, and the cylinder of Freon®-12 is opened carefully to allow the contents to distill into the flask until 300 ml. has been collected (Note 9). The cylinder valve is closed, and helium is admitted through *D* as far as the cylinder valve. Then *B* is closed, and Freon®-12 freezes, the system is evacuated to the propylene cylinder and then filled with helium. Stopcock *B* is closed, and 70 ml. of propylene is distilled from the cylinder into the graduated trap; then the propylene is cooled with a liquid nitrogen bath. Stopcock *B* is opened, the system is evacuated to about 0.4 torr, and stopcock *C* closed. The cooling bath is removed from the propylene trap; as the trap warms, propylene distills into the polymerization flask. The system is again filled with helium, stopcock *B* is closed, the propylene cylinder and graduated trap are removed, and a trap containing 35 ml. of thiocarbonyl fluoride is connected at *A*. The thiocarbonyl fluoride is frozen with liquid nitrogen; then stopcock *B* is opened, and the system is evacuated, filled with helium, and evacuated again. Stopcock *C* is closed and the thiocarbonyl fluoride is allowed to warm, whereupon it distills into the polymerization flask. Stopcock *B* is closed, trap *F* is immersed in liquid nitrogen, stopcock *C* is opened, and the system is evacuated. The liquid nitrogen bath is removed from the polymerization flask, and a stream of acetone is directed at the outside of the flask to warm the contents and bring about melting as quickly as possible (Note 10). When melting is complete, the polymerization flask is reimmersed in liquid nitrogen, and the bath surrounding trap *F* is removed. Material collected in the trap redistills into the polymerization flask (Note 11). The system is again filled with helium, the liquid nitrogen bath surrounding the polymerization flask is replaced with a solid carbon dioxide-acetone bath, the clamp is removed from the stirrer joint at *I*, and the rubber sleeve at *H* is raised to free the stirrer shaft. As soon as the flask contents melt, the stirrer is started.

One and five-tenths milliliters of 0.5 *M* diethyl (ethylperoxy)borane (7.5 ×

10^{-4} mole) in heptane (7) (*Caution!*) (Note 12) is added with a syringe through the serum stopper at G, followed by similar addition of 0.7 ml. of 1.86 *M* triethylborane (*Caution!*) (Note 12) (13×10^{-4} mole) in heptane. Polymer formation begins almost immediately and is complete at the end of 2.5 hours. Cold methanol is then added to the reaction mixture, and the whole mass is allowed to warm to room temperature. Freon®-12 and excess propylene are removed by distillation. The solid polymer is removed from the flask, dried, dissolved in 300 ml. of chloroform, and reprecipitated by pouring the chloroform solution into methanol. After drying, the precipitated polymer weighs 50-51 g. It has a sulfur content of 32.02%, corresponding to a 2.34:1 mole ratio of thiocarbonyl fluoride to propylene.

2. Characterization

High molecular weight polythiocarbonyl fluoride (8) is soluble in chloroform and tetrahydrofuran. Inherent viscosities of 4-6 correspond to $\overline{M}_n$ of well over 500,000 and perhaps over 1,000,000. Inherent viscosities of 0.5% in chloroform solutions are preferred because the polymer is stable in chloroform but degrades quite rapidly in tetrahydrofuran. The degradation can be overcome if the tetrahydrofuran is saturated with dry hydrogen chloride. The polymer has a crystalline melting point of about 35°, at which temperature it changes from an opaque plastic to a very resilient elastomer. In the amorphous elastomeric form, it can be molded but only at a relatively high temperature. Typical film-forming conditions are to heat the platens of a Carver press to 150° and press the polymer between aluminum sheets in the platens under a ram pressure of 10,000 lb. for several hours. As removed from the press, the films are elastomeric. At room temperature, they slowly crystallize to a non-elastomeric form. Reheating above 35° reconverts these films into the elastomeric form. An outstanding characteristic of this elastomer is high resilience (95% as measured by the Yerzley method, ASTM-D945). The T_g of the polymer measured by the torsion pendulum method is −118°.

The thiocarbonyl fluoride-propylene copolymer (8) is also soluble in chloroform. Inherent viscosities in 0.5% chloroform solutions typically fall between 2 and 3 dl./g. The compositions of these polymers range from about 2.2 molecules of thiocarbonyl fluoride per propylene molecule to much higher ratios. With a very large excess of propylene, copolymers having very nearly a 2:1 ratio can be prepared. The products are all soft polymers that stiffen only to a small extent at temperatures as low as −55°.

3. Notes

1. *Caution! Thiophosgene is very toxic. Because it can cause permanent injury by contact, ingestion, or inhalation, it should be used only in a forced-draft hood.* If it comes into contact with the skin, it should be immediately neutralized with ammonia, and the affected area washed with soap and water. If the eyes are affected, they should be washed with sodium bicarbonate solution.

2. If agitation is too vigorous, the dimer will remain suspended in the liquid and will block the entrance of ultraviolet light.

3. *Caution! Because ultraviolet light damages eyes, the apparatus should be shielded to prevent experimenter exposure. When working behind the shield, he should wear glasses that filter ultraviolet light.*

4. The product is colorless, has a refractive index of 1.3902^{26}-1.3919^{24}, and does not develop color when several drops are mixed with an alcohol solution of tripropylphosphine.

5. Stainless steel ball joints are friction-fitted to each end of the platinum tube, and the fit is sealed by silver soldering. A female joint on the upper end is connected to an appropriate glass joint on the T-tube, and the male joint on the lower end is connected to its opposite number on the stainless steel trap. The platinum tube was obtained under special agreement from Engelhard Industries, 700 Blair Road, Carteret, New Jersey 07008.

6. High inherent viscosities are obtained only if anhydrous conditions are maintained during polymerization.

7. The rubber sleeve is wired to both the stirrer shaft and the inner spherical joint for a tight fit.

8. This operation is repeated because it is important to remove all atmospheric oxygen. Any residual oxygen will react with the triethylborane that is to be used later and thereby will unbalance the stoichiometry of the redox initiating system.

9. It is convenient to mark the flask ahead of time at a level that coresponds to 300 ml.

10. This operation serves to degas the polymerization mixture and thus remove traces of atmospheric oxygen.

11. This material is presumed to be mostly thiocarbonyl fluoride. It is lower boiling than either Freon®-12 or propylene.

12. This solution can be made by cooling a 0.5 *M* solution of triethylborate (*Pyrophoric! Handle with extreme care.*) in heptane in a solid carbon dioxide-acetone bath and then carefully adding oxygen. Only 1 mole of oxygen reacts at this low temperature. After reaction, excess oxygen over the solution is replaced

with helium or other inert gas. These solutions should be standardized by iodimetric titration.

4. Methods of Preparation

A number of methods have been developed for the preparation of thiocarbonyl fluoride (9). In one method (10), thiophosgene is chlorinated to trichloromethanesulfenyl chloride, followed by fluorination to obtain chlorodifluoromethanesulfenyl chloride, which is then dechlorinated. The reaction of bis(trifluoromethylthio)mercury and iodosilane gives trifluoromethylthiosilane, which decomposes to thiocarbonyl fluoride and fluorosilane (11). Thiocarbonyl fluoride is formed when tetrafluoroethylene (5) or chlorofluoromethane (12) reacts with sulfur at high temperatures. Polymerization through the carbon-sulfur double bond, including thiocarbonyl fluoride, has been reviewed (13).

5. References

1. Central Research Department, E. I. du Pont de Nemours and Company, Wilmington, Delaware 19898.
2. Contract Research Division, PCR, Gainesville, Florida 32601.
3. A. Schonberg and A. Stephenson, *Chem. Ber.*, **66B**, 567 (1933).
4. This relatively large-scale preparation of 2,2,4,4-tetrachloro-1,3-dithiethane was worked out with the help of Messrs. H. D. Carlson and S. P. Gauntt of this laboratory.
5. W. J. Middleton, E. G. Howard, and W. H. Sharkey , *J. Org. Chem.*, **30**, 1375 (1965).
6. W. J. Middleton, H. W. Jacobson, R. E. Putnam, H. C. Walter, D. G. Pye, and W. H. Sharkey, *J. Polym. Sci.*, A, **3**, 4115 (1965).
7. A. L. Barney, J. M. Bruce, Jr., J. N. Coker, H. W. Jacobson, and W. H. Sharkey, *J. Polym. Sci.*, A-1, **4**, 2617 (1966).
8. W. J. Middleton, U.S. Pat. 3,240,765 (March 15, 1966).
9. W. Sundermeyer and W. Meise, *Z. Anorg. Allgem. Chem.*, **317**, 334 (1962).
10. N. N. Yarovenko and A. S. Vasil'eva, *J. Gen. Chem. USSR*, **29**, 3754 (1959).
11. A. J. Downs and E. A. V. Ebsworth, *J. Chem. Soc.*, 3516 (1960).
12. D. M. Marquis, U.S. Pat. 2,962,529 (September 29, 1960).
13. W. H. Sharkey, *Adv. Polym. Sci.*, **17**, 73 (1975).

Cyanoethylcellulose

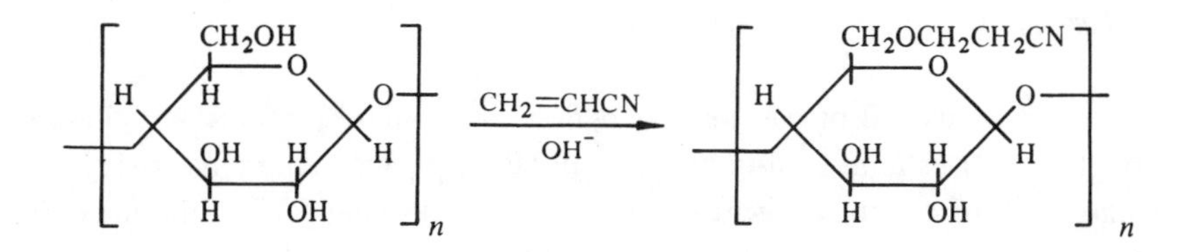

Submitted by N. M. Bikales (1a, b)
Checked by R. E. Leonard and J. E. Kiefer (2)

1. Procedure

A. Highly Cyanoethylated Cellulose (Note 1)

A. 1-1. flask equipped with condenser, 100-ml. additon funnel, thermometer and propeller stirrer is charged with 25 g. of regenerated cellulose (Note 2) and 600 ml. of acrylonitrile (*Caution!* Note 3). Over a 5-min. period 1.2 g. of sodium hydroxide is added as an aqueous solution to the rapidly stirred reaction mixture. The concentration of the alkaline solution is such as to provide a total water content of 25 g. (Note 4). The reaction mixture is then heated to 50° for 4 hr. with continued good agitation. The cellulose gradually dissolves in the acrylonitrile. To the resultant slightly yellow-brown (*Caution!*) (Note 5) and highly viscous solution a 5% aqueous acetic acid solution (50 ml.) is added, and stirring is continued for 10 min. The hot solution is filtered through a coarse fritted glass filter (Note 6) and transferred to an addition funnel.

An apparatus like that described for the cyanoethylation but equipped also with a Dean-Stark trap is charged with 500 ml. of deionized water. The water is heated to boiling, and the cyanoethylcellulose solution is added dropwise with vigorous stirring. To precipitate the cyanoethylcellulose, all excess acrylonitrile is removed by distillation. The mixture is cooled and filtered (Note 7). The slightly off-white fibrous product is reslurried in hot deionized water, filtered, washed with about 1 l. of hot deionized water, and dried at 60°. The yield is about 40 g. The nitrogen content (by Kjeldhal analysis) is 12.6%, which corresponds to an average substitution of 2.75 of the 3 hydroxyl groups of each anhydroglucose unit (Note 8).

B. Partially Cyanoethylated Cellulose

The apparatus and precautions are as described under Procedure A. Cellulose, 50 g. (Note 9), is immersed in 500 ml. of 2.0% aqueous sodium hydroxide containing 0.1% of a wetting agent (Note 10). After steeping for 30 min. the cellulose is pressed or centrifuged until it retains about an equal weight of the caustic solution (Note 11). The alkali-impregnated cellulose is then stirred in 600 ml. of acrylonitrile at 55° for 45 min. Addition of 5% aqueous acetic acid (50 ml.) neutralizes the suspension. After filtration, the cellulose is washed repeatedly with deionized water and then dried at 105°. The product weighs approximately 50 g. and has a nitrogen content of about 3.6%, corresponding to a degree of substitution of about 0.5.

2. Characterization

Highly cyanoethylated cellulose is readily soluble in such polar solvents as acetonitrile (5), dimethylformamide, dimethyl sulfoxide, and pyridine. Properties and uses are described in ref. 3.

3. Notes

1. The equation of p. 611 shows, for simplicity, only one of the three hydroxyl groups of the cellulose molecule being substituted with a cyanoethyl group. Depending on reaction conditions, the average degree of substitution can be varied from 0.0 to 3.0.
2. Finely chopped rayon is especially suitable.
3. Commercial grade acrylonitrile can be used without further purification. Because acrylonitrile is toxic and flammable, adequate precautions should be observed, such as those outlined in Chemical Safety Data Sheet SD-31 of the Manufacturing Chemists Association, Washington, D.C.

4. The total water content is the sum of the water in the rayon, the acrylonitrile, and the caustic solution. The water content of rayon can be determined by weighing a sample before and directly after heating in an oven at 110° for 1 hr. The water content of acrylonitrile is available from the manufacturer; most commercial grades are anhydrous.

5. *Caution! The appearance of a deep orange color and/or a rapid rise in temperature indicate the possible onset of anionic polymerization of acrylonitrile. Because this polymerization is violent, the reaction mixture should be promptly treated with about 100 ml. of 5% aqueous acetic acid.* The acetic acid can be placed in the addition funnel, after the addition of the caustic has been completed. However, we never observed anionic polymerization when we followed this procedure.

6. The filtration can be omitted if slight amounts of insolubles can be tolerated.

7. An alternative method of precipitation is to add slowly the cyanoethylcellulose solution in acrylonitrile to about 600 ml. of ethanol with vigorous stirring.

8. For the relationship between nitrogen content, degree of substitution, and weight increase, see ref. 3.

9. The cellulose can be cotton fibers, linters, or wood pulp (α-cellulose). Manila, jute, and sisal react more slowly and therefore require more drastic conditions.

10. The wetting agent must be stable in alkali; Aerosol OS (American Cyanamid Company) has worked satisfactorily.

11. The immersion in alkali can be omitted by a slight change in procedure (6). Forty grams of 2.4% sodium hydroxide is added dropwise at room temperature to the cellulose-acrylonitrile mixture with vigorous stirring. After 15 min., the temperature is raised to 50-55°. The procedure is then the same.

4. Methods of Preparation

The procedure for highly cyanoethylated cellulose is based on ref. 4. Other types of cellulose, such as cotton linters or wood pulp, can be substituted for rayon but give very viscous solutions that are more difficult to handle. Other hydroxylic polymers, such as poly(vinyl alcohol) or hydroxyethylcellulose, can be used with some adjustments in the procedure.

Numerous modifications of the procedure for partially cyanoethylated cellulose have been reported (7). The steeping in alkali can be omitted (6); a more concentrated alkaline solution (ca. 9%) can be used at room temperature (6); the reactivity of cellulose can be increased by swelling (8); the medium can be essentially aqueous with a small amount of acrylonitrile (9); or reaction can take place in cellulose impregnated with small amounts of both alkali and acrylonitrile (10).

The rate of cyanoethylation of cellulose is increased by preactivation with sodium hydroxide or ammonia or by using dimethyl sulfoxide in the cyanoethylation bath (11). A substitution of 2.86 has been obtained using dimethyl sulfoxide in the presence of NaOH solution vs. 0.70 without it (12).

5. References

1a. Consulting Chemist, 8 Trafalgar Drive, Livingston, New Jersey 07039.
1b. Current Address: Rutgers University, New Brunswick, New Jersey 08901.
2. Tennessee Eastman Company, Division of Eastman Kodak Company, Kingsport, Tennessee 37662.
3. N. M. Bikales and L. Segal, eds., *Cellulose and Cellulose Derivatives (High Polymers)*, (Vol. V), Wiley-Interscience, New York, 1971, pp. 811-833.
4. N. M. Bikales and W. O. Fugate (American Cyanamid Company), U.S. Pat. 3,067,141 (December 4, 1962).
5. K. W. Saunders and N. M. Bikales (American Cyanamid Company), U.S. Pat. 3,097,956 (July 16, 1963).
6. A. H. Gruber and N. M. Bikales, *Textile Res. J.*, **26**, 67 (1956).
7. For a review, see N. M. Bikales and L. Segal, eds., *Cellulose and Cellulose Derivatives (High Polymers)*, (Vol. V), Wiley-Interscience, New York, 1971, pp. 811-833, 1169-1223.
8. N. M. Bikales, *Ind. Eng. Chem.*, **50**, 87 (1958).
9. N. M. Bikales and L. Rapoport, *Textile Res. J.*, **28**, 737 (1958).
10. J. L. Morton and N. M. Bikales, *Tappi*, **42**, 855 (1959).
11. H. Schleicher, B. Lukanoff, and B. Philipp, *Faserforsch. Textiltech.*, **25** (5), 179 (1974).
12. B. Lukanoff and H. Schleicher, Ger. (East) Pat. 109,640 (November 12, 1974).

Squaric Acid-1,3-Polyamides

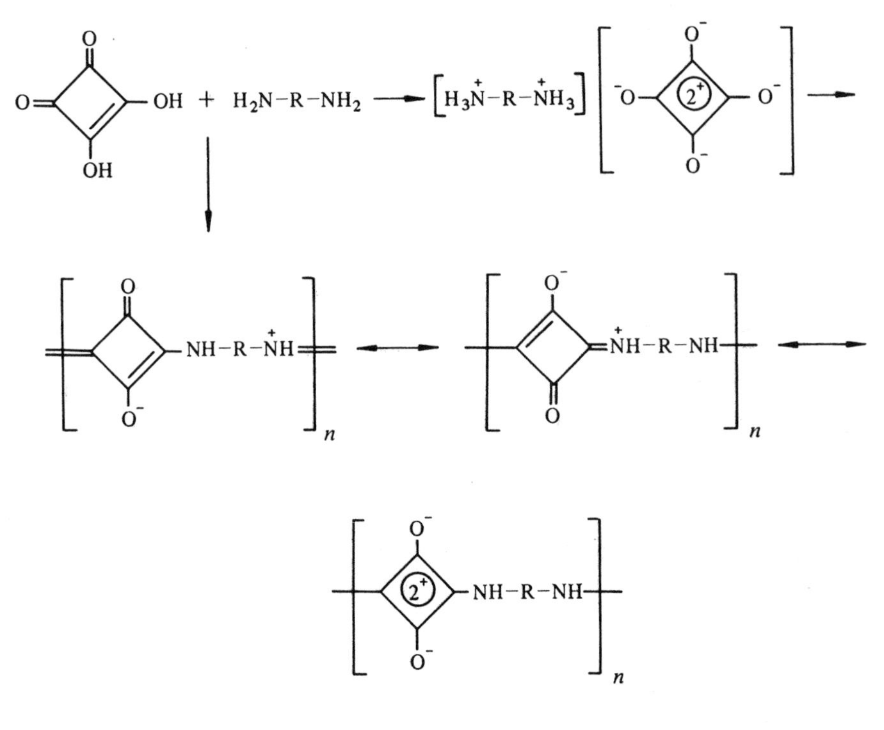

$$R = -(CH_2)_6- \text{ and } -C_6H_4-$$

Submitted by G. Manecke and J. Gauger (1)
Checked by Z. Janovic and N. G. Gaylord (2)

1. Procedure

A. 1,6-Diaminohexane-Squaric Acid Salt

1,2-Dihydroxycyclobutenedione (squaric acid) (1.14 g., 0.01 mole) is dissolved in 15 ml. of hot water. A solution of 1.16 g. (0.01 mole) of 1,6-diaminohexane in 200 ml. of acetone is then added, dropwise, with stirring. After cooling the squarate salt is removed by filtration and recrystallized twice from water-acetone, by dissolving the product in 5-10 ml. of hot water and adding 150 ml. of acetone in portions. There is obtained 2.1 g. (90.8% yield) of salt (Note 1).

B. Polyamide from Squaric Acid-1,6-Diaminohexane Salt: Poly(azi-hexamer-alt-quadratyl-1,3-amer)

After 1 g. of the salt is refluxed for 30 min. with 5 ml. of *m*-cresol (Note 2), nearly all of the solvent is removed by distillation at atmospheric pressure for about 15 min. (about 4 ml.). After cooling, the viscous liquid is stirred with 200 ml. of acetone, and the polyamide crystallizes. The product is collected on a filter and recrystallized by dissolving in 2 ml. of formic acid under reflux. After the volume has been reduced to 1 ml., the solution is placed in a refrigerator. A pale yellow polycondensate is obtained in 83-100% yield, m.p. above 350°. The polyamide shows low solubility in common nonpolar organic solvents (Notes 3 and 4). The viscosity (η) = 0.06-0.163 dl./g. in concentrated H_2SO_4 as solvent at 20°.

C. Polyamide from Squaric Acid and p*-Phenylenediamine: Poly[azi-phenylene-(1,4)-azmer-alt-quadratyl-(1,3)-amer]*

1,2-Dihydroxycyclobutenedione (squaric acid) (1.14 g., 0.01 mole) is refluxed with 1.08 g. (0.01 mole) of *p*-phenylenediamine in 40 ml. of glycerine for 10 min. About 10 ml. of the glycerine is removed by distillation at atmospheric pressure for 15 min. to remove the water eliminated during the condensation. After cooling, the dark brown polycondensate is removed by filtration and extracted with pyridine for 8 hr. followed by extraction with methanol for 8 hr. (Soxhlet apparatus) to yield 98-100% of product, m.p. above 350°. The viscosity of this aromatic polyamide could not be measured because of its very low solubility in concentrated sulfuric acid.

2. Characterization

The infrared spectra of squaric acid-1,3-polyamides resemble those of the monomer model compounds except that the bands are very diffuse and not as well resolved.

Carbonyl stretching frequencies appear at 1770-1841 and 1680-1730 $cm.^{-1}$ (7, 8) in the case of the squaric acid-1,2-bisamides, while the isomeric squaric acid-1,3-bisamides show no absorption in this region because of the quasi-aromatic character of the cyclobutene-diyliumdiolate system. [Ed. Note: Recent work has produced evidence that 1,2 as well as 1,3 orientations of the substituent links may be present in these systems (9).]

Thermogravimetric measurements show that the polyamides with relatively long carbon chains between the cyclobutene-diyliumdiolate units are the most thermostable products of the squaric acid-polyamides which have been investigated. At a temperature of 350°, after heating for 2 hr. these polyamides lost 20-30% of their weight.

3. Notes

1. Preparation from the squarate salts is preferable to direct mixing of squaric acid and diamine because equimolar amounts of the components are essential for the attainment of high molecular weights. Also, purification of the components is easier by recrystallization of the squarates.
2. Because of their high boiling points and good solvent properties for both the squarates and the polycondensation products, the most suitable solvents for the preparation of the polyamides are cresols and glycerine. The condensation in cresol (with azeotropic distillation of water) yields nearly colorless polycondensates.
3. By dissolving the polyamides formed with 1,6-diaminohexane in formic acid and precipitating into dimethylformamide, in which the yellow, low molecular weight by-products are soluble, the purification can also be carried out with sulfuric acid and water.
4. From a formic acid, inorganic acid, or cresol solution transparent films can be formed. When heated in the presence of air, the polycondensation products darken when the temperature is higher than 250°.

4. Methods of Preparation

Condensation of 1,2-dihydroxycyclobutenedione (squaric acid) with primary and secondary monoamines leads to the mesomeric squaric acid-1,3-bisamides derived from 1,3-bisaminocyclobutene-diyliumdiolate, not the squaric acid-1,2-bisamides (3, 4, 5). (Ed. Note: However, see ref. 9) Analogously, the free squaric acid forms the polycondensates described in Procedures B and C (with bifunctional aliphatic or aromatic amines) (4). This procedure has been described in the literature (6).

In an alternative polymerization, the squarate salt of 1,6-diaminohexane in formic acid is heated under nitrogen, and the solvent is slowly distilled to yield a nearly colorless precondensate. This can be further condensed under nitrogen and then a vacuum by raising the temperature. Polyamides so prepared are colorless to pale yellow.

5. References

1. Fritz-Haber-Institute der Max-Planck-Gesellschaft, Berlin-Dahlem.
2. Gaylord Research Institute, Newark, New Jersey 07014.
3. G. Manecke and J. Gauger, *Tetrahedron Letters,* 3509 (1967).
4. G. Manecke and J. Gauger, *Tetrahedron Letters,* 1339 (1968).
5. J. Gauger and G. Manecke, *Chem. Ber.*, **103**, 2696 (1970).
6. G. Manecke and J. Gauger, *Makromol Chem.*, **125**, 231 (1969).
7. S. Cohen and S. G. Cohen, *J. Amer. Chem. Soc.*, **88**, 1533 (1966).
8. G. Maahs and P. Hegenberg, *Angew. Chem.,* 78, 927 (1966); *Angew. Chem. Internat. Ed.*, **5**, 888 (1966).
9. E. W. Neuse and B. R. Green, *Polymer* **15**, 339 (1974).

Poly[4(5)-vinylimidazole]

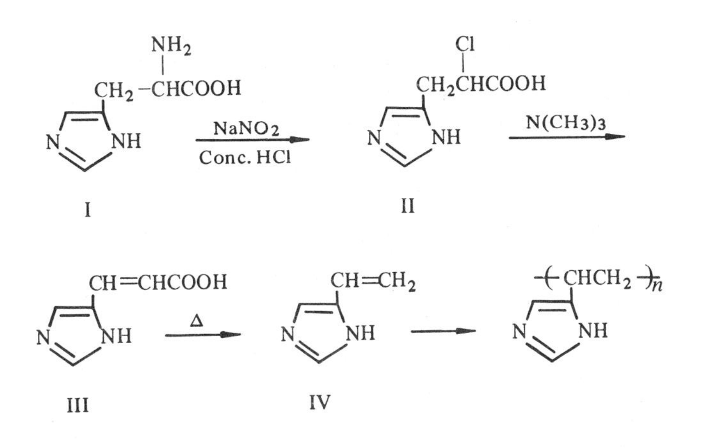

Submitted by C. G. Overberger, (1a) R. C. Glowaky, (1a, b)
T. J. Pacansky, (1a, c) and K. N. Sannes (1a, d)
Checked by H. Kaye (2a, b) and J. S. Hwang (2a)

1. Procedure

A. 2-Chloro-3(4-Imidazolyl)-Propionic Acid (11) (3)

A 3-1., three-necked, round-bottomed flask equipped with a thermometer, mechanical stirrer, and dropping funnel is charged with 1000 ml. of concentrated hydrochloric acid. With stirring, 150 g. (0.78 mole) of *l*-histidine (I) monohydrochloride (Note 1) is added portionwise. The fine, white dispersion is cooled in an ice-water bath, followed by dropwise addition, with stirring, of a solution of 150 g. of sodium nitrite in 300 ml. of water. The addition is controlled to keep the reaction temperature below 10°. Heavy precipitation of sodium chloride occurs, and the reaction mixture turns red. After the addition of the sodium nitrite solution is complete, the reaction mixture is stirred for an additional 3 hr. without adding more ice.

The reaction mixture is filtered to remove sodium chloride (*Caution!*) (Note 2), and the filtrate is evaporated *in vacuo*. During evaporation, more sodium chloride precipitates and is filtered. Evaporation is continued until a syrup free of sodium chloride is obtained. The syrup is dissolved in about 500 ml. of water and reevaporated. This process is repeated twice (Note 3). The syrup is then dissolved in about 100 ml. of water and stirred in an ice-water bath with about 500 ml. of ice-cold 2 *N* ammonium hydroxide. This solution is evaporated almost to dryness, leaving a precipitate of 2-chloro-3(4-imidazolyl)propionic acid (II) (3). The crude product is recrystallized from water with Norit; with concentration of the mother liquor, yields of 40-50 g. of chloro compound (~35% yield) are obtained. This material may contain a small amount of ammonium chloride, but it can be used in the next step.

B. *Urocanic Acid (III) (4)* (Note 4)

A solution of 90 g. (0.511 mole) of crude II in 500 ml. of 25% trimethylamine is placed in a bomb (Note 5), which is heated at 65° for at least 2 days. The reaction mixture is refluxed in a 1000-ml. flask until the odor of trimethylamine is no longer detected. After refluxing, the solution is cooled in an ice-water bath, and crude urocanic acid crystallizes immediately as long needles. The crystals are separated by filtration and recrystallized twice from water with Norit. Concentration of the mother liquor gives pure urocanic acid dihydrate as colorless needles, m.p. 224° (dec.).

The urocanic acid dihydrate crystals are dehydrated by heating overnight at 100° in a vacuum oven, giving 28 g. (40% yield) of III as a powder, which is stored in a desiccator.

C. *4(5)-Vinylimidazole (IV)*

Anhydrous urocanic acid (III) is decarboxylated in a simple distillation unit as shown in Fig. 1. In the 10-ml. distilling flask A is placed 2.5 g. (0.018 mole) of III (Note 6), and vacuum (~0.1 torr) is applied to the system. An air bath (Note 7) preheated to 230° is then applied to the flask. Decarboxylation takes place immediately, and crude 4(5)-vinylimidazole (IV) distills slowly as a viscous liquid. When only black residue remains in the distilling flask, the air bath is removed and the reaction is stopped. At this time the vacuum has returned to

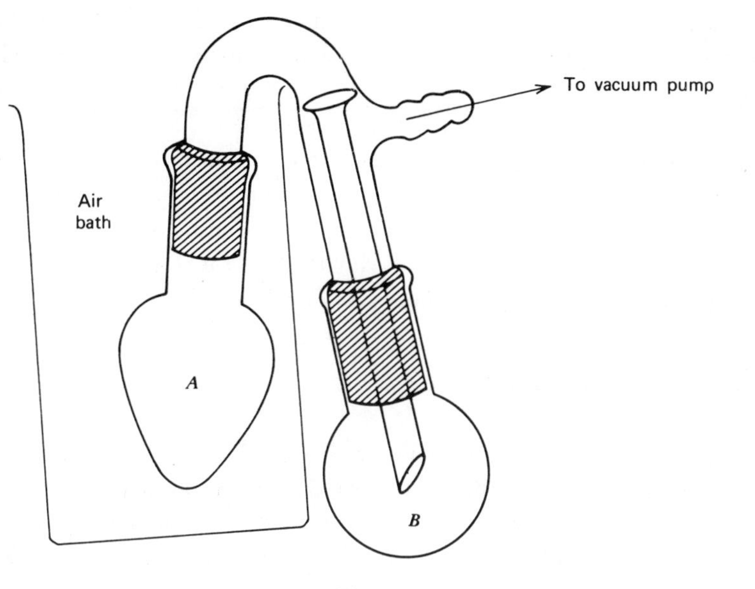

Fig. 1.

~0.1 torr. The product (1.01 g., 60% yield) slowly crystallizes in the receiver when stored in the refrigerator overnight. The crude product containing a small amount of polymerized material can be purified by recrystallization from water, but it is best purified by subliming at 50-55° under high vacuum (10^{-4} torr). Pure IV melts at 84° and exhibits an infrared band for the vinyl group at 1650 $cm.^{-1}$.

Urocanic acid (III) can be decarboxylated in batches of 10 g., and probably larger (Note 8), when the pressure rise that attends decarboxylation is reduced

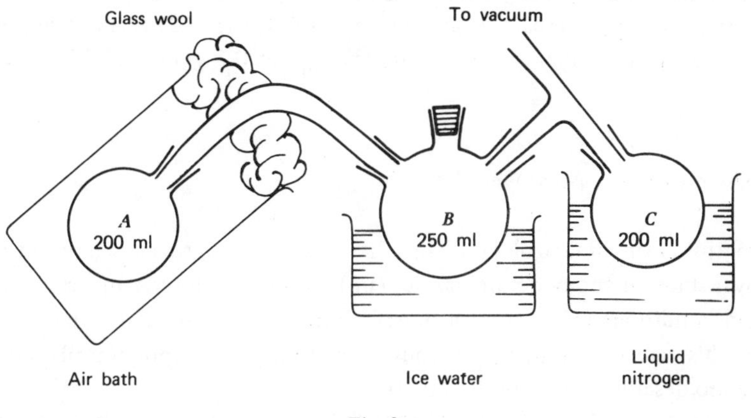

Fig. 2.

(Note 9) by freezing the liberated carbon dioxide with liquid nitrogen. Flask *A* (see Fig. 2) is charged with 10 g. (0.072 mole) of urocanic acid (III) and heated to 100° ± 10°, at 0.01 torr for 48 hr. to ensure dryness (Note 10). The temperature of the air bath surrounding flask *A* is then raised to effect decarboxylation. At 170-180° the pressure in the system, which is open to the vacuum pump, increases, and flask *C* is cooled to the temperature of liquid nitrogen to freeze carbon dioxide as it is evolved (Note 11). The temperature is raised to, and maintained at 210° ± 15° (note 12), until vinylimidazole no longer distills into flask *B*. The yield of crude 4(5)-vinylimidazole (IV) is 4.7 g. (69%).

D. *Polymerization of 4(5)-Vinylimidazole*

Freshly sublimed 4(5)-vinylimidazole, 0.5 g. (5.3 mmoles), and 0.9 mg. of azobis(isobutyronitrile) (AIBN, 0.1 mole %) are dissolved in 10 ml. of dry benzene. The solution is then introduced with a hypodermic syringe into a 25-ml. polymerization tube (Note 13). The clear solution is frozen in a Dry Ice-acetone mixture, evacuated to 0.1 torr closed to the vacuum, and allowed to warm to room temperature. After this procedure has been repeated three times, the tube is sealed while frozen and under vacuum. Polymerization is accomplished by placing the sealed tube in a refluxing methanol bath (65°) (Note 14). The polymer begins to precipitate after approximately 1 hr. of reaction. After 24 hr. at 65°, the tube is opened (*Caution!*) (Note 15) and the contents are added to 30 ml. of benzene and stirred for 1 hr. The polymer is isolated on a 15-ml. medium-porosity, sintered glass funnel and reprecipitated (Note 16) from methanol into benzene, acetone, and then benzene again to yield 0.4 g. of poly-[4(5)-vinylimidazole] (80% recovery). The polymer is dried over P_2O_5 in an Abderhalden drying apparatus at 100° and 0.1 torr for 10 hr. and then powdered for use in the kinetic experiment. The η_{sp}, at 0.5 g./dl. and 25.0°, is 0.50, dl./g.

E. *Hydrolysis of* p-*Nitrophenyl Acetate*

Poly[4(5)-vinylimidazole] (PVIm) has been shown to be a more efficient catalyst then monomeric imidazole (Im) in the hydrolysis of esters, under certain conditions (6). A typical kinetic experiment will be described to illustrate this rate enhancement. A convenient though non-optimum pH has been chosen to describe the catalytic activity.

Solutions of PVIm and Im, 5.33×10^{-4} *M* (Note 17), were prepared in 28.5%

ethanol-water (v/v), which was buffered to pH 7.96 with 0.02 *M* tris (hydroxymethyl)aminomethane (TRIS) and hydrochloric acid with sufficient KCl to adjust the ionic strength to 0.0213 (Note 18). The substrate, *p*-nitrophenyl acetate (PNPA), was dissolved with stirring in 28.5% ethanol-water (v/v) to give an 8×10^{-4} *M* solution. In a 1-cm. quartz cell, 200 μl. of substrate solution was added to 3 ml. of catalyst solution (Note 19). The final solution in the cell was 5×10^{-4} *M* in catalyst, 5×10^{-5} *M* in substrate, and 0.02 *M* in buffer: $\mu = 0.02$. The rate of formation of *p*-nitrophenol at 26° was followed by measuring the increase in optical density ($O.D._t$) at 400 nm as a function of time (t) (Note 20). After at least 10 half-lives the optical density was measured for complete reaction ($O.D._\infty$). Catalysis by buffer only (blank) was measured in the same fashion.

The measured data were treated as pseudo-first-order kinetics by plotting ln ($O.D._\infty - O.D._t$) versus t, the slope being $-k_{measd}$. The solvolysis of substrate (k_{measd}) is the composite of the catalyzed (k_{obs}) and uncatalyzed reactions: therefore, $k_{obs} = k_{measd} - k_{blank}$. Furthermore, the second-order rate constant, k_{cat}, can be obtained via the following relation:

$$k_{cat} = \frac{k_{obs}}{[\text{catalyst}]}$$

if the catalyst is present in sufficient excess. The data obtained under the experimental conditions described above produce the following results;

$$k_{measd}\ (\text{PVIm}) = 1.52 \times 10^{-3}\ \text{min}^{-1}$$

$$k_{measd}\ (\text{Im}) = 1.09 \times 10^{-3}\ \text{min}^{-1}$$

$$k_{measd}\ (\text{blank}) = 0.43 \times 10^{-3}\ \text{min}^{-1}$$

and, using the above definitions for k_{obs} and k_{cat},

$$k_{cat}\ (\text{PVIm}) = 21.8\ M^{-1}\ \text{min}^{-1}$$

$$k_{cat}\ (\text{Im}) = 13.5\ M^{-1}\ \text{min}^{-1}$$

2. Notes

1. This material is commercially available from Pierce Chemical Company, TLC homogeneous, $[\alpha]_D^{21} = -40.3°$ in H_2O.

2. *Caution! The filtration should be carried out in a hood because of the escape of noxious gases.*

3. The hydrogen chloride should be removed as completely as possible to minimize ammonium chloride contamination in the next step.

4. Urocanic acid is commercially available from Koch-Light and Company, Ltd., England. It can also be prepared from histidine by an enzymatic process (4).

5. A stainless steel tube (5.5 cm I.D. × 24 cm. high) with a Teflon® O-ring seal in the top was found to be suitable. Its purpose is to keep the trimethylamine in the reaction mixture. A pressure bottle can also be used.

6. For best results, the urocanic acid should be freshly dehydrated and ground into a fine powder. For this apparatus, increasing the amount of urocanic acid decreased the efficiency and yield in the decarboxylation reaction.

7. An oil bath can also be used.

8. The problems caused by foaming, which may accompany larger batches, can be avoided by simply using a large reaction flask (flask *A*, Fig. 2).

9. This reduction in pressure is necessary to distill the 4(5)-vinylimidazole from flask *A* at the decarboxylation temperature.

10. If previously dried urocanic acid is used, this drying step is probably unnecessary.

11. During the reaction, the rate at which the carbon dioxide could be removed became too low. The ensuing interruption of the reaction could be avoided by incorporating an alternative trap into the system.

12. It is necessary to heat the tube leading from flask *A* to a temperature close to that of flask *A* to distill the product.

13. The polymerization tube has a bulb 8 cm. in length and 2 cm. in diameter. A constricted neck 8 cm long is connected to the bulb.

14. A sealed polymerization tube represents a potential bomb. See ref. 5, pp. 8-12, for appropriate precautions in handling.

15. *Caution! The tube should be wrapped in a towel to prevent flying glass as a result of implosion or explosion.*

16. The non-solvent is used in a volume ten times that of the solvent (methanol) to ensure complete precipitation and removal of traces of monomer. The polymer should be dissolved in a minimum of the solvent, in this case 5-10 ml.

17. The concentration of catalyst is molar in pendent imidazole residues.

18. The following recipe was used to prepare 1 1. of buffer solution: 11.3 ml. of 1 *M* KCl, 2.422 g. of TRIS, 105.2 ml. of 0.1 *N* HCl, 285 ml. of ethanol

(abs.); dilute to 1 l. with glass-distilled water. The pH of the buffer solution so prepared was 7.96 at 26°.

19. The solution was mixed by inverting the stoppered cell several times before placing it in the spectrophotometer.

20. The rates were determined using a Beckman model DU-2 ultraviolet spectrophotometer and a "Time-It" electric stopwatch (Precision Scientific Company).

3. References

1a. University of Michigan, Ann Arbor, Michigan 48104.
1b. Current Address: Sherwin Williams Research Center, 10909 Cottage Grove Ave., Chicago, Illinois 60628.
1c. Current Address: Xerox Corp., Rochester, New York 14580.
1d. Current Address: General Electric Co., Schenectady, New York 12345.
2a. Department of Chemistry, Texas A and M. University, College Station, Texas 77843.
2b. Current Address: Howard Kaye & Associates, P. O. Box 39086, Houston, Texas 77039.
3. H. von Bidder, *Z. Physiol. Chem.*, **276**, 129 (1943); *J. Amer. Chem. Soc.*, **85**, 951 (1963).
4. A. H. Mehler, H. Tabor, and O. Hyaishi, *Biochem. Prep.*, **4**, 50 (1955).
5. W. R. Sorenson and T. W. Campbell, *Preparative Methods of Polymer Chemistry*, 2nd ed., Wiley-Interscience, New York, 1968.
6. C. G. Overberger and J. C. Salamone, *Acc. Chem. Res.*, **2**, 217 (1969).

Poly(methyl 2-cyanoacrylate)

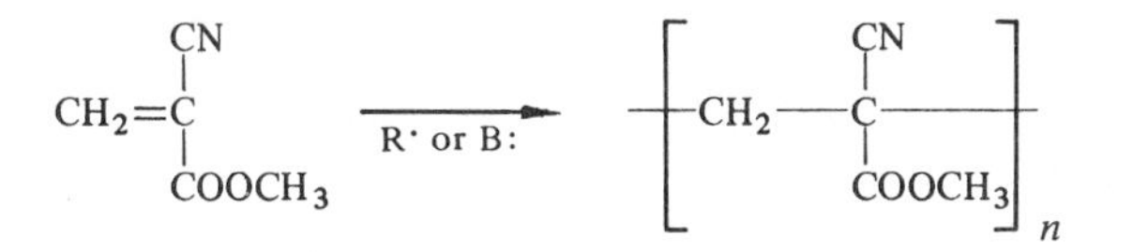

Submitted by H. W. Coover, Jr., and J. M. McIntire (1)
Checked by J. C. Salamone and P. Taylor (2)

1. Procedure

A. Solution Polymerization by Free-Radical Initiation

A clean, dry, 7-oz. beverage bottle (Note 1) is flushed with dry nitrogen for 10 min. and then is charged with 50.00 g. of isobutyronitrile (Note 2), 10.00 g. of freshly distilled and degassed methyl 2-cyanoacrylate (Note 3), and 0.10 g. of 2.2′-azobis(2-methylpropionitrile) (Note 4). After a final 30 sec. purge with dry nitrogen, the bottle is capped with a Teflon®- or polyethylene-lined crown closure. The monomer polymerizes as the bottle tumbles in a 60° water bath for 20 hr. (Note 5).

Poly(methyl 2-cyanoacrylate) precipitates when the solution is poured into a compressed-air-driven blender containing 200 ml. of anhydrous ethyl ether (Note 6). The precipitated polymer is filtered and washed with an additional 400 ml. of anhydrous ethyl ether (Note 7). After drying overnight in a vacuum oven at 50°, the poly(methyl 2-cyanoacrylate) weighed 10.00 g. and had an η_{inh} = 0.32 dl./g. at 20° in a 0.25 wt. % *N,N*-dimethylformamide solution (Note 8).

B. Anionic Polymerization

A 250-ml. beaker is charged with 180 ml. of methanol and 20 ml. of distilled water. This solution is mechanically stirred while 10.00 g. of freshly distilled and degassed methyl 2-cyanoacrylate (see Procedure A, Note 3) is added dropwise (Note 9). After the addition is complete, the mixture is stirred for an additional 15 min. The precipitated poly(methyl 2-cyanoacrylate) is filtered (Note 10), washed with methanol, washed thoroughly with ethyl ether, and dried overnight at 50° in a vacuum oven. The fluffy white polymer weighs 10.00 g. and has an η_{inh} = 0.04 dl./g. at 20° in a 0.25 wt. % *N,N*-dimethylformamide. solution (Note 11).

2. Notes

1. The 7-oz. beverage bottle must be thoroughly cleaned and dried. A soap and water wash, a distilled water rinse, a 75:25 sulfuric acid-nitric acid rinse, and at least three thorough rinses with distilled water, followed by drying for 24 hr. in a 70° circulating air oven, constitutes a suitable cleaning and drying procedure.

2. The isobutyronitrile is fractionated, purged with dry nitrogen, and dried over molecular sieves or other suitable drying agents before use.

3. Methyl 2-cyanoacrylate is prepared by methods described in refs. 3-7, or it can be obtained by distillation of Eastman 910® adhesive. For consistent results, the methyl 2-cyanoacrylate monomer should be distilled before use. Conventional vacuum distillation equipment can be employed. A fairly short path distillation head will give the most trouble-free service. The distillation apparatus should be washed with acid, rinsed thoroughly with distilled water, rinsed with acetone, and dried in a 50° oven under high vacuum (<1.0 torr). After charging the distillation pot with monomer, 0.1% hydroquinone is added. The distillation pot is fitted with a sulfur dioxide gas inlet adapter. The methyl 2-cyanoacrylate monomer is distilled under vacuum (2-5 torr). A typical distillation temperature is 48-49°. (2.5-2.7 torr).

After the distillation is completed, the system is allowed to equilibrate under sulfur dioxide. The receiver containing the distilled monomer is then fitted with a nitrogen gas inlet adapter, and a high vacuum (<1.0 torr) is applied for 1 hr. to remove sulfur dioxide. The system is allowed to equilibrate under dry nitrogen. Then the monomer is ready for polymerization.

Caution! Methyl 2-cyanoacrylate monomer will rapidly polymerize when it comes into contact with basic moieties, including water. Care must also be taken to avoid spreading the monomer into a thin film between two substrates (including fingers!) because it will rapidly form a strong adhesive bond.

4. 2,2′-Azobis(2-methylpropionitrile) is available from Eastman Organic Chemicals, catalog number 6400. Material from other sources is also suitable.

5. The beverage bottle is attached to a 2-foot-dia. wheel inside a 60-gal. water bath. The wheel is mechanically turned at 10-15 r.p.m. The temperature is maintained at 60 ± 0.5°. Other means for moderate agitation at a controlled temperature should also be satisfactory.

6. It is also possible to concentrate the solution *in vacuo* and then precipitate the polymer by adding the concentrated solution to 600 ml. of magnetically stirred ethyl ether.

7. The poly(methyl 2-cyanoacrylate) often develops a static charge during filtration and washing. Care should be taken to avoid a static spark that might cause an ether fire.

8. The checkers obtained a 75% conversion with an η_{inh} = 0.35 dl./g. at 25° in a 0.25 wt. % *N,N*-dimethylformamide solution.

9. If the methyl 2-cyanoacrylate is added too rapidly, the heat of polymerization will boil the methanol-water solution. If the mixture begins to boil the addition should be stopped and the mixture allowed to cool before the dropwise addition is resumed.

10. It is possible to obtain finely suspended polymer. This can be separated by centrifugation.

11. The checkers obtained an 82% conversion with an η_{inh} = 0.057 dl./g. at 25° in a 0.25 wt. % *N,N*-dimethylformamide solution.

3. Methods of Preparation

The preparation of poly(methyl 2-cyanoacrylate) and other poly(alkyl 2-cyanoacrylates) has been reported (8, 9). Solution polymerization techniques similar to the procedure described previously were employed. Benzoyl peroxide was used as the free-radical initiator in one study (9). Free radical and anionic polymerizations in solvents having solubility parameters of about 6.9 to 9.0 and dielectric constants <10 have been reported (12). These solvents effect precipitation polymerizations with methyl and higher alkyl 2-cyanoacrylates.

Copolymerizations of methyl 2-cyanoacrylate with methyl acrylate, methyl methacrylate, styrene, α-methylstyrene, and vinyl acetate have been reported by Kinsinger and coworkers (10). They obtained Q and e values of 4.5 and 2.1. Otsu and Yamada (11) reported Q and e values of 17 and 2.48 for methyl 2-cyanoacrylate.

The use of ^{14}C and H^3-labeled benzoyl peroxide showed methyl α-cyanoacrylate to be relatively unreactive towards BzO· and polymerized by Ph· (13). The polymerization of poly(alkyl α-cyanoacrylates) by anionic and free radical polymerization has been studied and properties of the polymers have been compared (14).

4. References

1. Research Laboratories, Tennessee Eastman Company, Division of Eastman Kodak Company, Kingsport, Tennessee 37662.
2. Polymer Science Program, Department of Chemistry, University of Lowell, Lowell, Massachusetts 01854.
3. F. B. Joyner and G. F. Hawkins (Eastman Kodak Company), U.S. Pat. 2,721,858 (1955).
4. F. B. Joyner and N. H. Shearer, Jr. (Eastman Kodak Company), U.S. Pat 2,756,251 (1956).
5. C. G. Jeremias (Eastman Kodak Company), U.S. Pat. 2,763,677 (1956).
6. G. F. Hawkins and H. F. McCurry (Eastman Kodak Company), U.S. Pat. 3,254,111 (1966).
7. G. F. Hawkins (Eastman Kodak Company), U.S. Pat. 3,465,027 (1969).
8. A. J. Canale, W. E. Goode, J. B. Kinsinger, J. R. Panchak, R. L. Kelso, and R. K. Graham, *J. Appl. Polym. Sci.*, **IV** (11), 231 (1960).
9. M. Yonezawa, S. Suzuki, H. Ito, and K. Ito, *Yuki Gosei Kagaku Kyokai Shi,* **27** (3), 280 (1969).
10. J. B. Kinsinger, J. R. Panchak, R. L. Kelso, J. S. Bartlett, and R. K. Graham, *J. Appl. Polym. Sci.*, 9, 429 (1965).
11. T. Otsu and B. Yamada, *Makromol. Chem.*, **110**, 297 (1967).
12. J. M. McIntire and T. H. Wicker, Jr. (Eastman Kodak Co.) U.S. Pat. 3,654,239 (1972).
13. J. C. Bevington and J. A. L. Jemmett, *J. Chem. Soc. Faraday Trans.* **1**, 69 (Pt 10), 1866 (1973).
14. R. K. Kulkarni, H. J. Porter, and F. Leonard, *J. Appl. Poly. Sci.*, **17** (11) 3509 (1973).

N,N-Divinylaniline Polymers

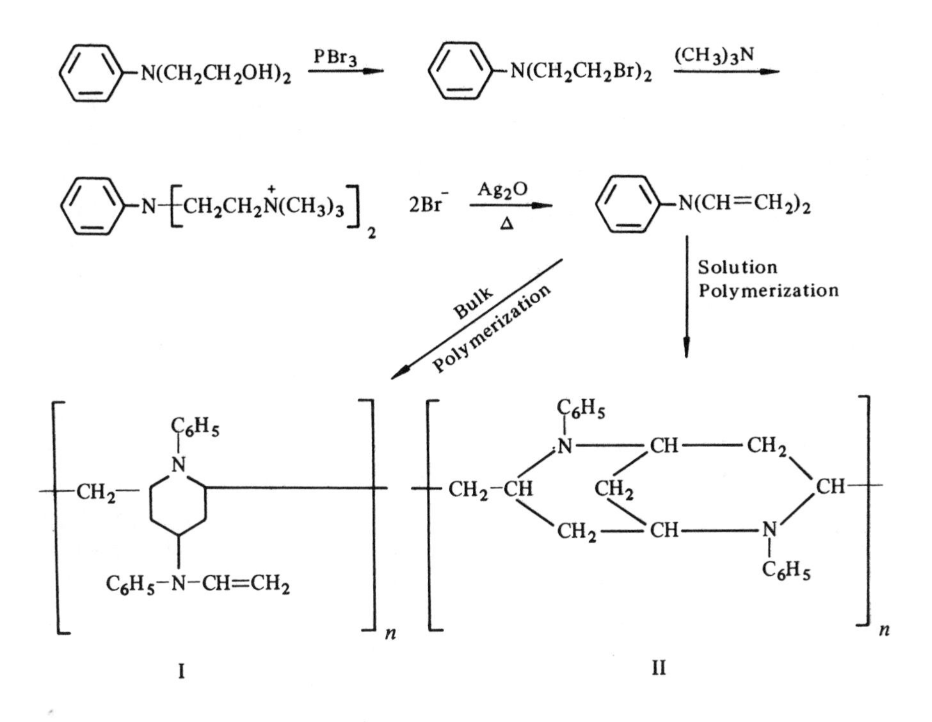

Submitted by E. Y. C. Chang and C. C. Price (1)
Checked by H. Höcker, (2a) H. Schnecko, (2a, b) and W. Kern (2a)

1. Procedure

A. Monomer Preparation

N,N-Bis(2-hydroxyethyl)aniline (111 g., 0.61 mole) is melted and dropped slowly into 500 g. of phosphorus tribromide (1.84 moles) with stirring and gentle warming (Note 1). The mixture is further heated for 3 hr. on a steam bath. An orange precipitate, probably a hydrobromide salt, appears during the reaction. The reaction mixture is diluted in 500 ml. of benzene and then added in portions, with stirring, to 500 ml. of benzene, 500 g. of crushed ice, and 1 l. of 10% caustic soda solution, After the organic layer is separated, the aqueous layer is extracted with two additional 200-ml. portions of benzene. The combined extracts are dried and passed through a column of activated alumina. After evaporation, the product is recrystallized from light petroleum ether to yield 125-135 g. (67-72%) (Note 2).

A mixture of 130 g. (0.422 mole) of *N,N*-bis(2-bromoethyl)aniline (*Caution!*) (Note 3) with 88 g. (1.49 moles) of anhydrous trimethylamine and 175 ml. of absolute ethanol is placed in a screw-capped flask and warmed to dissolve the contents. The reactants are allowed to stand at room temperature for 3 or 4 days with occasional shaking. The precipitate that forms is washed with ether and absolute ethanol to yield 120-140 g. (67-78%) of *N,N*-bis(2-trimethylammonioethyl)aniline dibromide. It is recrystallized from ethanol; m.p. = 245-246° (dec.).

A solution of 33 g. (0.078 mole) of the quaternary ammonium salt in methanol is shaken with freshly prepared, moist silver oxide, added in portions until excess silver oxide persists (Note 4). After the silver bromide precipitate is removed, the methanol solution is evaporated to a very viscous syrup, which is then dropped into a distilling flask kept at 150°. The decomposition products are entrained with a stream of nitrogen to a cooled receiver while the whole system is kept under a reduced pressure of 25 torr. A 25-ml. portion of benzene is added to the distillate, and the organic layer is separated, dried over $CaCl_2$, and distilled first through a 30-cm. packed column and then through a Todd column to yield 5.5-7.0 g. (49-62%) of *N,N*-divinylaniline as a colorless liquid, b.p. 81-81.7° at 10 torr., n_D^{20} 1.5734, d_4^{20} 0.955. The liquid becomes very viscous at −70° without crystallization.

B. *Bulk Polymerization*

A 1.06-g. sample of *N,N*-divinylaniline and 18.6 mg. of azobis(isobutyronitrile) (Note 5) are weighed into an ampoule, cooled to −70°, evacuated, and sealed. After heating in a 60° bath for 8 days, the ampoule is cooled and then opened, and the contents are dissolved in benzene. The solution is filtered through a fritted glass funnel (Note 6) into petroleum ether (b.p. 30-60°), and the precipitate is then redissolved and reprecipitated to give 0.30 g. (28%) of polymer after vacuum drying overnight. The white powdery polymer softens in a capillary at 160-170° and has a molecular weight of 5000-7000 cryoscopically in *p*-nitrotoluene, an inherent viscosity (η_{rel}/c of 0.065-0.09 (benzene, 30°, 1 g./100 ml.), and, by hydrogenation (Note 7), ca. 0.6 double bond/monomer unit.

C. *Solution Polymerization*

A mixture of 0.91 g. of monomer, 3 g. of benzene, and 24 mg. of azobis(isobutyronitrile catalyst in an ampoule is cooled to −70°, evacuated, and sealed. After heating for 7 days at 60°, the solution is diluted with 5 ml. of benzene and poured into 500 ml. of petroleum ether. After a second reprecipitation, the yield is 0.25 g. (28%) of polymer. The white powdery polymer shows a capillary softening point of 150-160°, a cryoscopic molecular weight of 4000-5000, an η_{inh} of 0.045-0.05 (benzene, 30°, lg./100ml), and, by hydrogenation (Note 7), ca. 0.03 double bond/monomer unit.

D. *Copolymerization of* N,N-*Divinylaniline* (M_1) *with Diethyl Fumarate* (M_2)

A mixture of 0.42 g. of M_1 and 1.60 g. of M_2 (1/3.2) with 95 mg. of azobis-(isobutyronitrile) catalyst is sealed in an ampoule and heated to 60° for 6 days. The ampoule is cooled and then opened, and the mixture is dissolved in 5 ml. of benzene. On pouring into petroleum ether (b.p. 30-60°), 0.30 g. (15%) of copolymer precipitates. Elemental analysis (4) shows this to contain two molecules of diethyl fumarate for each molecule of divinylaniline, and catalytic hydrogenation reveals only 0.15 double bond/divinylaniline unit.

The product had a softening point of 80-90°, an η_{inh} of 0.03-0.04 (benzene, 30°, lg./ml), and a number average molecular weight of 6000-7000, (Note 8).

2. Polymer Structure

The polymer prepared in bulk has unsaturation in accord with Structure 1 for two adjacent monomer units, while that prepared in solution corresponds to II. Evidence supporting these formulations is present in ref. 4.

The diethyl fumarate copolymer composition ratio of 1:2 and the substantial absence of double bonds in the copolymer suggest the following structural unit:

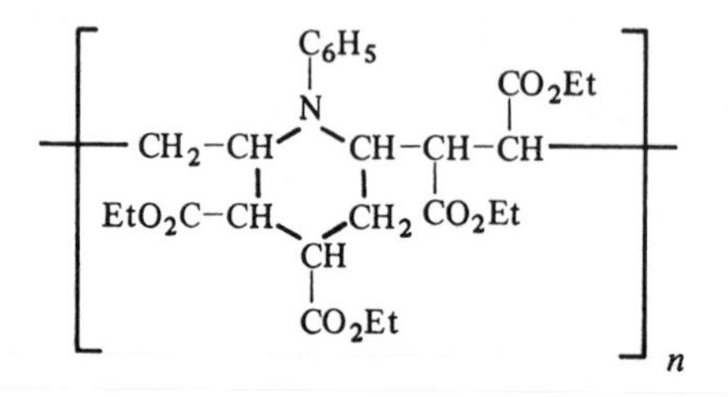

This, of course, amounts to a 1:1 alternating copolymerization of the vinyl double bonds and the fumarate double bonds.

3. Notes

1. The addition should be interrupted when brownish vapors are liberated to avoid excessively vigorous reaction.

2. This procedure is according to Ross (3), who reports the melting point as 53-55°.

3. *Caution!* N,N-*Bis(2-bromoethyl)aniline is a vesicant, and exposure to the skin must be avoided!*

4. Silver oxide was prepared by mixing equimolar amounts of aqueous silver nitrate and aqueous sodium hydroxide. The precipitate was collected on a filter, washed thoroughly with distilled water, and used promptly.

5. Benzoyl peroxide and hydroperoxides are not suitable catalysts, because they cause rapid discoloration of the monomer.

6. The bulk polymer is likely to contain a small amount of gel particles, particularly at high conversion.

7. The hydrogenation is carried out in a volumetric microhydrogenation apparatus in absolute ethanol or diglyme as solvent and with Raney nickel, 10% palladium on charcoal, or platinum oxide as catalyst.

8. Other ratios of M_1 to M_2 all tend to give polymers of nearly 1:2 ratio (4).

4. References

1. Department of Chemistry, University of Pennsylvania, Philadelphia, Pennsylvania 19104.
2. University of Mainz, D-65 Mainz, Germany.
3. W. C. J. Ross, *J. Chem. Soc.*, **183**, (1949).
4. E. C. Y. Chang and C. C. Price, *J. Amer. Chem. Soc.*, **83**, 4650 (1961).

Optically Active Polybenzofuran

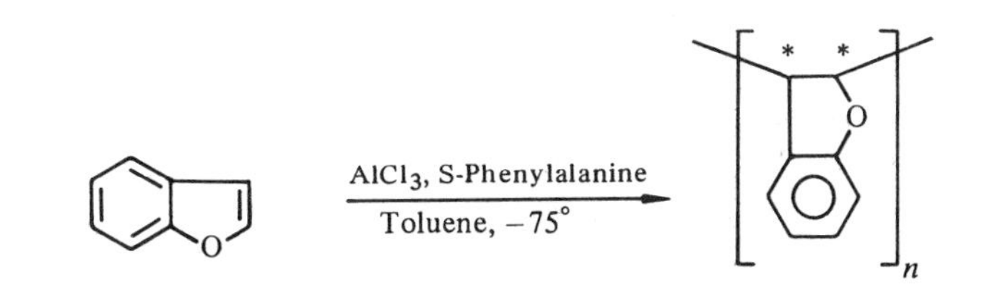

Submitted by G. Bressan (1a) and M. Farina (1b)
Checked by D. C. Kaufman and W. H. Daly (2)

1. Procedure

A. *Catalyst*

In a thoroughly dried 500-ml., three-necked flask provided with a mechanical stirrer and a nitrogen inlet is placed 0.3 g. (1.8 mmole) of S-phenylalanine (Notes 1 and 2). After further drying *in vacuo*, nitrogen is readmitted and 250 ml. of toluene is distilled into the flask (Note 3). Aluminum chloride (2 g., 15 mmoles) (Note 4) is then added with an airless solids addition funnel. After stirring for 90 min. at 20°, the undissolved solids are allowed to settle and a clear yellow-green solution is obtained that contains approximately 15 mmoles/l. of aluminum chloride and 5 mmoles/l. of S-phenylalanine, probably in the form of a 3:1 complex (Note 5).

B. *Polymerization*

The clear supernatant catalyst solution (220 ml.) is siphoned into a 500-ml., three-necked flask equipped with a stirrer and maintained under nitrogen and is cooled to $-75°$ in a Dry Ice-methanol bath. A solution of 2.36 g. (20 mmoles) of benzofuran (Note 6) in 20 ml. of toluene (Note 4) is slowly added from a dropping funnel while maintaining the reaction mixture at $-75°$. After 3 hr. of stirring at this temperature, a large excess of methanol is added, and the precipitated polymer is collected on a filter and dried to yield 1.9-2.1 g. (81-89%) of polybenzofuran (Note 7).

2. Characterization

The proposed structure for polybenzofuran is in accord with IR and NMR spectra. The intrinsic viscosity of the polymer was 2.2 dl./g. ($CHCl_3$ at 30°), and $[\alpha]_D^{25}$ was $-68°$ ($CHCl_3$, $C = 0.2\%$). Usually these values range from 0.8 to 2.5 dl./g. and from -25 to $-75°$. Dextrorotatory polybenzofuran with similar values of optical activity may be obtained in the presence of R-phenylalanine as cocatalyst (Note 7).

The polymer prepared as described does not show crystallinity by x-ray diffraction. However, crystallization may be induced by heating a 2% solution of the polymer in diethylene glycol diethyl ether to 130° (Note 8); separation of crystalline polybenzofuran starts after a few hours and is complete within 25-30 hr. During this treatment, the intrinsic viscosity decreases to 1.2, dl./g. but the specific rotation remains nearly the same; the polymer melts with some decomposition at 355-360°.

Amorphous polybenzofuran dissolves easily in a variety of organic solvents. The crystallized polymer, on the other hand, is generally insoluble except in chloroform, which causes slow dissolution. Precipitating the polymer from this solution (or melting crystalline polybenzofuran) produces an amorphous product (Note 9).

3. Notes

1. The action of optically active phenylalanine in the asymmetric synthesis of polybenzofuran is specific. Examination of other common optically active amino acids showed that only α-phenylglycine has a limited asymmetric induction power. Low-rotatory-power polymers were obtained when brucine or camphorsulfonic acid was used as cocatalyst.

2. S-Phenylalanine was dried under vacuum for several hours. In the prepara-

tion of the catalytic complex and in subsequent treatment, it is necessary to carefully exclude traces of water and to employ pure reagents. Otherwise S-phenylalanine is replaced as cocatalyst by other substances, resulting in formation of a polymer with low or zero rotatory power.

3. Commercial toluene was treated with conc. H_2SO_4 and washed several times with water; it was then purified by chromatography on active alumina and allowed to reflux over Na-K alloy for at least 20 hr.

4. Commercial aluminum chloride was purified by sublimation under dry nitrogen.

5. The composition of the catalytic complex depends on the initial concentration of the reagents: molar ratios of $AlCl_3$ to amino acid near 2 are obtained by allowing equimolar amounts of the two substances to react. The power of asymmetric induction of this catalyst is essentially the same as that described above, but the polymerization rate is lower.

6. Commercial benzofuran was fractionated at reduced pressure; gas chromatographically pure fractions were used. For a successful asymmetric synthesis, low molar ratios of monomer to catlaytic complex are required. The checkers preferred to purify commercial benzofuran on a preparative scale gas chromatograph, using a Carbowax 20M column at 140°. Benzofuran was also synthesized by decarboxylation of coumarilic acid (8).

7. It was observed that, as the polymerization goes beyond a certain conversion, the rotatory power of the polymer decreases, probably because of a progressive alteration of the counterion caused by termination or transfer reactions (4). On the contrary, by operating at low monomer and catalyst concentrations and removing polymer samples at low conversions, an increase of the rotatory power with conversion is obtained. This was attributed to an autocatalytic effect resulting from an interaction between the polymer and the catalytic system. The optical purity of polybenzofuran is not known.

8. This method of crystallization, which probably involves a specific polymer-solvent interaction, is effective with polybenzofuran obtained with the $AlCl_3$-S-phenylalanine catalyst. Samples obtained using other stereospecific catalyst systems, however, may crystallize by annealing with a non-solvent aliphatic ether such as diisoamyl ether. The checkers report that the polymer from the preparation described here could be induced to crystallize in diethylene glycol diethyl ether at 70° with no degradation. However, longer times were required.

9. Such behavior did not allow x-ray analysis or determination of the type of structure (threo- or erythrodiisotactic).

4. Methods of Preparation

A procedure based on a different catalyst containing (−) methoxytriethyltin as the asymmetric component has yielded results similar to those described here (optical activity: +73.8°) (9).

5. References

1a. Montedison-DIPI, Milan, Italy.
1b. Istituto de Chimica Industriale, Universita, degli Studi, Milan, Italy.
2. Department of Chemistry, Louisiana State University, Baton Rouge, Louisiana 70803.
3. G. Natta, M. Farina, M. Peraldo, and G. Bressan, *Makromol. Chem.*, **43**, 68 (1961).
4. M. Farina and G. Bressan, *Macromol. Chem.*, **61**, 79 (1963).
5. M. Farina, G. Natta, and G. Bressan, *J. Polymer Sci.*, C **4**, 141 (1963).
6. G. Bressan and R. Broggi, *Chim. Ind. (Milan)*, **50**, 1326 (1968).
7. G. Bressan, *Chim. Ind. (Milan)*, **51**, 705 (1969).
8. L. A. Paquette, *Principles of Modern Heterocyclic Chemistry*, Benjamin, New York, 1968, p. 160.
9. Y. Takeda, Y. Hayakawa, T. Fueno, and J. Furukawa, *Makromol. Chem.*, **83**, 234 (1965).

Copoly (acylhydrazones)

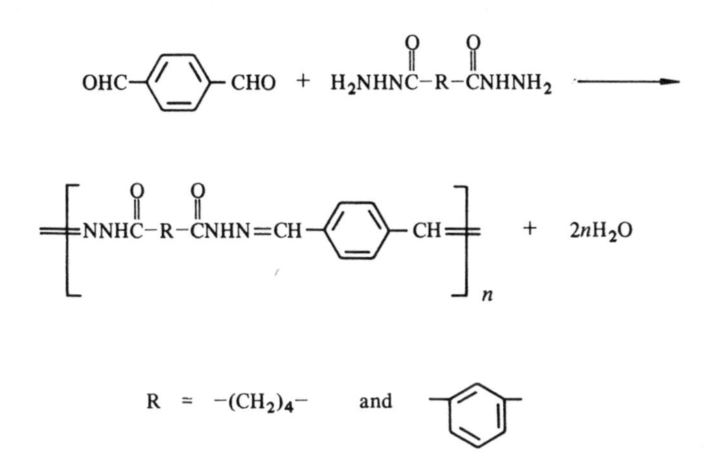

Submitted by R. H. Michel (1)
Checked by I. K. Miller (2)

1. Procedure

A 250-ml., three-necked, round-bottomed flask equipped with a stirrer and two glass stoppers is charged with 100 ml. of dimethyl sulfoxide (Note 1), 3.48 g. (0.02 mole) of adipic acid dihydrazide (Note 2), and 3.88 g. (0.02 mole) of isophthalic acid dihydrazide (Note 2). When all the hydrazide has dissolved, 5.36 g. (0.04 mole) of terephthalaldehyde (Note 3) is added. The solution is stirred at room temperature for 48 hr.

The poly(acylhydrazone) is isolated (Note 4) by pouring the viscous solution into methanol. The light yellow polymer is washed with methanol (three 200-ml. washes) in an explosion-proof blender, and is then dried in a vacuum oven at 60°. The yield of polymer is 90-95%, with an inherent viscosity of 1.2 dl./g. in dimethyl sulfoxide at 30° (conc. 0.25%). The polymer melt temperature on a hot metal surface is 280° (Note 5).

2. Notes

1. Dimethyl sulfoxide is purified in vacuum distillation from molecular sieves, type 4A (Linde).

2. The dihydrazides are prepared by condensation of the acid dimethyl esters with hydrazine hydrate in benzene (3). They are sufficiently pure after one recrystallization from water. The melting points are as follows: adipic acid dihydrazide, 180-181°; isophthalic acid dihydrazide, 227°.

3. Terephthalaldehyde with a melting point of 111-112° is needed to obtain the reported inherent viscosity. In checking the synthesis, Eastman terephthalaldehyde with a melting point of 108-110° gave an η_{inh} of 0.7. The melting point was raised to 112° by recrystallizing from 10% methanol in water.

4. The solution can be cast into films without isolation.

5. The polymer is essentially amorphous according to x-ray diffraction.

3. References

1. Film Department, Experimental Station, E. I. du Pont de Nemours and Company, Wilmington, Delaware 19898.
2. Textile Fibers Department, Experimental Station, E. I. du Pont de Nemours and Company, Wilmington, Delaware 19898.
3. R. H. Michel and W. A. Murphey, *J. Polym. Sci.*, 7, 623 (1963).

Mechanochemical Polymerization: Natural Rubber - Methacrylic Acid Block - Graft copolymer

$$\left[CH_2-\overset{\overset{\displaystyle CH_3}{|}}{C}=CH-CH_2 \right]_n + CH_2=\underset{\underset{\displaystyle COOH}{|}}{\overset{\overset{\displaystyle CH_3}{|}}{C}} \longrightarrow \text{Block and Graft Copolymer}$$

Submitted by R. Shuttleworth and W. F. Watson (1)
Checked by R. T. Morrissey, (2a) H. E. Diem, (2a) D. J. Harmon, (2a) R. W. Smith, (2a) M. R. Walters, (2a) and N. C. Hess (2a, b)

1. Procedure

Caution! Methacrylic acid is toxic and should be used only in a well-ventilated hood.

To conduct a mechanochemical polymerization (Note 1), suitable operating conditions and the quantities required for the available laboratory plastics extruder must first be determined. Each experiment requires approximately three times the volume of material needed to fill the extruder. A sample size of 200 g. is adequate for an extruder with a 1-in. diameter barrel and a 12:1 length: diameter screw (Note 2).

Natural rubber (650 g.) of RSS1 grade if formed to approximately 0.5-cm. thickness and 25-cm. width with minimum milling on a laboratory mill with 6-in. diameter rolls at between 50° and 90°. The sheet is loosely rolled in drawing linen and covered with acetone in a 5-1. beaker covered with a watch glass. After 24 hr. in the dark at room temperature, the liquid is decanted, and the rubber is covered with fresh acetone and left in the dark for an additional 24 hr. (Note 3). The sheet is then unrolled on its backing and allowed to dry in air for 6 hr. The loosely rolled sheet is then placed in a vacuum desiccator, and the residual acetone collected in a liquid nitrogen trap by continuously evacuating with a vacuum pump for 16 hr.

Freshly distilled methacrylic acid (150 g.) (Note 4) is poured slowly and as evenly as possible over 600 g. of the rubber. The sheet of rubber and monomer is then rolled up and left in à closed desiccator in air for 2 hr. to homogenize the mixture. A 200-g. sample is then passed through the extruder, running at a speed of approximately 20 r.p.m. Temperature is maintained as low as possible by cooling the extruder. Between one and ten passes through the extruder may be required to effect substantial polymerization, which will be evident by an increase in toughness and opacity of the mixture and a decrease in methacrylic acid odor (Note 2). If the extrudate smells strongly of monomer, it should be passed through the extruder again. Viscosity increase should be confirmed on a rubber plastimeter (Note 5). The speed of the screw should be varied, if necessary, in further experiments to achieve the greatest extent of polymerization on one pass through the machine, as indicated by a decrease in the odor of monomer.

The extrudate is sheeted to ½-cm. thickness on a laboratory mill, weighed, and evacuated in a vacuum desiccator for 24 hr. The decrease in weight corresponds to the amount of reacted monomer. Yields vary from 53 to >80%, depending on the extruder and the number of passes.

2. Characterization

Two samples, each of approximately 25 g., are weighed to 0.01 g. They are placed in 500-ml. conical flasks, covered with approximately 250 ml. of petroleum ether fraction of 60-80° boiling range, and left in the dark for 24 hr. at room temperature. The liquid is poured off and replaced by another 250-ml. batch of petroleum ether, and cold extraction again carried out for 24 hr. (Note 6). The polymer is then removed and dried overnight to constant weight in a vaccum desiccator. The loss in weight represents the amount of homopolymeric rubber present.

Similarly, two 25-g. samples are extracted wtih 50% ethanol-water mixture to remove the homopolymeric poly(methacrylic acid). Approximately 20% of natural rubber and 3% of poly(methacrylic acid) are removed, leaving 77% block copolymer.

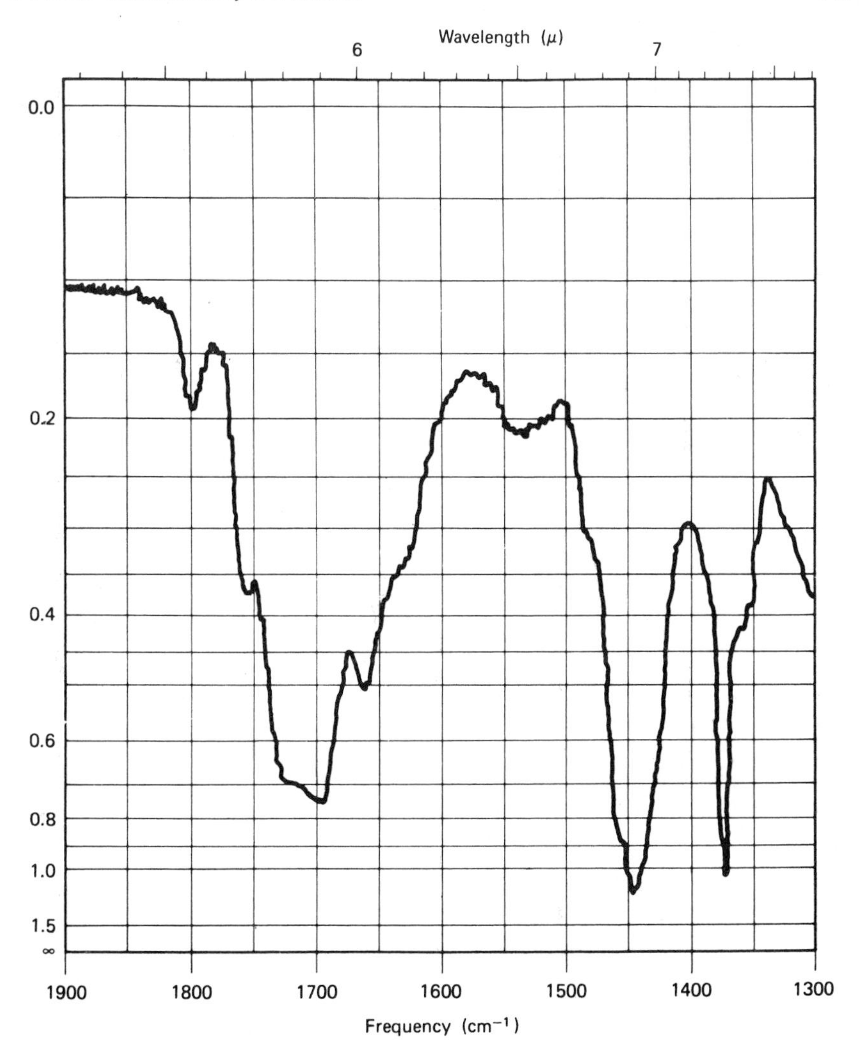

Fig. 1. Reaction product, natural rubber-MMA.

The following additional observations were made by the checker.

The Mooney viscosity (Note 7) showed a marked rise for the natural rubber-MAA polymer versus a marked reduction for natural rubber passed through the extruder under the same conditions. The data are indicated in the following table:

Mooney viscosity at 100°, large rotor	1′ML	4′ ML
Original natural rubber (RSS)	153	96
MAA–modified natural rubber	175	130
Natural rubber after passing through extruder (no MAA)	52	37

Infrared spectra (Figs. 1 and 2) show a strong absorption at peak in the 1700-

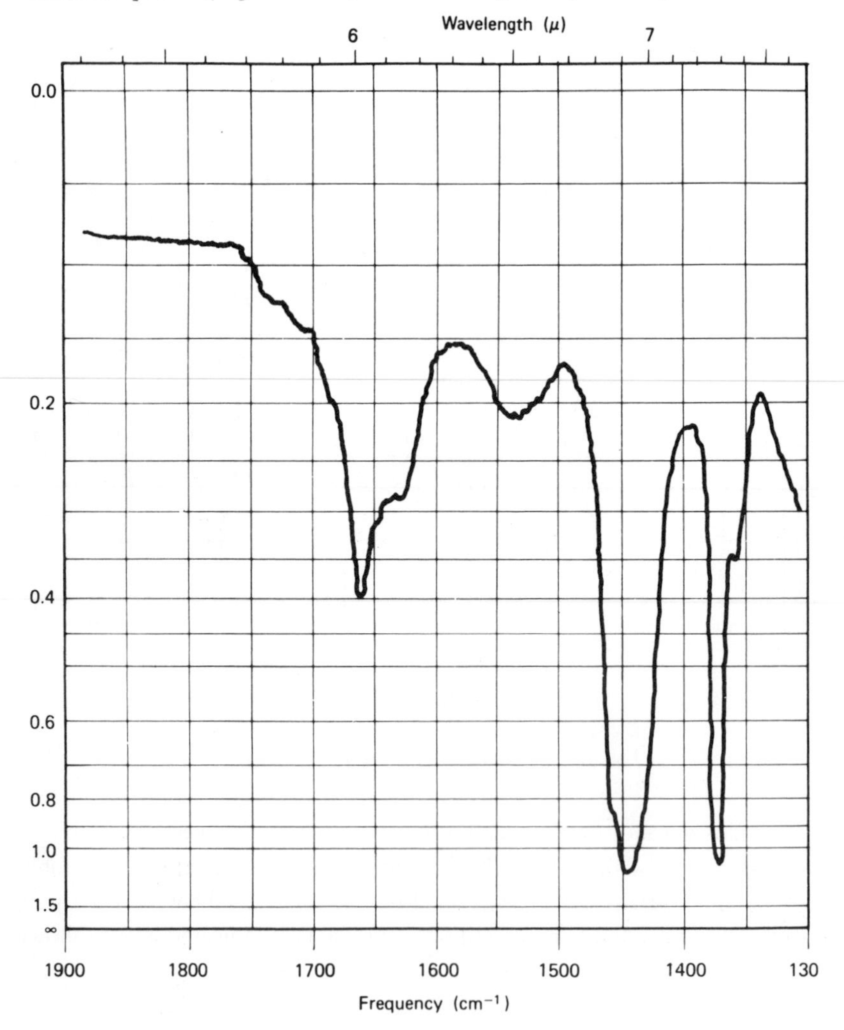

Fig. 2. Natural rubber.

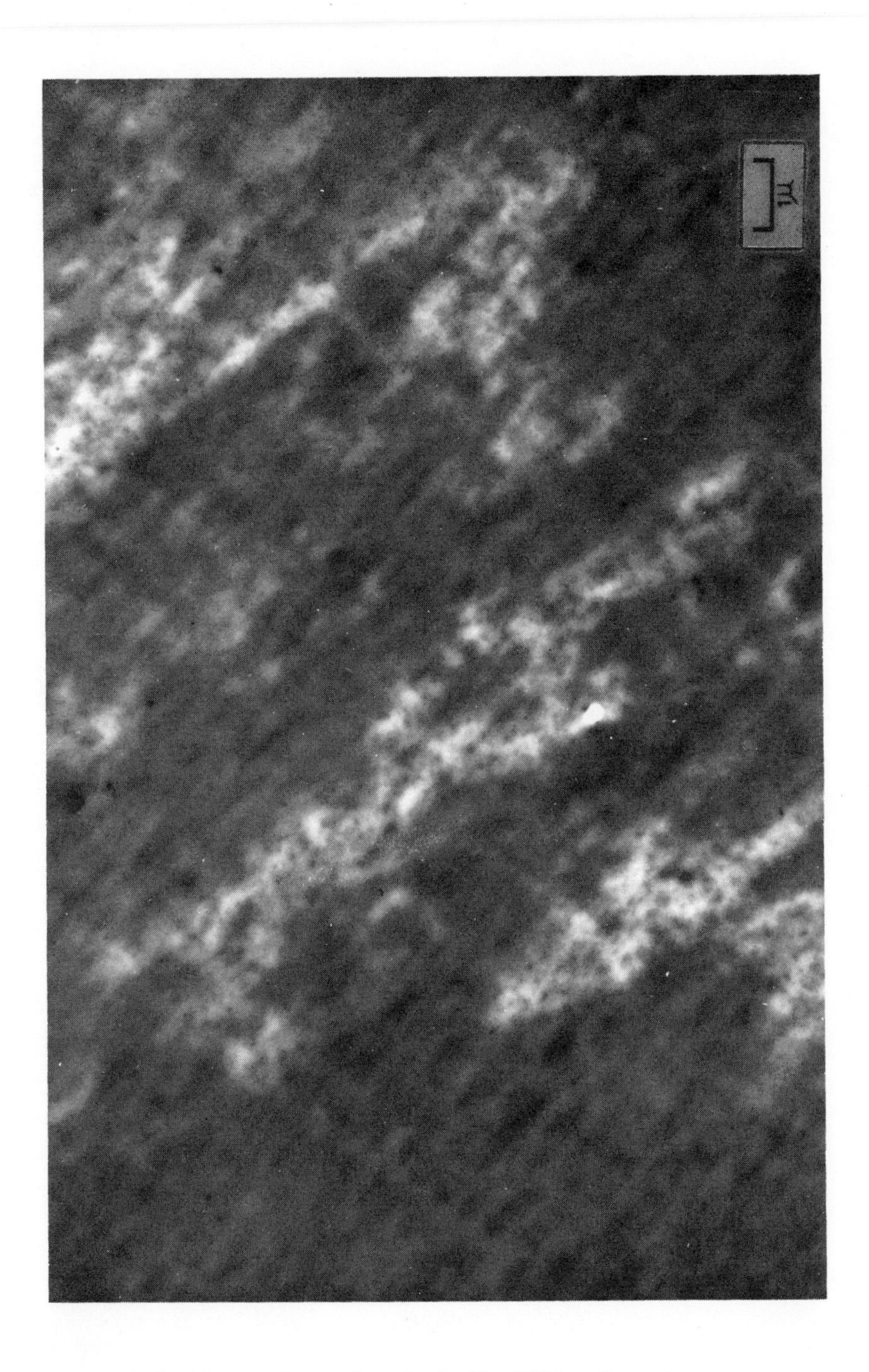

Fig. 3. Electronmicrograph: natural rubber-MMA graft copolymer.

cm^{-1} that indicates the presence of carboxyl groups in the modified polymer. The spectrum of Hevea rubber is strongly evident, suggesting blocks of isoprene units.

Electron microscopic examination of the polymer domains reveals a relatively coarse dispersion of the methacrylic acid graft. The electron micrograph in Fig. 3 shows the two polymeric phases. The gray areas are natural rubber, and the white areas are the blocks or grafts of methacrylic acid.

Mill mixing of the natural rubber-MAA polymer has revealed dry, non-tacky rubber with very poor processing characteristics. Attempts to cure the polymer with sulfur and accelerator and relatively high amounts of zinc oxide, 10-15 parts, have failed to produce a cured sheet with any strength. It may be concluded that the polymer is already crosslinked or highly branched. GPC data (Figs. 4 and 5) indicate a lowering of molecular weight distribution similar to that occurring in a natural rubber extruded in the same manner without MAA. If the polymer were completely a block polymer, very little change or actually an

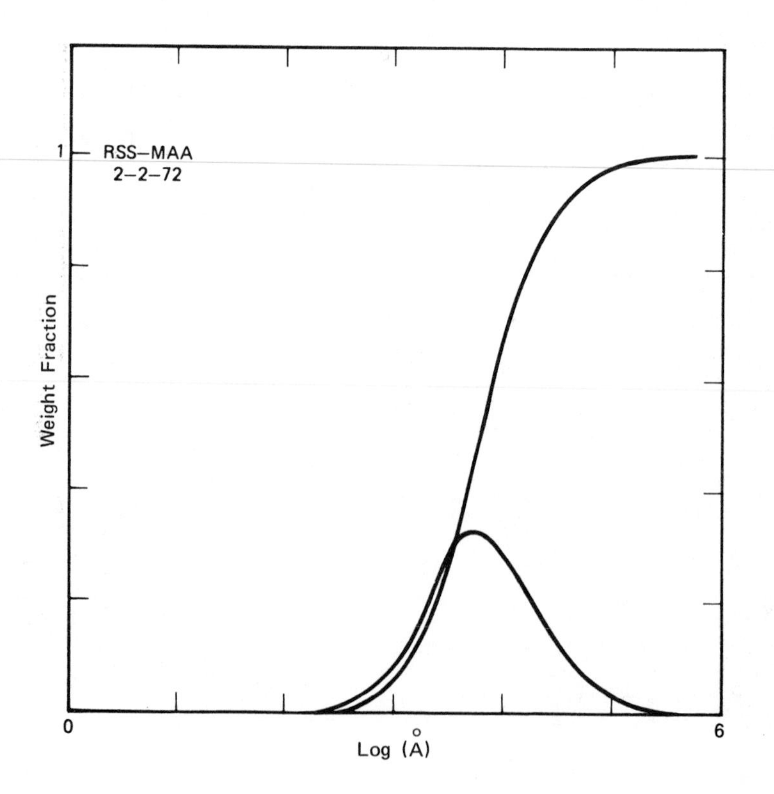

Fig. 4.

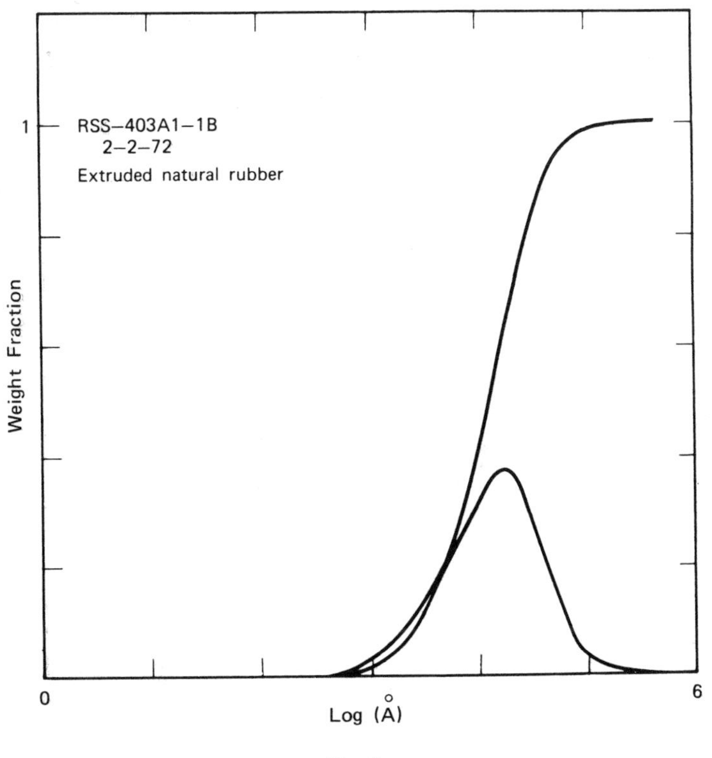

Fig. 5.

increase in molecular weight would be expected. Also, no change has been observed in the crystallization rate of the natural rubber-MAA reaction product, as compared to the untreated natural rubber. One may conclude from characterization of the polymer that it is not solely a block copolymer but rather a natural rubber-methacrylic acid graft that is highly branched.

3. Notes

1. This preparation is an example of mechanochemically induced polymerization. Conditions can be found for the scission of the molecular chains of almost all polymers to produce free radicals and for their initiation of polymerization of monomers of suitable reactivity (3, 4). The experiments described here use a typical laboratory plastics extruder. Systematic laboratory studies of mechanochemical reactions should preferably be carrried out employing a small masticator of special design (5).

2. The volume of the extruder is determined by feeding with strips of natural rubber until the screw is seen to be filled and extrusion is continuous from the die. Feeding is stopped and the extrudate cut off when all the rubber is just taken up by the screw. The weight then extruded as the extruder empties gives the required weight for filling the machine. Multiplication of this weight by 3 indicates the weight of rubber-monomer mixture for each experiment. A circular die of 1 $cm.^2$ aperture was used, but any available die to this general dimension can be employed.

The extruder should have maximum possible cooling during the experiment. A continuous flow of cold water will normally be the most convenient method. The rate of screw rotation will usually be the slowest speed available, 20 r.p.m. in the case of the extruder used. The important measurement for obtaining the best conditions is the temperature of the extruded rubber, which should be as low as possible to permit mechanochemical scission—certainly not above 70°.

When natural rubber of grade RSS1 is used to determine the volume of the extruder, it also cleans the machine. If there is a delay in carrying out the reaction, the machine should be cleaned with natural rubber shortly before use.

3. Natural rubber contains approximately 2% of non-rubber constituents that retard polymerization; these are removed by cold extraction with acetone.

4. Methacrylic acid is distilled from the stabilizer at 25-30 torr, discarding the first and last 10-g. fractions.

5. A convenient machine is the Wallace Rapid Plastimeter (10), which requires only a 1-g. sample for test. Otherwise, a Mooney viscometer can be used.

6. The characterization method employed here is selective elution of homopolymers from a mixture with block polymer. It is necessary to adopt two independent methods to characterize the polymer mixture, because cosolution and coprecipitation frequently occur (11).

7. This viscosity is a dimensionless value read directly from the Mooney viscometer. The denotations "1′ML" and "4′ML" represent, respectively, the readings obtained 1 min. and 4 min. after the start of rotation of the Mooney Large Rotor.

4. Methods of Preparation

A review of the mechanical syntheses of block and graft copolymers, including natural rubber-methacrylic acid has been published (12).

5. References

1. Rubber and Plastics Research Association, Shawbury Shrewsbury, SY4 4NR, England.
2a. Research Center, The B. F. Goodrich Company, 9921 Brecksville Road, Brecksville, Ohio 44141.
2b. Current Address: B. F. Goodrich Company, Cleveland Ohio 44131.
3. E. M. Fettes, ed., *Chemical Reactions of Polymers,* Wiley-Interscience, New York, 1964, Ch. 14.
4. N. K. Baramboim, *Mechanochemistry of Polymers* (translation), Maclarens Press, London, 1964.
5. P. Arnaud, *Off. Act. Plast. Caout.*, **16**, 510 (1969).
6. N. K. Baramboim and N. L. Maxmudbekova, *Bys. Soed.,* **6**, 418 (1972).
7. E. Z. Eckareva, *Plast. Massy,* **12**, 28 (1973).
8. A. Casale and R. S. Porter, *Kaut. u. Gummi,* **28**, 536 (1975).
9. W. F. Watson and D. Wilson, *Rubber Plastics Age,* **38**, 982 (1957). The Unirotor mixer is marketed by Baker Perkins, Ltd., Peterborough, England.
10. H. W. Wallace and Company, Ltd., Croydon, England.
11. R. J. Ceresa, *Block and Graft Copolymers,* Butterworths, London, 1962.
12. A. Casale and R. Porter, *Adv. Polym. Sci.,* **17**, 1 (1975).

5-Nylon

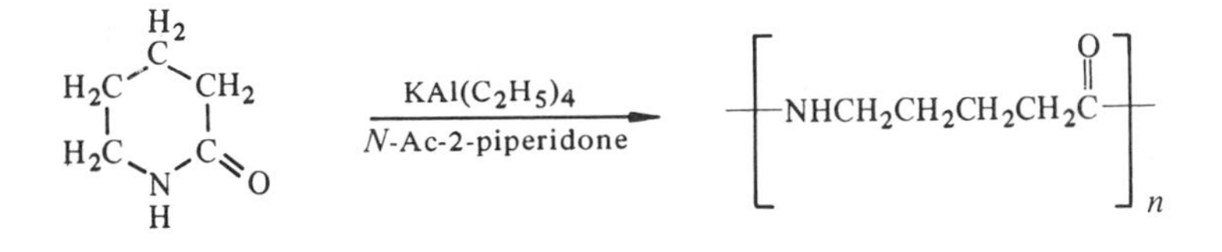

Submitted by T. Konomi (1) and H. Tani (2)
Checked by H. Fujii, R. Kawase, M. Motoi, and T. Saegusa (2a)

1. Procedure

A. Catalyst

$$3K + 4Al(C_2H_5)_3 \rightarrow 3KAl(C_2H_5)_4 + Al$$

A 200-ml., four-necked flask equipped with a mechanical stirrer, a reflux condenser, a thermometer, a nitrogen inlet, and a stopper (Fig. 1) is flushed with dry nitrgen and then is charged with 70 ml. of freshly distilled tetrahydrofuran (Note 1), 5.85 g. (0.15 g. atom) of freshly cut potassium turnings (Note 2), and 30 ml. (25 g., 0.219 mole) of triethylaluminum (Note 3), using a hypodermic syringe under a dry nitrogen atomsphere.

Caution! Triethylaluminum is pyrophoric in air, extreme care should be exercised in handling it.

After the reaction mixture has been refluxed with vigorous stirring for 6 hr. the supernatant solution is transferred to another flask, previously flushed with dry nitrogen. Tetrahydrofuran and triethylaluminum are removed from the solution at 65-75° under a pressure of 10^{-1} torr and the residue is crystallized

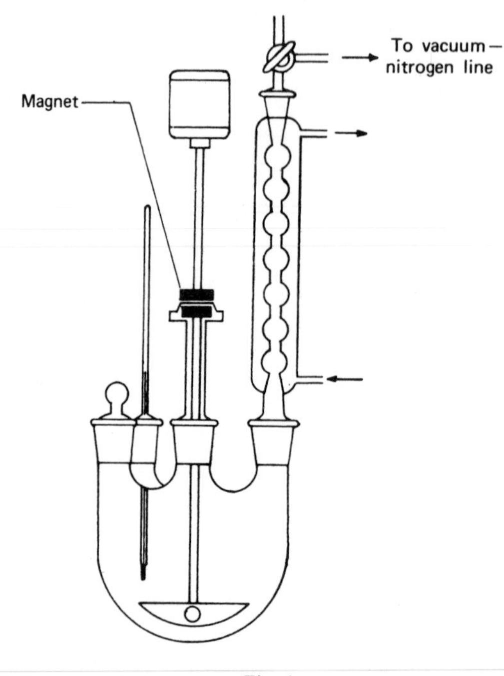

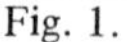
Fig. 1.

from tetrahydrofuran solution by adding *n*-hexane (3, 4). Crystalline $KAlEt_4$ having a melting point of 78° is obtained.

A tetrahydrofuran solution (*Caution!*) (Note 4) containing 0.5-2.0 moles/l. of $KAlEt_4$ is used as the catalyst for the polymerization. The concentration of $KAlEt_4$ is determined by back-titration (indicator: phenolphthalein) or gasometry after hydrolysis with 5% hydrochloric acid-tetrahydrofuran (1:1, v/v).

B. *Polymerization*

All the operations are done under a dry nitrogen atomsphere. In one compartment (*A*) of the polymerization apparatus (Fig. 2) flushed with dry nitrogen is charged 2.0 ml. (20 mmoles) of 2-piperidone (Note 5). To the 2-piperidone, the solution of tetrahydrofuran containing 0.44 mmole of $KAlEt_4$ is added, with stirring, with a hypodermic syringe at 40-50°. The temperature should not be allowed to exceed 50° in this operation, otherwise the exothermic reaction decreases the activity of catalyst. After the evolution of gaseous ethane has ceased (Note 6), tetrahydrofuran is removed at 45-50° under reduced pressure. In compartment *B* is placed 0.080 g. (0.57 mmole) of freshly distilled *N*-acetyl-

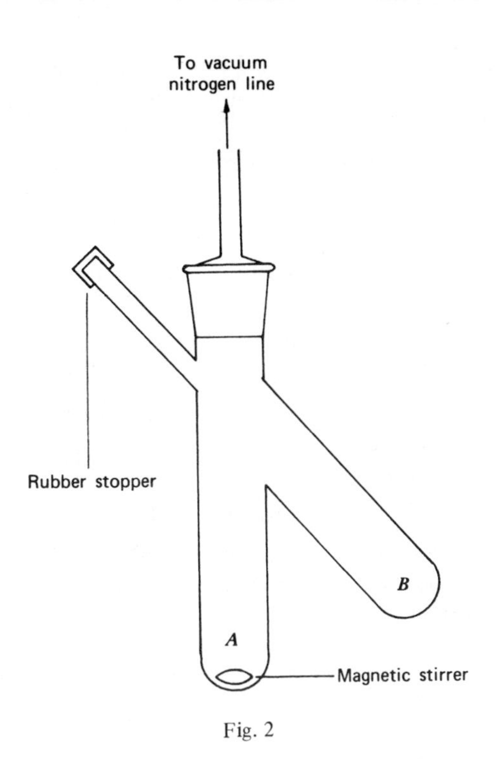

Fig. 2

2-piperidone (Note 7). After the contents of both compartments have reached 45°, they are mixed together. After a specified time, the polymerization is terminated by adding water or methanol. The catalyst is removed by washing thoroughly with water the methanol, and the polymer is dried at 60-70° under reduced pressure. To reach high molecular weight, quite a long time is required:

Polymerization time (days)	1	5	10	25	30
Yield of polymer (%)	18	21	31	43	50
$\eta_{sp/c}$ (Note 8)	0.16	0.25	0.40	0.79	0.91
Melting point of polymer	253	259	264	267	–

Polymer having reduced viscosity higher than 0.8 dl./g. forms films by casting from formic acid solution.

2. Notes

1. Tetrahydrofuran was previously treated by refluxing with sodium metal for 5 hr.

2. There is danger here; the dispersed form is preferable.

3. Triethylaluminum (Ethyl Corporation) was used without further purification: b.p. at 760 torr., 186.6°; density at 25°, 0.8324 g./ml.

4. *Caution! Stepwise addition of tetrahydrofuran to the crystalline $KAlEt_4$ and cooling the flask with cold water may be necessary as heat is released because of complex formation.*

5. 2-Piperidone is synthesized from cyclopentanone oxime, which is derived from cyclopentanone obtained from adipic acid (5): b.p., 112.4° at 5 torr., 138° at 10 torr., 196° at 115 torr; m.p. 40° 2-Piperidone is purified by repeated vacuum distillation under a dry nitrogen atomsphere. The water content of the monomer should be less than 0.010 wt.% by the Karl Fisher method. Water is a powerful inhibitor of anionic polymerization.

6. This reaction requires about 40 min.

7. *N*-Acetyl-2-piperidone was prepared by refluxing 2-piperidone with excess acetic anhydride for about 8 hr., b.p. 72° at 0.1 torr (6).

8. Reduced viscosity of the polymer ($\eta_{sp/c}$) is measured in *m*-cresol (conc., 0.5 g. of polymer/100 ml. of *m*-cresol) at 30° (7).

3. Method of Preparation

Alkali metal (5, 8), alkali metal hydroxide (5, 8), and $KAlEt_4$ (piperidone) (9) are used as catalysts for polymerization. The catalytic activity of triethylaluminum is low (10).

Initiators which may be used for polymerization include *N*-acyllactam (5, 8), diphenyl ketene (11), and lactone (11).

4. References

1. Katata Research Institute, Toyo Spinning Company, Ltd., Katata, Shiga, 520-20, Japan.
2. Department of Polymer Science, Faculty of Science, Osaka University, Toyonaka, Osaka, 560, Japan.

2a. Department of Synthetic Chemistry, Kyoto University, Kyoto, Japan.

3. L. I. Zakharkin and V. V. Gavrilenko, *U. Gen. Chem. USSR*, **32**, 688 (1962).
4. T. Konomi and H. Tani, *J. Polym. Sci.*, A-1, 7, 2269 (1969).
5. N. Yoda and A. Miyake, *J. Polym. Sci.*, **43**, 117 (1960).
6. H. K. Hall, Jr., M. K. Brandt, and R. M. Mason, *J. Amer. Chem. Soc.*, **80**, 6420 (1958).
7. H. Tani and T. Konomi, *J. Polym. Sci.*, **6**, 2281 (1968).
8. W. O. Ney, Jr., and M. Crowther, U. S. Pat. 2,739,959 (1956).
9. H. Tani and T. Konomi, *J. Polym. Sci.*, A, **6**, 2295 (1968).
10. H. Tani and T. Konomi, *J. Polym. Sci.*, A, **4**, 301 (1966).
11. T. Konomi and H. Tani, *J. Polym. Sci.*, A, 7, 2255 (1969).

Poly(allyl phenylphosphonate)

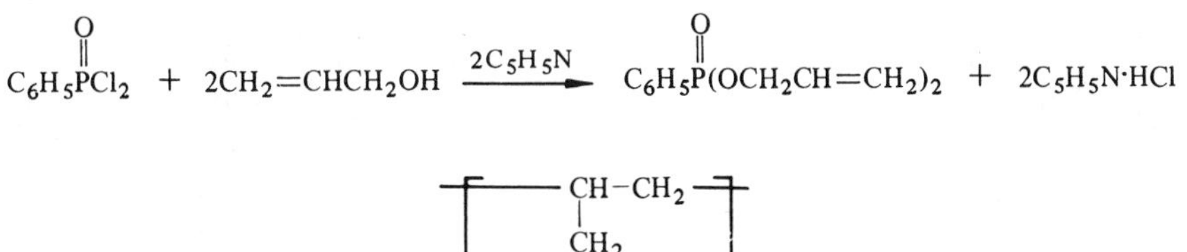

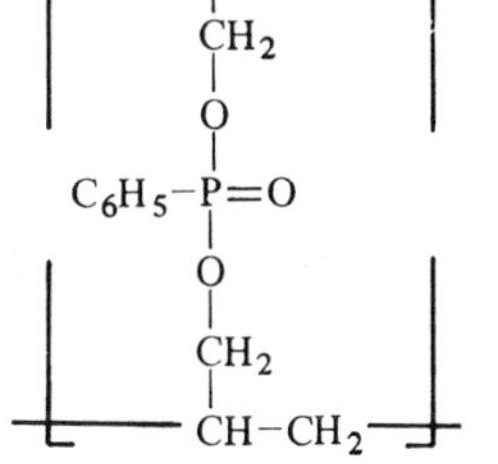

Submitted by A. D. F. Toy (1)
Checked by A. G. Bachman and J. R. Elliot (2)

1. Procedure

A. Diallyl Phenylphosphonate

A 3-1., three necked, round-bottomed flask equipped with a Precision Bore stirrer, a thermometer, and a 1-1. dropping funnel is charged with 464 g. (8 moles) of allyl alcohol (Note 1) and 632 g. (8 moles) of dry pyridine (Note 2). To this mixture is added, with stirring, 780 g. (4 moles) of phenylphosphonyl dichloride (Notes 3 and 4). The temperature of addition is maintained at 2-5° with an ice-salt bath. About 6 hr. is required for the addition. Upon its completion, the reaction mixture is allowed to come to room temperature with continuous stirring overnight; 400 ml. of water is then added to dissolve the amine hydrochloride. The oily layer is separated and is then distilled immediately under reduced pressure (Notes 5 and 6). The distillation flask should be equipped with a thermometer for measuring the liquid temperature. The residue should not be heated above 150° (*Caution!*) (Note 7). The yield of the distilled product, diallyl phenylphosphonate, is 777 g. (81.6%); b.p.= 128° at 1 torr n_D^{25} = 1.5128.

B. Polymerization

In a 2-oz. screw-cap glass bottle is placed 10 ml. of redistilled (Note 8) diallyl phenylphosphonate. To it is added 0.2 g. of benzoyl peroxide. (*Caution! Benzoyl peroxide is a strong oxidant and should be used with care.*) The bottle is purged with nitrogen (Note 9), and the cap is quickly and tightly closed. After shaking to dissolve the benzoyl peroxide, the bottle is three-quarters suspended in a water bath at 78-80° (Note 10). After 16 hr. the bottle is removed from the bath, cooled, and broken (after wrapping in a towel) to obtain a disk of slightly yellowish, clear, hard, glassy solid. This solid is self-extinguishing when ignited and then removed from a flame. Being a crosslinked polymer, it is insoluble.

2. Notes

1. The allyl alcohol is commercial grade from Shell Chemical Company.
2. The pyridine is dried over KOH. An equivalent quantity of anhydrous triethylamine may be used.
3. Phenylphosphonyl dichloride is from Stauffer Chemical Company, Specialty Chemical Division, New York, It should be freshly redistilled before use (b.p. 137-138° at 15 torr, n_D^{25} = 1.5581).
4. The top of the dropping funnel containing the phenylphosphonyl dichloride should be protected from atmospheric moisture with a tube containing calcium chloride.
5. If the crude, wet ester cannot be immediately distilled, it should be dried over $MgSO_4$, because the ester will undergo slow hydrolysis in the presence of water.
6. In effecting the distillation, a high-capacity vacuum pump (e.g., Precision Model D-150 two-stage rotary mechanical pump with free air capacity of 150 l./min.) should be used, and a Dry Ice/acetone-cooled trap should be connected to the system to condense water and other volatiles. When the crude ester is heated under reduced pressure, volatiles evolve that make it difficult to maintain a low pressure without a high-capacity pump. The pressure should be less than

20 torr up to 50° (liquid temperature), less than 10 torr from 50 to 100°, and less than 5 torr to 120°. The preferred pressure for distillation is less than 1-2 torr. If a high-capacity vacuum pump is not available, the crude ester, after drying over $MgSO_4$, may be distilled in portions of 50-100 ml.

7. *Caution! A safety shield should be used during the distillation in case accidential overheating causes rapid decomposition.*

8. Diallyl phenylphosphonate, redistilled (discarding about the first 5-10%), gives a light-colored, hard, glassy solid.

9. When the polymerization is carried out in the presence of atmospheric oxygen, the top surface of the solid polymer obtained is soft and tacky.

10. The polymerization reaction is exothermic during the gelling stage. If the temperature as this stage exceeds 100°, only a soft, art gum-like, clear solid is obtained.

3. Methods of Preparation

Preparation procedures for diallyl phenylphosphonate and poly(allyl phenylphosphonate) are described in the literature (3, 4).

4. References

1. Stauffer Chemical Company, Dobbs, Ferry, New York, 10522.
2. Loctite Corporation, Newington, Connecticut 06111.
3. A. D. F. Toy, *J. Amer. Chem. Soc.*, 70, 186 (1948).
4. A. D. F. Toy and Lee V. Brown, *Ind. Eng. Chem.*, **40**, 2276 (1948).

Poly[methylene bis(4-phenylene) sebacamide]

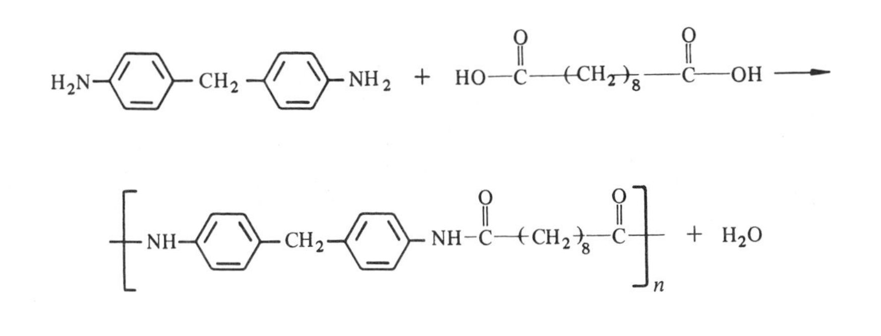

Submitted by D. A. Holmer, O. A. Pickett, Jr., and J. H. Saunders (1)
Checked by J. A. Moore (2a) and J. P. Wesson (2a, b)

1. Procedure

A heavy-wall Pyrex test tube (Note 1) approximately 300 mm. × 35-40 mm. I.D.) is fitted with a Neoprene rubber stopper. Holes are bored through the stopper to accept a glass capillary (350 mm. × 6 mm., ¼-¾ mm. bore), a glass thermocouple well (350 mm. × 6 mm. O.D.), and a gas exit tube (6 mm. O.D.). The capillary and thermocouple well are inserted through the stopper so they extend to within ¼ in. of the bottom of the test tube. The gas exit tube is bent at a 135° angle, and one end of the tube is inserted just to the bottom of the stopper. A mixture of 9.91 g. (0.0500 mole) of 4,4′-diaminodiphenylmethane (*Caution! The National Institute for Occupational Safety and Health has included this chemical in a list of 1500 substances shown to cause cancer in animals. A possible link between this compound and toxic hepatitis has also been cited.*) and 10.62 g. (0.0525 mole) of sebacic acid is added to the test tube (Notes 2 and 3). The reactor is then assembled (Note 4) and connected to the

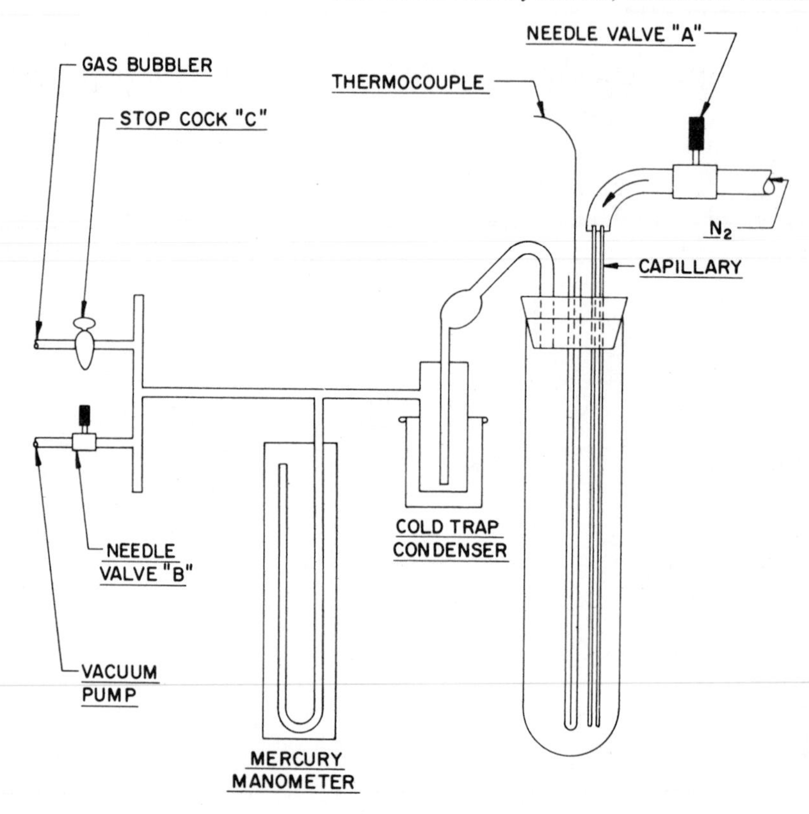

Fig. 1. Polymerization apparatus for synthesis of poly[methylene bis(4-phenylene)sebacamide].

manifold system as shown in Fig. 1 (Note 5). Valve *A* and stopcock *C* are closed, and valve *B* is opened slowly to evacuate the system. A check is made for leaks, particularly where the tubes are inserted through the stopper. When the apparatus is free from leaks, the system is alternately evacuated and purged with nitrogen in the following manner. Valve *B* is closed, and valve *A* is slowly opened (Note 6) to admit nitrogen. When atmospheric pressure is reached in the system, valve *A* is again closed and valve *B* is opened until a pressure of 5 torr or less, is attained. After the fourth evacuation, the system is returned to atmospheric pressure and stopcock *C* is opened. The nitrogen flow is adjusted (valve *A*) to give a gentle sweep through the reactor. A heating bath (preferably a heated, fluidized bed of sand) is mounted around the lower half of the test tube (Note 7).

The contents of the reactor are heated rapidly (40-50 min.) (Note 8) to 285°. The temperature of the polymer melt is held at 285° for 30 min. at atmospheric pressure. Stopcock *C* is then closed and valve *B* is opened slowly, while allowing valve *A* to admit a very slow flow of nitrogen to sweep through the molten polymer. The pressure is reduced to 5-15 torr over a period of 25 min. (Note 9). The melt temperature (285°) and pressure (5-15 torr) are held constant for 15-30 min. to finish the polymerization while continuing to sweep nitrogen through the melt. Valve *B* is then closed, and by regulating valve *A* the system is slowly brought to atmospheric pressure. Stopcock *C* is opened, and the heating bath is removed. The polymer is allowed to cool to room temperature by removing the stopper and introducing a blanket of nitrogen directly into the mouth of the test tube (Note 10). The polymer is removed after cooling by wrapping the test tube with a heavy towel and breaking the glass tube.

2. Characterization

The opaque, cream-colored polymer (Note 11) has an inherent viscosity of 0.70-0.90, determined at 30° on a 0.5% solution in *N,N*-dimethylacetamide containing 5% dissolved lithium chloride (Note 12).

Differential thermal analysis (Note 13) shows the polymer to have a crystalline melting point of 270°. In addition, polymer which has been rapidly quenched from the melt shows a crystallization exotherm at 140°.

Fibers and films may be obtained by melting the polymer in an inert atmosphere and spinning or extruding through a suitable apparatus.

3. Notes

1. This test tube should be fabricated, by a competent glass blower, from tubing with a minimum wall thickness of 2 mm.
2. 4,4'-Diaminodiphenylmethane was obtained from Allied Chemical Company. It was recrystallized twice from toluene (100 g./500 ml.), decolorized with charcoal, and dried for 48 hr. in a vacuum oven at 50° to remove the last traces of solvent. The capillary melting point was 93-94°.
3. Sebacic acid (reagent grade) was obtained from Matheson, Coleman, and Bell and used without further purification. A 5 mole % excess of sebacic acid served to inhibit gel formation during polymerization.
4. Care must be exercised in assembling the reactor to prevent plugging the capillary with the reactants. The tube for introducing nitrogen can also be just

above the surface of the polymer. If this is done and the thermocouple well is raised from the melt at the end of the polymerization, it is not necessary to remove the stopper during cooling. The chance of oxygen contacting the polymer and causing color is reduced.

5. A Dry Ice-methanol bath was used to cool the cold trap. Care must be taken not to plug the condenser during the reaction. A safety shield mounted in front of the reactor and bath is recommended.

6. Care should be taken during purging and evacuation to prevent reactants from being swept from the test tube.

7. It is generally desirable to secure a second thermocouple to the lower outside wall of the test tube (with a high-temperature glass tape) to prevent possible degradation from overheating.

8. The heating rate is rather critical. If the mixture is heated too slowly, the prepolymer will solidify ("phase out"). If the reaction mixture is heated too rapidly, excessive foaming will result. A crust of prepolymer frequently forms on the surface of the polymer melt because of cooling by the refluxing water of condensation. This crust will generally melt when the reaction temperature reaches 285°.

9. Care should be exercised during the pressure reduction to prevent excessive foaming of the melt.

10. On cooling, the polymer frequently pulls glass from the walls of the tube. Hence the test tube should be wrapped with a towel to prevent loss of product or injury from flying glass.

11. Color varies from white to cream to light yellow, depending on the quality of the 4,4′-diaminodiphenylmethane used and the content of oxygen in the polymerization system.

12. This value was determined with a Cannon-Fenske Series 100 viscometer.

13. A Du Pont Model 900 Differential Thermal Analyzer was used. The heating rate was 20°/min., in nitrogen.

4. Methods of Preparation

In addition to this procedure for poly[methylene bis(4-phenylene)sebacamide] (3, 4), the polymer may be prepared from dimethyl (or diethyl) sebacate and 4,4′-diaminodiphenylmethane, using a similar melt polycondensation scheme (5, 6).

The polymer may also be prepared by the Schotten-Baumann reaction, using sebacoyl chloride and 4,4′-diaminodiphenylmethane in a water-tetrahydrofuran system (7).

5. References

1. Monsanto Textiles Company, Pensacola, Florida 35202.
2a. Department of Chemistry, Rensselaer Polytechnic Institute, Troy, New York 12181.
2b. Current Address: Union Carbide Corp., Tarrytown, New York 10591.
3. D. A. Holmer, O. A. Pickett, Jr., and J. H. Saunders, *J. Polym. Sci.,* A-1, **10**, 1547 (1972).
4. D. A. Holmer and O. A. Pickett, Jr., U.S. Pat. 3,651,022 (1972).
5. O. Y. Fedotova and A. S. Kurochkin, *Vysokomol. Soedin.,* **2**, 1688 (1960).
6. D. A. Holmer, Monsanto Textiles Company, unpublished results.
7. H. W. Hill, S. L. Kwolek, and P. W. Morgan, U.S. Pat. 3,066, 899 (1961).

Poly[N-(1,1-dimethyl-3-hydroxylbutyl) acrylamide]

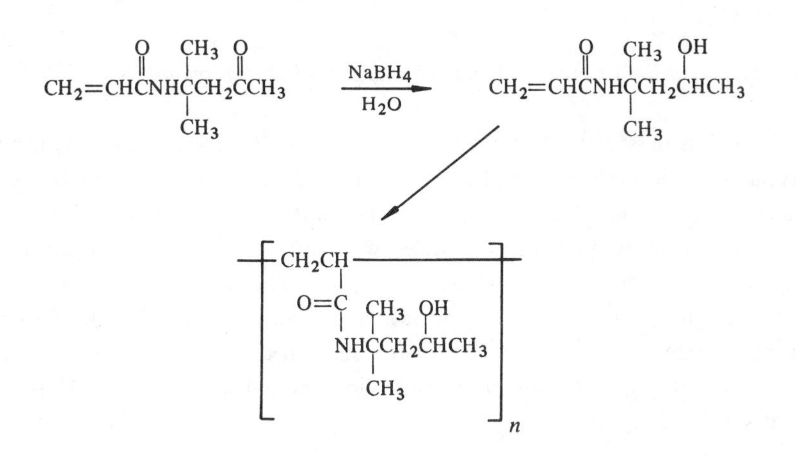

Submitted by D. I. Hoke (1)
Checked by R. Kalnajs and E. S. Poklacki (2)

1. Procedure

A. N-*(1,1-Dimethyl-3-hydroxybutyl)acrylamide*

A solution of 37.8 g. (1.0 mole) of sodium borohydride (Note 1) in 1 l. of water is added to a stirred solution of 338 g. (2.0 moles) of diacetone acrylamide (Notes 2 and 3) in 1 l. of water. The temperature is maintained between 0° and 10° throughout the reaction and the neutralization. The solution is stirred for 2 hr. after addition of the sodium borohydride is completed. The excess sodium borohydride is then destroyed by dropwise addition of 5% aqueous sulfuric acid until the pH of the solution is 4 (*Caution! Hydrogen gas is evolved*). The pH of the solution is adjusted to 8; at this point 2-3 g. of a gummy precipitate forms that adheres to the flask. After the aqueous solution is decanted from the gum, the product is extracted with five 200-ml. portions of chloroform. The extracts are dried over Drierite and filtered. The chloroform is removed on a rotary evaporator, and the residue is distilled (using an oil bath) through an 8-in. heated Vigreux column in the presence of 1 g. of 4,4′methylenebis-2,6-di-*t*-butylphenol. A colorless liquid, boiling at 116° (0.3 torr), n_D^{25} 1.4782 is collected as product (Note 4). Yields of 65-87% are obtained. The purity by gas chromatographic analysis is 95-98% (Note 5).

B. Polymerization of N-*(1,1-Dimethyl-3-hydroxybutyl)acrylamide*

Nitrogen is passed through a solution of 50 g. of *N*-(1,1-dimethyl-3-hydroxybutyl)acrylamide in 200 g. of distilled water for 30 min. A solution of 0.250 g. of ammonium persulfate in one ml. of distilled water and a solution of 0.25 g. of sodium metabisulfite in 1 ml. of distilled water are added while stirring at room temperature. Stirring is continued for 4 hr. The polymer, which precipitates as it forms, is separated by filtration and washed with 500 ml. of water. The precipitate is then dried to constant weight in a vacuum oven at 40°. The intrinsic viscosity of the product, measured in dimethylformamide at 30°, is 1.1. Yields are 72-95%.

C. Reduction of Poly(diacetone acrylamide)

Poly(diacetone acrylamide) (50 g.), prepared by solution or emulsion polymerization (3), is dissolved in 1 l. of ethanol (Note 6). A suspension of 5.6 g. (0.15 mole) of sodium borohydride in 40 ml. of ethanol is added over a period of 1-2 hr. The mixture is stirred at room temperature overnight. Glacial acetic acid (24 g., 0.4 mole) is then added dropwise during a 3-hr. interval (*Caution! Hydrogen gas is evolved and causes foaming.*) The polymer is precipitated by slowly pouring the ethanol solution into 8-10 times its volume of well-stirred water. The precipitate is collected on a filter, washed with two 250 ml. portions of water, and dried to constant weight in a vacuum over a 40°. The yield is 65-95% (Note 7).

2. Characterization

Poly[*N*-(1,1-dimethyl-3-hydroxybutyl)acrylamide] is soluble in methanol,

ethanol, wet acetone, and dimethylformamide. It is insoluble in water, dry acetone, and aliphatic hydrocarbons. The polymer has a glass transition temperature of 110° and decomposes at 227° (determined with the differential scanning calorimeter, an accessory for the Du Pont Thermal Analyzer: heating rate 20°/minute, nitrogen atmosphere). Films prepared by casting from solution or by melt pressing are colorless, transparent, and brittle.

3. Notes

1. When the stoichiometric quantity of sodium borohydride is used, a product containing unreacted diacetone acrylamide is obtained. A product free of diacetone acrylamide is obtained by use of a 100% excess of sodium borohydride.

2. Diacetone acrylamide, *N*-(1,1-dimethyl-3-oxobutyl)acrylamide, may be prepared by the reaction of acetone with acrylonitrile in concentrated sulfuric acid (3).

3. Diacetone methacrylamide (4) may be substituted for diacetone acrylamide in the preparation of *N*-(1,1-dimethyl-3-hydroxybutyl)methacrylamide.

4. The product should be stored in a refrigerator in the presence of about 500 p.p.m. *N*-phenyl-2-naphthylamine or 4,4′-methylenebis-2,6-di-*t*-butylphenol or used immediately after distillation.

5. The packing for the column for gas chromatographic analysis was 20% FFAP on Aeropak 30, both available from Varian Corporation, Walnut Creek, California.

6. Ethanol denatured with methanol may be used. The quantity of ethanol required is dependent on the molecular weight of the poly(diacetone acrylamide). The amount specified is satisfactory for polymers having an intrinsic viscosity of 1.5-3.0 dl./g.

7. Carbonyl group analysis (5) shows that the reduction is complete. Hydroxyl group analysis (6) gives a high value because of residual water in the polymer.

4. Methods of Preparation

N-Hydroxyalkylacrylamides and methacrylamides have been prepared by the reaction of acrylyl chloride or methacrylyl chloride with a hydroxyalkylamine (7) and by amide-ester interchange with an acrylate or methacrylate ester and a hydroxylalkylamine (8). *N*-Hydroxymethylacrylamide is prepared by the reaction of acrylamide with formaldehyde (9). *N*-(1,1-Dimethyl-3-hydroxybutyl)acrylamide has also been prepared by an indirect reduction of diacetone acrylamide (4). The product of conjugate addition of methanol to diacetone acrylamide is subjected to catalytic hydrogenation; then methanol is eliminated by heating in ther presence of a basic catalyst.

5. References

1. Lubrizol Corporation, Cleveland, Ohio 44117.
2. Borg-Warner Corporation, Des Plaines, Illinois 60018.
3. L. E. Coleman, J. F. Bork, D. P. Wyman, and D. I. Hoke, *J. Polym. Sci.,* A, **3**, 1601 (1965).
4. D. I. Hoke, *J. Polym. Sci.,* A-1, 9, 2949 (1971).
5. J. S. Fritz, S. S. Yamamura, and E. C. Bradford, *Anal. Chem.,* **31**, 260 (1959).
6. C. L. Ogg, W. L. Porter, and C. O. Wilts, *Ind. Eng. Chem.,* Anal. Ed., **17**, 394 (1945).
7. G. D. Jones (General Aniline and Film Corporation), U.S. Pat. 2,593,888 (April 22, 1952).
8. P. L. DeBenneville, L. S. Luskin, and H. J. Sims, *J. Org. Chem.,* **23**, 1355 (1958).
9. H. Feuer and U. E. Lynch, *J. Amer. Chem. Soc.,* **75**, 5027 (1953).

Poly(cetyl vinyl ether)

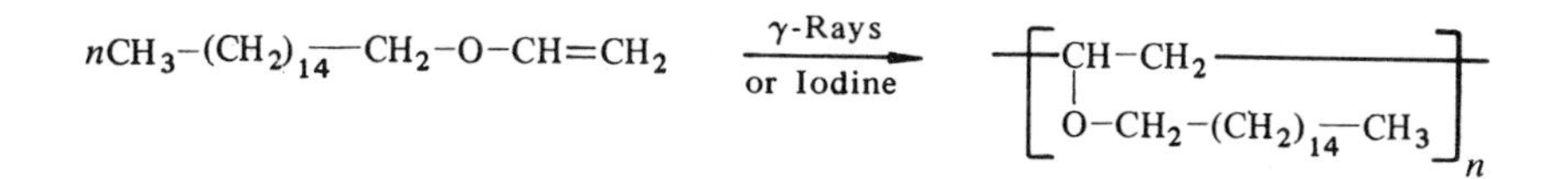

Submitted by Gy. Hardy, K. Nyitrai, and F. Cser (1)
Checked by K. N. Sannes (2a, b) and C. G. Overberger (2a)

1. Procedure

A. Radiation-Initiated Polymerization

Cetyl vinyl ether is polymerized at two temperatures. In each of two ampoules (Note 1) is placed 1 g. of purified (Note 2) cetyl vinyl ether. Each sample is deoxygenated (Note 3) and sealed under 600 torr of nitrogen. The polymerizations are initiated by a Co^{60} radiation source (Note 4). During polymerization one ampoule is maintained at 12° and the other at 0°; both samples are irradiated for 120 hr. at a dose rate of 5.26×10^4 rad./hr. The contents of each ampoule are dissolved in 5 ml. of *n*-heptane at its irradiation temperature, and each polymer is then precipitated by adding each *n*-heptane solution to 500 ml. of acetone stirred at −20°. The polymers are filtered on fine fritted glass filters and dried at room temperature in a vacuum chamber to constant weights to afford 0.63 g. (63%) from the 12° polymerization and 0.38 g. (38%) from the 0° polymerization.

B. Iodine-Initiated Polymerization

The inner surface of a 14-mm.-diameter glass ampoule is coated uniformly (to an approximate thickness of 0.02 mm.) with 0.5 g. of cetyl vinyl ether (Note 2).

This is accomplished by melting the monomer at 40°, followed by rapid rotation of the tube at a 45° angle in an ice bath, so that an even, thin layer is obtained. The thickness of the layer is calculated from the surface area and the weight and density of the monomer. A separate glass vial containing 1 g. of iodine is introduced into the polymerization ampoule at 0°, which is the polymerization temperature. The iodine is removed after 50 hr. and the unreacted monomer is removed from the polymer by treating the reaction mixture with acetone at 0°. The product is filtered on a fritted glass filter and dried to a constant weight in a vacuum chamber to afford 0.26 g. (52%) of poly(cetyl vinyl ether).

2. Characterization

Poly(cetyl vinyl ether) is soluble in *n*-heptane benzene, petroleum ether, and tetrahydrofuran.

The viscosities of the polymers are measured with an Ubbelohde viscometer at 30° in tetrahydrofuran in the 0.1-0.4 g./dl. range of concentration. The intrinsic viscosities of poly(cetyl vinyl ether) formed at 12° and 0° are 0.072 and 0.070 dl./g. respectively; for the iodine-initiated polymerization at 0° the intrinsic viscosity is 0.078 dl./g. (Note 5).

Procedures A and B both yield polymers with melting points of 39°. Until reaching the melting point, the polymers have a spherulitic structure. They exist as smectic double layers.

3. Notes

1. The ampoule is 200 mm. long, has a 14-mm. I.D., and is equipped with a ground glass joint.

2. Before using the cetyl vinyl ether, it is advisable to distill it from metallic sodium at 115° (0.03 mm.). The cetyl vinyl ether has a melting point of 16.5° and is in a smectic liquid crystalline state within the 16.5 to -2.5° temperature range. Further data concerning its crystalline structure are available (3, 4).

3. The contents of the ampoule are frozen in ice. The ampoule is evacuated and blanketed with nitrogen, and the monomer is allowed to melt. This procedure is repeated three times before sealing.

4. A 500-Ci. ^{60}Co source with panoramic construction has been used as an irradiation facility.

5. For numerical estimation of the $\overline{DP}$ values, the constants determined for poly(octadecyl vinyl ether) can be used (6). The reported intrinsic viscosities correspond to $\overline{DP}$ values of about 60-70.

4. Methods of Preparation

The iodine-initiated polymerization method has been adapted from the procedure used by Gy. Hardy, K. Nyitrai, and F. Cser (5).

5. References

1. Research Institute for the Plastics Industry, Budapest.
2a. University of Michigan, Ann Arbor, Michigan 48104.
2b. Current Address: General Electric Co., Schenectady, New York 12345.
3. Gy. Hardy, K. Nyitrai, and F. Cser, *Magy. Kem. Folyoirat,* 74, 602 (1968).
4. Gy. Hardy, K. Nyitrai, and F. Cser, *Magy. Kem. Folyoirat,* 47, 608 (1968).
5. Gy. Hardy, G. Kovacs, Gy. Kosterszita, and F. Cser, *Kinetics and Mechanism of Polyreactions,* IUPAC Symposium, Budapest, 1969, Preprints, Vol. 4, p. 147.
6. J. G. Fee, W. S. Port, and L. P. Witnaner, *J. Polym. Sci.,* 33, 95 (1958).

Isotactic Polypropylene

$$CH_2{=}\underset{\displaystyle CH_3}{\underset{|}{CH}} \xrightarrow[TiCl_3]{Et_2AlCl} -\!\!\left(CH_2-\underset{\displaystyle CH_3}{\underset{|}{CH}}\right)\!\!-_n$$

Submitted by E. J. Vandenberg (1)
Checked by J. A. Turner (2)

1. Procedure

A. $TiCl_3$ *Catalyst Component*

$$\tfrac{1}{3}Et_3Al + TiCl_4 \xrightarrow[\text{Heat}]{} TiCl_3 \cdot \tfrac{1}{3}AlCl_3$$

A 1-1., 3-necked, round-bottomed flask is fitted with a stirrer (half-moon Teflon® blade, mercury seal, about 600 r.p.m.), thermometer, nitrogen inlet (over the surface of the liquid), nitrogen outlet through a reflux condenser to a bubbler containing *n*-heptane, and a rubber serum stopper for injecting ingredients with hypodermic equipment (Note 1). The air in the flask is replaced by passing a fast stream of nitrogen (Note 2) through it for 30 min. The nitrogen is passed through a bubbler containing *n*-heptane before entering the flask. Then 320 g. of *n*-heptane (Note 3) is added, and nitrogen sweeping is continued for 15 min. While stirring and using a slow stream of nitrogen, 76.8 ml. of $TiCl_4$ (132.8 g., 0.70 mole) is added (Note 4) with hypodermic equipment (Note 5). Then, with the flask in a cold water bath to maintain the reaction temperature at 18-26°, 154 ml. (110.5 g., 0.236 mole) of 1.53 *M* Et_3Al in *n*-heptane (*Caution!*) (Note 8) is added dropwise during 1 hr (Note 5). After an additional 30 min. at 25-33°, the water bath is replaced with a heating mantle and the temperature is raised slowly to reflux (about 95°) over 1.3 hr. Considerable gas is evolved during this period. Refluxing is continued until gas evolution ceases (about 3.5

hr.). The reaction mixture is cooled to room temperature. The product, a dark brown to purple dispersion, is transferred to a nitrogen-filled 28-oz. pressure bottle (Note 5) with a long (18-24 in.) hypodermic tube (13-15G) having a needle point on each end (Note 9). The tubing is first introduced into the free space of the flask through the rubber stopper and then, after the air is swept from it with nitrogen, it is inserted into the nitrogen-filled 28-oz. bottle. The hypodermic tubing in the flask is then placed near the bottom of the flask, and the transfer of product is accomplished by applying vacuum to the bottle. When the transfer is completed, the free space of the bottle should be filled with nitrogen to 15 p.s.i.g. (Note 10). This catalyst is stable to room-temperature storage. It is 1.02 *M* with respect to Titanium [all Ti(III)], density 0.81 (Note 11).

B. Polymerization

An 8-oz. pressure bottle (Note 6) is charged with 97.8 ml. of *n*-heptane (Note 3). After the free space is swept with nitrogen, the bottle is topped with a metal cap having two 1/8-in. holes and fitted with a special Buna N rubber liner (Note 7). The bottle is then placed in a protective metal shield (Note 12). Air is further excluded by alternately applying vacuum and nitrogen pressure through a hypodermic needle (20G), which is inserted through the cap liner and is attached by pressure tubing through a three-way stopcock to an oil pump and to a source of nitrogen at 15-p.s.i.g. pressure. Vacuum is applied for 1 min. and is followed by nitrogen pressure until the pressure reaches 15 p.s.i.g. (15-30 sec.). The evacuation is repeated once more; thereafter, the bottle is tared and is then filled with propylene (Note 13) through a hypodermic needle (20G) attached to a reducing valve, fitted to the propylene cylinder and set at 40 p.s.i.g. (Note 14). Propylene is added to saturation; this requires about 1 min. if the bottle is shaken. The bottle is reweighed to determine the exact amount of propylene added (approximately 14 g.) with a hypodermic syringe-stopcock-needle assembly, 1.4 ml. (2 mmoles) of 1.46 *M* Et_2AlCl in *n*-heptane (Note 8) is injected. The bottle is then tumbled in a 50° constant temperature water bath for 30-60 min. The pressure is approximately 68 p.s.i.g. (Note 15). The $TiCl_3$ component (1.0 ml., 1 mmole) described above is then injected to start the polymerization.

After 2.5 hr. the pressure drops to 0, and 10 ml. of anhydrous ethanol is injected. After the reaction mixture changes from salmon-brown to essentially colorless, the bottle is vented in a hood with a hypodermic needle and is uncapped. The very viscous reaction mixture is transferred to a separatory funnel with sufficient additional *n*-heptane to make it fluid. It is washed by being shaken twice with 100 ml. of 10% methanolic-HCl (2 l. of methanol plus 612 ml. of concentrated HCl) and then with water until neutral. The *n*-heptane layer is filtered to recover the heptane-insoluble material. It is washed three times with an equal amount of *n*-heptane and dried for 16 hr. at 80° in a water-pump vacuum. There is thus obtained an 88% conversion of isotactic polypropylene (Note 16), $\eta_{sp/c}$ 11.3 dl./g. (0.1% Decalin, 135°) (Note 17), m.p. 170° (Note 18). isolated (Note 19).

Alternatively, the preparation can be carried out at lower temperatures (Note 20) or in stirred reactors at lower pressures (Note 21).

2. Notes

1. Apiezon N lubricant (Apiezon Products, Ltd., England) is used for all joints and stopcocks. Rubber serum stoppers are available from Arthur H. Thomas, Philadelphia, Pennsylvania, catalog number 8826.

2. High-purity, dry nitrogen from Linde Company, Division of Union Carbide Corporation, is satisfactory.

3. Phillips Petroleum Company pure grade *n*-heptane may be used as received.

4. Matheson, Coleman, and Bell 9425 $TiCl_4$ (99.5%) is satisfactory.

5. It is convenient to store $TiCl_4$, aluminum alkyl solutions, and reactive catalysts under nitrogen in 8-oz. or 28-oz. pressure bottles (Note 6) fitted with crown caps having two ¹/₈-in. holes and Buna N self-sealing liners (Note 7). Aliquots are then withdrawn with an appropriate nitrogen-filled hypodermic syringe (10-100 ml.) fitted with a metal stopcock (no. FL/Z, Propper Manufacturing Company, Long Island City, New York; a better but more expensive stopcock is the No. 1LF1 valve, part number 86570, from Hamilton Company, Whittier, California) and hypodermic needle (20G) and are injected into the reaction flask through the rubber stopper. Larger quantites are more conveniently measured in a pressure bottle and then forced under nitrogen pressure into the flask through a stopcock (or needle valve as in Note 15) fitted with a hypodermic needle on each end.

6. Standard, crown-capped beverage bottles of the returnable type are satisfactory if they are not exposed to more than 50° thermal shock. Scratched bottles should not be used.

7. The general procedure for using puncture-sealing gaskets has been

described (3), particularly as a convenient method for polymerizing in the absence of air. Some additional specific details are given in the polymerization section (Procedure B). The evacuation time for 28-oz. bottles should be increased to 4 min. Molded Buna N liners from Firestone Rubber and Latex Products Company, 1620 S. 49 Street, Philadelphia, Pennsylvania 19143 are satisfactory. They are extracted for 3 days in a Soxhlet apparatus with benzene and then dried to their original size.

8. *Caution! Extreme care must be exercised in handling aluminum alkyls.* Heptane solutions, approximately 25%, from Texas-Alkyls, Inc., Weston, Michigan, are satisfactory. Such solutions are still very reactive, although not pyrophoric, as are pure aluminum alkyls. *Contact with the skin must be avoided.* Futher details are available in the Texas-Alkyls brochure "Aluminum Alkyls–Safety and Handling Information."

9. Polyethylene tubing fitted on each end with a Becton, Dickinson and Company (Rutherford, New Jersey) No. 5208A adapter and an appropriate hypodermic needle is conveniently used in place of hypodermic tubing, especially for viscous materials.

10. For long-term storage, the bottle should be recapped with a new Buna N liner under nitrogen. Liners which have been punctured with needles larger than 18G are not reliable.

11. A small quantity of this catalyst at 0.125 *M* Ti is conveniently prepared in a 6-8 oz., capped, nitrogen-filled pressure bottle (Notes 6 and 7) containing 101.9 ml. of *n*-heptane by injecting 3.3 ml. of 1.53 *M* Et_3Al in *n*-heptane and 15.0 ml. of 1 *M* $TiCl_4$ in *n*-heptane. The $TiCl_4$ is added all at once, followed by shaking. The catalyst is aged for 2 hr. at room temperature and then heat-treated for 16 hr. in a 90° bath (no agitation). *Caution! The bottle must be warmed in hot water before placing it in the 90° bath!* This catalyst gives polymerization results similar to those obtained with the larger-scale concentrated catalyst.

An analogous catalyst, prepared by the Al reduction of $TiCl_4$ is available from Stauffer Chemical Company, 299 Park Avenue, New York, New York, 10017, as titanium trichloride AA (4). It will give similar results in this procedure. The reaction time is conveniently adjusted by pressure measurements to allow for differences in catalytic activity.

12. For pressures above 15 p.s.i.g., it is recommended that the bottles be used in protective metal cans. These are conveniently made from an appropriate size of brass tubing which has been drilled with enough ¼-in. holes near the top and bottom to permit water circulation when placed in the polymerization bath.

13. A pure grade of propylene from Phillips Petroleum Campany or Matheson, Coleman, and Bell is satisfactory.

14. The outlet on a standard reducing valve for propylene is fitted with a metal coupling, prepared from solid hexagonal brass rod, to a No. 478/D hypodermic needle adapter from Becton, Dickinson and Company.

15. The pressure in the bottle is conveniently measured by a pressure gauge (Ashcroft No. 1004, 2-in. dial, 100 p.s.i. to 30 in., from Manning, Maxwell, and Moore, Inc., Stratford, Connecticut) fitted with metal couplings to a needle valve (No. 323, stainless steel, 1/8-in. N.P.T., Teflon® packing, Hoke Inc., Englewood, New Jersey) and to a hypodermic needle adapter (No. 478/D, Becton, Dickinson and Company). This pressure gauge assembly, attached to a hypodermic needle (20G), is filled with nitrogen or propylene to almost the expected bottle pressure (higher if convenient) by attaching to a pressure bottle (through the usual self-sealing liner) containing either nitrogen or propylene at or above this pressure. After several fillings and ventings, the gauge assembly, is ready to measure the reaction pressure. As the needle is inserted through the cap liner of the reaction bottle, the valve is opened slightly to sweep out the free space in the needle. Such pressure measurements permit the polymerization to be followed readily.

16. This product is hghly isotactic, on the basis of a hot heptane extraction test similar to that used by Natta, Corradini, and Allegra (5). Thus, the product is 98.5% insoluble after extracting for 16 hr. in a Soxhlet extractor with *n*-heptane (temperature $> 85°$). The product also proved to be 3% soluble in decahydronaphthalene isotactic content.

17. Lower molecular weight polymer is conveniently prepared by using hydrogen as a chain transfer agent (7). For example, a partial pressure of 6 p.s.i. of hydrogen, in 2 hr. at 50°, gives the same conversion and a 95% yield of isotactic polypropylene, $\eta_{sp/c}$ 2.0 dl./g. A convenient method of adding the hydrogen is first to add propylene to a pressure of 2 p.s.i.g. and then to apply 8 p.s.i.g. of hydrogen pressure by the same general procedure. Thereafter the rest of the propylene (12.2 g.) is introduced.

18. The crystalline melting point is conveniently determined by measuring the temperature at which birefringence due to crystallinity disappears (8). Other methods such as differential thermal analysis and differential scanning calorimetry may also be used.

19. The total heptane-soluble atactic polymer (2.3% conversion) is determined by drying a 20-g. aliquot of the combined filtrates from the isotactic polymer isolation, first on a steam bath and then for 1 hr. at 80° *in vacuo*.

20. At 30°, the same high conversion and a slightly greater yield of isotactic polymer, $\eta_{sp/c}$ 30 dl./g. are obtained in 19 hr. A hydrogen partial pressure of about 12-17 p.s.i. is required to decrease the $\eta_{sp/c}$ to about 2 dl./g.

21. A stirred pressure vessel (autoclave or magnetically stirred 8- or 28-oz. pressure bottle) is conveniently used at lower propylene pressures (about 15 p.s.i.g.). With such equipment, the propylene is advantageously fed to the reactor at constant pressure until the desired solids are achieved. Under these conditions, approximately 5-10 hr. are required to duplicate the results shown. This procedure has obvious safety advantages for larger scale preparations.

3. Methods of Preparation

A low-yield method of preparing isotactic polypropylene, analogous to that used in its early synthesis (9, 10) and based on an *in situ* El_3Al-$TiCl_4$, catalyst, has been described (11).

A high-yield synthesis based on ball-milled, violet $TiCl_3$ (prepared by hydrogen reduction of $TiCl_4$) plus Et_3Al or Et_2AlCl (12) gives very low rates of polymerization. High rates of polymerization with high yields were first achieved by using $TiCl_3$ prepared by the aluminum alkyl reduction of $TiCl_4$ (13). A particularly effective $TiCl_3$, prepared by reducing $TiCl_4$ with a stoichiometric amount of Et_3Al, is described in this report. A similar process for preparing this catalyst and using it to prepare isotactic polypropylene is described in various U.S. Patents (14). A preparation of isotactic polypropylene with a similar catalyst but under less favorable conditions has also been described (15). An analogous $TiCl_3$ catalyst has been prepared by the reduction of $TiCl_4$ with aluminum metal and is available commercially (4). The use of this catalyst is described for preparing isotactic polypropylene (4) and isotactic poly(1-butene) (16).

The $TiCl_3$ catalyst discussed here is useful for preparing other isotactic polyolefins. Its use in preparing isotactic polystyrene has been described (17). The pressure bottle, self-sealing liner technique described in this preparation is very useful for conducting polymerizations and reactions in the absence of air and under low pressures. Pyridine has been used in conjunction with titanium trichloride-diethylaluminum chloride to increase the per cent of isotactic polymer (18). A vanadium trichloride-diethylaluminum catalyst for the production of isotactic polypropylene has been reported (19).

4. References

1. Hercules Research Center, Wilmington, Delaware 19899.
2. Hercules Research Center, Wilmington, Delaware 19899.

3. S. A. Harrison and E. R. Meincke, *Anal. Chem.*, **20**, 47 (1948).
4. *Preliminary Bulletin* No. 56-6, July 9, 1957, and *Manual of Procedure for the Evaluation of Olefin Polymerization Catalyst, Ziegler-Natta Type,* 2nd ed., Stauffer Chemical Company, 380 Madison Avenue, New York, 17, New York.
5. G. Natta, P. Corradini, and G. Allegra, *J. Polym. Sci.*, **51**, 39 (1961).
6. *Federal Register,* **25**, 3320 (April 16, 1960), and **26**, 6429 (July 18, 1961), CFR 121.2501.
7. E. J. Vandenberg, Brit. Pat. 807,204 (January 7, 1959); U.S. Pat. 3,051,690 (August 28, 1962), assigned to Hercules Incorporated.
8. C. W. Hock and J. F. Arbogast, *Anal. Chem.*, **33**, 462 (1961).
9. A. D. Ketley, *The Stereochemistry of Macromolecules,* Vol. 1, Marcel Dekker, New York, 1967, Chs. 1 and 6.
10. B. C. Repka, "Polypropylene," in *Kirk-Othmer Encylcopedia of Chemical Technology,* Vol. 14, Wiley, New York, 1967, p. 282.
11. D. Braun, H. Cherdron, and W. Kern, *Techniques of Polymer Syntheses and Characterization,* Wiley-Interscience, New York, 1972, p. 158 ("Preparation of Isotactic Polypropylene").
12. G. Natta, *Chim. Ind. (Milan),* **42**, 1207 (1960).
13. E. J. Vandenberg (Hercules Powder Company), U.S. Pat. 2,954,367 (filed July 29, 1955, issued September 27, 1960).
14. L. W. Gamble, A. W. Langer, Jr., and A. H. Neal (Esso Research and Engineering Company), U.S. Pat. 2,951,045 (August 30, 1960); D. E. Winkler and K. Nozaki (Shell Oil Company), U.S. Pat. 2,971,925 (February 14, 1961); E. V. Fasce and R. J. Fritz (Esso Research and Engineering Company), U.S. Pat. 2,999,086 (September 5, 1961); E. J. Vandenberg (Hercules Incorporated), U.S. Pat. 3,261,821 (July 19, 1966).
15. W. R. Sorenson and T. W. Campbell, *Preparative Methods of Polymer Chemistry,* 2nd ed., Wiley-Interscience, New York, 1968.
16. W. Kern, H. Schnecko, W. Lintz, and I. Kollar, "Preparation of Isotactic Poly(1-butene)," this volume, p. 425.
17. E. J. Vandenberg, "Isotactic Polystyrene", *Macromolecular Syntheses,* J. E. Mulvaney, ed. Vol. 6, p. 39.
18. T. Setoguchi and S. Mizuno, Ger. Pat. 2,313,533 (Sept. 27, 1973).
19. I. L. Dubnikov. . .Vysokomol. Soedin., Ser. A**15**, (7), 1635 (1973).

Preparation and Polymerization of N-Vinylpyridinium Fluoborate

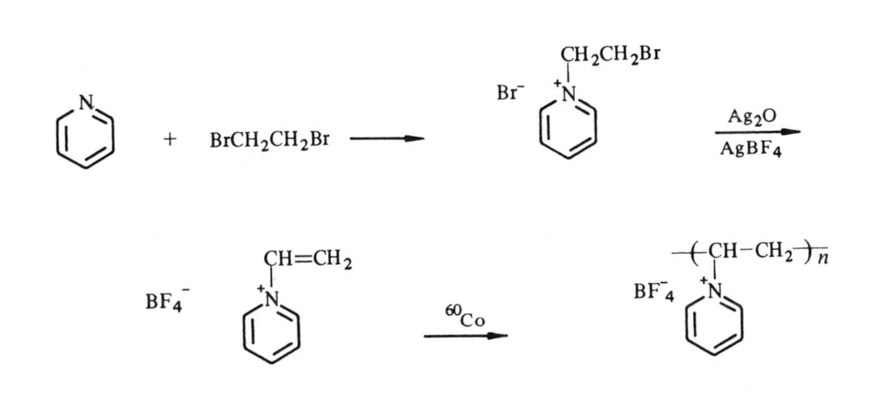

Submitted by I. N. Duling and C. C. Price (1)
Checked by E. Gipstein and O. U. Need (2)
G. B. Butler, and K. B. Wagener (3)

1. Procedure

A. *Monomer Preparation*

A mixture of 190 g. (0.01 moles) of 1,2-dibromoethane and 39.5 g. of pyridine, 0.5 mole, is allowed to stand in a stoppered flask for 4-5 weeks. The tan, crystalline mass (Note 1) is crushed, washed two or three times with anhydrous ether by decantation, and transferred to a Büchner funnel. The filter cake is washed several times with anhydrous ether and dried under vacuum over phosphorus pentoxide, yielding 106 g. (79%) of *N*-(2-bromoethyl)pyridinium bromide (Note 2).

To a solution of 44.4 g. (0.166 mole) of this crude salt in 100 ml. of water is added 0.175 mole of freshly prepared silver oxide (Note 3). After 3 min. of vigorous stirring, the pH is adjusted to 7 with hydrobromic acid and the precipitate of silver bromide is filtered. A concentrated aqueous solution of silver fluoborate (Note 4) is added carefully to the red-yellow filtrate until no more silver bromide precipitates. The silver bromide is again filtered and the solution is evaporated to dryness, leaving 23.2 g. (73%) of crude *N*-vinylpyridinium fluoborate. After five recrystallizations from methanol, followed by vacuum drying over P_2O_5, 36-47% of pure product is obtained as hygroscopic colorless needles, m.p. 71-75°.

B. Solid-State Polymerization by Gamma Irradiation

About 0.5 g. of crystalline monomer is weighed into a 15-cm. test tube evacuated to 10^{-4} torr, and sealed. The sample is exposed to a 300-Ci. ^{60}Co source, the rod being 10 mm. from the center of the tube. Irradiation is carried out for 96 hr. at ca 50°. The crystals become light tan on irradiation, but otherwise their appearance is unaltered. The melting point, however, is changed from 75.5-76.5° to 192-215°. Removal of monomer by thorough washing with methanol leaves a yield of 0.43-0.47 g. (86-94%) of polymer. The polymer is soluble in hot water (25 ml./g.), and a 0.05-0.065 g. (10-13%) yield precipitates on cooling (Note 5). The polymer is soluble in DMF and nitromethane, and insoluble in ethanol, methanol, chloroform, THF, and pyridine. It turns red when exposed to bases. As a polyelectrolyte the polymer displays anomalous viscosity behavior, that is, the reduced specific viscosity rapidly increases as zero concentration is approached. Values for $\eta_{sp/c}$ at 30.0° in water are as follows:

Concentration (g./100) ml.)	Viscosity (dl/g.)
0.163	0.779
0.01	2.53
8.5×10^{-4}	5.41

2. Characterization

The x-ray diffraction pattern of the methanol-extracted polymer is very different from that of monomer but does show some degree of order (sharp lines corresponding to spacings of 3.83 and 3.46 Å.), whereas the reprecipitated polymer showed only two amorphous bands (4).

3. Notes

1. If the usual commercial "pure" grades of reagents are prepurified before use, the product is nearly colorless.

2. Recrystallization from ethanol gives colorless, hygroscopic needles, m.p. 126-128° (4), although the checkers report a melting point of 135-137° (3).

3. The silver oxide is prepared by adding an aqueous solution of 7.02 g. (0.175 mole) of sodium hydroxide (100 ml.) to an aqueous solution of 29.8 g. (0.175 mole) of silver nitrate (500 ml.). The precipitate is washed four or five times by decantation (until the wash water is neutral to pH paper) and used promptly.

4. The solution is prepared by adding concentrated fluoboric acid to a suspension of freshly prepared silver oxide (Note 4) until it just dissolves.

5. Undoubtedly much low molecular weight polymer remains soluble in the cold water.

4. References

1. Department of Chemistry, University of Pennsylvania, Philadelphia, Pennsylvania 19104.
2. IBM, Monterey and Cottle Roads, San Jose, California 95114.
3. Department of Chemistry, University of Florida, Gainesville, Florida 32601.
4. I. N. Duling and C. C. Price, *J. Amer. Chem. Soc.*, **84**, 580 (1962).

Synthesis of Poly(isophthaloyl-adipolyhydrazide) by Phosphorylation

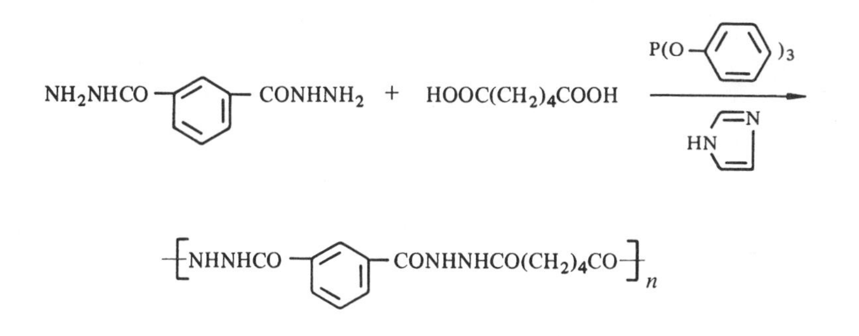

Submitted by N. Ogata and G. Suzuki (1)
Checked by W. Memeger, Jr., and A. H. Frazer (2)

1. Procedure

In a 200-ml. conical flask containing 50 ml. of dimethylformamide and equipped with a mechanical stirrer are dissolved 20.6 g. (0.067 mole) of triphenyl phosphite and 4.53 g. (0.067 mole) of imidazole (Note 1). To this solution, 4.85 g. (0.025 mole) of isophthalic dihydrazide and 3.65 g. (0.025 mole) of adipic acid are added with stirring (Note 2). The flask is immersed in a constant temperature bath at 30° (Note 3). The solution gradually transforms from a suspension into a clear homogeneous solution over a period of 2 hr. After stirring for 5 hr., the solution is poured into 500 ml. of acetone and the resulting polyhydrazide is filtered. The polymer is washed once with water, followed by repeated washings with hot water and then with acetone. It is dried under reduced pressure over phosphorus pentoxide. The yield of polyhydrazide reaches 95-98%, and the inherent viscosity in dimethyl sulfoxide (lg./100 ml.) is 0.20-0.30 dl./g. (Note 4). The polymer is also soluble in formic acid and concentrated H_2SO_4 and slightly soluble in dimethyl acetamide, *m*-cresol, and *N*-methyl pyrrolidone.

2. Notes

1. It is not necessary to replace air with an inert gas. The optimum molar ratio of triphenyl phosphite to the monomers used to yield polyhydrazide is in the range of 2-3; any additional excess amount of triphenyl phosphite gives poor results. Equal moles of imidazole and triphenyl phosphite should be used. Dimethyl formamide was dried by azeotropic distillation with benzene followed by vacuum distillation from barium oxide. The other reagents were purified by recrystallization according to conventional methods. The checkers used reagents as received from the manufacturers. The DMF used by the checkers was du Pont Technical Grade and was used without purification, while isophthalic hydrazide was recrystallized from water, m.p. 229.5-231.5°.

2. Stirring is a critical variable in the production of polymers with high inherent viscosities because the reaction proceeds at first in a heterogeneous phase. When aliphatic dihydrazides are used, the solution remains heterogeneous and ultimately becomes a gel after the polycondensation is completed.

3. The rate of the polycondensation reaction increases with rising temperatures, and the reaction is completed within 1 hr. at 60°.

4. When the reaction is carried out at 60° in dioxane in place of dimethyl formamide, the inherent viscosity of the polymer increases up to 0.30-0.40, although the polymer yield decreases to 80-90%. In this case the reaction proceeds in a heterogeneous phase throughout the reaction, The checkers obtained an 80% yield of polymer with an inherent viscosity of 0.17 dl./g. (.5 g/100 ml of DMSO at 30°) using the 30° procedure.

3. Methods of Preparation

This preparation has been reported (3).

Polyhydrazides are usually prepared by interfacial or solution polycondensation of a diacid chloride with a dihydrazide in the presence of an inorganic or organic base as an acid acceptor (4, 5, 6).

4. References

1. Sophia University, 7 K101CHO, Chiyoda-Ku, Tokyo, Japan.
2. Textile Fibers Department, Pioneering Research Laboratory, Experimental Station, E. I. du Pont de Nemours and Company, Wilmington, Delaware 19898.
3. N. Ogata, H. Tanaka and G. Suzuki, *J. Polym. Sci.,* **11**, 675 (1973).
4. P. W. Morgan, *Condensation Polymers: by Interfacial and Solution Methods,* Wiley-Interscience, New York, 1965, p. 508.
5. W. R. Sorenson and T. W. Campbell, *Preparative Methods of Polymer Chemistry,* 2nd ed., Wiley-Interscience, New York, 1968, p. 171.
6. A. H. Frazer and T. A. Reed, "Alternating Copolyhydrazide of Terephthalic and Isophthalic Acids," in *Macromolecular Syntheses,* this volume, p. 299.

Author Index

Subject Index